The Role of the Polymeric Matrix in the Processing and Structural Properties of Composite Materials

The Role of the Polymeric Matrix in the Processing and Structural Properties of Composite Materials

Edited by
JAMES C. SEFERIS

University of Washington
Seattle, Washington

and
LUIGI NICOLAIS

University of Naples
Naples, Italy

PLENUM PRESS • NEW YORK AND LONDON

Library of Congress Cataloging in Publication Data

Joint U.S.–Italy Symposium on Composite Materials (1981: Capri, Italy)
   The role of the polymeric matrix in the processing and structural properties of composite materials.

   "Proceedings of a Joint U.S.–Italy Symposium on Composite Materials, held June 15–19, 1981, in Capri, Italy"—P.
   Includes bibliographical references and index.
   1. Composite materials—Congresses. 2. Polymers and polymerization—Congresses. I. Seferis, James C. II. Nicolais, Luigi. III. Title.
TA418.9.C6J584 1983   620.1'92   82-18895
ISBN 0-306-41134-2

Proceedings of a Joint U.S.–Italy Symposium on Composite Materials,
held June 15–19, 1981, in Capri, Italy

©1983 Plenum Press, New York
A Division of Plenum Publishing Corporation
233 Spring Street, New York, N.Y. 10013

PREFACE

The state of development of composite materials is quite unique in the scientific world with simultaneous advances being made both in their usage and basic understanding. The complexity and high technology required in manufacturing structural parts with these materials as well as the need for fundamental description of their processing and property characteristics necessitates a close collaboration between industrial and academic researchers. This collaboration has become significant not only in solving specific technical problems, but in providing a much needed supply of scientists with training and background focused on anticipated demand for further advances in composite usage.

The fact that the transportation industry with its current international character has a vital interest in composite materials for weight savings applications has provided a strong incentive for extending these developments beyond national boundaries. An excellent example of an established international venture is the building of the new generation commercial aircraft by the Boeing Company with composite parts manufactured by Aeritalia in Italy. Accordingly, we organized a Joint U.S.-Italy Symposium on Composite Materials in Italy which was successfully held on June 15-19, 1981, under the primary sponsorship of NSF in the U.S.A. and CNR in Italy. The strong support we also received from industrial co-sponsors, both from Italy and the U.S.A., as well as our respective academic institutions gave us confidence that we were addressing a timely and important area in Science and Engineering with a unique concept. After many discussions with colleagues from Boeing, Aeritalia, the U.S. Air Force, as well as from academia, we centered on the symposium theme that addressed specifically the Role of the Polymeric Matrix on Processing and Structural Properties of Composite Materials.

The symposium was structured in such a way as to bring together active scientists both from industry and academia as well as management from industry and government who have a direct interest in seeing work in the field develop beyond existing boundaries. Ample time for discussion periods was incorporated in the program as an

v

integral part of the symposium. Although we can not capture the
spirit and important points that were discussed during the five full
days of the meeting, the reader has now before him in this volume
the work that was presented by the participants and forms a coherent
and thorough coverage of the subject matter from different perspec-
tives.

As for the organization of the symposium, we wish to thank the
following colleagues who acted as discussion leaders and made an
important contribution to its success:

G. Astarita, F. Betz, E. Colombo, P. Corradini, R. Coulehan,
C. Denson, A. DiBenedetto, J. Fitzgerald, H. Hopfenberg,
P. Laconte, R. Landel, J. Manson, G. Marrucci, E. Meinecke,
G. Ombra, M. Pegoraro, S. Piccarolo, A. Valvassori

We are indebted to our students, co-workers and colleagues who
were recruited for helping in this endeavor and responded to the
challenge with enthusiasm supplementing our own.

James C. Seferis
Luigi Nicolais
November, 1981
Seattle, Washington, U.S.A.

ACKNOWLEDGMENTS

Support from the following Institutions is gratefully acknowledged:

Consiglio Nazionale delle Ricerche, Progetto Finalizzato
    Chimica Fine e Secondaria

National Science Foundation

Aeritalia

Alitalia

Azienda Autonoma di Soggiorno, Cura e Turismo di Capri

Boeing

Centro Richerche Fiat

Comitato Nazionale per l'Energia Nucleare

Costruzioni Aeronautiche "G. Agusta"

Hooker Chemical

Industrie Pirelli

Istituto Donegani (Montedison)

Owens-Corning Fiberglas

Regione Campania

Regla Fiberglass

Universita' di Napoli

University of Washington

CONTENTS

CHEMICAL AND ENVIRONMENTAL EFFECTS

SHORT FIBER REINFORCEMENT EFFECTS

INDUSTRY–ACADEMIA AND INTERNATIONAL
COLLABORATIVE RESEARCH EFFORTS

INTRODUCTORY REMARKS

J.C. Seferis * and L. Nicolais**

*University of Washington
 Seattle, Washington, USA
**University of Naples
 Naples, Italy

Composite materials with polymeric matrices are emerging as strong candidates for load bearing structural applications in the commercial airplane and automotive industries. Furthermore, in view of the energy shortage, the composite's low weight coupled with energy efficient operations during their processing makes them one of the most desirable materials for use in the future. However, unlike the metals which the composites are replacing in structural applications, their processing characteristics and their final properties can be greatly affected by chemical composition, the load conditions and the various environments under which they perform. Consequently, the problem of understanding and predicting the influence of these effects on the composite material performance has attracted attention from both academic and industrial researchers due to its complexity and relatively prominent role in providing short- and long-term solutions to many needs of the transportation industry.

The polymeric matrix and its interaction with a reinforcing phase, in the form of continuous or discontinuous high strength and stiffness fibers, is one of the major controlling factors in the processing and property characteristics of composites. For high performance load bearing applications, the epoxy polymeric matrix has resulted as the dominant form of matrix material. Although traditionally the matrix has been thought to play a passive role in composite use and development with major emphasis being placed on the fiber performance alone, the more demanding recent uses of composites require that polymeric matrix plays an increasingly important role in composite performance. The wide variety of epoxy polymers available, as well as the numerous other polymeric materials both of thermoplastic and thermosetting types that can be tailormade

1

to meet specific performance characteristics, necessitate a clear
identification of property and processing requirements of the
material to be used as a matrix in composites. Accordingly, the
papers in this volume address explicitly the effects of processing
and properties of the polymeric matrix on a variety of topics providing an unusual comprehensive coverage of the composite material
field.

The papers in this volume have been organized in six major sub-
sections. First, the introductory section provides an excellent
overview of composite research, its successes, accomplishments,
current needs, and sets the stage for specific contributions grouped
within four major topics within our overall theme:

        Chemical and Environmental Effects
        Short Fiber Reinforcements
        Interfacial Effects
        Continuous Fiber Reinforcements and Design

Finally, as was stated earlier, one of the major strengths of
the Symposium was that it effected an interaction between industry,
academia, and government across international boundaries in dealing
with composite material research and development. Fittingly, the
last two contributions grouped as the fifth topic provide coverage
of the presentations made in this area as an integral part of our
theme in composites.

We hope that collectively this volume will provide and stimu-
late further research in this area, as well as serve as useful
reference for ongoing activities in the composites field.

THE ROLE OF THE MATRIX

IN FIBROUS COMPOSITE STRUCTURES

John C. Halpin

ASD Engineering
Aeronautical Systems Division
Wright-Patterson AFB OH 45433

INTRODUCTION

The purpose of this communication is to provide an overview to advanced composite structural technology and the role played by the matrix material system in this technology. In developing this overview the topic discussed will address: why composites; and what are some of the technical issues unique or different about composites which affect structural design and/or their processing or fabrication via traditional metallic structure.

Why Composites?

Advanced composite systems consist of fibers with moduli (stiffness) and strength properties (generally graphite) superior to glass fiber technology embedded in matrix system; usually a resin. The technology became available to the engineering community about 1972 and has demonstrated a weight savings in aircraft structures of 20 to 25% as illustrated in Figure 1. The data in Figure 1 is for individual components which when assembled would constitute an aircraft. In their design these components were designed to the specification (including geometrical size) of existing aluminum structures. A new aircraft designed would experience this weight savings plus additional weight savings of 15% due to resizing of the vehicle. Resizing results because you need less fuel to travel the same distance, a smaller engine to propel the reduced mass of the airframe, smaller landing gear, etc. The combined effect of weight savings and a resizing effect is a total system mass or weight reduction of 28 to 33%. This weight reduction is within the current materials system and design capability. In fact the LEAR FAN Corporation is currently designing and certifying an all composite aircraft

3

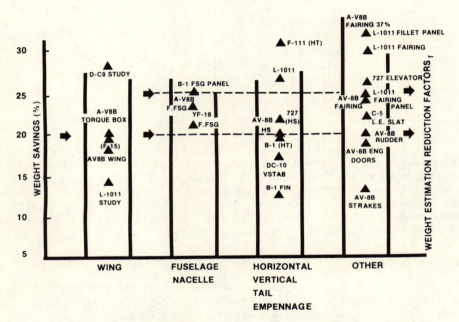

Fig. 1.   Illustrated Component Weight Experience

with these characteristics. Current studies suggest a resulting operation fuel savings of 7.5 to 10% in the weight of fuel burned to accomplish a flight requirement. The importance of fuel savings can be appreciated from Figure 2 which illustrates the impact of fuel costs escalating faster than general inflation. Note that fuel costs were about 11% of total life cycle costs of a fleet of aircraft. By the year 2000, with current technology that cost will increase to 45 to 50% of the total life cycle costs. Current estimates suggest that an all composite aircraft (45 to 60% composite material by weight) can be designed and produced for a cost comparable to conventional aluminum airframes. The cited fuel savings then become life cycle operating cost savings. Our society has reached a condition in which it is more economical to build the aircraft from hydrocarbon technology than burn the hydrocarbons in the engines of metallic vehicles.

The role of the matrix in the composite is: (a) transfers stresses between the fibers, (b) provides a barrier to contain or exclude an environment (pipe wall contains a fluid), (c) protects the surface of the fibers from mechanical abrasion, etc. This role will be illustrated through discussion related to moisture effects, mechanical properties including toughness and/or fracture resistance, and processing and quality control.

Moisture
<u>Moisture</u>

Water is a universally present compound capable of being absorbed into and diffusing through the epoxy based composite laminates. The primary mechanism for moisture absorption in epoxy composite laminates appears to be instantaneous surface absorption followed by diffusion through the resin phase. The rate of the apparent diffusion process and magnitude of the equilibrium weight gain is accelerated by the presence of cracks and voids in a laminate. A direct consequence of moisture absorption is a hydrostatic dilation of the matrix phase in the composite. The induced dilation of the matrix produces internal stresses which appear to produce some debonding of the resin from the fibers, a reversible change in the glass transition temperature which in turn produces a sensitivity to rapid heating rates (supersonic thermal spikes) which exceed the local glass transition temperature within the laminate.

The current data base may be summarized as follows:

1. <u>Moisture Induced Dilation of the Resin</u>

If additivity of volumes of the absorbed swelling agent and the matrix are realized[1] the induced dilation would be:

$$\Delta = \frac{\Delta V}{V_o} = \rho_r M_t \bar{V}_D \tag{1}$$

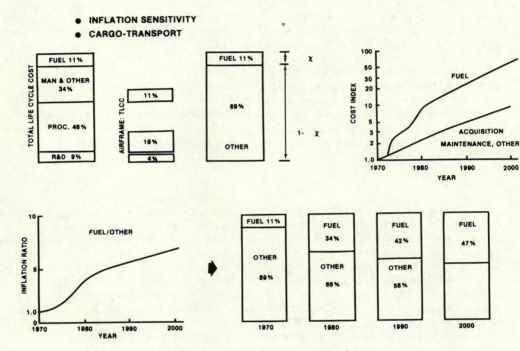

Fig. 2.   Total life cycle cost scaling for a cargo-
transport type of aircraft as a function of
the relative costs of fuel to other cost items.

within the restriction that the moisture distribution is uniform and
that the material is isotropic, the corresponding engineering strains
would be:

$$e_m = \frac{1}{3}\frac{\Delta V}{V_o} = \frac{1}{3}\rho_r M_t \overline{V}_D \tag{2}$$

The current experience for bulk resins is that the volumetric change
is less than the additivity of volumes.

Material	Δ Equals	$M_r(\infty)$
3501-5	$1.265M^{1.12}$	.067
3501-6	$1.265M^{1.18}$	.067
5208	$1.265M^{1.17}$	.060
5505 (2373)	$1.265M^{1.18}$	.100
934	$1.265M^{1.12}$	.075

While the induced dilation is somewhat less than originally expected
the maximum dilation strains are still quite large: 4 to 8%. The
Hercules 3501, Fiberite 934 and Narmco 5208 are of similar chemical
composition: tetraglyciclyl methylene dianiline (TGMDA) epoxy with
a diaminodiphenyl sulfone (DDS) curative.

2. Swelling Stresses

The induced dilation can produce significant internal
stresses on the bonds between the resin and fiber as well as either[2]
edge or internal stresses in the presence of concentration gradients.
For example, the bond between fiber and matrix will experience a[3]
hydrostatic tension, p, induced by both swelling and temperature:

$$p \sim \frac{E_m E_f}{(1 + V_m)E_f + (1 + V_f)E_m}\;[(\alpha_m - \alpha_f)\Delta T + e_m]$$

This process produces hydrostatic tensile stresses on the order of
the wet transverse strengths of typical graphite or glass epoxy
lamina. When these materials are taken to saturation, dried and re-
tested at room temperature there appears to be an irreversible loss
(25 to 50%) in the interlaminar shear strength as well as the transverse strain allowables. It is also generally observed that the
modulus properties are unaffected. Apparently the induced stresses
at high degree of swelling fractures the resin bond to produce the
degradation in shear and transverse properties. It is generally
observed that the stiffness properties of the resins are reversibly

altered by moisture absorption and redrying.

3. The Glass Transition Temperature

The induced dilation accompanying moisture absorption depresses the glass-transition temperature (the temperature at which a resin goes from a hard-to-soft material) of the resin in both bulk resin samples and laminates. This effect is well known in the physics of polymers[4] and may be evaluated from the following expressions:

$$Tg(wet) = \frac{\alpha_r Tg(resin/dry)v_r + \alpha_{H_2O}Tg(H_2O)v_{H_2O}}{\alpha_r v_r + \alpha_{H_2O}v_{H_2O}} \tag{3}$$

where

$$v_r = \frac{V_o}{V_o + \Delta H_2O} = \frac{1}{1 + \rho_r M_t \bar{v}_D}$$

$$v_{H_2O} = 1 - v_r$$

$Tg(H_2O)$ - assume approximately 4°C

$\alpha_{H_2O} \simeq 4 \times 10^{-3}/°C$

$\alpha_r \simeq 3.78 \times 10^{-4}/°C$ (Epoxy)

$M_t \simeq$ wt fraction moisture gain

$\rho_r \sim 1.265 - 1.28$ g/cm^3

$\bar{v}_{H_2O} \sim 1.0$ g/cm$^3 = \bar{V}_D$

All of the above cited resins follow this relationship. In general, these resins have dry glass transition temperature on the order of 400-470°F. C. Browning[5] has observed a hundred-fold change (E_R = 300,000 psi → 3,000 psi) in modulus at 300°F as the resin Tg was reduced from 420°F to 260°F. The general experience in the industry is that the matrix sensitive allowables at elevated temperature (ε_{12}, $-\varepsilon_{11}$, etc.) are translated down the temperature in proportion to Tg (dry) - Tg (wet @ M_t). The engineering impact is that the current resin based materials are still adequate for low mach number applications at current design levels (ε[max fatigue spectrum] $\sim$ 0.002), but the engineering design margins are smaller than hoped for at temperature. The situation is moderated by the relatively low load levels incurred at temperature and the infrequent occurrence of peak temperature. High temperature curing systems, such as cited above, are very adequate for subsonic application. The usage of 250°F curing resins does produce a sensitivity to environment for subsonic usage comparable to the supersonic-high temperature cure resins and should be utilized with caution. The narrowing of the engineering margins due to these effects is placing additional pressure on engineering data development during the design verification procedures.

4.   Absorption

The assessment of the engineering impact of the moisture sensitivity of epoxy materials and adhesives is dependent upon our understanding of the rate of in-service moisture accumulation. It is generally accepted[1,6-8] that moisture absorption is described by Fick's first and second laws for diffusion and a statement regarding the equilibrium moisture absorption: $M(\infty)$. In general, the material is described by a temperature dependent diffusion coefficient, D, and a solubly term $M(\infty)$. The equilibrium moisture content is a function of two variables: (1) the fraction of resin, voids and fiber in the composite laminate; and (2) the vapor pressure of the moisture. The weight gain is primarily due to the moisture absorption of the resin matrix and is given as:

$$M_t(comp) = \frac{\rho_r X_r}{\rho_{comp}} M_t \ (resin) \qquad\qquad (4)$$

@t $\rightarrow \infty$

$$M_\infty(comp) \cong A(RH)^b.$$

where ρ_r is the density of the fibers (approximately 1.75 gm/cm^3), x_r is mass fraction of the resin and RH is the relative humidity divided by 100 at the specified temperature; A is M_t (t = ∞, composite) using above cited equation with $M_t(\infty)$ data of 1. The power of b is generally found by experiment to vary between 1 and 2 with most laminates. The physical cause of this is uncertain at this time but it appears to be related to the degree of resin cure. Typical data is contained in Table 1.   Table 1 also includes the associated permeability D_o (mm^2/sec) and activation energies E (cal/gm) for the transverse diffusivity characteristics.

$$D = D_o \ exp - (E/RT) \qquad\qquad (5)$$

of the cited systems. R is the gas constant and T is the temperature in degrees Kelvin. The transverse diffusivities of the composite are related to the resin characteristics through the expression[9].

$$D(comp) = D(resin)(1-2) \ \pi/v_r) \qquad\qquad (6)$$

J. C. HALPIN

TABLE 1. Comparative Properties of Current Material Systems (9-12)

QUANTITY MATERIAL	A, Absorbed Moisture Concentration @ RH = 1.0	b, exponent	D_o, Average Permeability Index (mm^2/sec)	E, Activation Energy (cal/gm)
RESIN (Neat)				
3501-5	0.065	1.5	16.1	11,700
3501-6	0.065	1.5		
5208	0.063	1.5	3.5	10,530
934	0.075	1.5	10.6	11,027
5505(2373)	0.092	1.5		
PREPREG				
T300/5208	0.0055 to 0.0070	∿ 2.0	6.15	∿ 10,000
LAMINATE				
AS/3501-5	0.0185	1.75	17.66	12,082
AS/3501-6	0.016	1.2	1.01	10,200
T300/5208	0.0155	1.0 to 1.5	0.49	10,143
Boron/5505	0.0225	1.5	3.4	9,190

The above cited relationships must be modified for porous laminates. In general, current fabrication procedures utilize high pressure, 75-100 psi, autoclave cure techniques which yield void contents of less than 1%. Current studies on reducing or eliminating autoclave pressure generally produce laminates which may contain 5 to 10% void content. The presence of these voids has a dramatic effect on increasing the equilibrium moisture uptake $M(\infty)$ and the diffusion coefficient, D_y. An analysis of the existing data base supports the following relationships:

$$\frac{M}{M_o} = \frac{M\ (voids)}{M\ (zero\ voids)} = $$

$$1 + \frac{\rho_{H_2O}\overline{v}_r}{M_t(resin)}\ [\frac{1}{v_R} - 1] \tag{7}$$

where v_R is the volume fraction of resin in the three phase composite of fiber, resin and voids:

$$v_R = \frac{1 - v_f}{1 - v_f(1 - v_v)}$$

with v_v being the void fraction. A two to threefold moisture increase is expected. The essential assumption in the derivation is that the voids accumulate liquid water. For water immersion experience, the equation does correlate with the data. Complementing this increase in apparent solubility is an enhanced diffusion coefficient:

$$\frac{D \text{ (void)}}{D \text{ (zero voids)}} = [\frac{M\text{(void)}}{M_o}]1.4 \tag{8}$$

which implies a three to fivefold increase in the apparent diffusion coefficient. This increase is larger than expected from the micromechanics estimates for closed pore structure, implying that some of the voids in laminates are probably open to the surface. This would promote a more rapid convection diffusion mechanism.

5.   In-Service Expectations

        Advanced composite structures will be used in a variety of geographical locations with varying temperature and relative humidities. The climatic variations have been studied in the following manners at various geographical locations:

        (1)   Historical mean monthly relative humidities $[\overline{RH}]_M$ and temperatures were evaluated;

        (2)   Average annual relative humidities $[\overline{RH}]_A$ and temperatures were established based on step (1);

        (3)   The delta monthly change, $\Delta[\overline{RH}]_M$ from the high monthly average $[\overline{RH}]_{MH}$ to the lowest monthly average $[RH]_{ML}$ relative humidity and temperature was established;

        (4)   For each month the average daily high $[\overline{RH}]_{MDH}$, and the average daily low $[RH]_{MDL}$ relative humidity and temperature was established; and

        (5)   The average delta change $\Delta[\overline{RH}]_{MDA}$, between the month's highs $[RH]_{MDH}$ and lows $[RH]_{MDL}$ relative humidity and temperature was established.

(6) The diurnal cycle was then modeled as a harmonic
variation.

$$[RH(t)] = [\overline{RH}]_A - \frac{1}{2} \Delta[\overline{RH}]_M \cos 2\pi(\frac{t}{8640})$$

$$+ \frac{1}{2} [\overline{RH}]_{MDA} \cos \pi(\frac{t}{24}). \tag{9}$$

where t is defined as day in-service with t equal to zero on
January 1. A similar expression was also developed for temperature
variation, [T(t)]. The values for [RH(t)] and [T(t)] were inserted
into expressions (4) and (5); which were in turn used with Fick's
second law to establish variations in absorbed moisture as a function
of laminate thickness, years in-service and the expected geopherical
history of the structures. Based upon this analysis a "design"
moisutre weight gain has been established:

$$M_L(t \sim 20 \text{ years}) \sim 1.1 \text{ to } 1.2 \text{ percent}$$

which suggests

$$M_R(t \sim 20 \text{ years}) \sim 4.0 \text{ to } 4.5 \text{ percent}$$

and a glass transition temperature (3) of the resin at the end of
service of 250° (121°C) to 270°F (132°C) for a dry initial Tg of
400°F (204°C). These conditions are comparable to a steady state
condition of 70 percent relative humidity and 75°F (24°C). The
Tg(wet) exceeds the maximum temperature requirements for a subsonic
aircraft and is compatible with low Mach number supersonic aircraft
applications. The current TGMDA-DDS formulation of Table 1 represents
the best hot wet capability of current epoxy technology. Higher
temperature application, that the class of structures illustrate in
Figure 1 will require new or alternative matrix systems as discussed
in subsequent presentations. The general approach for alternative
matrix systems is to seek either higher Tg material or material which
absorps less moisture and/or other solvents to yield a higher in
service Tg(t≃20 years service).

Mechanical Properties

In addition to the thermal limits the matrix dictates, at least,
four essential mechanical characteristics of a unidirectional
lamina and consequently the structural laminates:

(1) The in-plane shear stiffness G_{12};

(2) The longitudinal compression strength $-\sigma_{11}$ or strain $-\varepsilon_{11}$;

(3) The transverse strength σ_{22} or strain ε_{22}; and

(4) The fracture, fatigue and impact properties on the lamination plane (similar to 90° lamina). Clearly the resin must maintain a high shear modulus, G_r, to provide an adequate G_{12}. Coupled with this is the correlation of compression capability with the shear stiffness of the resin.

$$-\sigma_{11}(max) \approx 0.63 \frac{G_r (T \sim Tg-100\%)}{V_M} \tag{10}$$

for current fiber-matrix systems. As temperature increased compression strength or strain is lost at the same rate as the in-plane shear stiffness or resin shear stiffness

$$\frac{-\varepsilon_{11}(T)}{-\varepsilon_{11}(T_o)} \sim \frac{-\sigma_{11}(T)}{-\sigma_{11}(T_o)} \sim \frac{G_{12}(T)}{G_{12}(T_o)} \sim \frac{G_r(T)}{G_r(T_o)} \tag{11}$$

The consequence of this relationship is that longitudinal lamina compression strength properties which are greater than the tension strength properties at room temperature are significantly lower than the tension strength at elevated temperatures dry or wet:

TABLE 2 Comparative Longitudinal Properties (T300/52308)

Temperature (°F)	Strain to Failure		T To	Variation of Properties From Room Temp.				
	Tension $+\varepsilon_{11}$(Dry)	Compression $-\varepsilon_{11}$(Dry)		Compression			Tension	
			D	W(0.6%)	W(1.2%)	D	W(0.6%)	
-65			1.1	1.2		>1.2	0.90	0.87
75	0.0095	0.0105	1.0	1.0	1.0	1.0	1.0	1.0
220			0.95	0.70		0.55	1.08	1.1
250		.0077						

This effect impacts those designs in which symmetrical design loads require comparable tension and compression properties. Also note that the tension side of the design becomes the limiting factor below 0°F. This result suggests the matrix volume changes (shrinkages) from curing influence longitudinal tensile strength. Two points are to be made here:

(1) A high Tg resin is required to maintain balanced design properties, and

(2) Temperature – environmental limitations are recognized in establishing design "allowable" properties; such as Table 2 hot-wet reduced properties. This design procedure limits the stress and/or

loads and design temperature regions such that viscoelastic effects
associated with the resin glass transition region are not a factor
in the design.

Other attributes of the structural operating environment for
composite structures is the events associated with low velocity
transverse impacts (tool drop, foreign objects, hail, etcetra. These
impacts, which are generally perpendicular to the lamination plane,
result in delamination damage to the laminate as well as transverse
matrix cracking in the lamina constituting the laminate. The damage
occurs because of the low toughness properties on the lamination
plane, G_{1c} (INTERLAMINAR) is roughly 0.5 to 1.5 in lbs/in^2, where
as the in-plane toughness, across the fiber direction, is G_{1c} (in-
plane) $\sim$ 200 in lbs/in^2 for a $0_2/+45$ for a typical graphite-epoxy
laminate. Two hundred in lbs/in^2 toughness is typical of aluminum
properties which are the same in-plane and through the thickness
direction. The low inter-plane toughness properties of the laminated
composites are a reflection of the brittle character of high Tg
epoxy formulations. Low velocity impact damage, results in reduced
compression strength. This damage may also grow as delamination
fatigue cracks in the normal service usage to impare compression at
some time later than the impact event. The objective is to be able
to withstand normal service low velocity impacts (Kinetic Energy loads
of 7 to 8 foot pounds, 0.5 to .75 inch diameter indentations)
without inspection and repair being required. The systems summarized
in Table 1 are limited in this capability. An improvement of
interlaminar toughness from 0.75 to 3 to 10 in lbs/in^2 would insure
robust durability for composite structures. This improvement would
also benefit the delamination growth resistance to the manufacturing
variability which results in delaminations in the as-manufactured
structure. Associated with that toughness would be improved resin
elongation capabilities which would compliment increased strength
or strain-to-failure (1% to 1.3 to 1.5% strain) fibers under development.

These high strain fibers imply the potential for increased
unidirectional tensile and compression (3) strength in the lamina.
The expectations of increased compression strength are ill-founded
as the matrix must have sufficient stiffness to utilize the increased
compression capability of the fibers. Significantly increasing the
glassy modulus of resin systems is beyond the technical nature of
the molecular potential fields which dictate glass-state properties.
The higher strength fiber would benefit bolted joint design (improved
bolt bearing-tension interactions) but that topic is beyond this
discussion.

This section is summarized as follows:

(1) Improved resins with a significant improvement in toughness
while retaining the hot-wet properties of Moisture are desired;

(2)   Traditional procedures (rubber-toughening, etcetra) conflict with the high glass temperature requirement; and

(3)   The material selection criteria should be expanded to include:

   (a)   Transverse laminar tensile strain; ε_{22};

   (b)   Interlaminar shear strength;

   (c)   Interlaminar toughness G_{1c} and G_{2c} or K_{1c} and K_{2c}; and

   (d)   Interlaminar fatigue crack growth rate characteristics (for example da(half crack size)/dn(number of load cycles) versus ΔG_1 and ΔG_2; or ΔK_1, and ΔK_2.

Processing and Quality Control

   Epoxy resin technology is utilized because it has relatively robust and tolerant processing characteristics. The basic approach in the industry is a prepreg concept wherein a partially cured resin-fiber form is fabricated and shipped to a primary manufacturer where the final structural part is manufactured including shaping and curing of the prepreg. In effect the manufacturer, who matured in the metallic technology, is now required to manage and complete a chemical reaction in the facility. The processing and manufacturing performance at the manufacturer's facility is a function of:

   (1)   the starting monomeric stoichiometry,

   (2)   matrix formulation/mixing procedures,

   (3)   the "controlled" amount of oligomeric material (partial reaction at the material vendor's site called "B-staging"),

   (4)   the time-temperature history of the prepreg until the material is worked in the manufacturer's facility; and

   (5)   the manufacturer's facilities environment (humidity, temperature) and work procedures (out time, etcetra). For example; the manufacturer's task is to consolidate prepreg lamina into a cured laminate with minimum porosity, delimination and uniform ply or lamina thickness through the thickness direction. The prepreg material must be pliable and "tacky" at room temperature to facilitate shaping: this requires Tg (prepreg) less than room temperature. Since the basic unreacted formulation has a glass transition tempera-ture of about 5°F(-15°C) typical prepreg reaches the manufacturer with about 10% reaction completion. During layup; the uncured resin is exposed to a warm moist environment. The chemical reaction

continues; if it reaches 30% or so the prepreg Tg is above room
temperature and the working characteristics are lost. While the
prepreg is being worked it is absorbing moisture. For a facility
with 85% RH the prepreg weight gain would be about 0.4% in twelve
hours. Typical working times for large parts are 48 to 96 hours.
This means that for each pound of material going into the autoclave
about 2200 cubic centimeters of water vapor (at STD temperature and
pressure) would be given off as the cure temperature exceeds 212°F
(100°C). For thin parts, this is not much of a problem. For thick
parts say 60 to 100 to 200 plies of material, the associated mass
transport and rheology characteristics lead to cured products with
porosity and delamination. It is for this reason that extensive
work is currently going on in the chemical composition, for epoxy
based systems. Very basic work is currently being done to understand the combined exothermic chemical kinetics, mass transport
and resin rheology characteristics of resins. The results of these
studies will not only guide new resin development, but establish
a scientific basis for cure cycle selection including the possibility
of nonautoclave cure processes, specifications on manufacturing
environments; and a rational basis for material specification and
control.

CONCLUSION

     Advanced composite technology is a technology whose time has
come. Subsonic aircraft application is within the current technological capability. The utilization of this technology is renewing
the polymer scientist's interest in thermosetting technology. Significant improvements are expected in matrix and fiber characteristics
with the continued maturing of the advanced composite structures
technology. These improvements will be the product of the research
community such as that represented in this symposium.

REFERENCES

1. C. D. Shirrell, J. C. Halpin, "Moisture Absorption and Desorption in Epoxy Composite Laminates," ASTM STP 617, 514, ASTM 1977.

2. T. Alfrey, E. F. Gurnee, and W. G. Lloyd, "Diffusion in Glassy Polymers," J. Polym. Sci, C12 249 (1966).

3. L. J. Broutman and F. J. McGarry, Mod. Plastics, 40(1), 161 (1962).

4. F. Bueche, Physical Properties of Polymers, Interscience, N.Y. (1962)

5. C. E. Browning, "The Mechanisms of Elevated Temperature Property Losses in High Performance Structural Epoxy Resin Matrix Material After Exposures to High Humidity Environments," PhD Thesis, University of Dayton, 1976.

6. G. T. vanAmerongen, Rubber Chemistry Techn., 37, 1065 (1964).

7. E. L. McKague, et al, "Moisture in Composites: The Effect of Supersonic Service on Diffusion," J. Comp. Mats. 9, 2 (1975).

8. E. L. McKague, J. D. Reynolds and J. E. Halkais, "Moisture Diffusion in Fiber Reinforced Plastics," ASME J. Eng. Materials and Technol., (1976).

9. C. H. Shen, G. S. Springer, "Moisture Absorption and Desorption of Composite Materials," J. Comp. Materials, 10, 2 (1976).

10. R. DeIasi and J. B. Whiteside, "Effect of Moisture on Epoxy Resins and Composites," Advanced Composite Materials-Environmental Effects, ASTM STP 658, J. R. Vinson, Ed., American Society for Testing and Materials, 1978, p.2.

11. C. D. Shirrell, "Diffusion of Water Vapor in Graphite/Epoxy Composites," Advanced Composite Materials-Environmental Effects, ASTM STP 658, J. R. Vinson, Ed., American Society for Testing and Materials, 1978, p.21.

12. G. S. Springer, "Environmental Effects on Epoxy Matrix Composites," Composite Materials: Testing and Design, ASTM STP 647, S. W. Tsai, Ed., American Society for Testing and Materials, 1979, p. 291.

13. B. W. Rosen, in Fiber Composite Materials, American Society for Metals, Metals Park, Ohio, (1966); Chapter 3, L. B. Greszczuk, "Compressive Strength and Failure Modes of Unidirectional Composites," Analysis of the Test Methods for High Modulus Fiber and Composites, ASTM STP 521, ASTM, 192-217 (1973).

COMPOSITES IN COMMERCIAL AIRCRAFT

D. T. Lovell

The Boeing Company

Seattle, Washington

This paper reviews (1) the present usage of advanced composites on Boeing airplanes, (2) the future importance of composites and why Boeing is interested in motivating basic research on them, and (3) the kind of research that we believe has the greatest benefit and payoff to industry and the research community. Specific topics include application of advanced composites to the Boeing 767, the international nature of the 767 program, key developments that we feel are required in advanced composites, and examples of research conducted jointly by industry and universities, such as Boeing and the University of Washington.

Figure 1 shows the application of advanced composites on the Boeing 767 airplane, including Kevlar, graphite, and hybrid (Kevlar and graphite) composite structures. Of key significance is the use of graphite composites for primary flight-control structures. The 767 is the first Boeing airplane on which composites have replaced aluminum alloys in a major production application. These composite applications have resulted in 20 to 25% weight savings.

Figure 2 demonstrates that the 767 is a world airplane. The 767 is an excellent example of national and international production collaboration, in that major components are made all over the world. For example, certain body sections are made by Commercial Transport Development Company (CTDC) at Nagoya, Gifu, and Utsunomiya, Japan, and the empennage structure and flight-control surfaces are made by Aeritalia (AIT) at Pomigliano D'Arco and Torino, Italy. Figure 3 shows the 767 graphite-epoxy composite rudder made at AIT. The part is 34 feet long and is the largest graphite-epoxy airplane structure currently in production.

The programs with AIT and CTDC have been very successful, and we are proud of these efforts. AIT and CTDC are program participants, not subcontractors, and operate under a special risk-sharing agreement. The initial

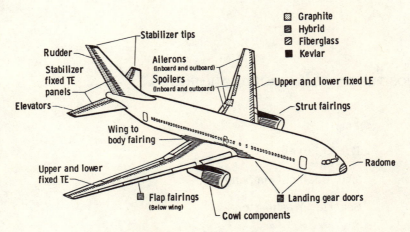

Fig. 1. Composite applications—767.

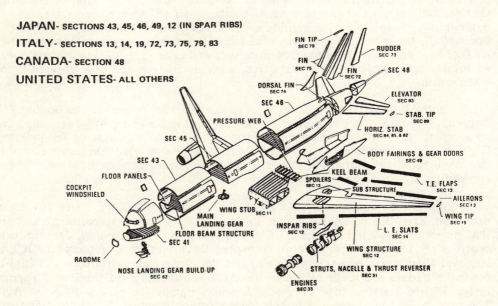

Fig. 2. Fabrication sources—767.

technical program and development effort included direct engineering interface with AIT and CTDC engineers located at Boeing facilities. In this technology transfer phase, a preproduction program was conducted at Boeing. The program then was transferred to Italy and Japan with complete technical integration by Boeing engineers. All material and processing requirements are in accordance with Boeing specifications.

Fig. 3. Graphite-epoxy composite rudder—767.

A summary of composite usage on Boeing airplanes, starting with fiberglass on the B52, is shown in Figure 4. Boeing greatly increased fiberglass usage from the B52 to the 747, from approximately 1 to 28% of total wetted (exterior) area. As can be seen in Figure 4 for the 767 and 757, there is significant use of graphite and graphite-Kevlar hybrid structure and relatively little use of fiberglass. Figure 5 illustrates structural material usage on the 747 and 767 as a percentage of structural weight. The 747 structure is 82% aluminum, 13% steel (e.g., landing gear), 4% titanium (we use a lot of titanium fasteners, as well as titanium landing-gear beams, which are the largest titanium forgings in the world), and 1% fiberglass. On the 767, the material relationships are similar, except that the use of composites increased from 1 to 3%, with about 2.5% being advanced Kevlar and graphite systems.

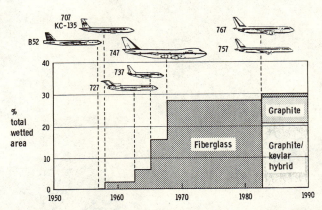

Fig. 4.   Advanced composite materials—structural usage trends.

Figure 6 shows a comparative distribution of major direct operating costs for commercial aircraft in 1973 and 1980. The illustration reveals that fuel is now the primary factor in total direct operating costs, due to the fuel cost portion increasing from 21 to 55%. This creates a requirement for airplanes with higher performance and lower weight to reduce fuel burn. Advanced composite structure offers the greatest potential for weight savings in future airplanes, thus accounting for the rapidly increasing interest in composite design, development, and application. It is estimated that advanced composites could comprise up to 65% of the structural weight of the next generation of commercial airplanes (Fig. 7).

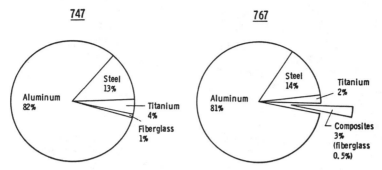

Fig. 5. Structural materials usage—747 and 767.

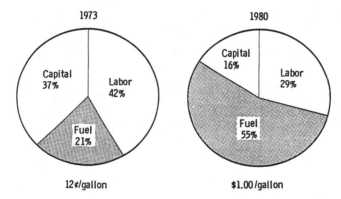

Fig. 6. Direct operating cost elements—1,000-nmi trip.

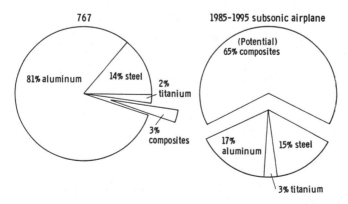

Fig. 7. Advanced technology airplane—materials weight distribution.

We have three major goals for structural composite applications. First, we want to be able to use composites in primary structure such as wings, empennage, and fuselage. In doing this, we can increase airplane performance by reducing weight. For primary structural applications, the most significant improvement desired is increasing the strain to fracture (elongation) of advanced composite materials. This is the key parameter for structural design, equivalent to the strength parameters in metals. As shown in Figure 8, the objective is to develop an improved fiber and matrix system with a 2% minimum elongation. The second goal is to develop new processing methods for composite structure to reduce manufacturing costs. Today, the processing of advanced composites is very labor intensive. New, automated processing methods will overcome this drawback and also will offer the designer greater flexibility. New processing methods include filament winding, mechanized and/or automated layup, pultrusion, and various forms of molding. It would be very desirable to have the same matrix material system available for all these processing methods, but we will probably need combinations of other systems such as thermoplastic or hybrid systems. The third and final goal is to use composites for high-temperature applications, up to and possibly greater than 500°F (260°C). To achieve these three major goals, improvement of the basic matrix resin system is the most crucial development required.

The key technical research challenges that we see from the engineering standpoint are shown in Figure 9. The first is consistency of properties. We need to establish an approach similar to that used for metals (i.e., a metal model), where we have consistent industry testing standards and industry-accepted data parameters. Improvements in environmental durability, toughness, impact and fatigue, and interfacial integrity of the fiber and matrix are other key challenges. Finally, process control, specifications, inspectability, and repairability are vital for cost-effective manufacturing. Each of these challenges will be discussed briefly below.

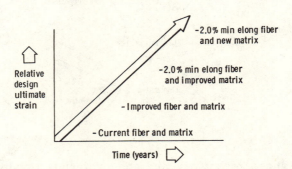

Fig. 8.   Advanced composites—fiber and matrix developments.

• Consistency of properties

• Environmental durability

• Durability of matrix and fiber

• Interfacial integrity

• Processing control

• Inspectability (NDT)

• Repairability

Fig. 9.   Technical challenges—key areas.

In terms of consistency of properties, the principal considerations are to improve basic strength and strength under high-temperature and high-moisture conditions, reduce data variation, and improve properties in zones with undetected damage. Figure 10 schematically illustrates the effect of these factors in reducing the allowable design strain in present advanced composites. All of these factors are matrix dominated and require improvements in the matrix system to increase the design values.

In terms of toughness and impact properties, there has already been significant improvement over currently used composite systems. Figure 11 shows stress-strain curves developed during impact testing of composite structure. The P_f point represents the maximum load carried by the specimen during impact. By this point, visible damage has occurred. In the "A" system, P_i, incipient damage load, is approximately equal to P_f. "A" systems typically have high modulus and are glassy matrix systems in which a crack formed at the P_i load propagates relatively easily, giving P_f values approximately at the P_i level. Some tougher systems, such as the "B" system, have P_f values higher than the P_i load. The "B" system shown here is an elastomer-modified form of a high-modulus, glassy matrix. The P_i values in the "B" system occur at the same load as do P_i values in the "A" system due to crack formation in the continuous, glassy parent matrix phase. Increased P_f values are due to increased resistance to crack propagation permitted by the elastomer modifications. Ideally, we would like to have a system with a higher P_i to reduce the likelihood of incipient damage and also with a much higher P_f to be able to tolerate damage. Achieving such a system will require use of thermosets with elastomer modification, thermoplastics, and hybrid thermoplastic/thermoset systems. What is desired is a system that will respond similarly to metal systems, where we have the phenomenon of yielding.

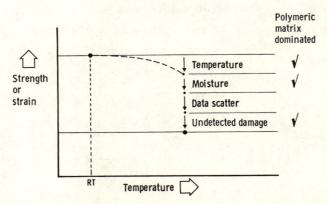

Fig. 10.   Advanced composites—variability of properties—reduction factors.

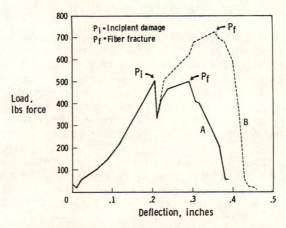

Fig. 11.   Durability of matrix and fiber—impact testing.

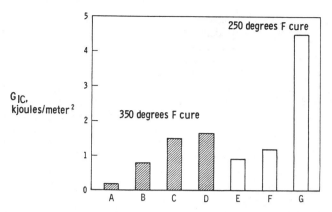

Fig. 12.　Composite materials—interlaminar fracture toughness.

A major structural concern is preventing interply delamination. Figure 12 shows a comparison of the delamination resistance of various matrix systems on graphite fibers. The delamination resistance is determined by measuring the energy required to move a delamination crack one unit of area in a composite specimen. Figure 12 shows the mode I (or opening mode) values. The various matrix systems use both 250°F (121°C) and 350°F (177°C) curing systems. In general, elastomer-modified systems have higher mode I fracture toughness (G_{IC}) values than do nonmodified systems. However, systems such as the "G" system shown in Figure 12 often can be developed with high G_{IC} values but with unacceptable mechanical properties (e.g., compression strength). Thus, one must always look at the balance of properties of new systems. Also, one can see from Figure 12 that cure temperature per se does not determine system toughness. We do not know the limits of achievable toughness with acceptable mechanical properties.

Interfacial bond integrity is another basic area that requires development. Again, a balanced system is desired. We do not necessarily want a brittle resin with excellent adhesion, since this creates stress concentration in the fiber elements. Likewise, we do not want a system where adhesive failure occurs and the resin slips over the fiber. The desirable compromise is where the resin can carry the fiber to its ultimate strength capability. When failure occurs in this case, the resin will contain the damage and allow other fibers to pick up the load.

Key engineering requirements are included in material processing specifications to control processing and inspection. In composite specifications of the 1970s, the major control requirements were mechanical properties. In the 1980s, we anticipate that chemical characterization, behavior characterization (viscoelastic/rheological performance), environmental testing, and quality demonstration will be the major specification requirements. The most recent Boeing specifications include many of these requirements.

Boeing has several cooperative advanced composite research programs with the University of Washington. Figure 13 shows one such program for evaluating the environmental durability of advanced composites by measuring the change in moisture and the viscoelastic moduli. Dr. Seferis (University of Washington) and Dr. Nicolais (University of Naples) are working on these tests with two different types of apparatus. By understanding the effects of dynamic moisture ingress and egress on molecular structures, we hope to design a system with a new molecular structure that will have better moisture resistance.

In summary, advanced composites are now successfully used in production airplane components and have the potential for future application in a major portion (approximately 65%) of commercial airplane structure. We believe that an improved resin matrix is the most critical development required for the successful application of advanced composites to airplane primary structure. This effort would encompass two key elements: (1) basic research and (2) applied development. Basic research is considered to cover chemistry/morphology of the matrix, compatibility of the matrix with different processing systems, and other areas such as environmental, rheological, and physical characteristics and fiber/matrix integrity. Applied development would include the actual development of specification data for production control, the service and structural parameters required for a particular design, the actual design techniques and methods, and the actual production "know-how." There is a definite need for government, industry, and universities to work together in basic research on resin matrix systems. In the applied development area, however, collaboration will be much more restricted, mainly due to the proprietary nature of applied development information generated by each company.

Problem: Environmental durability of graphite/epoxy composites

Approach: Measure changes in H_2O gain/loss and
visco-elastic moduli

Investigator: Dr. J. C. Seferis and Dr. G. Nicolais

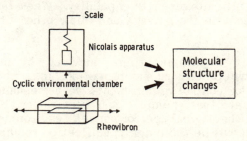

Fig. 13.   Advanced composites—Boeing/University of Washington program.

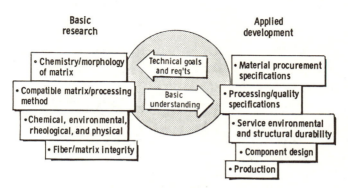

Fig. 14. Advanced composites—research and development summary.

We believe that basic research has the greatest potential benefit to industry and the research community. As a key link between basic research and applied development, industry needs to establish technical goals and requirements and then communicate these to the research community (Fig. 14). The research community, in turn, can conduct basic research in the desired areas with support from industry funding and establish fundamental understandings that industry can use directly for applied development.

Today, we develop specification requirements and run many tests and statistical analyses to project design and processing controls to ensure that materials will be durable for the life of an airplane (approximately 20 years). For the future use of advanced composites in primary structure, we need a more basic understanding of these materials. We need to understand, for example, molecular structure versus bulk properties. A more fundamental understanding of the interrelationship of matrix properties also is required. This is one major area where industry needs more input from the research community (Fig. 15).

Figure 16 shows the structural material usage trend in commercial airplanes over the last 50 years and projected usage for the next 20 years. The Boeing goal for the next 10 years is to increase the rate of materials improvement over 10 times the rate that was obtained during the last 50 years. We hope to accomplish this primarily through increased use of advanced composite structure.

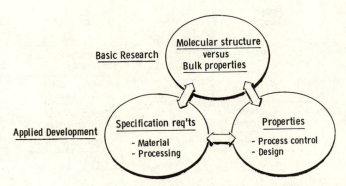

Figure 15.   Advanced composites—proposed future R&D relationships.

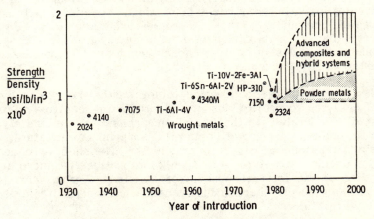

Fig. 16.   Competitive materials—structural materials trend.

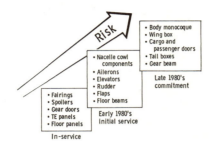

Fig. 17.   Advanced composites—increasing risk and payoff.

Finally, it must be realized that as the payoff for use of advanced composites becomes greater, the risk also increases (Fig. 17).   Therefore, industry and research communities need to generate viable procedures for the mutual assessment of these risks. In doing this work, a major thrust will be to develop new matrix and fiber systems.   It will be crucial for academia, government, material producers, and the aircraft industry to work together effectively without compromising proprietary concerns. The challenge is great, but the 1980s promise to be an exciting decade for material engineers.

CURRENT MATERIAL RESEARCH NEEDS

IN AIRCRAFT STRUCTURES

Claudio Voto

Material and Process Engineering
Aeritalia - Via Privata
80038 Pomigliano d'Arco - Napoli - Italy

Abstract

To-day's structural materials have in fact been available to the aircraft industry for the last twenty-forty years.

Only lately the definition of more sophisticated test methods determined substantial improvements in their structural efficiency.

We're referring to a better detail design, better corrosion protection, improved fatigue and damage resistance analysis, and refined non-destructive tests.

These are the main parameters that justify the need for a continuous, future development of a new generation of structural materials: let's read composites.

The goal is the improvement of the productivity and the economy of the operative life.

Introduction

The fatigue resistance and the damage tolerance of the structures are the main design parameters of an aircraft.

Environmental durability goes along with the compliance with the strength and stiffness requirements, while aiming at a structure as light as possible.

That's essential for the fuel economy and the maintenance costs.

It appears evident the role of structural materials
in reducing the aircraft weight while ensuring safe, durable and costeffective structures.

Graphite/epoxy composites

The concept of composites materials explains their
potential over other materials in meeting this goals and
the specific characteristics of graphite-epoxy make them
particularly attractive for the aerospace applications.

They are characterized by a continuous phase (the
matrix) embedding the fibrous phase (the graphite fibers).
The matrix stabilizes the fibers, keeping them all together, allowing to get the required shape of the structural components. The fibers are responsible for the
mechanical properties of the material. The specific
strength and stiffness of graphite fibers offer the potential for major weight savings in aircraft design.
The loads go along the fibers rather than going across
them. The possibility of tailoring the material to the
structural needs, highly increases the efficiency of the
composite part.

Research objectives

The application of graphite-epoxy composites in the
primary structures of commercial airplanes needs further
development in three main areas of research:

    °   Material technology
    °   Structural durability
    °   Quality control

To-day's material limits are constituted by both
the matrix and fibers strain characteristics.Formulations
improvements and additional data on microcracking,impact
damage,fatigue flaw growth and environmental effects
will lead to a material with higher toughness and environmental durability, improved allowables and increased
weight savings.

Another important parameter is the fabrication costs.
A higher volume of production will be attained with au-
tomatization and implementation of innovative techniques,
like pultrusion and filament winding.

The consistency and uniformity of incoming material and final products are characteristics as important as well to ensure the application of composites in the area of primary structures.

Material technology

The versatility of epoxy resins is a really valid reason for concentrating on their development as a tougher matrix for composites.

A wide spectrum of resins, hardeners and modifiers are available; the viscosity can vary from very low to high.

The cure latitude is ample, from room temperature to 350 F.

The toughness can vary, depending upon the distance between cross link points. The low shrinkage during cure allows a good adhesion to fibers. Mechanical properties are higher than most other cast resins. The chemical resistance to alkalies, acids and solvents is good, as are the thermal and dimensional stabilities.

The main disadvantage is in the long term durability because of the degradation due to ultraviolet radiation and humidity. Ultraviolets react with epoxy with a photoxidation reaction, while the humidity absorption causes the plasticization of the matrix.

During and after the exposure there are changes in the composition. Microcracking originated in the matrix is increased by thermal cycles with subsequent increase in water absorption. The consequence is a drop of the mechanical characteristics in the matrix and at the interface.

Composite specifications for primary structure require the control of material and processing.

The goal is the consistency of chemistry and the capability of foreseeing mechanical properties and environmental durability.
There are several variables in the material that impact the final properties of the composite part. We refer to the catalyst level, molecular weight distribution cross-linking density, formulation and formulation sequence, fiber property variation, fiber finish, fabric weave.

Final mechanical properties are dependent as well on the process variables: different "B" staging, storage conditions, fabrication environment, cure parameters. The key factors to this investigation is the chemical and rheological characterization.

The subject of chemical characterization is the determination of the prepreg components, their quantitative evaluation and their correlation with the mechanical characteristics. The analysis objective is the control of the product consistency and long term durability.

Strictly tied to this must be the rheological characterization of the material, to get a basic understanding of the kinetic of reaction, the changes in viscosity and the aging behavior of the resin system.

Another important area of investigation is that of corrosion.

Altough resin matrix composites are considered organic in nature, graphite-epoxy have a cathodic behavior towards the airframe material (aluminum mostly) coming in contact with them. The phenomenon is dependent on the amount of moisture which penetrates the joint interface and is made worse by the unfavorable ratio of the cathod/anod surfaces. Not enough data are available to-day to make prediction on life expectancy of assemblies. We only know that protection is required and further development is needed for evaluating the effectiveness of coatings.

The best approach is the definition of inhibiting systems allowing to eliminate the cathodic surfaces of graphite-epoxy and drop down the extra weight due to the protective systems.

Structural durability

Graphite-epoxy structures have an elastic behavior to the failure with a fragile fracture. Microcracking in the resin determines limiting allowable strains.

Today's design is based on the criteria of damage tolerance. This implies the understanding of fracture toughness and fatigue crack growth characteristics. These properties are dependent upon several parameters, like form, thickness, chemical composition, environment. The evaluation of the effects of these variables on fracture properties, will allow to set up the requirements of minimum fracture toughness.

Analytical procedures supported by test data will characterize the fatigue and crack propagation behavior of composites and the residual strength after fatigue cycling and cracking as well. It is essential to certify the environmental durability that is the effects of temperature and humidity.

In fact the graphite-epoxy impact resistance is low. There are several sources of damage, like punctures during handling, or tool drop impact, or runway debris. The extension of damage is dependent on several parameters, like the geometry of impactor, the energy of impact and its location. The consequence is the degradation of mechanical characteristics, like tension, compression and fatigue.

Then there is on one hand the need of determining the extension of this degradation (i.e. matrix damage, fiber damage, penetration), on the other one the requirement of developing a resin more resistant to minor impact damage. Other kinds of damages can be generated on the production line, like delaminated skins, misaligned holes, misaligned edges, voids, crushed core, foreign object inclusions.

On one hand acceptance/rejection criteria must be set up, evaluating the type of defects, their limits to the purpose of ensuring the structural integrity with the support of test data.

On the other one, repair methods must be defined for different types of structures (sandwiches, laminates, joints). Static and fatigue characteristics must be evaluated to the purpose of ensuring the part reliability.

Quality control

Today's receiving inspection of incoming prepregs is based on the evaluation of physical and mechanical characteristics of the material. More stringent controls are required for the primary structure applications. Only chemical fingerprinting can make sure of the consistency and quality of the resin matrix, through the control of type, purity, concentration of ingredients and homogeneity of the resin mix. This consistency control is the best warranty of the final product quality.

On the other hand, better non destructive inspection techniques are required to fully evaluate the type and extension of defects in the cured parts (like flaw,voids, delaminations).

Overall, there is the final need of testing standardization, to get generally accepted, reliable test methods, like for metal parts.

Conclusion

The graphite-epoxy materials have the potential of bringing a real strong improvement in the structural efficiency of airplanes. The data that are going to be cumulated out of the secondary structure application,together with a development effort in the indicated areas of research will allow the application in major airframe that only will significantly improve the aircraft performances.

CHARACTERIZATION OF THE MATRIX GLASS TRANSITION IN CARBON-EPOXY

LAMINATES USING THE CSD TEST GEOMETRY

S. S. Sternstein and P. Yang

Rensselaer Polytechnic Institute

Troy, New York 12181

INTRODUCTION

The mechanical properties of carbon-epoxy laminates display limited viscoelastic behavior when tested in the plane of the laminate for most stacking sequences. While there are certain stacking sequences which display strong matrix dependence, such as a zero degree layup tested in the 90° direction, these layups are not generally used for structural components. It would appear that a need exists for a test geometry which is sensitive to matrix behavior and which can be performed on standard (structural) stacking sequences.

In this paper, a newly developed test geometry is described which has several useful attributes: (1) the geometry is very sensitive to matrix behavior; (2) both transient (creep and stress relaxation) and dynamic mechanical tests are readily performed over a wide range of timescales; (3) the test geometry provides viscoelastic dispersion data which are independent of angular orientation of the sample; consequently sample alignment problems are eliminated; (4) the same test geometry may be used to provide information on delamination strength; (5) relatively small test specimens are required; (6) the test geometry is sufficiently sensitive to matrix changes to allow its use for postcuring, humidity, crosslink density and other matrix change studies.

TEST GEOMETRY

The test geometry used to obtain the data presented here will

be referred to as the centro-symmetric deformation (hereafter CSD)
test geometry. The sample consists of a thin disc, typically 6
to 14 plys thick, having a nominal diameter of 30 mm. The disc
is freely supported on a circular anvil around its entire perimeter at a diameter of 25 mm. The load is applied to the center
of the disc using a ball bearing nosepiece having a diameter of
8 mm. The sample is not clamped in anyway; consequently, the
sample may be mounted on the test fixture in seconds. The lack
of clamping also provides for a high degree of reproducibility
in the test method since variations of clamping stresses and jaw
slippage are a major source of errors and lack of reproducibility,
especially for stiff samples.

The absence of clamping requires that the load must always
be applied in the same direction as shown in Figure 1. This presents no difficulty in transient testing; however, in dynamic
tests it is necessary to apply a static load at least as large
as the peak to center amplitude of the dynamic sinusoidal load,
thereby insuring that the ball bearing nosepiece remains in contact with the sample. This posed no problem in the apparatus
used since the static and dynamic loads are independently pro-
grammed when in closed loop load control.

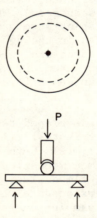

Figure 1.   Schematic of the CSD Test Geometry.

With reference to Figure 1, the sample is subjected to a
circularly symmetric set of boundary conditions. If R is the
radius of the anvil support, then there are no deflections or
mements (freely supported) at R and W (R) = 0 and M (r) = 0
where W (r) is the out of plane deflection of the disc and M (r)
is the out of plane moment. An analysis of the CSD geometry has

been performed by Taggart (1) who shows that to a good first approximation the centerline deflection W (0) is independent of stacking sequence and depends only on ply properties and the applied load P.

Regardless of the validity of the independence of stacking sequence, it is clear from the circularly symmetric boundary conditions that the orientation of a given sample (given stacking sequence) can have no effect on the test results. This is in marked contrast to the more conventional 3-point bend test where the strip orientation with respect to the bend axis is a primary variable. This is a decided advantage of the CSD geometry since it is not necessary to align or even keep track of the direction of the first ply fiber direction when mounting the sample.

APPARATUS

The instrument used to obtain the data reported here was developed in this laboratory over a number of years and is now available commercially. Briefly, the main viscoelastic test device is a closed-loop linear motion apparatus that is capable of both dynamic and transient tests in either load or displacement control. This instrument known as a "Dynastat" contains several patented (2) components including a D.C. coupled electromagnetic motor circuit. Temperature programming and control to 0.1 C were obtained using a "Dynatherm". Dynamic moduli information were obtained using a Fourier transform analyzer known as a "Dynalyzer" which enabled accurate in-phase and out-of-phase moduli to be obtained at displacement amplitudes as low as 10 microns. Scaling of the data, plotting of graphs and storage of data on a floppy disc were all performed using a "Datalyzer". Additional details of the instruments are given elsewhere (3).

SAMPLE PREPARATION AND PROCEDURE

Laminated plates approximately 30 cm x 30 cm were prepared using prepreg manufactured by the Fiberite Co. and designated HYE-1048A1E. This prepreg contains T-300 fibers in a 948 epoxy matrix. Twelve ply symmetric laminates were prepared using three stacking sequences each of which had a fixed angle increment from ply to ply, for example, the so-called 90 degree laminates have the stacking sequence $[0,90,0,90,0,90]_s$ whereas the 45 degree laminates have the stacking sequence $[0,45,90,135,0,45]_s$.

The layups were vacuum bagged and cured in a press using the following history: 60 minutes at 105°C; 120 minutes at 150°C and 0.69 MPA. The CSD test pieces were cut from the cured plates using a diamond coring bit in a milling machine but without the use of coolant. All samples were postcured prior to testing.

The data presented in this paper were obtained on the 90 degree stacking sequence samples only. Additional data on the effects of stacking sequence, moisture, postcuring and the behavior of neat resin will be cited later.

All samples were tested using the CSD test geometry. A static load of 3 Kg was superposed on the dynamic load of 1.0 Kg. Dynamic frequency was swept between 0.1 Hz and 25 Hz at all test temperatures. Variations in both the static and dynamic load levels have been used to verify that the data are in the so-called linear viscoelastic regime.

PREPRESENTATION OF RESULTS

Linear viscoelastic dynamic data may be represented in a variety of ways. For the reader's convenience, the interrelationships are presented here. For linear viscoelastic behaviro, a sinusoidal stress σ produces a sinusoidal strain ε, the ratio of which is designated here by M*.

$$\frac{\sigma}{\varepsilon} = M* = M' + i M'' \tag{1}$$

where M' and M'' are the in-phase and out-of-phase components of the dynamic modulus, respectively. The magnitude of the modulus $|M*|$ is given by

$$|M*| = [(M')^2 + (M'')^2]^{\frac{1}{2}} \tag{2}$$

and the loss factor tan δ by

$$\tan \delta = M''/M' \tag{3}$$

Wedgewood and Seferis (4) have proposed that the components of the complex compliance are more sensitive to certain changes in the viscoelastic behavior of composites than the components of the complex modulus. Without commenting on the validity of this proposal for the time being, the complex compliance is defined by

$$\frac{\varepsilon}{\sigma} = S* = S' - iS'' \tag{4}$$

which when combined with equation 1 and 2 gives

$$S' = \frac{M'}{|M*|^2} \quad \text{and} \quad S'' = \frac{M''}{|M*|^2} \tag{5}$$

The complex viscosity $\eta*$ is defined by the ratio of dynamic stress to dynamic strain rate:

$$\frac{\sigma}{\dot{\varepsilon}} = \frac{\sigma}{i\omega\varepsilon} = \eta* = \eta - i\eta'' \tag{6}$$

Combining equations 6 and 1 gives

$$\eta' = \frac{M''}{\omega} \quad \text{and} \quad \eta'' \quad \text{and} \quad \frac{M'}{\omega} \tag{7}$$

where $\omega = 2\pi f$ and f is the frequency in Hertz.

RESULTS

A temperature sweep of the dynamic properties of a typical 12 ply carbon-epoxy laminate at a fixed frequency of 6.5 Hz is shown in Figure 2 where the magnitude of the modulus $|M*|$ and the loss factor (tan δ) are given. The pronounced loss factor peak at ca. 160°C is the glass transition of the eopxy matrix. It is obvious that the CSD test geometry is quite sensitive to matrix softening. Additional evidence will be cited later.

An alternate representation of these data would be to plot the components M' and M''. For example, the loss factor is compared with the out-of-phase modulus M'' in Figure 3. It is interesting to note that the peak in loss factor occurs at a higher temperature than does the peak in M''. Thus, one must be careful in comparing glass transition temperatures obtained by different dynamic methods. For example, a damping measurement (e.g. torsion pendulum) generally gives the loss factor and not M''.

The same data may also be represented in terms of the components of the complex compliance as shown in Figure 4. Note that S'' is plotted on a log scale whereas S' is on a linear scale. Again, it is interesting to note that the peak in S'' occurs at a different temperature (ca. 170°C) than the peak in M'' (Figure 3). The "correct" dynamic mechanical glass transition temperature depends therefore on one's definition and preference for plotting of the data.

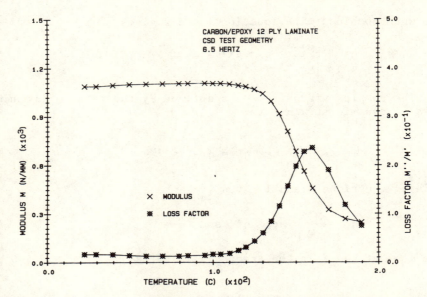

Figure 2.   Complex modulus magnitude and loss factor versus
             temperature at 6.5 Hz for a 12 ply carbon–epoxy
             laminate tested in the CSD mode.

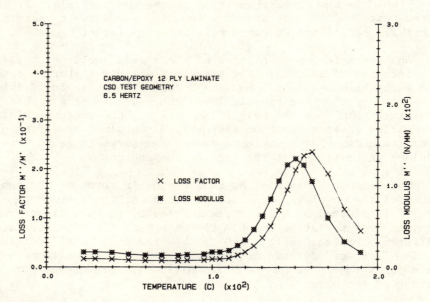

Figure 3.   Comparison of loss factor and loss modulus peaks.

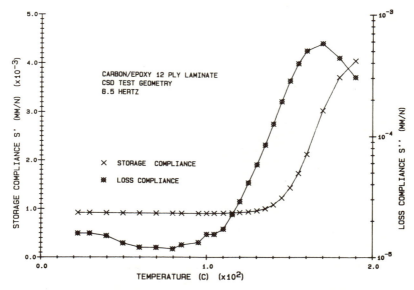

Figure 4.   The representation of the data in Figure 2 in terms
            of complex compliance components.

     It should be noted that the data in Figures 2-4 were ob-
tained on a postcured sample which was then exposed to moisture
and subsequently dried. Our experience indicates that such a
sample history gives a glass transition temperature which slowly
increases. Thus, the M'' peak temperature in Figure 3 will increase from 160°C to 165°C after several repetitions of the thermal sweep shown in the figure. A very detailed study of the
effects of moisture, both prior to and after postcuring will be
published in the near future (5).

     The data shown in Figures 5-10 were obtained on a postcured
sample which was kept dry at all times.

     A series of frequency sweeps at a number of temperatures
are shown in Figure 5 where the in phase (or storage) compliance
is given and in Figure 6 where the out of phase (or loss) compliance is given. When these data are shifted horizontally on
the frequency axis it is found that they superpose to form master
curves of the storage compliance (Figure 7) and loss compliance
(Figure 8). The shift factors required to obtain the superposition are shown in Figure 9. The master curves shown in Figures 7 and 8 are plotted for a reference temperature of 165°C
which is, by definition, the temperature where the shift factor
(Figure 9) is unity.

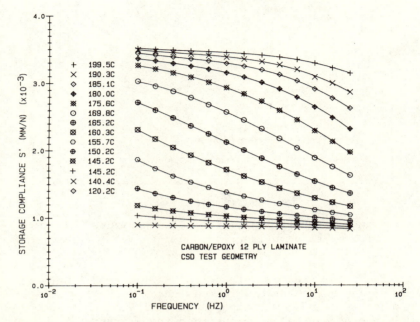

Figure 5.   Storage compliance versus frequency for several
            temperatures.

        We have shown previously (see ref. 6) that the storage and
loss components of the complex modulus form master curves and
that these master curves are the same for the three stacking
sequences investigated when scaled vertically by a simple scaling
factor.   Thus, it would appear that the viscoelastic behavior of
the matrix is independent of stacking sequence.   Furthermore, it
was shown that the master curves of neat resin also coincided
with the master curves for the composite laminates but that the
glass transition temperature of the in situ resin was about 12°C
higher than that of the neat resin.

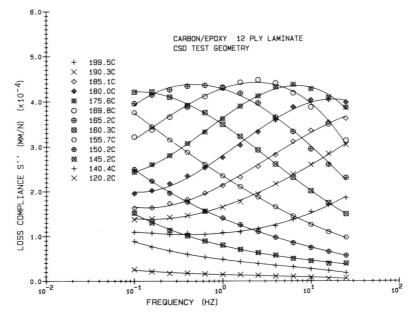

Figure 6.   Loss compliance versus frequency for several
            temperatures.

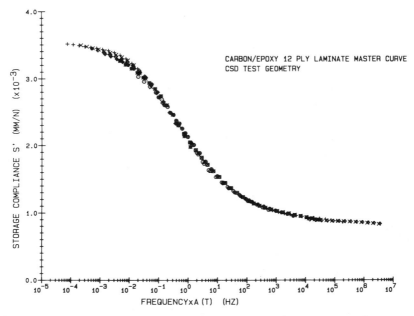

Figure 7.   Master curve for storage compliance.   Reference
            temperature is 165°C.

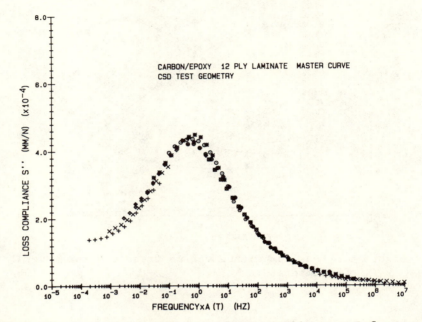

Figure 8.   Master curve for the loss compliance.   Reference
temperature is 165°C.

The results shown in Figures 7 and 8 are no better or worse
than our previous master curves of the complex modulus components.
The major difference in the two methods of presenting the present
data (moduli vs. compliances) is the pronounced broadening of the
low frequence side of the loss compliance peak (Figure 8) relative
to the master curve of loss modulus (ref. 6).   Insofar as the in
phase (storage) compliance master curve is concerned (Figure 7),
its low and high frequency limits display somewhat more slope

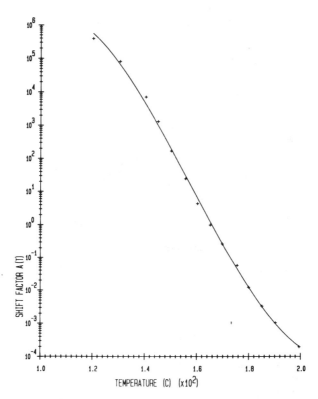

Figure 9.   Shift factor versus temperature.

than does the master curve of storage modulus which was virtually flat at low and high frequencies. It would appear then
that presentation of the data as compliance components rather
than as moduli components serves to emphasize the data extremes
(low and high temperature or frequencies) of the master curves.
Further consideration of the Wedgewood and Seferis (4) suggestion is justified.

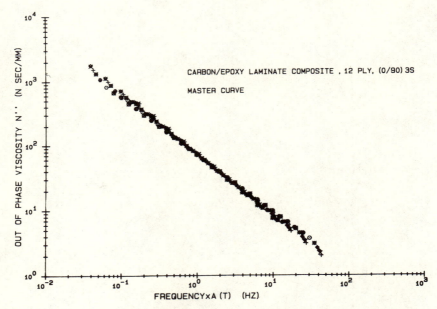

Figure 10.   Out of phase viscosity master curve.   Reference
temperature is 165°C.

Additional insight into the nature of the viscoelastic
dispersion of in-situ epoxy resin is gained by presenting the
data of Figure 5 as the out of phase viscosity (see equations
5 and 7). The master curve is shown in Figure 10. This curve
is virtually a straight line with a log–log slope of –0.9. A
relaxation time spectrum analysis of Figure 10 is currently in
progress; however, the linearity of the curve allows a very quick
estimate of the shape of the relaxation time spectrum, namely a
wedge shaped spectrum of very flat slope approximating a "box",
that is

$$H(\tau) \propto \tau^{-0.1} \tag{8}$$

We close this paper by noting that the CSD geometry has also
been used to study the delamination failure of the carbon-epoxy
laminates described here (7,1).

ACKNOWLEDGMENTS

      This work was supported by NASA/AFOSR under grant NGL-33-
018-003 and is a contribution from the Composites Program at
Rensselaer Polytechnic Institute.

REFERENCES

1. David Taggart, "Delamination Failure of Laminated Composite
        Materials: A Theoretical Analysis", M.S. Thesis,
        Rensselaer Polytechnic Institute, Troy, N.Y. (May, 1981),
        Advisor: S. S. Sternstein.
2. S. S. Sternstein, U.S. Patent 4,096,741.
3. S. S. Sternstein, "Transient and Dynamic Characterization of
        Viscoelastic Solids", Submitted to A.C.S. Advances in
        Chemistry No. 203.
4. A. R. Wedgewood and J. C. Seferis, Polymer, 22, 966 (1981).
5. P. Yang and S. S. Sternstein, to be published.
6. P. Yang, L. Carlsson and S. S. Sternstein, "Dynamic Mechani-
        cal Response of Graphite/Epoxy Composite Laminates and
        Neat Resin", Submitted to Polymer Composites.
7. Carl Altman, "Delamination Failure of Laminated Composite
        Materials: An Experimental Evaluation", M.S. Thesis,
        Rensselaer Polytechnic Institute, Troy, N.Y. (August,
        1981), Advisor: S. S. Sternstein.

NETWORK STRUCTURE DESCRIPTION AND ANALYSIS OF AMINE-CURED EPOXY MATRICES

Helen S. Chu[*] and James C. Seferis[**]

Department of Chemical Engineering
University of Washington
Seattle, Washington 98195

INTRODUCTION

Epoxy polymers have emerged as the dominant form of matrix material employed in the manufacture of carbon fiber reinforced composite parts for load bearing structural applications. The variety of epoxy systems that are available, as well as the fact that their properties depend strongly on the network structure they form, has provided a strong motivation for studies on the description of the network.

High performance epoxy systems that retain their properties to high temperatures are usually made up of one or more multifunctional epoxies and are cured in the presence of multifunctional amines. The chemical structure, functionality, and composition of the constituent components, as well as the extent to which they react, can all influence the resulting network structure. The variety of reactions that may be involved during the cure cycle of the epoxy system makes the detailed theoretical determination of the network structure quite complex and in most cases intractable. Furthermore, the fact that the majority of high performance epoxy systems are quite rigid and insoluble in most solvents eliminates experimental procedures which could be used to provide a direct method for the determination of fundamental network parameters like molecular weight, crosslink density, and molecular weight between crosslinks. These difficulties have led to a variety of

---

 * Present Address: Boeing Commercial Airplane Company
** Author to whom correspondence should be addressed

experimental and theoretical techniques that can provide indirectly
information about the network structure of these epoxy systems.
The most widely used techniques have dealt with dynamic mechanical
(1,2,3,4,5) and swelling (4,6) measurements on samples made under
carefully controlled and known conditions. These experiments, when
coupled with theoretical information about the network structure,
have the potential of becoming extremely useful for the characterization of the epoxy networks. The work of Bell and Murayama (6,7)
has demonstrated the feasibility of this approach. Bell's formulation (7), making simplifying experimentally verifiable assumptions about the reactions that may take place during the cure of
the epoxy system, successfully estimated the molecular weight
between crosslinks of the network, based on the stoichiometry of
the curing reaction and the amount of primary amino and epoxide
groups remaining in the system at a given time. This modeling
approach was developed specifically to describe an epoxy system made
up of a difunctional epoxy and a tetrafunctional diamine when the
curing reaction is carried to completion. The model was capable of
providing a description of the resulting network structure for the
cases where stoichiometric amounts of epoxy and amine were used or
when excess epoxy or excess amine was used. The total molecular
weight and molecular weight between crosslinks for such a network
could then be simply calculated from stoichiometric considerations
alone. However, for high performance epoxy systems which are made
up of two or more multifunctional epoxies and cured with a bulky
tetrafunctional amine, stoichiometric considerations alone are not
sufficient to describe their resulting network structure. Dynamic
mechanical experiments coupled with a first approximation modeling
approach by Keenan and Seferis (5,8) have shown that for high performance epoxy system the cure may not be carried to completion due
to steric and diffusional limitations that may arise as a result of
the developed epoxy network structure. Accordingly, a probabilistic
model for the description of the network structure for a tetrafunctional epoxy cured with a tetrafunctional amine was developed
(8). This modeling approach provided qualitative agreement with
the observed dynamic mechanical response of a commercial system (5)
as well as demonstrated the potential to take into account other
structural considerations that have been observed for the same epoxy
system by Morgan and O'Neal (10).

Our previously reported dynamic mechanical experiments (5)
were performed specifically on a high performance epoxy system which
was made up of a mixture of two different epoxies: tetraglycidyl
4,4'-diaminodiphenyl methane (TGDDM) and polyglycidyl ether of Bisphenol A Novalac. The epoxy mixture was cured with a tetrafunction-
amine: 4,4'-diamino diphenyl sulfone (DDS). Consequently, our system contained the primary constituents of several commercially
important resins including the NARMCO 5208 epoxy (14). These resin
systems have also attracted the attention of numerous investigators

(9,11,12,13,14), providing a large data bank for network structure and property comparisons. Accordingly, the present study is focused on describing with a probabilistic approach, the network structure of an epoxy system which is made up of one or two distinct epoxies with functionalities of two or larger and is cured with a tetra-functional diamine. For clarity in the development that follows, the probabilistic model is presented specifically for the TGDDM-Novalac-DDS epoxy system. However, the model is also presented in a generalized form so that its results can be applied and com-pared to other epoxy systems.

CHEMICAL, STRUCTURAL AND EXPERIMENTAL CONSIDERATIONS

The structure of each component in the TGDDM-Novalac-DDS epoxy system that is modeled in this study is shown schematically in Figure 1 along with an abbreviated symbolic representation that is used in the model development for clarity. Each of these com-ponents is available commercially. The TGDDM (Araldite MY 720) and DDS (Eporal Hardener 972) are available from CIBA-GEIGY while the Novalac (EPI-REZ-SU-8) is available from Celanese. The epoxy equivalent weight for each component (weight per epoxide group) provided by the manufacturer as well as the theoretical and equi-valent molecular weights and functionalities are listed in Table I. It is apparent from the values listed in the table that the com-mercially available epoxy components are not pure and may contain higher molecular weight species or, more likely, that not all of the epoxide groups indicated in the theoretical structural formula may be available for reaction. Furthermore, it should be noted that the structural formula provided for the Novalac component in Figure 1b is at best an averaged representation of its structure. The $-CH_2-$ group linking the diglycidyl ether of Bisphenol A units could be positioned on any of the available positions of the phenyl rings, resulting in a staggered arrangement of the Bisphenol A units. In addition, although detailed information is not disclosed by the manufacturer, the functionality depicted by the structural formula in Figure 1b may be high since it has been observed by gel permeation chromatography that the number of epoxide groups per Novalac molecule may range from four to eight (23). From the above considerations, it is essential to recognize the importance of pro-viding a general model whose input parameters can routinely accom-modate variable functionalities, molecular weights and epoxy equivalents for its constituent epoxy components. For simplicity, however, during the initial stages of the model development, the theoretically assumed structures of Figure 1 for the epoxy com-ponents will be utilized.

It will also be advantageous at this point to define the range of compositions of interest as well as the conditions under which

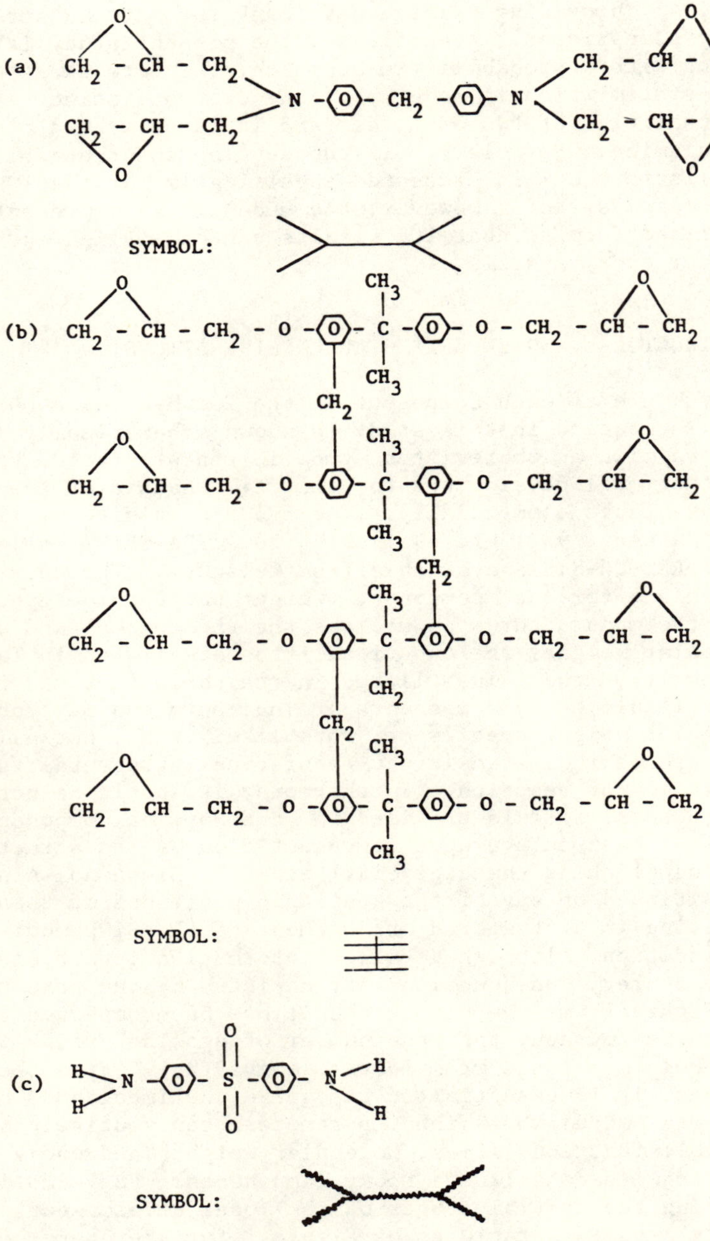

Fig. 1. Components of the TGDDM-Novalac-DDS system: (a) Tetragly-cidyl 4,4'-diaminodiphenylmethane (TGDDM); (b) Polyglycidyl ether of Bisphenol A Novalac; (c) 4,4'-diaminodiphenyl sulfone (DDS). An abbreviated symbolic representation for the structure of each component is also shown for later reference.

TABLE I

Molecular Weights and Epoxy Equivalents for
the Components of the TGDDM-Novalac-DDS System

Component	Theo. Epoxy Equiv.	Theo. M. W.	Exp. Epoxy Equiv.	Calcu- lated Equiv. M. W.	Theo. Function- ality	Symbol
TGDDM	105.5	422	125	500	4	
Novalac	175.3	1402	210	1722	8	
DDS	-	248	-	248	4	

the cure for this epoxy system is usually performed. These considerations will be useful when the model is compared with experimental results and/or when the various simplified assumptions that are introduced are critically examined.

To approximate a composition of commercial interest, a mixture of the epoxies consisting of 88.5% TGDDM and 11.5% Novalac by weight may be assumed to react with the DDS. Although the DDS content in such commercial systems is estimated (14) to range between 22 PHR and 25 PHR (parts by weight of DDS per hundred parts by weight of the two epoxies combined), it will be of interest to examine systems with greater variations in DDS content. It has been observed experimentally that the steric and diffusional limitations exhibited during the cure of this system are strongly dependent on the DDS content used in the formulation (5,8,10). Accordingly then, when the model is developed and presented for the TGDDM–Novalac–DDS system, variations with DDS content in the range of 19 PHR to 40 PHR would be of particular interest since epoxies with formulations in this range have been examined experimentally in other studies as well (5,8,10,28).

To be able also to place the modeling concepts within the experimental realm, the conditions under which the various components are reacted to produce the final crosslink network are also specified here. Furthermore, we have found that for ease of testing a variety of properties of these epoxies, they should be shaped in a film form. Accordingly, the following procedure was employed for making reproducible films: First, the TGDDM, Novalac components were premixed in the desired proportions, with the aid of a motor stirrer. The mixing temperature was set at 135°C. The two epoxies were mixed for 20 minutes first before DDS was added slowly. As soon as mixing was complete (usually in an additional 20 minutes) the reacting mixture was cooled to 0°C and stored at that temperature to prevent any additional polymerization reaction. When a film was desired, the premixed frozen mixture was melted and degassed in a vacuum oven at 121°C (250°F) for about 30 minutes. Subsequently, it was poured on to a film mold that was preheated to a temperature of 121°C. The mold consisted of two polished stainless steel plates perforated with tightening bolts at the edges and was spaced with shim stock so that the resulting epoxy film had a 0.01 cm thickness. After the resin was poured on the preheated mold, it was sealed by tightening the bolts and placed into an oven which was preheated at 177°C oven temperature. The mixture in the mold was allowed to cure for two additional hours at the 177°C temperature. The mold was then taken out of the oven, and allowed to cool to room temperature. The resulting films were stored in an ambient environment for two weeks prior to any testing.

The above procedure provides a general reference framework of conditions in which the components react. Although different curing procedures may produce different network structures with distinct property behavior, a model which is capable of describing the network structure of the epoxy should be able to accomplish this task independent of the processing conditions employed. These considerations are addressed explicitly, after the model development, in the discussion section where experimental results are compared with the model predictions.

PROBABILISTIC MODEL FORMULATION

For a model to provide a description of the network structure of a complex polymeric system, such as the epoxy system considered in this study, it must be based on simplifying assumptions in order to be tractable. The most important simplifying assumptions on which the model development is based are summarized below along with a brief description of the model formulation that follows. The validity as well as the applicability of the assumptions within the realm of experimental and theoretical evidence are discussed in detail in a subsequent section.

The major assumption in the formulation of the model concerns the possible reactions that may occur when an epoxy molecule is reacted with the amine. In the present model, as well as in Bell's (7) development, only the reaction between an epoxide group and an amino hydrogen is considered to be of importance. Homopolymerization of the epoxy molecules or the reaction between a hydroxyl and an epoxide group are assumed to occur at a negligibly small extent. Accordingly, the network structure is calculated on a probabilistic fashion by a simple calculation of the probability of the epoxide groups to react with an amino hydrogen. This probability, however, depends on the stoichiometry of the reactants used and the extent of completion of the epoxy-amine reaction.

Step-wise reactions of the primary and secondary amino hydrogens with the epoxide groups are assumed with the implication that the secondary amino hydrogen reaction can occur only after completion of the primary amino hydrogen reaction. However, the secondary amino hydrogen reaction is allowed to remain incomplete and its extent is denoted by a parameter $\gamma (0 \leq \gamma \leq 1)$. Thus, for an amine that has been reacted with an equal or excess stoichiometric composition of epoxies at a fixed reaction extent that is less than one, only a certain fraction of the total possible functional groups of the epoxy and amine molecules will have reacted. If three or more epoxide groups of the epoxy molecules or a secondary amino hydrogen of the amine molecules have reacted, a crosslink or junction point will have been created. From the probabilities of

the epoxide groups to react with an amino hydrogen, the fraction
of epoxy and amine molecules having one or more epoxide groups or
amino hydrogens reacted is calculated. This calculation then leads
to network parameter descriptions such as, for example, the number
of crosslinks and the molecular weight between crosslinks.

The primary and secondary amine reactions considered to occur
in a stepwise fashion to produce the final epoxy network structure
are shown in Figure 2. However, for an epoxy network containing a
mixture of different epoxies, some assumptions concerning their
relative reactivity with respect to the amine must also be made.
Chapman, et al (9) concluded that electron withdrawing groups would
increase the rate of reaction with nucleophilic curing agents, such
as amines, and decrease the rate of reaction with electrophilic
curing agents. In the TGDDM-Novalac-DDS system, the DDS is an
aromatic nucleophilic amine. However, the Novalac epoxy has an
electron withdrawing oxygen attached to its glycidyl group, while
the TGDDM has an electron donating nitrogen attached to its glycidyl
group. Consequently, the Novalac may be expected to be more reactive
than the TGDDM. In the development that follows, equal reactivity
for these two epoxies will be assumed initially for simplicity.
This assumption will be subsequently relaxed and differences in
reactivity will be accounted for in the general formulation of the
model.

Finally, it will be shown that the present probabilistic development is restricted to epoxy formulations in which stoichiometric
or excess amount of epoxies with respect to the amount of amine are
used. Although this restriction could be easily removed by incorporating additional considerations in the description of the
reaction extent, the fact that commercial formulations contain
excess amounts of epoxies make this restriction of the model a
desirable feature rather than a hindrance.

Primary Amine Reactions

For the TGDDM-Novalac-DDS epoxy system under consideration,
assume in general that there are a moles of DDS, b moles of TGDDM,
and c moles of Novalac epoxy. These amounts will result in $4a$ moles
of amino hydrogens, $4b$ moles of TGDDM epoxide groups and $8c$ moles
of Novalac epoxide groups to be present in the mixture before the
reactions begin. If attention is restricted to mixtures with
stoichiometric or excess epoxide groups (i.e., $8c+4b \geq 4a$), the probability of any epoxide group to have reacted with a primary amino
hydrogen (when this type of reaction has proceeded to completion)
is equal to the number of available primary amino hydrogens divided
by the total number of epoxide groups present. If equal reactivity
for the two epoxies (TGDDM and Novalac) is assumed, this probability
is given by:

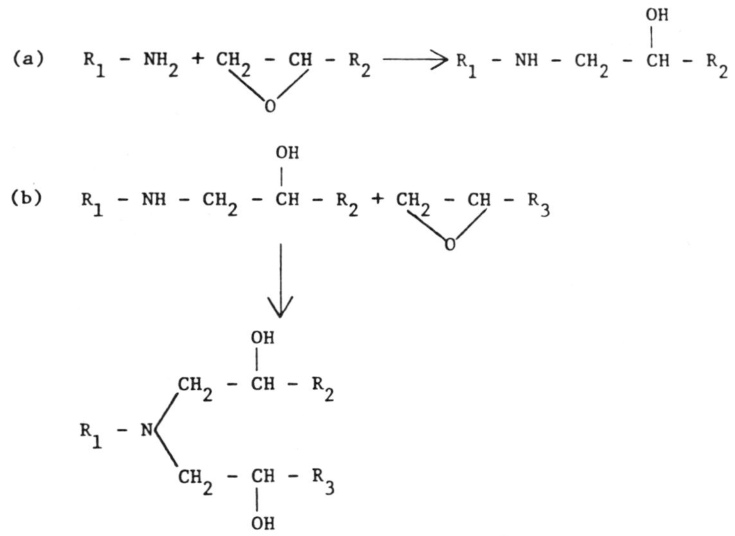

$$\text{(a)} \quad R_1 - NH_2 + \underset{O}{CH_2 - CH} - R_2 \longrightarrow R_1 - NH - CH_2 - \overset{OH}{\underset{|}{CH}} - R_2$$

$$\text{(b)} \quad R_1 - NH - CH_2 - \overset{OH}{\underset{|}{CH}} - R_2 + \underset{O}{CH_2 - CH} - R_3$$

$$R_1 - N \begin{cases} CH_2 - \overset{OH}{\underset{|}{CH}} - R_2 \\ CH_2 - \underset{|}{CH} - R_3 \\ OH \end{cases}$$

Fig. 2. Possible primary and secondary amine reactions of the DDS
molecules with epoxide groups of the TGDDM and Novalac
molecules considered by the model in forming and resulting
network structure of the TGDDM–Novalac–DDS epoxy system.

$$P' = \frac{2a}{4b+8c} \qquad (0 \le P' \le 1) \qquad\qquad (1a)$$

Consequently, the probability of having no epoxide groups reacted with primary amino hydrogens is given by:

$$Q' = 1 - P' \qquad (0 \le Q' \le 1) \qquad\qquad (1b)$$

As shown above, all probabilities are restricted to values in the range from 0 to 1. At any instant during the course of the curing reaction, an epoxy molecule may already have one, two, three .... or all of its epoxide groups reacted with an amino hydrogen. Each of these occurrences is denoted as the reaction state of the epoxy molecule. For each molecule, when no epoxide group has reacted, its reaction state is denoted as zero; when one epoxide group has reacted, its reaction state is denoted as two. Thus, reaction states on the epoxy molecules can be defined up to their functional-ities. The fraction of epoxy molecules present in a reaction state of i is expressed by N_i, where i can assume values from zero up to the functionality of the epoxy molecule under consideration. For example, N_0 would represent the fraction of epoxy molecules in a reaction state of zero while N_3 would represent the fraction of epoxy molecules in a reaction state of three. In general, the fraction of epoxy molecules in a reaction state of i is identical to the probability of existence of the molecule in such a state i. For a multifunctional epoxy molecule, this probability is simply the cumulative probability of the epoxide groups that belong to the epoxy molecule, to either have or have not reacted with the amino hydrogens. To clarify the above definitions, it will be instructive at this point to consider reaction states for the TGDDM and Novalac epoxy molecules separately when reaction with only a primary amino hydrogen has taken place. If we denote N_i' as the fraction of all the epoxy molecules (TGDDM and Novalac) in a reaction state of i, then N_i' expressed as the sum of individual fractions of the TGDDM and Novalac epoxy molecules in the same reaction state i is given by the simple relation:

$$N_i' = N_i^{T'} + N_i^{B'} \qquad\qquad (2)$$

where $N_i^{T'}$ and $N_i^{B'}$ denote the fraction of TGDDM and Novalac epoxy molecules, respectively, based on the total number of epoxy mol-ecules available in the TGDDM-Novalac mixture (i.e., $\sum_i N_i' = 1$).

   If a TGDDM epoxy molecule in the TGDDM-Novalac mixture has no epoxide groups that have reacted with a primary amino hydrogen, its reaction state by definition would be zero and denoted by $N_0^{T'}$. The probability for this occurrence will be the cumulative probability of all four epoxide groups that belong to the TGDDM epoxy molecule not to have reacted. From equation (1b) the probability of one

epoxide group not reacting was defined as Q', so the cumulative
probability for an epoxy molecule with a functionality of four in
a zero reaction state would be Q'^4. Since the probabilities defined by equations 1a and 1b were based on the total epoxide groups
available, the cumulative probability Q'^4 should be weighted by the
mole fraction of TGDDM epoxy molecules present in the TGDDM-Novalac
mixture in order to calculate the TGDDM fraction with a zero
reaction state (i.e., $N_0^T = \frac{b}{b+c} Q'^4$). Similarly, fractions for higher reaction states can be calculated. However, the number of possible arrangements by which reacted epoxide groups may distribute
themselves in the epoxy molecule must also be considered in the
calculation of the cumulative probability. This will be properly
taken into account if the cumulative probability is calculated
explicitly through the binomial distribution, viz.:

$$\text{Cumulative Probability} = \frac{n!}{i!(n-i)!}\, Q'^{n-i}P'^{i} \tag{3}$$

where n denotes the functionality of the epoxy molecule. Thus for
TGDDM molecules, which have a functionality of four (n=4), in a
reaction state of two (i=2) the fraction of TGDDM molecules in the
mixture will simply be $N_2^T = \frac{6b}{b+c} Q'^2P'^2$.

The same procedure is used to calculate the fraction of Novalac epoxy molecules in their i^{th} reaction state. However, in the
Novalac case, the cumulative probabilities should be weighted by a
mole fraction of the Novalac epoxy molecules in the mixture (i.e.,
by $\frac{c}{b+c}$). A summary of the fractions of epoxy molecules in the
various reaction states after reaction with a primary amino hydrogen for the TGDDM-Novalac epoxy mixture is given in Table II along
with a symbolic representation for each reaction state. The advantage of basing the probabilities on the total epoxide groups available in a mixture of TGDDM-Novalac epoxies is evident from this
table which demonstrates explicitly that the calculation for a
single epoxy system may be performed as a special case of a mixture
for which either the TGDDM (b=0) or the Novalac (c=0) epoxy is not
present.

To provide a feel for the fractions of epoxy molecules in the
various reaction states, the N_i^T and N_i^B were calculated as a
function of DDS content for a TGDDM-Novalac epoxy mixture made up
of 88.5% TGDDM and 11.5% Novalac by weight. The results are
plotted in Figure 3 as a function of DDS content from 12 to 58
PHR DDS. The significance of the compositions used for these
calculations was noted earlier inasmuch as they approximate systems
of commercial importance. It is noteworthy to observe from Figure 3
that within the range of 19 to 40 PHR DDS, the primary amine reaction can only account for a very low degree of polymerization.
Furthermore, in this range of DDS content the N_i^T and N_i^B values

TABLE II. Fractions of Epoxy Molecules in the Various Reaction States after Primary Amine Reaction of a TGDDM-Novalac Epoxy Mixture (o on symbolic representation signifies reactions of epoxide group with primary amino hydrogen).

Reaction State	TGDDM Fractions $N_1^{T'} = \frac{4!}{1!(4-1)!} Q'^{4-1} P'^1 \times \frac{b}{b+c}$	Symbol	Novalac Fractions $N_1^{B'} = \frac{8!}{1!(8-1)!} Q'^{8-1} P'^1 \times \frac{c}{b+c}$	Symbol	Fraction of Both Epoxies $N_1' = N_1^{T'} + N_1^{B'}$
0	$Q'^4 \times \frac{b}{b+c}$		$Q'^8 \times \frac{c}{b+c}$		$N_1' = N_1^{T'} + N_1^{B'}$
1	$4Q'^3 P' \times \frac{b}{b+c}$		$8Q'^7 P' \times \frac{c}{b+c}$		
2	$6Q'^2 P'^2 \times \frac{b}{b+c}$		$28Q'^6 P'^2 \times \frac{c}{b+c}$		
3	$4Q' P'^3 \times \frac{b}{b+c}$		$56Q'^5 P'^3 \times \frac{c}{b+c}$		$N_3' = N_3^{T'} + N_3^{B'}$
4	$P'^4 \times \frac{b}{b+c}$		$70Q'^4 P'^4 \times \frac{c}{b+c}$		$N_4' = N_4^{T'} + N_4^{B'}$
5	0		$56Q'^3 P'^5 \times \frac{c}{b+c}$		$N_5' = N_5^{B'}$
6	0		$28Q'^2 P'^6 \times \frac{c}{b+c}$		
7	0		$8Q' P'^7 \times \frac{c}{b+c}$		
8	0		$P'^8 \times \frac{c}{b+c}$		$N_8' = N_8^{B'}$
$\sum\limits_1 N_1'$	$\frac{b}{b+c}$	+	$\frac{c}{b+c}$	−	$1 - \frac{b+c}{b+c}$

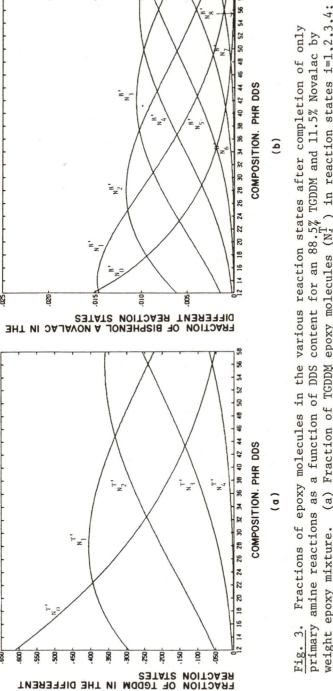

Fig. 3. Fractions of epoxy molecules in the various reaction states after completion of only primary amine reactions as a function of DDS content for an 88.5% TGDDM and 11.5% Novalac by weight epoxy mixture. (a) Fraction of TGDDM epoxy molecules (N_i^T) in reaction states $i=1,2,3,4$; (b) Fraction of Novalac epoxy molecules (N_i^B) in reaction states $i=1,\dots,8$.

are quite small for i > 2 which is indicative of a low degree of
crosslinking. Finally, it should be noted from Figure 3 that
$\sum\limits_{i} N'_i = 1$ for all DDS contents examined, as was indicated by the

expressions provided in Table II.

Secondary Amine Reactions

It was mentioned initially that step-wise reactions for the
primary and secondary amino hydrogens are assumed, that is, the
secondary amine reaction will proceed only after the completion
of the primary amine reaction. For most epoxy systems, the
secondary amine reaction may be incomplete even though epoxide
groups are present in excess, because the epoxy matrix after cure
had been found to be steric and diffusion limited (5,8,10). A
reaction parameter, γ, may be introduced to define the extent of
the secondary amino hydrogens that have reacted to the total number
of moles of secondary amino hydrogens present, viz.,

$$\gamma = \frac{\text{no. of moles of secondary amino hydrogens that have reacted}}{\text{total no. of moles of secondary amino hydrogens}} \quad (4)$$

According to the above definition, the parameter γ is bounded in
the range $0 \leq \gamma \leq 1$. Furthermore, since there are 2a miles of sec-
condary amino hydrogens after completion of the primary amine reaction, at any instant, the number of secondary amino hydrogens that
have reacted is equal to $2a\gamma$. The extent of secondary amine
reaction, γ, may also be viewed as the probability of a secondary
amino hydrogen of the DDS molecule to have reacted with an epoxide
group. Thus the probability of having an unreacted secondary amino
hydrogen is equal to $(1-\gamma)$.

The probability of the remaining epoxide groups, after completion of the primary amine reaction, to undergo reaction with
the secondary amino hydrogen, is denoted by P", and can be calculated simply as:

$$P'' = \frac{2a\gamma}{4b+8c-2a} \qquad (0 \leq P'' \leq 1) \qquad (5a)$$

The probability of an epoxide group not to have reacted with a
secondary amino hydrogen (denoted by Q") is simply:

$$Q'' = 1 - P'' \qquad (0 \leq Q'' \leq 1) \qquad (5b)$$

In the above equations, the double primes denote probabilities of the secondary amine reactions to distinguish them from
the probabilities for the primary amine reactions, which were
denoted by single primes.

At this point, the validity of the model to account only for systems
containing a stoichiometric or excess amount of epoxide groups
should be noted. In view of Eq. (5a), P" could be greater than 1
for $0 < \gamma < 1$ if excess amine was present in the formulation. This
situation will lead to a physically unrealistic description of the
probability and a model breakdown.

Definition of Total Reaction States

After the onset of secondary amine reactions, the reaction
state of an epoxy molecule in a TGDDM-Novalac mixture that describes
possible reactions of epoxide groups with amino hydrogens will be
dependent on both the primary and secondary amine reactions.
Therefore, it will be advantageous at this point to define total
reaction states for both amine reactions, primary and secondary.
The probability of existence then, of an epoxy molecule in a
reaction state of i for the TGDDM-Novalac mixture, will be the sum
of the cumulative probabilities describing three possible ways by
which the i epoxide groups belonging to the epoxy molecule have
reacted with the amino hydrogens. The three possible ways are
a) for all i epoxide groups to have reacted with only primary amino
hydrogens, or b) for all i epoxide groups to have reacted with only
secondary amino hydrogens, or c) for some of the i epoxide groups
to have reacted with primary amino hydrogens and some of the i
epoxide groups to have reacted with secondary amino hydrogens. As
was done for the primary amine reaction, the fraction of all the
epoxy molecules (TGDDM and Novalac) in a total reaction state of i
will be denoted by N_i (i.e., after both primary and secondary amine
reactions). The $N_i's$ may again be expressed as the sum of the individual fractions of the TGDDM and Novalac epoxy molecules in the
same total reaction state i, viz.,

$$N_i = N_i^T + N_i^B \tag{6}$$

where now N_i^T and N_i^B denote the fractions of TGDDM and Novalac epoxy
molecules, respectively, in the total reaction state i. Again it
should be noted that these fractions are based on the total number
of epoxy molecules available in the TGDDM-Novalac mixture (i.e.,
$\sum_i N_i = 1$). Consequently, the fractions N_i^T and N_i^B may be calculated
directly by weighing the appropriate sum of cumulative probabilities which describe the possible reactions of the epoxide groups,
with the mole fraction of TGDDM epoxy molecules ($\frac{b}{b+c}$) or the mole
fraction of Novalac epoxy molecules ($\frac{c}{b+c}$), respectively.

As an example, let us consider the fraction of TGDDM epoxy
molecules with a total reaction state of two (i=2). The cumulative
probabilities that two epoxide groups of the TGDDM molecule have

reacted with an amino hydrogen may be calculated for the three possible combinations as follows: a) For both epoxide groups to have
reacted with only primary and no secondary amino hydrogens, the
cumulative probability will be $6P'^2Q'^2 \cdot Q''^2$. b) For both epoxide
epoxide groups to have reacted with only secondary and no primary
amino hydrogens, the cumulative probability will be $Q'^4 \cdot 6P''^2Q''^2$.
c) For primary and secondary amino hydrogens to have reacted with
the two epoxide groups, the cumulative probability will be
$4P'Q'^3 \cdot 3P''Q''^2$, reflecting the event that one epoxide group has
reacted with a primary amino hydrogen while the other epoxide group
has reacted with a secondary amino hydrogen. The sum of the cumulative probabilities given in Steps a, b, and c above when multiplied by the mole fraction of the epoxy molecules present in the
mixture ($\frac{b}{b+c}$) will provide the fraction of TGDDM epoxy molecules
in a total reaction state of two (i.e., N_2^T). Following the same
procedure as for the example above, the expressions for the fractions of TGDDM molecules in their various total reaction states were
developed and are provided below:

$$N_0^T = \frac{b}{b+c} [Q'^4 \cdot Q''4] \tag{7a}$$

$$N_1^T = \frac{b}{b+c} [Q'^4 \cdot 4P''Q''^3 + 4P'Q'^3 \cdot Q''^3] \tag{7b}$$

$$N_2^T = \frac{b}{b+c} [6P'^2Q'^2 \cdot Q''^2 + Q'^4 \cdot 6P''^2Q''^2 + 4P'Q'^3 \cdot 3P''Q''^2] \tag{7c}$$

$$N_3^T = \frac{b}{b+c} [Q'^4 \cdot 4P''^3Q'' + 4P' \, Q'^3 \cdot 3P''^2Q''$$
$$+ 6P'^2Q'^2 \cdot 2P''Q'' + 4P'^3Q' \cdot Q''] \tag{7d}$$

$$N_4^T = \frac{b}{b+c} \quad [Q'^4 \cdot P''^4 + 4P'Q'^3 \cdot P''^3 + 6P'^2Q'^2 \cdot P''^2$$
$$+ 4P'^3Q' \cdot P'' + P'^4 \cdot 1] \tag{7e}$$

On a similar basis, the fraction of Novalac molecules existing in the various total reaction states, which are denoted by N_i^B,
can be obtained as follows:

$$N_0^B = \frac{c}{b+c} \quad [Q'^8 \cdot Q''^8] \tag{8a}$$

$$N_1^B = \frac{c}{b+c} \quad [8P'Q'^7 \cdot Q''^7 + Q'^8 \cdot 8P''Q''^7] \tag{8b}$$

$$N_2^B = \frac{c}{b+c} \quad [28P'^2Q'^6 \cdot Q''^6 + Q'^8 \cdot 28P''^2Q''^6 + 8P'Q'^7 \cdot 7P''Q''^6] \tag{8c}$$

$$N_3^B = \frac{c}{b+c} \quad [56P'^3Q'^5 \cdot Q''^5 + 28P'^2Q'^6 \cdot 6P''Q''^5$$
$$+ 8P'Q'^7 \cdot 21P''^2Q''^5 + Q'^8 \cdot 56P''^3Q''^5] \tag{8d}$$

$$N_4^B = \frac{c}{b+c} \; [70P'^4Q'^4 \cdot Q''^4 + 56P'^3Q'^5 \cdot 5P'Q''^4$$

$$+ 28P'^2Q'^6 \cdot 15P''^2Q''^4 + 8P'Q'^7 \cdot 35P'^3Q''^4$$

$$+ Q'^8 \cdot 70P''^4Q''^4] \tag{8e}$$

$$N_5^B = \frac{c}{b+c} \; [56P'^5Q'^3 \cdot Q''^3 + 70P'^4Q'^4 \cdot 4P''Q''^3$$

$$+ 56P'^3Q'^5 \cdot 10P''^2Q''^3 + 28P'^2Q'^6 \cdot 20P''^3Q''^3$$

$$+ 8P'Q'^7 \cdot 35P''^4Q''^3 + Q'^8 \cdot 56P''^5Q''^3] \tag{8f}$$

$$N_6^B = \frac{c}{b+c} \; [28P'^6Q'^2 \cdot Q''^2 + 56P'^5Q'^3 \cdot 3P''Q''2$$

$$+ 70P'^4Q'^4 \cdot 6P''^2Q''^2 + 56P'^3Q'^5 \cdot 10P''^3Q''^2$$

$$+ 28P'^2Q'^6 \cdot 15P''^4Q''^2 + 8P'Q'^7 \cdot 21P''^5Q''^2$$

$$+ Q'^8 \cdot 28P''^6Q''^2] \tag{8g}$$

$$N_7^B = \frac{c}{b+c} \; [8P'^7Q' \cdot Q'' + 28P''^6Q'^2 \cdot 2P''Q''$$

$$+ 56P'^5Q'^3 \cdot 3P''^2Q'' + 70P'^4Q'^4 \cdot 4P''^3Q''$$

$$+ 56P'^3Q'^5 \cdot 5P''^4Q'' + 28P'^2Q'^6 \cdot 6P''^5Q''$$

$$+ 8P'Q'^7 \cdot 7P''^6Q'' + Q'^8 \cdot 8P''^7Q'' \tag{8h}$$

$$N_8^B = \frac{c}{b+c} \; [P'^8 \cdot 1 + 8P'^7Q' \cdot P' + 28P'^6Q'^2 \cdot P''^2$$

$$+ 56P'^5Q'^3 \cdot P''^3 + 70P'^4Q'^4 \cdot P''^4 + 56P'^3Q'^5 \cdot P''^5$$

$$+ 28P'^2Q'^6 \cdot P''^6 + 8P'Q'^7 \cdot P''^7 + Q'^8 \cdot P''^8] \tag{8i}$$

In equations (7) and (8), the advantage of basing the probabilities on the total epoxide groups available in a mixture of
TGDDM-Novalac epoxies is again evident for calculations of a single
epoxy system. As it was shown previously for the primary amine
reactions, a single epoxy reacting with the DDS may be treated as a
special case of a mixture for which either TGDDM (b=0) or the
Novalac (c=0) epoxy is not present. Furthermore, as expected,
equations (7) and (8) demonstrate that if no secondary amine reactions have taken place in the system (i.e., $\gamma=0$) the fractions of
epoxy molecules in the various total reaction states, N_i^T and N_i^B,
become identical to the fractions of epoxy molecules in the reaction
states N_i^T and N_i^B, which were obtained for primary amine reactions
only. This identity may be verified by setting $P'' = 0(Q'' = 1)$ in
equations (7) and (8) and comparing the resulting simplified expressions with the ones provided in Table II.

It will be advantageous at this point to also define fractions
of DDS amine molecules in total reaction states. These fractions
or probabilities are denoted as N_i^A and are based on the total number
of DDS amine molecules present in the system. According to the
definition of reaction states, each N_i^A describes a fraction or probability of the DDS molecules that have i amino hydrogens reacted
(i=1,2...4). However, since stepwise reactions have been assumed
for the amino hydrogens, when the cure of the epoxy system has proceeded, no DDS molecules may remain completely unreacted or containing unreacted primary amino hydrogens. Thus the stepwise amine
reactions imply that our description is restricted to systems where
all the primary amino hydrogens of the DDS molecules have reacted
with epoxide groups (i.e., $N_0^A = N_1^A = 0$). Consequently, we only
need to consider DDS molecules that have available for reaction
only secondary amino hydrogens. Since we have denoted the probability of having a secondary amino hydrogen reacted as γ and the
probability of having no secondary amino hydrogen reacted as $1-\gamma$,
the fractions of DDS molecules in their various reaction states
may be described as a cumulative probability by the following
expressions:

$$N_0^A = 0 \tag{9a}$$

$$N_1^A = 0 \tag{9b}$$

$$N_2^A = (1 - \gamma)^2 \tag{9c}$$

$$N_3^A = (1 - \gamma)\gamma \tag{9d}$$

$$N_4^A = \gamma^2 \tag{9e}$$

where N_2^A denotes the fraction or probability of DDS molecules that
have no reacted secondary amino hydrogen, N_3^A denotes the fraction
or probability of DDS molecules that have one reacted secondary
amino hydrogen and N_4^A denotes the fraction or probability of DDS
molecules that have both of their secondary amino hydrogens reacted.
Finally, it should be noted from equation (9) that the N_A fractions pertain only to the DDS molecules present in the system.
Furthermore, as expected, their sum adds to unity ($\sum_i N_i^A = 1$) for all
γ's; while at $\gamma=0$ equation (9) predicts the obvious fact that all
the DDS molecules will have only their primary amino hydrogens
reacted.

The expressions provided by equation (9) describing the influence of γ on the fractions of DDS molecules in their various reaction states are straightforward and need no illustration. However,
the dependence of the fractions of epoxy molecules on γ provided by
equations (7) and (8) cannot be easily elucidated by the analytical
expressions provided. Accordingly then, numerical evaluations of
the fractions of epoxy molecules in the various total reaction states

were performed as a function of γ and DDS content for the 88.5%
TGDDM-11.5% Novalac epoxy mixture. Specifically, the TGDDM (N_i^T) and
Novalac (N_i^B) fractions are plotted as a function of the reaction
parameter γ for DDS contents of 19 and 31 PHR in Figure 4 and
Figure 5, respectively. Values of N_i^T and N_i^B as a function of DDS
content after completion of the secondary amine reaction (i.e., γ=1)
are also plotted in Figure 6 to elucidate in a continuous fashion the
effect of DDS on the epoxy reaction states. In these figures, it is
clearly shown that as the extent of secondary amine reaction is
increased, more epoxide groups that belong to either TGDDM or Novalac
molecules are reacted further with amino hydrogens, thus increasing
the fraction of epoxy molecules in higher reaction states. At a
higher DDS content, it is seen that increasing extent of secondary
amine reaction provides even further increases to the number of
epoxide groups that have reacted with amino hydrogens. The model
calculations then imply that the DDS content and extent of reaction,
γ, will have similar effects on the fractions of epoxy molecules in
a given reaction state.

The ability to provide a quantitative measure of the fractions
of epoxy and amine molecules in their various reaction states form
the basis by which more familiar descriptors of the epoxy-amine
network structure can be calculated. In the sections that follow,
basic network parameters like molecular weight between crosslinks
and crosslink density will be shown to have a direct relation to
the reaction states. Furthermore, from the reaction states,
descriptions that are specific to the epoxy system used can also be
inferred providing further elucidation of the network structure.

STRUCTURAL INFERENCES

The crosslinked network structure of the epoxy system considered here can be inferred if the distribution of the reaction
states of the constituent epoxy and amine molecules is known.
Before the onset of secondary amine reactions, or when the majority
of the epoxy molecules are in the reaction states of one or two,
the resulting network will be made up primarily of linear chains.
However, when secondary amine reactions have occurred or when the
majority of the epoxy molecules are in reaction states of three or
higher, 3-way and 4-way crosslinks are possible resulting in a
crosslinked network as shown in Figure 7 for the TGDDM-Novalac-DDS
epoxy system. As was noted by Bell (7) and shown in Figure 7, a
crosslink may be difficult to define without any reference to the
individual molecules involved in the formation of the network.
Accordingly, for the epoxy system considered, a crosslink is visual-
ized as a junction point from which three or more reacted molecular
chains radiate from it. Specifically, following Bell's (7) definition, the nitrogen atoms of the TGDDM or DDS molecules will be
identified as junction points when at least three of the epoxide

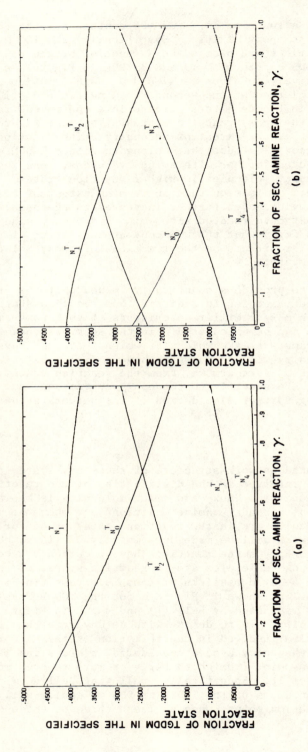

Fig. 4. Fractions of TGDDM epoxy molecules (N_i^T) in reaction states $i=1,2,3,4$ as a function of the reaction extent for an 88.5% TGDDM and 11.5% Novalac by weight epoxy mixture. The DDS content is 19 PHR for (a) and 31 PHR for (b).

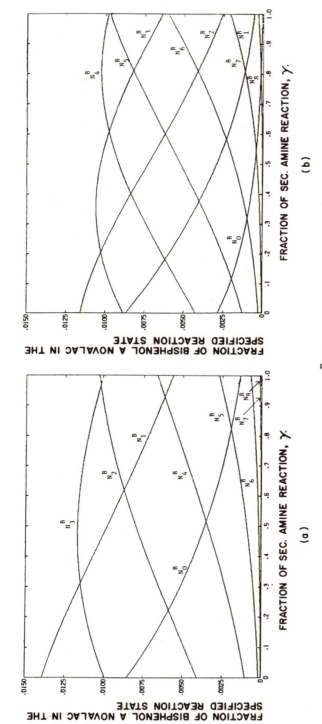

Fig. 5. Fractions of Novalac epoxy molecules (N_i^B) in reaction states i=1,...,8, as a function of the reaction extent for an 88.5% TGDDM and 11.5% Novalac by weight epoxy mixture. The DDS content is 19 PHR for (a) and 31 PHR for (b).

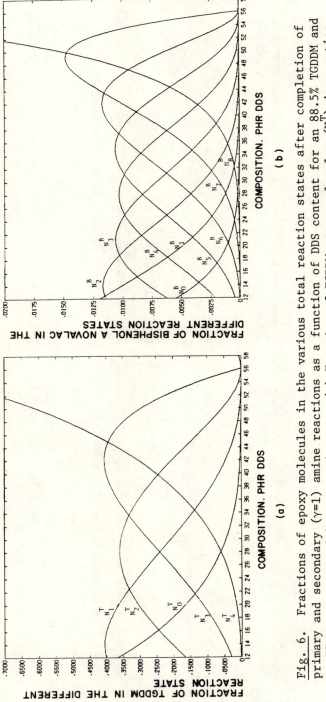

Fig. 6. Fractions of epoxy molecules in the various total reaction states after completion of primary and secondary ($\gamma=1$) amine reactions as a function of DDS content for an 88.5% TGDDM and 11.5% Novalac by weight epoxy mixture. (a) Fraction of TGDDM epoxy molecules (N_i^T) in reaction states i=1,2,3,4. (b) Fraction of Novalac epoxy molecules (N_i^B) in reaction states i=1,...,8.

end groups or amino hydrogens of the molecule have reacted. Similarly, for the Novalac molecules, the carbon atoms of the phenyl
rings of the Bisphenol A that are attached to the $-CH_2-$ linking
group will be identified as junction points, when at least three
of the epoxide groups of the molecule have reacted (see Figure 1
and Figure 7). It should be clear then, from the above definition
of a crosslink, that the TGDDM amd DDS molecules in a reaction state
of three or four can yield 3-way crosslinks. However, the Novalac
molecule, having a functionality of eight, can accommodate both
3-way and 4-way crosslinks. These definitions of crosslinks, along
with an indication that the molecular weight between crosslinks
(M_c) may be calculated from the molecular weight of the chains that
connect two adjacent junction or crosslink points are also depicted
in Figure 7. It should be further noted from Figure 7 that a junction or crosslink point may be connected to a molecular chain whose
other end may not be attached to another crosslink point. These
molecular chains are identified as dangling chains and the crosslinks that they originate from are dangling crosslinks. By identi-
fying the number of dangling crosslinks that may exist in the network and substracting this number from the total number of crosslinks that may be formed, the number of dangling free junctions or
effective crosslinks for the network is obtained. The crosslink
density of the network may then be calculated by dividing the number
of effective crosslinks by the total moles of reactants used in the
TGDDM-Novalac-DDS epoxy formulation.

    The terms that have been defined above, i.e., junction or
crosslink points, crosslink density, molecular weight between
crosslinks and the number and length of the dangling chains are but
a few parameters that may be used to infer the structure of the
epoxy network. However, since these parameters may be considered
fundamental for any network description, we will concentrate on
them for developing quantitative relations with the previously
defined reaction states of reaction extent γ. Qualitatively, the
relations to be developed can be easily understood. For example,
when the reaction state of the epoxy molecules that make up the
network are low and the extent of secondary amine reactions is
also low (i.e., $\gamma \ll 1$), only a small number of crosslinks will
be present in the network. Consequently, the crosslink density
will be low and the molecular chains between crosslinks will be
long, accounting for large molecular weight between crosslinks.
Furthermore, in this case, the number and lengths of the dangling
chains will be quite large. Conversely, the values assumed by these
parameters will be reversed when the reaction states of the epoxy
molecules are high and the extent of secondary amine reaction
approaches completion $(\gamma \cong 1)$. The quantitative description for
each network parameter based on the probabilistic terms previously
developed is considered in the following sections.

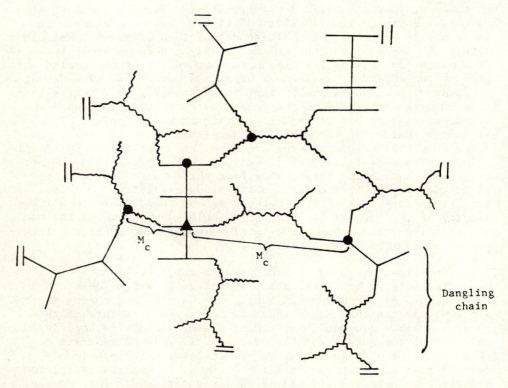

Fig. 7.   Schematic representation of the network structure for
          a TGDDM-Novalac-DDS epoxy system. The components and
          the possible reactions considered were shown in Figs.
          1 and 2. (0) denotes 3-way crosslinks; (Δ) denotes
          4-way crosslinks and (||) the continuation of the net-
          work. Chains between crosslinks, whose molecular
          weight allows calculation and dangling chains are al-
          so identified.

Junction Points

The junction or crosslink points are terms used interchange-ably in this development, and as was discussed previously, only epoxy and amine molecules in reaction states of three or higher need to be considered when calculating the number of crosslinks.

For the TGDDM molecules, only 3-way junction points are pos-sible as a result of the chemical structure of this epoxy. Thus the TGDDM junction points in the TGDDM-Novalac-DDS epoxy system under consideration may be calculated from the fraction of TGDDM molecules in a total reaction state of three plus twice the frac-tion of TGDDM molecules in a total reaction state of four, since two 3-way junctions exist in a TGDDM molecule when all of its epoxide groups have reacted. For an epoxy mixture containing b moles of TGDDM and c moles of Novalac that has reacted with a moles of DDS, the total number of 3-way junctions attributed to the TGDDM, J_3^T, may be explicitly calculated, viz.,

$$J_3^T = A(b+c) \ (N_3^T + 2N_4^T) \tag{10}$$

where the constant A denotes Avogadros number. For the DDS molecules that have similar functionality to the TGDDM molecules the total number of 3-way junctions contributed to the network by the DDS, J_3^A, may be similarly expressed, viz.,

$$J_3^A = Aa \ (N_3^A + 2N_4^A) \tag{11a}$$

which may also be expressed in terms of γ with the aid of equation (9) as:

$$J_3^A = 2Aa\gamma \tag{11b}$$

Unlike the TGDDM and DDS molecules, for the Novalac epoxy molecules, both 3-way and 4-way junctions are possible as a result of the functionality of eight that was assumed for this epoxy. Con-sequently, the calculation of the number of crosslinks that may be attributed to the Novalac molecules becomes quite cumbersome, since there are numerous ways by which 3-way or 4-way crosslinks may be accommodated in a molecule. Accordingly, for each total reaction state of a Novalac molecule, an average number of 3-way and 4-way crosslinks was calculated. The procedure that was used for this calculation was as follows: For each total reaction state of Novalac molecules, the total number of ways in which the reacted epoxide groups could be arranged was calculated by the binomial coefficient. The number and different configurations of 3-way and 4-way junctions that could be accommodated in each arrangement was then calculated which, upon division by the binomial coefficient, provided for an average number of 3-way and 4-way crosslinks per Novalac molecule in each reaction state. For brevity, the procedure

is summarized schematically in Table III for each possible total
reaction state. From the table, it should be noted that for a
Novalac epoxy molecule in a reaction state greater than three, 3-way
and 4-way junctions are possible. Furthermore, Table III also shows
that for the Novalac molecule, whenever an inner Bisphenol A chain
has its epoxide groups reacted (i.e., positions 3 and 4, or 5 and 6
in reference to the legent of Table III) in conjunction with two
other reacted epoxide groups on either side of the inner unit, a
4-way junction results. Multiplication of the average number of
crosslinks (3 or 4-way) per molecule in each reaction state i
given in Table III by its corresponding fraction N_i^B and summation
over all the terms gives the total number of Novalac junction
points. For an epoxy mixture containing b moles of TGDDM and c
moles of Novalac that has been reacted with a moles of DDS, the
total number of 3-way junctions attributed to the Novalac molecules,
J_3^B , may be explicitly calculated with the results of Table III as:

$$J_3^B = A(b+c)\left[N_3^B + \frac{108}{70} N_4^B + \frac{104}{56} N_5^B + 2N_6^B + 2N_7^B + 2N_8^B\right] \qquad (12a)$$

Similarly, from Table III the total number of 4-way junctions attributed to the Novalac molecules, J_4^B , is calculated as:

$$J_4^B = A(b+c) \left[\frac{16}{70} N_4^B + \frac{32}{56} N_5^B + N_6^B + \frac{3}{2} N_7^B + 2N_8^B\right] \qquad (12b)$$

Finally, the total number of 3-way junctions formed, J_3, in the
TGDDM-Novalac-DDS system from a moles of DDS, b moles of TGDDM and
c moles of Novalac is simply calculated as:

$$J_3 = J_3^T + J_3^B + J_3^A \qquad (13a)$$

while the total number of 4-way junctions formed, J_4 is simply obtained from the 4-way junctions of the Novalac molecules, viz.,

$$J_4 = J_4^B \qquad (13b)$$

     In the calculation of the number of junction points provided
above for each component, it is clear that only for the DDS molecules with unreacted secondary amino hydrogens, the number of 3-way
junctions have been properly accounted for through γ. However, in
all other situations, the above procedure for calculating the number
of junction points does not take into account the fact that some of
the chains attached to the junction points may be dangling (i.e.,
some chains originating from an epoxy or amine junction point may
end with an unreacted portion of an epoxy molecule rather than into
another junction point). Thus, these dangling junctions that are
associated with dangling chains must be substracted from the total
number of junction points to obtain the effective number of junctions or crosslinks for the network.

TABLE III   Network Junction Points from the Novalac
Epoxy Molecules in Their Reaction States

○ on symbolic representation depicts a reacted epoxide group site
● on symbolic representation depicts a 3-way junction point
▲ on symbolic representation depicts a 4-way junction point
The positions of the epoxide groups on the symbolic representation of the Novalac molecules are keyed
with the following numbering system:

$$1 \ {+}\ 2$$
$$3 \ {+}\ 4$$
$$5 \ {+}\ 6$$
$$7 \ {+}\ 8$$

Reaction State, i	Fract. of Molec. in i[th] State	Symb. Rep. of Molecule	No. of Possible Arrange. of React. Epoxides	No. of Possible Arrange. for 3-way Junctions	No. of Possible Arrange. for 4-way Junctions	Average No. of Junctions/Molecule 3-way	4-way	Comments
3	N_3^B		$\frac{81}{3!5!} = 56$	56	–	$\frac{56}{56} = 1$		Every possible arrangement of epoxide groups results in a 3-way junction.
4	N_4^B	a)	$\frac{81}{4!4!} = 70$	–	16	–	$\frac{16}{70}$	If 3,4 or 5,6 positions are reacted they result in a 4-way junction.
		b)		2(70-16)=108	–	$\frac{108}{70}$	–	When no Bisphenol unit is reacted structures result in to two 3-way junctions.
			70	108	16	$\frac{108}{70}$	$\frac{16}{70}$	Total
5	N_5^B	a)	$\frac{81}{5!3!} = 56$	32	32	$\frac{32}{56}$	$\frac{32}{56}$	a) One 3-way and 4-way junction per Novalac molecule are possible
		b)		3(56-32)=72	–	$\frac{72}{56}$	–	or b) Three 3-way junctions are possible.
			56	104	32	$\frac{104}{56}$	$\frac{32}{56}$	Total
6	N_6^B	a)	$\frac{81}{6!2!} = 28$	–	2x8=16	–	$\frac{16}{28}$	a) Two 4-way junctions are possible
		b)		2x12=24	12	$\frac{24}{28}$	$\frac{12}{28}$	or b) Two 3-way and 4-way junctions
		c)		4x8=32	–	$\frac{32}{28}$	–	or c) Four 3-way junctions per Novalac molecule are possible.
			28	56	28	2	1	Total
7	N_7^B	a)	$\frac{81}{7!1!} = 8$	4	2x4=8	$\frac{4=1}{8\ 2}$	1	a) If 3 and 4 positions along with 5 and 6 are reacted, one 3-way and two 4-way junctions result
		b)		3x4=12	4	$\frac{12=3}{8\ 2}$	1/2	or b) If only 3 and 4 or 5 and 6 positions are reacted three 3-way and one 4-way junctions result.
			8	16	12	2	3/2	Total
8	N_8^B		$\frac{81}{8!0!} = 1$	2	2	2	2	For fully reacted Novalac molecules two 3-way and two 4-way junctions can be accommodated per molecule.

Dangling Chains and Junctions

     According to the definition provided earlier, a dangling chain
may consist of TGDDM, Novalac, or DDS molecules in any series
arrangement ending with unreacted epoxide groups. However, the
epoxy molecules that need to be accounted as part of the dangling
chains may only be in reaction states of one or two since these
states do not contain any junction points or crosslinks. Similarly,
the DDS molecules that need to be accounted as part of dangling
chains must be in reaction state of two. Since the epoxy system
under consideration contains excess amount of epoxy molecules, all
the primary amine hydrogens are expected to have reacted. No dangling chain therefore would end with a singly-reacted DDS molecule.

     A dangling chain may be of variable length depending on the
number of epoxy and amine molecules that make it up. The probability of the occurrence of a particular molecular sequence of a dangling chain may be simply calculated as the product of the probabilities (N_i's) describing the reaction states of the constituent molecules. Furthermore, since each dangling chain originates from a
junction point of the network, the calculated probability for the
occurrence of a dangling chain is also the probability that the
junction of its origin will be a dangling junction. Consequently,
the dangling junctions may be associated with the DDS, TGDDM, or
Novalac molecules in reaction states of three or higher that contain junction points. In what follows, we consider the probability
of occurrence of dangling chains and then we associate them with the
junctions of the molecules from which they originate.

     If we consider a dangling chain at some distance from its ori-
ginating junction, it is evident from the restrictions of the model
that this chain will be made up of some repeated sequence of
(TGDDM-DDS) and/or (Novalac-DDS) molecules, all in reaction state
of two and ending with a TGDDM or Novalac molecule in a reaction
state of one. Examples of the repeat sequence of dangling chains
are schematically illustrated in Table IV along with their associated probabilities. If we define the length of the dangling
chain, ℓ, as the number of Epoxy-DDS pair of molecules that make
up the chain and j as the number of Novalac-DDS pair of molecules
so that the number of TGDDM-DDS pair of molecules will be $\ell-j$, then
the probability of occurrence of a particular repeating portion of
the dangling chain, $P_{\ell,j}$ may be simply defined by the binomial distribution (25) as:

$$P_{\ell,j} = \frac{\ell!}{j!(\ell-j)!} \ (N_2^T \ N_2^A)^{\ell-j} \ (N_2^B \ N_2^A)^j \tag{14a}$$

which accounts for all possible arrangements by which the j (Novalac-
DDS) pair of molecules can be distributed among the ℓ positions of

TABLE IV

Examples of Repeat Sequence of Dangling Chains
(Representations of molecules in the sequence are: T = TGDDM:
B = Novalac; D = DDS)

Chain Sequence	Probability of Occurrence of Chain	Molecular Weight of Chain
	TGDDM-DDS SYSTEM	
-T	N_1^T	M_T
-T-D-T	$N_2^T \, N_2^A \, N_1^T$	$M_T + M_A + M_T$
-T-D-T-D-T	$(N_2^T \, N_2^A)^2 \, N_1^T$	$2 \, (M_T + M_A) + M_T$
-T-D-T-D-T-D-T	$(N_2^T \, N_2^A)^3 \, N_1^T$	$3 \, (M_T + M_A) + M_T$
$\underbrace{-T-D-T-D.....}_{m \text{ pairs}}.T$	$(N_2^T \, N_2^A)^m \, N_1^T$	$m \, (M_T + M_A) + M_T$
	NOVALAC-DDS SYSTEM	
-B	N_1^B	M_B
-B-D-B	$N_2^B \, N_2^A \, N_1^B$	$M_B + M_A + M_B$
-B-D-B-D-B	$(N_2^B \, N_2^A)^2 \, N_1^B$	$2 \, (M_B + M_A) + M_B$
-B-D-B-D-B-D-B	$(N_2^B \, N_2^A)^3 \, N_1^B$	$3 \, (M_B + M_A) + M_B$
$\underbrace{-B-D-B-D.....}_{n \text{ pairs}}.B$	$(N_2^B \, N_2^A)^n \, N_1^B$	$n \, (M_B + M_A) + M_B$
	TGDDM-NOVALAC-DDS SYSTEM	
-B-D-T	$N_2^B \, N_2^A \, N_1^T$	$M_B + M_A + M_T$
-B-D-B-D-T	$(N_2^B \, N_2^A)^2 \, N_1^T$	$2 \, (M_B + M_A) + M_T$
-B-D-B-D-T-D-T	$(N_2^B \, N_2^A)^2 \, (N_2^T \, N_2^B) \, N_1^T$	$2 \, (M_B + M_A) + (M_T + M_A) + M_T$
-B-D-B-D-T-D-T-D-T	$(N_2^B \, N_2^A)^2 \, (N_2^T \, N_2^B)^2 \, N_1^T$	$2 \, (M_B + M_A) + 2 \, (M_T + M_A) + M_T$
$\underbrace{-B-D-B...}_{j \text{ pair}}\underbrace{T-D-T...}_{\ell-j \text{ pair}}T$	$(N_2^B \, N_2^A)^j \, (N_2^T \, N_2^A)^{\ell-j} \, N_1^T$	$j \, (M_B + M_A) + (\ell-j) \, (M_T + M_A) + M_T$

$\underbrace{\qquad\qquad\qquad}_{\ell \text{ pair}}$

*Note: B-D and T-D pairs exist in the chain in random. For a chain ℓ containing
j pairs of B-D, the probability of occurrence is:

$$\frac{\ell!}{j!\,(\ell-j)!} \, (N_2^B \, N_2^A)^j \, (N_2^T \, N_2^A)^{\ell-j} \, (N_1^T + N_1^B)$$

For a chain ℓ containing any No. of B-D pairs, the probability of occurrence is:

$$\sum_{j=0}^{\ell} \frac{\ell!}{j!\,(\ell-j)!} \, (N_2^B \, N_2^A)^j \, (N_2^T \, N_2^A)^{-j} \, (N_1^T + N_1^B)$$

Summation of all the possible arrangement and length of chains gives the probability of
having any dangling chain attached to the network is:

$$\sum_{\ell=0}^{\infty} \sum_{j=0}^{\ell} \frac{\ell!}{j!\,(\ell-j)!} \, (N_2^B \, N_2^A)^j \, (N_2^T \, N_2^A)^{\ell-j} \, (N_1^T + N_1^B)$$

the dangling chain. Multiplication of the probability $P_{\ell,j}$ by the fraction of epoxy molecules in reaction state of one, N_1, will account for the ending portion of the dangling chain in a general fashion, accommodating automatically a TGDDM or Novalac epoxy molecule as the ending molecule of the chain. Thus the probability of having a dangling chain of length ℓ with j (Novalac–DDS) units and ending with a TGDDM or Novalac epoxy molecule is simply:

$$P_{\ell,j} \, N_1 = P_{\ell,j}(N_1^T + N_1^B) \tag{14b}$$

Furthermore, it should be noted that equations (14a) and (14b) can also describe dangling chains that consist exclusively of either TGDDM or Novalac epoxy molecules. For example, for a system containing only TGDDM and DDS molecules (i.e., when c=0 moles of Novalac are present) recall that $N_1 = N_1^T$ and thus equations (14a) and (14b) reduce to:

Probability of having a
dangling chain of
length ℓ with no $\qquad = P_{\ell,j=0}(N_1^T) = (N_2^T \, N_2^A)(N_1^T)^\ell \,(15)$
Novalac units

It is also evident from equations (14a) or (14b) that dangling chains of large length will have a small probability of occurrence. In addition, for the TGDDM–Novalac–DDS systems of commercial interest where only a small amount of Novalac is used in their formulation, the makeup of the majority of dangling chains with a significant probability of occurrence will be from the TGDDM–DDS molecules. Furthermore, the summation of equation (14a) can account for the probability of occurrence of a dangling chain of length ℓ for all the possible sequences of TGDDM–DDS and Novalac–DDS molecules that may exist, viz.,

Probability of having a
dangling chain of
length ℓ with any $\qquad = P_\ell N_1 = N_1 \displaystyle\sum_{j=0}^{\ell} P_{\ell,1} \tag{16}$
sequence of (TGDDM–DDS)

where P_ℓ as defined in equation (16) denotes exclusively the frequency of the repeating portion of a dangling chain of length ℓ for all the possible Epoxy–DDS sequences that may exist. It is obvious from equation (16) that the frequency P_ℓ is based on the fraction of epoxy molecules in a reaction state of one by which the dangling chain may terminate. Finally, the probability of having a dangling chain irrespective of its length and makeup may be simply calculated by summing up equation (16) over all possible lengths, viz.,

Probability of having a
dangling chain in the
epoxy network (of any
length)

$$= P^d N_1 = N_1 \sum_{\ell=o}^{\infty} P_\ell \qquad (17)$$

where the frequency, P^d, for the occurrence of the repeating portion of the dangling chain has again been defined based on the
fraction of the ending epoxy molecules. Combining equations (14a),
(16), and (17), we can obtain a summary of the interrelated definitions for the frequencies of the repeating portion of the dangling chain, viz.:

$$P^d = \sum_{\ell=o}^{\infty} P_\ell = \sum_{\ell=o}^{\infty} \sum_{j=o}^{\ell} P_{\ell,j} = \sum_{\ell=o}^{\infty} \sum_{j=o}^{\ell} \frac{\ell!}{j!(\ell-j)!} (N_2^T N_2^A)^{\ell-j} (N_2^B N_2^A)^j \quad (18)$$

In order to obtain a better feel for the dangling chain description provided by equations (17) and (18), one can examine their
limits at $\ell=o$ and $\ell \to \infty$. For the case where $\ell=o$ as can be seen
from equation (18), $P^d = 1$. Consequently, for $\ell=o$ equation (17)
simply gives the probability of the shortest dangling chain that may
exist in the network, i.e., a dangling epoxy molecule in reaction
state of one. Conversely, for the case where $\ell \to \infty$ since the products
$N_2^T N_2^A$ and $N_2^B N_2^A$ are less than unity, then $\lim_{\ell \to \infty} P_\ell = 0$. However, depending on how much smaller than unity the products $N_2^T N_2^A$ and $N_2^B N_2^A$
are, the frequency P_ℓ may become insignificant at a finite and
relatively small ℓ value. Thus in practice the dangling chain
frequency, P^d obtained through the summation of P_ℓ's will reach a
constant value at some finite value of ℓ. For the TGDDM-Novalac-DDS
system, we have found that an $\ell = 5$ more than adequately accounts
for all the dangling chain lengths that may be encountered at all
compositions of interest.

Armed with the probability of having a dangling chain in the
epoxy network, we can now proceed to associate a dangling chain to
its originating junction or crosslink point, which will then become
a dangling junction. As it was stated earlier, the dangling junction may be on a DDS, TGDDM or Novalac molecule which is in a
reaction state of three or higher. For example, let us consider a
dangling chain which originates from a DDS molecule in a reaction
state of three. At this reaction state, the DDS molecule can
accommodate only one 3-way junction. The probability that a DDS
molecule in a reaction state of three has reacted with a dangling
chain originating from its 3-way junction may be simply calculated
as:

Probability of DDS molecule
in reaction state of $= N_3^A P^d N_1$ (19a)
three to have reacted
with a dangling chain

which is also the probability that the 3-way DDS junction will be
a dangling junction. A schematic representation of the dangling
junctions associated with the DDS molecules in reaction state of
three or higher is provided in Table (V) along with their corresponding probabilities. From Table (V), it should be noted that the
probability of a dangling junction calculated with equation (19a)
also accounts for the two configurations by which the unreacted
secondary amino hydrogen may be distributed in the DDS molecule.
Thus, the dangling junction can be located either on the nitrogen
atom where the dangling chain is attached or on the nitrogen atom
of the DDS molecule at the opposite side of where the dangling chain
begins. As it will be shown later, for purposes of assigning a
molecular weight to the dangling chain originating from a junction
point that may have more than one configuration, an average between
the various structures will be assumed to exist. Continuing with
the case that the dangling junction may also originate from a DDS
molecule in a reaction state of four, the probability then, of
having a 3-way dangling junction is:

Probability of DDS molecule
in reaction state of $= 2N_4^A P^d N_1$ (19b)
four having two dang-
ling 3-way junctions

which accounts for the fact that two 3-way junctions may be accommodated in a DDS molecule in a reaction state of four. The total
number of dangling 3-way junctions, $J_{3,D}^A$, that may be attributed
to the DDS molecules can now be calculated from equations (19a) and
(19b) to obtain:

$$J_{3,D}^A = Aa(N_3^A + 2N_4^A)\ P^d N_1$$ (20a)

which in view of equation (11a) can also be expressed as:

$$J_{3,D}^A = J_3^A P^d N_1$$ (20b)

Dangling junctions for the TGDDM and Novalac molecules can also
be obtained in a similar fashion. However, it should be recognized
that in the calculation of the dangling chain frequency P^d, the
dangling chain origin was assumed to be an epoxy molecule. Thus,
when the dangling chain originates from a junction point of an epoxy
molecule, the frequency P^d should be multiplied by the fraction of
DDS molecules in reaction state of two (N_2^A). As shown in the schematic representations provided in Table V, a DDS molecule in a reaction state of two must always be present to connect an epoxy junction

TABLE V

Examples of Dangling Junctions Associated with
the DDS, TGDDM and Novalac Molecules

Junction Molecule	No. of Junction Points

DDS:

$N_3^A \; P^d \; N_1$

$2N_4^A \; P^d \; N_1$

$\left.\begin{array}{c} \\ \\ \end{array}\right\} J_3^A \; P^d \; N_1$

TGDDM:

$N_3^T \; N_2^A \; P^d \; N_1$

$2N_4^T \; N_2^A \; P^d \; N_1$

$\left.\begin{array}{c} \\ \\ \end{array}\right\} J_3^T \; N_2^A \; P^d \; N_1$

Novalac:

See Table III
for Novalac
junction point
expressions

$\left.\begin{array}{c} \\ \\ \end{array}\right\} \begin{array}{l} J_3^B \; N_2^A \; P^d \; N_1 \\ J_4^B \; N_2^A \; P^d \; N_1 \end{array}$

Note: ⊢o denotes continuation of network

o— denotes dangling chain

point to the point where consideration of a dangling chain begins. Thus the number of 3-way dangling junctions that may be attributed to the TGDDM and Novalac epoxy molecules, $J_{3,D}^T$ and $J_{3,D}^B$, can be calculated from equations (10) and (12a), respectively, by simply multiplying them by $N_2^A P^d N_1$ to obtain:

$$J_{3,D}^T = J_3^T N_2^A P^d N_1 \qquad (20c)$$

and

$$J_{3,D}^B = J_3^B N_2^A P^d N_1 \qquad (20d)$$

Similarly, for the Novalac epoxy molecules the total number of 4-way dangling junctions $J_{4,D}^B$ may be obtained with the aid of equation (12b) as:

$$J_{4,D}^B = J_4^B N_2^A P^d N_1 \qquad (20e)$$

From the above expressions for the number of dangling junction the effective or non-dangling number of crosslinks or junction points can now be calculated. However, before we proceed with this calculation, it should be noted that only one dangling chain per junction has been accounted for with the above formulation. Although the analysis could be modified to account for more than one dangling chain per junction, the resulting expressions would become too complicated without adding any significant refinements to the calculation. Consequently, the effective number of non-dangling junctions formed by the reacting species DDS, TGDDM and Novalac are calculated by subtracting from J_3^A, J_3^T, J_3^B, and J_4^B, the corresponding dangling junctions calculated in equations (20b), (20c), (20d) and (20e) to obtain:

$$J_{3,E}^A = J_3^A - J_{3,D}^A = J_3^A (1 - P^d N_1) \qquad (21a)$$

$$J_{3,E}^T = J_3^T - N_{3,D}^T = J_3^T (1 - N_2^A P^d N_1) \qquad (21b)$$

$$J_{3,E}^B = J_3^B - J_{3,D}^B + J_{4,D}^B = J_3^B (1 - N_2^A P^d N_1) + J_4^B N_2^A P^d \qquad (21c)$$

$$J_{4,E}^B = J_4^B - J_{4,D}^B = J_4^B (1 - N_2^A P^d N_1) \qquad (21d)$$

From the above expressions for the effective crosslinks of the network, the effect of the one dangling chain per junction assumption can be clearly seen in the expressions for the Novalac junction points. In particular, equation (21c) shows that as a result of this assumption, the dangling 4-way junctions become effective 3-way junctions and must be added in the expression for $J_{3,E}^B$.

Crosslink Density

In principle, crosslink densities for the network can now be

defined with respect to any of the previously calculated effective
junctions. For reasons that will become clear as we proceed, we
have chosen to calculate the crosslink density of the network in
terms of the total number of effective 3-way and 4-way crosslinks
that may exist in the network irrespective of their molecular
origin (i.e., if they belong to a DDS, Novalac or TGDDM molecule).
Accordingly, the total number of effective 3-way crosslinks for the
network, $J_{3,E}$, was calculated by simple addition of the effective
number of 3-way crosslinks that had been attributed to the TGDDM,
Novalac, and DDS molecules, viz.,

$$J_{3,E} = J_{3,E}^{T} + J_{3,E}^{B} + J_{3,E}^{A} \qquad (22a)$$

However, the total number of effective 4-way crosslinks for
the network, $J_{4,E}$, is simply identical to the number obtained from
the Novalac molecules, viz.,

$$J_{4,E} = J_{4,E}^{B} \qquad (22b)$$

From equation (22a) and equation (22b), the number of 3-way and
4-way junctions per mole of reactants, i.e., the crosslink density
may be calculated as:

$$\hat{J}_3 = \frac{J_{3,E}}{a+b+c} \qquad (23a)$$

and

$$\hat{J}_4 = \frac{J_{4,E}}{a+b+c} \qquad (23b)$$

To illustrate numerically the feasibility of these calculations, the
crosslink densities $\hat{J}_3$ and $\hat{J}_4$ of a network resulting from the reaction of 88.5% TGDDM and 11.5% Novalac by weight of epoxy mixture
with various PHR's of DDS were calculated as a function of the
reaction parameter γ. The results are plotted in Figure 8a for $\hat{J}_3$
and Figure 8b for $\hat{J}_4$. It is interesting to note from these figures
that as expected, due to the small amount of Novalac epoxy present
in the system, the model predicts that the number of 4-way crosslinks per mole of reactants to be much smaller than the number of
3-way crosslinks.

The considerations involved in the calculation of junction
points and crosslink density can now be used to calculate the
molecular weights of the various chains that have been assumed to
form the network structure of the TGDDM-Novalac-DDS epoxy system.

Molecular Weight Considerations

Theoretically, when stoichiometric amounts of epoxy and amine
molecules are reacted, upon completion of the amine reactions that
have been considered, a single macromolecule will result. The

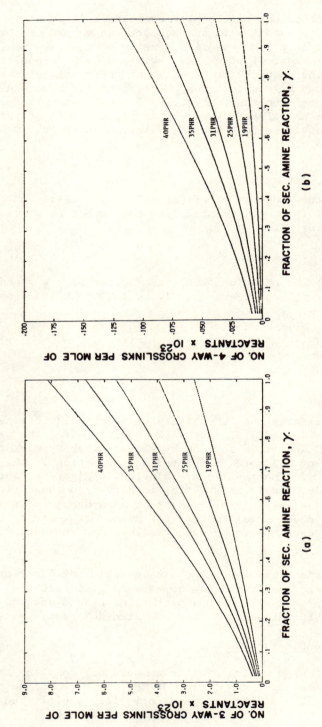

Fig. 8. Crosslink densities for the TGDDM-Novalac-DDS epoxy network as a function of the reaction parameter γ. Calculations were performed for the 88.5% TGDDM and 11.5% Novalac by weight epoxy mixture reacting with PHR DDS shown on each curve. (a) Crosslink density for 3-way crosslinks ($\hat{J}_3$); (b) Crosslink density for 4-way crosslinks ($\hat{J}_4$).

molecular weight of such network molecule will be quite large and can be simply calculated from Avogadros number and the molecular weights of the reacting components. However, for the TGDDM-Novalac-DDS epoxy system considered, where excess epoxies would be present along with incomplete amine reactions, the weight of unreacted epoxy molecules must be substracted from the weight calculated for the single network molecule. Accordingly, the total weight of a single network molecule, W, is obtained by the expression:

$$W = M_T[b-(b+c)N_o^T] + M_B[c-(b+c)N_o^B] + M_A a \tag{24}$$

where M_T, M_B, and M_A denote the molecular weights of the TGDDM, Novalac and DDS, respectively. It should be noted from equation (25) that the number average molecular weight of the network molecules may be simply calculated as AW, where A is Avogadros number. If the network molecules did not contain any dangling chains, dividing the number average molecular weight, AW, by the total number of molecular chains between crosslinks or junction points would have provided the number average molecular weight between crosslinks. However, since we have seen that each network molecule may contain dangling chains, their weight must be subtracted (equation 24) to give the total weight of chains between crosslinks.

The total weight of the dangling chains that must be subtracted will depend on their length and on the amount of the different kinds of junctions from which they originate. From the probabilities of occurrence of a dangling chain considered in the previous section, a number average molecular weight for the dangling chain population present in the TGDDM-Novalac-DDS system can be obtained. It is much easier, however, in view of the different probabilities, for the three parts of a dangling chain, to consider the average molecular weight of each part separately. The three parts of a dangling chain for which we can calculate an average molecular weight are the part of the originating molecule that includes the dangling junction, the repeating portion of the dangling chain and the end molecule of the dangling chain.

In the section where dangling chains were considered in detail, it was observed that the dangling chain must always end with an unreacted epoxy molecule in a reaction state of one. Thus, for the TGDDM-Novalac epoxy mixture under consideration, the number average molecular weight of the dangling chain ending part may be obtained as:

Molecular weight of end molecule of dangling chain
$$= \frac{N_1^T M_T + N_1^B M_B}{N_1^T + N_1^B} = \frac{N_1^T M_T + N_1^B M_B}{N_1} \tag{25}$$

which can account automatically for one component epoxy systems
(i.e., when c=0 or b=0).

In a similar fashion, the number average molecular weight of
the repeating portion of the dangling chain may be obtained with
the aid of appropriate frequencies that were summarized in equation
(28). For the repeating portion, the molecular weight for a dangling chain of length ℓ with j, Novalac-DDS units may be simply
obtained from the expression

$$\begin{array}{l} \text{Molecular weight} \\ \text{of repeating por-} \\ \text{tion of a dangling} \\ \text{chain with length } \ell \\ \text{and j Novalac-DDS units} \end{array} = (\ell-j)(M_T+M_A) + j(M_B+M_A) \quad (26)$$

which upon appropriate subsitution into equation (18) gives the
following expression for the number average molecular weight of
the repeating portion of the dangling chain, viz.,

$$\begin{array}{l} \text{Number average} \\ \text{molecular weight} \\ \text{of repeating por-} \\ \text{tion of a dangling} \\ \text{chain} \end{array} = \dfrac{\displaystyle\sum_{\ell=0}^{\infty} \sum_{j=0}^{\ell} (P_{\ell,j}\{(\ell-j)(M_T+M_A) + j(M_B-M_T)\}}{\displaystyle\sum_{\ell=0}^{\infty} \sum_{j=0}^{\ell} P_{\ell,j}} \quad (27)$$

which upon combination with equation (25) and (27) gives the number
average molecular weight of a dangling chain $\bar{M}^d$, which excludes the
originating molecule containing the dangling junction:

$$\bar{M}^d = \dfrac{\displaystyle\sum_{\ell=0}^{\infty} \sum_{j=0}^{\ell} (P_{\ell,j})\{\ell-j(M_T+M_A)+j(M_B-M_T)\}}{P^d} + \dfrac{N_1^T M + N_1^B M}{N_1} \quad (28)$$

Finally, the number average molecular weight of the originating
molecule of the dangling chain that contains the dangling junction
must be calculated according to the particular component and possible location of the junction. For example, as we have seen in
Table V, if a dangling chain originates from a DDS molecule in a
reaction state of three, the 3-way dangling junction may be located
in either of two nitrogen positions of the DDS molecules. As can
be seen from the schematic representation of Table V, depending on
which position the dangling junction is located, a different weight
contribution will be made to the molecular weight calculation of the
dangling chain. Thus, it can easily be seen that if the DDS molecule
is in a reaction state of three and the dangling junction is located
on the nitrogen atom which is nearest to the network and has both

secondary amino hydrogens reacted, the portion of the DDS molecule
that should be counted as part of the dangling chain should include
all constituent groups except the NH_2 group that has its amino
hydrogens reacted as part of the network. So for this case, the
molecular weight that should be added to $\overline{M}^d$ to obtain the originating molecular weight portion of the dangling chain, M_A^o , should
be:

$$M_A^o = M_A - 16$$

where 16 is molecular weight of the NH_2 group which has been
reacted as part of the network molecule. Conversely, if the dangling chain is located on the nitrogen atom which is farther away
from the network and has both of its secondary amino hydrogens
reacted, then

$$M_A^o = 1$$

where 1 is the molecular weight of the hydrogen atom that has
reacted with an epoxide group of the dangling chain. Thus for the
two cases considered above the average molecular weight for a DDS
originating dangling chain portion is:

$$\overline{M}_A^o = \frac{N_3^A(M_A-16) + N_3^A(1)}{2N_3^A} = \frac{M_A-15}{2}$$

If the DDS molecules in a reaction state of four are also considered, then an average molecular weight for all possible configurations of DDS from which dangling chains originate is obtained, viz.,

$$\overline{M}_A^o = \frac{N_3^A \dfrac{M_A-15}{2} + 2N_4^A(1)}{N_3^A + 2N_4^A} \tag{29a}$$

where we should recall that $N_3^A + 2N_4^A$ represent the total number of
3-way junctions per mole of DDS present in the system.

In a similar fashion, the number average molecular weight for
the TGDDM originating portion of the dangling chain $\overline{M}_T^o$ is obtained,
viz.,

$$\overline{M}_T^o = \frac{N_3^T(\dfrac{M_T-71}{2}) + 2N_4^T(M_T -365)}{N_3^T + 2N_4^T} \tag{29b}$$

where $\dfrac{M_T-71}{2}$ and M_T-365 represent average molecular weights for the
parts of the TGDDM molecule in reaction state of three and four from
which a dangling chain can originate.

For the Novalac epoxy where both 3-way and 4-way junctions are
possible in numerous configurations, the number average molecular
weight for the parts from which dangling chains may originate was
approximated as follows: For each reaction state, an average
weight for parts of the Novalac molecule that could belong to a
dangling chain was calculated by dividing the Novalac molecular
weight by the reaction state number. Then with the aid of Table III
or equations (12a) and (12b) the number average molecular weight for
Novalac originating portion of a dangling chain could be calculated
for each type of junction.

Thus, for dangling chains originating from 3-way Novalac
junctions, the number average molecular weight of the originating
part $\bar{M}^o_{B,3}$ calculated from the appropriate possible configurations
of Table III and equation (12a) is:

$$\bar{M}^o_{B,3} = \frac{N^B_3\,\dfrac{M_B}{3} + \dfrac{108}{70}\,N^B_4\,\dfrac{M_B}{4} + \dfrac{104}{56}\,N^B_5\,\dfrac{M_B}{5} + 2N^B_6\,\dfrac{M_B}{6} + 2N^B_7\,\dfrac{M_B}{7} + 2N^B_8\,\dfrac{M_B}{8}}{J^B_3/A(b+c)}$$

(29c)

while the number average molecular weight of the originating part
for dangling chains attached to 4-way Novalac junctions $\bar{M}^o_{B,4}$ was
calculated from the appropriate possible configurations of Table III
and equation (12b) by an analogous expression, viz.,

$$\bar{M}^o_{B,4} = \frac{\dfrac{16}{70}\,N^B_4\,\dfrac{M_B}{4} + \dfrac{32}{55}\,N^B_5\,\dfrac{M_B}{5} + N^B_6\,\dfrac{M_B}{6} + \dfrac{3}{2}\,N^B_7\,\dfrac{M_B}{7} + 2N^B_8\,\dfrac{M_B}{8}}{J^B_4/A(b+c)}$$

(29d)

From the above expressions for the molecular weight of a dangling chain, their total weight contribution, $W_{D.C}$, to the single
network molecule can now be obtained. Multiplication of the molecular weight for each dangling chain originating from a DDS, TGDDM or
Novalac molecule by the corresponding number of dangling chains (or
junctions) associated with each molecule and adding the resulting
weights, provides the following expression:

$$W_{D.C.} = \frac{1}{A} \{ J^A_{3,D}(\bar{M}^o_A + \bar{M}^d) + J^T_{3,D}(\bar{M}^o_T + \bar{M}^d)$$

$$+ J^B_{3,D}(\bar{M}^o_B + \bar{M}^d) + J^B_{4,D}(\bar{M}^o_B + \bar{M}^d) \}$$

$$= \frac{1}{A} \{ J^A_{3,D} \bar{M}^o_A + J^T_{3,D} \bar{M}^o_T + J^B_3 \bar{M}_{B,3}$$

$$+ J^B_4 \bar{M}_{B,4} + (J_{3,D} + J_{4,D})\bar{M}^d \} \qquad (30)$$

The weight between effective crosslinks for the network molecule, $W_{B.C}$, is obtained by substracting $W_{D.C}$ of equation (30) from the total weight of the network molecule, W, given in equation (24), viz.,

$$W_{B.C} = W - W_{DC} \qquad (31)$$

Finally, to obtain the number average molecular weight between crosslinks, consideration must be given to the number of chains that exist between the effective junctions or crosslinks of the network molecules. The number of these chains can be calculated simply from the number of 3-way and 4-way effective crosslinks obtained for the network molecules. By the definition of a 3-way crosslink or junction point, there are three chains radiating from each 3-way junction and there are two junction points at the end of each chain. Hence the number of non-dangling chains or chains between effective crosslinks is 3/2 the number of 3-way effective crosslinks. Similarly, since there are four chains radiating from each crosslink for a 4-way crosslinks, the corresponding number of non-dangling chains is 4/2 the number of 4-way effective crosslinks. Thus, the total number of chains between effective crosslinks, C, for the TGDDM-Novalac-DDS system can be calculated with the aid of equations (22a) and (22b) as:

$$C = \frac{3}{2} J_{3,E} + 2J_{4,E} \qquad (32)$$

Dividing equation (32) into (31) and remembering that the weight calculation was for a single network molecule while the number of junction points were calculated per mole of network molecules, the number average molecular weight between crosslinks, $\bar{M}_c$ for the epoxy systems can now be obtained as:

$$\bar{M}_c = \frac{A \ W_{B.C}}{C} \qquad (33)$$

where A, as was previously defined, denotes Avogadros' number.

From the above expressions, $\bar{M}_c$ was evaluated for the established epoxy and DDS compositions using the theoretical molecular weights of Table I. The calculated $\bar{M}_c$ values for systems containing 19,25,31,35, and 40 PHR DDS are plotted in Figure 9 as a function of the reaction extent γ. From this figure, as expected,

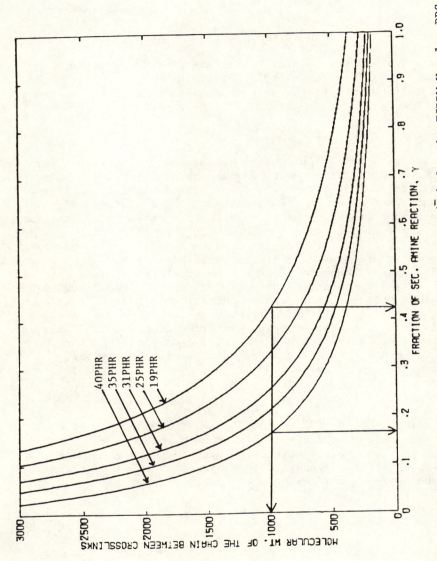

<u>Fig. 9.</u>   Number average molecular weight between crosslinks ($\overline{M}_c$) for the TGDDM-Novalac-DDS epoxy network as a function of the reaction parameter γ.   Calculations were performed for the 88.5% TGDDM and 11.5% Novalac by weight epoxy mixture reacting with PHR DDS shown on each curve.   The equivalent effect that the DDS content and the reaction parameter γ can have on $\overline{M}_c$ is noted on this figure.

it is evident that the model equations predict very large $\bar{M}_c$ values at small γ values (i.e., at small extent of secondary amine reaction). Furthermore, at small γ's, the variation of $\bar{M}_c$'s for the different DDS contents is quite large and small increases in DDS content result in large decreases of the molecular weight between crosslinks. However, for $\gamma > 0.5$ as can be seen from Figure 9, $\bar{M}_c$ variations with DDS content diminish. For example, the $\bar{M}_c$ value calculated at $\gamma = 0.1$ for the epoxy system containing 19 PHR DDS is 3817, while at 40 PHR, $\bar{M}_c$ is 1488. However, at $\gamma = 1.0$, the $\bar{M}_c$ value calculated for the 19 PHR DDS content is 359 while $\bar{M}_c$ at 40 PHR is 163. These calculations for $\bar{M}_c$ of the epoxy system also show quantitatively the equivalent effect that the DDS content and extent of reaction can have on the network structure of this epoxy system. As it is shown in Figure 9, for example, and $\bar{M}_c$ value of 1000 for the epoxy system under consideration can be achieved either with a DDS content of 40 PHR and $\gamma = 0.16$ or with a DDS content of 19 PHR and $\gamma = 0.42$.

The validity of the proposed model equations can now also be easily verified in terms of the average molecular weight results obtained for systems of stoichiometric composition. For an epoxy system containing stoichiometric compositions of Novalac and DDS at $\gamma = 1.0$, $\bar{M}_c$ can be calculated from the theoretical molecular weights of Table I and from the possible number of chains present for these two components:

$$\bar{M}_c^{BP} = (M_B + 2M_A)/13 = \frac{1898}{13} = 146$$

while for an epoxy system containing stoichiometric compositions of TGDDM and DDS at $\gamma = 1.0$, from the theoretical molecular weights of Table I and from the possible number of chains present, we obtain

$$\bar{M}_c^{TP} = (M_T + M_A)/6 = \frac{670}{6} = 112$$

Thus, for an epoxy system containing 88.5% TGDDM and 11.5% Novalac by weight that has reacted with a stoichiometric amount of DDS, the weighted $\bar{M}_c$ based on the mole fractions of each epoxy and amine present is calculated as:

$$\bar{M}_c = \frac{(0.02)(1898) + (0.47)(670)}{(0.02)(13) + (0.47)(6)} = 114.6$$

where 0.02 and 0.47 are the mole fraction of Novalac and TGDDM epoxies present respectively. To be able to compare the above $\bar{M}_c$ values on the basis of parts of DDS to one hundred parts of epoxy by weight that has been used in presenting the results of the model so far, the stoichiometric compositions were based on PHR of DDS content, and $\bar{M}_c$ values from the model equations were calculated for these compositions. For an epoxy network made up of stoichiometric compositions of a single epoxy–DDS mixture, the

stoichiometric DDS content is 58.8 PHR for the TGDDM-DDS system
while for the Novalac-DDS system it is 35.4 PHR. As expected,
the model equations for these systems gave an $\bar{M}_c^{TP}$ = 112 and an
$\bar{M}_c^{BP}$ = 146 when they were numerically evaluated for the TGDDM-DDS
and Novalac-DDS systems, respectively, at γ = 1. For the mixed
epoxy-DDS system containing 88.5% TGDDM and 11.5% Novalac by
weight of epoxies where the stoichiometric DDS content is 56 PHR,
the model equations gave, as expected, an $\bar{M}_c$ = 114 at γ = 1. Thus,
this numerical check for the stoichiometric compositions provides
added confidence to the model equations that have been developed
so far. For later comparison, values of $\bar{M}_c$ evaluated with the
model equations for the TGDDM-Novalac-DDS system at selected γ
and PHR values of DDS can also be found in Table VI.

The various descriptions of the epoxy network structure that
have been considered in the development of the probabilistic model
can now be combined in several different ways to calculate network
parameters that would be more descriptive for the system under consideration than the calculation of average molecular weight between
crosslinks. In particular, since the TGDDM-Novalac-DDS system is
made up with two different epoxies, it will be advantageous to
explore separately the molecular weight between crosslinks for each
type of epoxy to elucidate further the network structure of the
system.

The Dual, Average Molecular Weight Model

For the TGDDM-Novalac-DDS network system that we have been considering specifically so far, it is clear that the number of crosslink points will be distributed amongst the constituent molecular
species that make up the network. In particular, the number of
crosslinks formed on each type of epoxy or DDS molecules will depend
on the amount and distribution of the reaction states of each molecular species present in the system. For example, if a small amount
of Novalac epoxy is used in making up a TGDDM-Novalac-DDS network
formulation and a major portion of the Novalac molecules are in low
reaction states, then the average molecular distance from one Novalac
crosslink point to another will be large. Hence the average molecular weight between effective Novalac crosslinks will also be large
since the weight of TGDDM and DDS molecules that connect these
effective Novalac crosslinks will also be counted in the calculation. Conversely, if a large amount of Novalac epoxy is used in
making up a TGDDM-Novalac-DDS network formulation and the major
portion of these molecules are in high reaction states, then the
average molecular weight between effective Novalac crosslinks
will be small. Thus it becomes evident that systems constituted
with different relative amounts of the TGDDM-Novalac-DDS species
may exhibit the same overall average molecular weight between
crosslinks ($\bar{M}_c$) but quite different and distinct molecular weights
averages between the various types of crosslinks attributed to the

TABLE VI

Numerical Values from $\bar{M}_c$ Plots Using Theoretical Molecular Weights of the Components for TGDDM-DDS, Novalac-DDS, and TGDDM-Novalac-DDS Systems as a Function of DDS Content, Reaction Parameter γ and Relative Reactivity R.

Composition of DDS	Reaction Parameter γ	TGDDM-DDS System $\bar{M}_c^{TP}$	Novalac DDS System $\bar{M}_c^{BP}$	TGDDM-Novalac-DDS System $\bar{M}_c$		TGDDM-DDS of TGDDM-Novalac-DDS System $\bar{M}_c^{MT}$		Novalac-DDS Comp. of TGDDM-Novalac-DDS System $\bar{M}_c^{MB}$	
				R = 1	R = 20	R = 1	R = 20	R = 1	R = 2
19PHR	.1	3917	1244	3817	2930	4376	4428	29885	8662
	.5	841	445	826	768	914	929	8594	4422
	1.0	365	258	359	351	393	400	4092	2844
25PHR	.1	3338	769	3073	2551	3586	3649	21481	8474
	.5	652	334	618	596	687	702	6156	3959
	1.0	272	198	265	263	289	292	3131	2579
31PHR	.1	2676	576	2351	2083	2764	2839	15707	7816
	.5	497	270	466	459	516	527	4729	3552
	1.0	216	163	210	210	228	230	2672	2443
35PHR	.1	2246	500	1928	1769	2268	2341	12860	7240
	.5	420	241	393	390	435	442	4127	3342
	1.0	191	147	186	186	201	202	2491	2385
40PHR	.1	1765	433	1488	1413	1744	1807	10147	6479
	.5	348	214	327	327	360	365	3617	3145
	1.0	168	(132*)	163	164	176	176	2324	2318

*The stoichiometric composition of DDS content for the Novalac-DDS system is 35.4 PHR. At 40 PHR of DDS content, the model is no longer applicable to this system.

different components that make up these systems. Consequently,
it may be expected that certain physical properties of these
systems, that depend on the broadness of the distribution of $\bar{M}_c$,
may be quite different even though all these systems may have the
same molecular weight average between all effective crosslinks.
Although the description of a molecular weight distribution in
thermoset resins made up with components of different functionality
can be considered in detail by Flory's rigorous methods (25), it
will be advantageous at this point to consider a simplified description based specifically on the information provided by the probabilistic model for the TGDDM-Novalac-DDS system. In particular,
we can easily separate the average molecular weight between
effective crosslinks into components attributed to TGDDM-DDS and
Novalac-DDS molecular species that make up the network providing
a dual $\bar{M}_c$ description for the TGDDM-Novalac-DDS system. It should
be obvious, however, that such dual description of M_c should col-
lapse to a single $\bar{M}_c$ description when a single epoxy-DDS formulation is considered.

To accomplish this dual $\bar{M}_c$ description, the number of chains
between effective crosslinks attributed to each type of epoxy
molecular species and associated DDS molecules must be calculated
first. Division of the total weight between effective crosslinks
for the complete network molecules, AW_{BC}, by this number of
effective chains will provide the average molecular weight between
crosslinks for each type of epoxy-DDS species.

The number of chains between effective crosslinks attributed
to the TGDDM-DDS molecular species in the TGDDM-Novalac-DDS network
may be obtained from the effective number of TGDDM crosslinks and
an estimate of the effective number of DDS crosslinks that have
formed on DDS molecules after reaction with only the epoxide groups
of the TGDDM molecules. Remembering that there are 3/2 effective
chains formed per 3-way effective crosslinks, this number of
effective chains formed, C^T, can be expressed as:

$$C^T = \frac{3}{2}[J^T_{3,E} + \frac{\frac{b}{b+c} - N^T_o}{1 - N_o} \cdot J^A_{3,E}] \tag{34}$$

where the term $\dfrac{\frac{b}{b+c} - N^T_o}{1 - N_o}$ accounts for the fraction of TGDDM epoxy

molecules that have reacted out of all reacted epoxy molecules
(TGDDM and Novalac) in the mixture. It also provides the weighting
factor by which DDS effective crosslinks may be preferentially
associated with the TGDDM effective crosslinks.

In a similar fashion, the number of chains between effective crosslinks attributed to Novalac-DDS molecular species in the TGDDM-Novalac-DDS network may be obtained as:

$$C^B = \frac{3}{2} \left[J^B_{3,E} + \frac{\frac{c}{b+c} - N^B_o}{1 - N_o} \cdot J^A_{3,E} \right] + 2J^B_{4,E} \tag{35}$$

where the term $\dfrac{\frac{c}{b+c} - N^B_o}{1 - N_o}$ accounts for the fraction of Novalac epoxy molecules that have reacted out of all reacted epoxy molecules (TGDDM and Novalac) in the mixture.

The consistency of the above two expressions (equations 34 and 35) in providing an estimate for two different numbers of chains between effective crosslinks for the TGDDM-Novalac-DDS system can be verified by simply adding C^B and C^T which in view of equations 6 and 32 gives:

$$C^B + C^T = C = \frac{3}{2} J_{3,E} + 2J_{4,E} \tag{36}$$

Furthermore, equations 34 and 35 reduce to describing a singly epoxy system as a special case of the mixture for which either the TGDDM (b=o) or the Novalac (c=o) epoxy is not present in the formulation. The average molecular weight of such a single epoxy-DDS system, TGDDM-DDS or Novalac-DDS, can then be calculated by simply substituting c=o or b=o in all the expressions provided earlier for the average weight between effective crosslinks of the network molecules, $AW_{B.C}$, and dividing this value by either C^T (with b=o) or C^B (with c=o): This calculation then will automatically provide the average molecular weight of single epoxy-DDS systems, which for future reference we denote as $\bar{M}^{TPP}_c$ for the pure TGDDM-DDS system and $\bar{M}^{BP}_c$ for the pure Novalac-DDS system. For consistency, we can also carry forward these definitions of $\bar{M}_c$ as a description of the dual molecular weight average for the mixed epoxy TGDDM-Novalac-DDS system. We thus can simply calculate, as was stated earlier, the average molecular weight between TGDDM-DDS crosslinks, $\bar{M}^T_c$:

$$\bar{M}^T_c = \frac{A\,W_{B.C}}{C^T} \tag{37}$$

while the average molecular weight between Novalac-DDS crosslinks, $\bar{M}^B_c$:

$$\bar{M}^B_c = \frac{A\,W_{B.C}}{C^B} \tag{38}$$

The above definitions for $\bar{M}_c^T$ and $\bar{M}_c^B$ may be related to the overall molecular weight average, $\bar{M}_c$ for the TGDDM-Novalac-DDS system provided earlier by equation (33). As can be seen by combination of equations 37, 38, 33, and 36, the overall $\bar{M}_c$ for the TGDDM-Novalac-DDS network is obtained in terms of simple addition of the inverses of $\bar{M}_c^T$ and $\bar{M}_c^B$, viz.,

$$\frac{1}{\bar{M}_c} = \frac{1}{\bar{M}_c^T} + \frac{1}{\bar{M}_c^B} \tag{39}$$

The implications of equations 37 through 39 in describing the network arrangement between the epoxy-DDS species can be further elucidated if we consider a mechanical property of the system like the modulus of elasticity. It is well known that the shear modulus, G, of a polymeric network measured above its glass transition temperature may be approximated (6) by an inverse relation to $\bar{M}_c$, viz.,

$$G \cong \frac{\rho RT}{\bar{M}_c} \tag{40}$$

where ρ is the density, R the gas constant, and T the absolute temperature. Application of this relation to the TGDDM-DDS and Novalac-DDS components and substitution into equation (39) gives:

$$G = G^T \frac{\rho}{\rho^T} + G^B \frac{\rho}{\rho^B} \tag{41a}$$

where G^T and ρ^T (or G^B and ρ^B) denote the shear modulus and density of the TGDDM-DDS (or Novalac-DDS) component of the TGDDM-Novalac-DDS network. Furthermore, if it is recognized that $\rho/\rho^T = v^T$ and $\rho/\rho^B = v^B$ simply represent the volume fractions of these components, then equation 41a becomes

$$G = G^T v^T + G^B v^B \tag{41b}$$

where the volume fractions must also obey the law of mass conservation, i.e.,

$$v^T + v^B = 1 \tag{41c}$$

Equation 41b should be recognized as a parallel addition "Rule of Mixtures" (26) for a two component mechanical analog of the network. Thus the definitions of the $\bar{M}_c^T$ and $\bar{M}_c^B$ molecular weights average provided earlier have a direct modelling implication as to how the TGDDM-DDS and Novalac-DDS components reinforce each other in sharing loads that may be imposed on the TGDDM-Novalac-DDS network. Thus structurally, the description provided by the dual $\bar{M}_c$'s, i.e., $\bar{M}_c^T$ and $\bar{M}_c^B$ is to decouple the TGDDM-Novalac-DDS network into two equivalent hypothetical networks of TGDDM-DDS and Novalac-DDS acting in a parallel arrangement. It should be obvious, however, from the developed expressions that both $\bar{M}_c^T$ and $\bar{M}_c^B$ depend on the amount of each epoxy and DDS used in making up the network as well as the reaction extent parameter γ. To illustrate this point further, $\bar{M}_c^T$ and $\bar{M}_c^B$ were numerically evaluated for the established epoxy (88.5% TGDDM, 11.5% Novalac) and DDS compositions and the results are plotted as a function of reaction parameter γ in Figure 10. As expected, both $\bar{M}_c^T$ and $\bar{M}_c^B$ decrease with increasing γ but because of the relatively small amount of Novalac used in the formulation, $\bar{M}_c^B$ decreases more than $\bar{M}_c^T$. For comparison, $\bar{M}_c^{TP}$ and $\bar{M}_c^{BP}$ referring to the corresponding single epoxy systems, TGDDM-DDS and Novalac-DDS respectively, were also calculated for the same DDS compositions and as a function of reaction extent γ. These results are plotted in Figure 11, while selected numerical values for $\bar{M}_c^{TP}$ and $\bar{M}_c^{BP}$ from this figure along with values for $\bar{M}_c^T$ and $\bar{M}_c^B$ from Figure 10 appear in Table VI for comparison with the overall $\bar{M}_c$ values obtained earlier.

It will be advantageous at this point to compare the various $\bar{M}_c$ descriptors, starting with the single epoxy-DDS systems and proceeding to the mixed system with its overall and dual descriptors. As was discussed in the previous section and shown in Table VI, at stoichiometric composition for the single epoxy DDS system and at $\gamma=1$, $\bar{M}_c^{BP}$ for the Novalac-DDS system is higher than $\bar{M}_c^{TP}$ for the TGDDM-DDS system. At first observation, this difference implies that the Novalac molecules when mixed with the TGDDM molecules and reacted with DDS should act as chain extenders and that the overall $\bar{M}_c$ value for the TGDDM-Novalac-DDS system should be higher than the $\bar{M}_c^{TB}$ value of the TGDDM-DDS system evaluated at the same PHR DDS and γ values. However, as it is evident from the results of Table VI, contrary results are obtained. The apparent discrepancy can be explained by the fact that the stoichiometric composition of the two epoxies is different (i.e., 58.8 PHR of DDS for the TGDDM-DDS system versus 35.4 PHR of DDS for the Novalac-DDS system.) Therefore, at a particular composition, the DDS content with respect to the Novalac epoxy is high, whereas the DDS content with respect to the TGDDM epoxy is low. At a particular DDS content and at a specific γ value, the extent of reaction for the Novalac is far greater than that of the TGDDM, resulting in a larger fraction of the Novalac molecules being at higher reaction states. For example, at $\gamma = 1.0$ and at 35 PHR of DDS, i.e., near stoichiometric composition for the Novalac, nearly all of the Novalac epoxide groups have reacted, whereas only a fraction of the TGDDM epoxide groups have reacted. Consequently, the Novalac molecules serve the role of chain concentrators, rather than chain extenders. As can be seen from Table VI, at a given DDS content and γ value, $\bar{M}_c^{BP}$ of the Novalac-DDS system is always smaller than the corresponding $\bar{M}_c^{TP}$ values of the TGDDM-DDS system. This chain concentration effect of the Novalac in the TGDDM-Novalac-DDS system will be

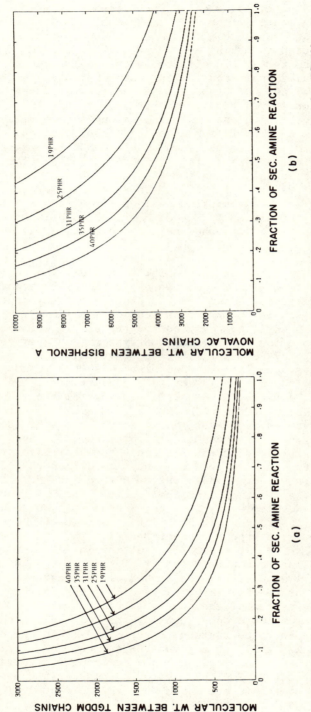

Fig. 10. Number average molecular weight between crosslinks for the two epoxies in the TGDDM–Novalac–DDS epoxy network as a function of the reaction parameter γ. Calculations were performed for the 88.5% TGDDM and 11.5% Novalac by weight epoxy mixture reacting with PHR DDS shown on each curve. (a) Average molecular weight between TGDDM–DDS crosslinks ($\overline{M}_C^T$); (b) Average molecular weight between Novalac–DDS crosslinks ($\overline{M}_C^B$). Crosslinks on DDS are also counted in these calculations according to equations (37) and (38) in the text.

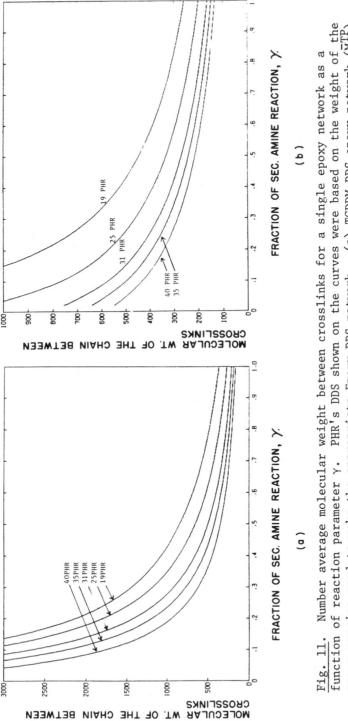

<u>Fig. 11.</u> Number average molecular weight between crosslinks for a single epoxy network as a function of reaction parameter γ. PHR's DDS shown on the curves were based on the weight of the epoxy species used to make the appropriate Epoxy-DDS network. (a) TGDDM-DDS epoxy network ($\overline{M}_c^{TP}$). (b) Novalac-DDS epoxy network ($\overline{M}_c^{BP}$).

seen in more pronounced form in the model results when the reactivity of the Novalac epoxide groups with the DDS will be assumed to be larger than that of the TGDDM epoxide groups.

The differences in the amine amount required to achieve a given degree of crosslinking for the TGDDM-DDS and Novalac-DDS systems can be seen for all values of γ and DDS content shown in Figure 11. At small γ values (i.e., $\gamma < 0.1$) $\bar{M}_c^{BP}$ are always less than the corresponding $\bar{M}_c^{TP}$ values at the same PHR DDS content. For example, at the DDS content of 19 PHR, the higher values obtained for $\bar{M}_c^{TP}$ imply that not enough crosslinks have been formed in comparison to the relative lower values obtained for $\bar{M}_c^{BP}$ which imply a relatively higher degree of crosslinking. Thus for this example, the Novalac-DDS system will have an appreciable number of crosslinks formed after completion of only a few secondary amine reactions. At the same time, however, in all these calculations, the stoichiometric amounts of the amine that are theoretically required to completely react with epoxide groups of Novalac should always be kept in mind since the model has been developed only for stoichiometric or excess epoxide groups to be present in any epoxy-DDS formulation to be considered. As it is shown in Table VI, at a DDS content of 40 PHR and at $\gamma = 1$, the $\bar{M}_c^{BP} = 132$ value of the Novalac-DDS system is smaller than the limiting value of 146 which was calculated stoichiometrically previously. Since 40 PHR DDS is greater than the stoichiometric amount of 35.4 PHR DDS needed to completely react all Novalac epoxide groups, the model is no longer applicable. (In fact, for this case, it can be shown from Equation 1a that $P' > 1$, thus violating the limits of probability.)

The dual M_c results of Table VI clearly show that for a given PHR DDS and γ, the molecular weight of the TGDDM-DDS component of the TGDDM-Novalac-DDS network, $\bar{M}_c^T$ has values only slightly greater than the $\bar{M}_c^{TP}$ values calculated for the single epoxy TGDDM-DDS system and the overall $\bar{M}_c$ of the network. According to the definition of $\bar{M}_c^T$, this result is to be expected since the accounted molecular chains between the TGDDM-DDS crosslinks will contain the few additional Novalac-DDS crosslinks present in the network. On the other hand, the molecular weight of the Novalac-DDS component, $\bar{M}_c^B$ has values much larger than the $\bar{M}_c^{BP}$ values calculated for the single epoxy-Novalac-DDS system and the overall $\bar{M}_c$ of the network. These results are consistent with the fact that only a small amount of Novalac was included in the numerical evaluation of the $\bar{M}_c$'s for the TGDDM-Novalac-DDS system shown in Table VI. As it was amply demonstrated in this section, the dual M_c model provides a basic description as to how the different types of chains made up from the different molecular species are distributed throughout the network structure. Further discussion on the usefulness of the dual M_c model is provided more appropriately when we later consider differences in the reactivity of the epoxide groups associated with the Novalac and TGDDM molecules when combining with the DDS amino hydrogens in forming the network structure.

EXTENSIONS OF THE BASIC FORMULATION

The network description presented in the previous sections was specifically developed by assuming equal reactivity of the epoxide groups, associated with each epoxy present, when reacting with an amino hydrogen. This assumption made possible the calculation of the basic probability of having these epoxide groups reacting with the amino hydrogens. A simple expression involving the functionality of each epoxy molecule and the number of moles of each distinct molecular species used for a given formulation of the final network was the starting point. All network descriptors specifically developed for the TGDDM-Novalac-DDS system were generated from these basic probabilities of the epoxide groups to undergo reaction with the primary and secondary amino hydrogens in forming the final network. The advantage of formulating these probabilities with no distinction between the epoxide groups associated with each epoxy molecule became obvious when calculations, for network systems containing only one of the epoxy species, were performed. By setting the number of moles of the epoxy molecules not present in the formulation, equal to zero, the same expressions could be used in generating a description of single epoxy-DDS systems (i.e., either Novalac-DDS or TGDDM-DDS systems). Furthermore, by accounting separately for each contribution of the distinct molecular species to the network description, it was also possible to elucidate the effect that each molecular component had on the final network structure. Thus, in accounting for different reactivities of epoxide groups associated with different epoxy molecules, we now need to only provide separate expressions for the probabilities of the distinct epoxide groups to react with a primary or a secondary amino hydrogen. With this distinction, chemical, as well as structural differences between the molecular constituents, will be incorporated into the network descriptors. Although the structural differences of the molecular constituents have been considered exclusively with the different functionalities that were assumed for the TGDDM-Novalac-DDS system, it should be clear from the model development that epoxy molecules of any functionality (obviously greater than two) could have been considered in reacting with a primary diamine to form a given network. However, in generalizing the model expressions to routinely accommodate any functionality of the epoxy species, some subtle modifications must be made to account for the different molecular structures that may be encountered. Accordingly, in the following two sections, the different reactivities of the epoxide groups and the functionality of epoxy molecules are incorporated as parameters into the model formulation.

Epoxide Groups with Different Reactivities

The development of the probabilistic model up to this point has been based on the assumption that the rate and mechanism of reacting an amino hydrogen with the TGDDM and Novalac epoxide groups were the same. However, as it was stated initially, based on the work of Chapman, et al (9), it is expected experimentally that the epoxide groups associated with the Novalac molecule would react at a faster rate with the amino hydrogens than the epoxide groups associated with the TGDDM molecule. Consequently, in the resulting TGDDM-Novalac DDS network, the probability of having a Novalac epoxide group reacted with an amino hydrogen would be greater than having a TGDDM epoxide reacted with an amino hydrogen. Thus, a distinction must be made for the probabilities of the primary and secondary amino hydrogens to react with the epoxide groups associated with TGDDM and Novalac molecules. These probabilities can then be substituted in the appropriate expressions for the reaction states of the TGDDM and Novalac molecules and thus provide with the previously developed descriptions an expansion of the model to account for the different reaction rates of the epoxide groups. Although the reaction rate of the epoxide groups with the amino hydrogens will depend on the amounts of each component used in formulating the network, through the reaction mechanism, for the probabilistic point of view, it will be more appropriate and simpler to consider only relative reaction rates of the different epoxide groups. For incorporating the different reaction rates into the probability terms, we can introduce a relative reactivity parameter, R, which we define as the rate of reaction with an amino hydrogen, of the more reactive Novalac epoxide groups divided by the rate of reaction of the less reactive TGDDM epoxide groups, i.e.

$$R= \frac{\text{React.rate of Novalac epoxide groups react. w/amino hydrogens}}{\text{React.rate of TGDDM epoxide groups react. w/amino hydrogens}} \quad (42)$$

with

$$R \geq 1$$

The values that can be assumed by this parameter as defined by equation 42 must always be greater than or equal to unity to reflect the expected higher reactivity of the Novalac epoxide groups.

The general procedure, then, for expanding the model to take the different reactivities of the epoxide groups into account, is to define the probability terms for each epoxide group to react with a primary and secondary amino hydrogen as a function of the parameter R. Thus two sets of primary (P^T and P^B) and secondary (P^T and P^B) probabilities must be defined as a function of R for the TGDDM and Novalac species, respectively. It should be recalled that these probabilities have already been defined for R=1 in the preceding development. Thus, from equation 1 we obtain:

$$P^{T'} (R=1) = P^{B'}(R=1) = P' = \frac{2a}{4b+8c} \qquad (43)$$

with $\quad 0 \le P' \le 1$

while from equation 5a, the probability of the various epoxide groups to react with a secondary amino hydrogen was defined again for R=1, viz.,

$$P^{T''} (R=1) = P^{B''}(R=1) = P'' = \frac{2a\gamma}{4b+8c-2a} \qquad (44)$$

with $\quad 0 \le P'' \le 1$

and $\quad 0 \le \gamma \le 1$

With a similar analogy to the one used in obtaining these probabilities at R=1, we can intuitively obtain expressions at the opposite extreme at $R \to \infty$, that is when the Novalac epoxide groups react at an extremely faster rate than the TGDDM epoxide groups. Denoting this value of R as R^∞, the probabilities at an intermediate relative reactivity value $(1 \le R \le R^\infty)$ can be obtained to a first approximation, by assuming a linear relationship between the various probability expressions and the relative reactivity. For example, for the TGDDM epoxide groups, the primary and secondary probabilities at any R can be obtained as:

$$P^{T'} (R) = \alpha^{T'} R + P^{T'} (R=1) \qquad (45a)$$

with

$$\alpha^{T'} = \frac{P^{T'} (R=R^\infty) - P^{T'} (R=1)}{R^\infty - 1} \qquad (45b)$$

and

$$P^{T''} (R) = \alpha^{T''} R + P^{T''} (R=1) \qquad (46a)$$

with

$$\alpha^{T''} = \frac{P^{T''} (R=R^\infty) - P^{T''} (R=1)}{R^\infty - 1} \qquad (46b)$$

where $P^{T'}$ (R) and $P^{T''}$ (R) are the probability of having a TGDDM epoxide group reacted with a primary and secondary amino hydrogens, respectively for a given relative reactivity. With similar expressions for the Novalac epoxide groups, we can define probabilities for a given relative reactivity R as $P^{B'}$ (R) and $P^{B''}$ (R) in having a Novalac epoxide group reacted with a primary and secondary amino hydrogens, respectively. By definition, for R > 1, the Novalac

epoxide group reactions are much faster than the corresponding TGDDM
epoxide group ones. Furthermore, in obtaining the probability
expressions at R^∞ in terms of the molar amounts used in making up the
TGDDM-Novalac-DDS network, it is clear that these expressions must
also assume values between the allowable limits of probability
(i.e., between zero and unity). Consequently, stoichiometric limits
of applicability for the TGDDM and the Novalac probability terms must
now be established. In addition, the total molar amount of epoxide
groups (from both TGDDM and Novalac) must always equal or exceed the
molar amount of amino hydrogens (both primary and secondary) in order
for the model formulation to be valid (i.e., $4b + 8c \geq 4a$). The
probability expressions at R^∞ can be naturally derived for two
distinct molar amounts of Novalac that will always guarantee that
the probabilistic limits of zero and one will never be violated.
These two cases correspond to two separate conditions that (a) the
number of Novalac epoxide groups in the mixture will always be equal
or less than the number of primary amino hydrogens or (b) the number
of Novalac epoxide groups in the mixture will always be equal or
greater than the number of primary amino hydrogens. The above cases
are considered separately in the following sections to obtain expressions for P^T, P^B, $P^{B'}$ and $P^{B''}$ at $r = R^\infty$. Based on the assumed
linearity of the probability expressions, values can then be obtained
for any relative reactivity in the interval $1 \leq R \leq R^\infty$.

(a)   Excess primary amino hydrogens ($8c \leq 2a$ and $4b + 8c \geq 4a$)

If the reactions of the Novalac epoxide groups with the amino
hydrogens are much faster than the ones with the TGDDM epoxide
groups (at $R = R^\infty$), then the reactions of the two kinds of epoxide
groups can be taken as stepwise. Furthermore, for this case, since
the amount of Novalac epoxide groups present is less than the amount
of primary amino hydrogens available for reaction, it is implied
that all the Novalac epoxide groups would be completely reacted with
only primary amino hydrogens. Thus for this case, the probability at
$R = R^\infty$ of the Novalac epoxide groups to react with primary amino
hydrogens is simply:

$$P^{B'} (R = R^\infty) = 1 \tag{47a}$$

Consequently, no more Novalac epoxide groups are left for reaction
with secondary amino hydrogens. Thus the probability of the Novalac
epoxide groups to react with secondary amino hydrogens is:

$$P^{B''} (R = R^\infty) = 0 \tag{47b}$$

After the reaction of the Novalac epoxide groups $(2a - 8c)$ moles
of primary amino hydrogen are still available for the TGDDM epoxide
groups to react, so that the probability of the TGDDM epoxide groups
to have reacted with primary amino hydrogens is:

$$p^{T'} \ (R = R^{\infty}) = \frac{2a - 8c}{4b} \tag{48a}$$

After completion of the reactions with the primary amino hydrogens $4b - (2a - 8c) = 4b + 8c - 2a$ moles of TGDDM epoxide groups are
left to react with the secondary amino hydrogens. Therefore, the
probability of the TGDDM epoxide groups to have reacted with secondary amino hydrogens is:

$$p^{T''} \ (R = R^{\infty}) = \frac{2a\gamma}{4b+8c-2a} \tag{48b}$$

The reaction parameter, γ, which has been defined previously
to be the fraction of secondary amine reactions, is used to account
for the incomplete reactions of the secondary amino hydrogens due
to steric and diffusion limitations that may be encountered as a
result of the network formation.

(b) Excess Novalac epoxide groups ($8c \geq 2a$ and $4b + 8c \geq 4a$)

In this case, during the reaction of the epoxide groups with
primary amino hydrogens, the limiting component is the amount of
primary amino hydrogens present in the formulation. Thus at $R = R^{\infty}$,
we simply obtain

$$p^{B'} \ (R = R^{\infty}) = \frac{2a}{8c} \tag{49a}$$

Since all the primary amino hydrogens have been consumed by the
Novalac epoxide groups, the probability of the TGDDM to react with
primary amino hydrogens is:

$$p^{T'} \ (R = R^{\infty}) = 0 \tag{50a}$$

In a similar fashion, all the remaining Novalac epoxide groups,
i.e., ($8c - 2a$) moles would react with the secondary amino hydrogens.
Thus the probability of the remaining Novalac epoxide groups to react
with the secondary amino hydrogens is:

$$p^{B''} \ (R = R^{\infty}) = \frac{2a\gamma}{8c-2a} \tag{49b}$$

Since none of the TGDDM epoxide groups have been consumed, a
total of $4b$ moles are available for reaction with secondary amino
hydrogens. Therefore, the probability of the TGDDM epoxide groups
to react with secondary hydrogens is obtained as:

$$p^{T''} \ (R = R^{\infty}) = \frac{2a\gamma}{4b} \tag{50b}$$

For the two cases that have been described above, it is relatively easy to show that the for TGDDM-Novalac-DDS system, we have
been describing so far (i.e., 88.5% by weight TGDDM, 11.5% Novalac,
and DDS content from 19 to 40 PHR) the first case is applicable.
Thus, to numerically obtain the network descriptors, equations 47

and 48 were used in appropriate linear expressions like equations 45
and 46 to obtain the appropriate probabilities which in turn were
used to obtain the fractions describing the various reaction states
as a function of R. In particular, the $\bar{M}_c$ descriptors of the
TGDDM-Novalac-DDS network were calculated for an R = 20 with R^∞ = 100.
The different values obtained for specific PHR DDS and γ values are
shown in Table VI for comparison with the results of the model at
R=1. The specific values for R = 20 was determined experimentally
by DSC experiments with the method described by Barrett (24) on
samples made up with different ratios for the constituent species
(27,28). Although R^∞ could also be determined experimentally from
these results (27), an R^∞ = 100 can be assumed to be sufficient in
providing a description for the condition of infinite relative
reactivity. As can be seen from the results of Table VI, the model
predicts a relatively minor decrease on the overall $\bar{M}_c$ value for the
TGDDM-Novalac-DDS system with increasing values of R from R = 1 to
R = 20. This result may be easily explained by the fact that only
a small amount of Novalac epoxy is assumed to be present in this
formulation so that the relative reactivity of the Novalac epoxide
groups will have little impact on the overall $\bar{M}_c$ values. Similarly,
only a slight increase is observed for the average molecular weight
between the TGDDM-DDS components ($\bar{M}_c^T$)with increasing R. However, a
large decrease is observed for the average molecular weight between
the Novalac-DDS component ($\bar{M}_c^B$) with increasing R especially for relatively low values of γ or DDS content. This large decrease may be
attributed directly to an increase of effective crosslinks on the
Novalac molecules as a result of the higher reactivity of the
associated epoxide groups. Thus as the model predictions indicate,
by incorporating the relative reactivity into the formulation, a
network structure can be described in which the effective crosslinks
will be preferentially located on the Novalac or more reactive epoxy
molecule, representing clusters of higher crosslink density material.
For the TGDDM-Novalac-DDS system for which such clusters of higher
crosslink denisty material have been observed experimentally, this
added capability of the probabilistic model can prove to be quite
useful in explaining some seemingly abnormal behavior that has been
observed in the dynamic mechanical data of these systems.

Epoxies with Different Functionalities

    To complete the probabilistic model development we need to ex-
plicitly examine the effect of the different functionalities of the
epoxy constituents on the final network structure. So far we have
examined the TGDDM-Novalac-DDS system by assuming a functionality
of four and eight to be applicable for the TGDDM and Novalac components respectively. However, we have seen earlier that not all
of the epoxide groups of the constituent molecules may be available
for reaction, or that specifically for the Novalac molecule the
functionality may vary from four to eight epoxide groups depending
on the actual molecular arrangement of this molecule. Furthermore, to

be able to apply the probabilistic model to other epoxy-diamine
systems, a generalized formulation incorporating the epoxy functionalities and all the features that were developed specifically
for the TGDDM-Novalac-DDS network descriptions must be provided.
In generalizing the probabilistic formulation, we will still desig-
nate A to be the diamine component, T to be one of the epoxy components and B to be a second epoxy component, However, now these
designations will not be specifically applicable to only the TGDDM-
Novalac-DDS system. If the functionality of the first and second
epoxy is f_T and f_B, the only changes that are required in the model
equations are to substitute these two parameters (instead of the
values 4 and 8 that were used) in the probability terms describing
the reactions of the epoxide groups with the amino hydrogens.
Furthermore if we define the relative reactivity parameter R by
always assuming that the B epoxide groups reactivity will be greater
than or equal to the T epoxide group reactivity, then the equations
considered in the previous section could also be directly employed
with only substituting f_T and f_B for the functionality of the
epoxies. For example, for R=1 the probabilities of having a given
epoxide group reacting with a primary amino hydrogen will be given
as:

$$P^{T'} \ (R = 1) = P^{B'} \ (R = 1) = \frac{2a}{f_T b + f_B c} \tag{51a}$$

while for reactions of secondary amino hydrogens these probabilities
at R = 1 may be expressed from equations (5) or (44) as:

$$P^{T''} \ (R = 1) = P^{B''} \ (R = 1) = \frac{2a\gamma}{f_T b + f_B c - 2a} \tag{51b}$$

The probability expressions at any R and epoxy functionality can be
substituted into the expressions for the reaction states (equations
7 and 8) to provide the fractions of each epoxy molecule existing
in the various total reaction states. In a general form, these
expressions for each epoxy molecule in the formulation become:

$$N_i^T = \frac{b}{b+c} \sum_{j=0}^{i} \left[\frac{f_T!}{(i-j)!(f_T-i+j)!} P(R)^{T'(i-j)} \cdot Q(R)^{T'(f_T-i+j)} \frac{(f_T-i+j)!}{j!(f_T-i)!} P^{T''j} Q^{T''(f_T-i)} \right] \tag{52}$$

where $0<i<f_T$ and $Q^{T'} \ (R) = 1 - P^{T'} \ (R)$ and $Q^{T''} \ (R) = 1 - P^{T''} \ (R)$ as
they were previously defined. A similar expression (by substituting
B for T in equation 54 can be derived for the second epoxy molecule
to provide the general expressions for the calculations of total
reaction states. From the above expression, it should be again clear

that the desired feature for calculating the total reaction states for a single epoxy-diamine system is still maintained and can be observed by setting either b or c equal to zero. The expressions for total reaction states which formed the basis of all the other network descriptions can then be used to obtain appropriately modified expressions as a function of functionality. Unfortunately, we were unable to provide a general compact form as in equation (52) for the number of crosslinks or junction points as an explicit function of the functionality. However, we were able to work out on a case by case basis individual expressions for a given functionality. We thus provide here, as an example, individual expressions for the crosslink or junction points for functionalities of epoxy molecules containing up to eight epoxide groups. Using the T epoxy molecule designation for this example, it is clear that if $f_T=2$, no crosslinks will be generated by this molecule in the final network. On the other hand for $f_T=3$, only 3-way crosslinks will be generated and the corresponding expressions will be:

$$J_3^T = A(b+c) \; N_3^T \qquad\qquad \text{for } f_T=3 \quad (53)$$

For a functionality of four ($f_T=4$) which was the functionality for the TGDDM epoxy, equation (10) will be directly applicable. However, for a functionality of five ($f_T=5$) both 3-way and 4-way crosslinks may be obtained with the following expressions:

$$J_3^T = A(b+c) \; [N_3^T + \frac{6}{5} N_4^T + N_5^T] \qquad\qquad (54a)$$
$$\text{for } f_T=5$$
$$J_4^T = A(b+c) \; [\frac{2}{5} N_4^T + N_5^T] \qquad\qquad (54b)$$

Similarly for a functionality of six and seven we obtained:

$$J_3^T = A(b+c) \; [N_3^T + \frac{22}{15} N_4^T + \frac{4}{6} N_5^T + 2N_6^T] \qquad\qquad (55a)$$
$$\text{for } f_T=6$$
$$J_4^T = A(b+c) \; [\frac{4}{15} N_4^T + \frac{4}{6} N_5^T + N_6^T] \qquad\qquad (55b)$$

and

$$J_3^T = A(b+c) \; [N_3^T + \frac{50}{35} N^{4T} + \frac{45}{21} N_5^T + \frac{10}{7} N_6^T + N_7^T] \qquad (56a)$$
$$\text{for } f_T = 7$$
$$J_4^T = A(b+c) \; [\frac{10}{35} N_4^T + \frac{9}{21} N_5^T + \frac{9}{7} N_6^T + 2N_7^T] \qquad\qquad (56b)$$

Finally, to complete the sequence, for a functionality of eight ($f_T=8$), which was the functionality considered for the Novalac epoxy molecule, equations (12a and 12b) are directly applicable.

The remaining expressions provided earlier for dangling chains, junctions, etc., can be used without modification since they do not

explicitly involve the epoxy functionality. The only caution that
needs to be exercised is to properly account for both 3-way and 4-way
crosslinks appropriately by the equations already provided. All
these conditions along with the generalized expressions have been
incorporated in a computer program (27), so any network system made
up with one or two epoxies with functionality of two or larger and
a tetrafunctional diamine can be routinely described. To test the
generality of the equations as well as the applicability of the proposed probabilistic model, the network system employed by Bell (6,7)
in his experimental and theoretical studies was evaluated. Bell's
system which was made with a diglycidyl ether of Bisphenol A (DGEBA)
epoxy and an m-phenylenediamine (MDA) provided us with a single
difunctional epoxy - tetrafunctional diamine network. However, due
to the limits of the probabilistic model, only compositions containing excess epoxy (or negative excess amine) were evaluated with our
model for Bell's system. In particular, Bell has reported two formulations of the DGEBA-DMA system: one containing -5.96% excess
amine from the stoichiometric formulation and another at -2.41%
excess amine. For these systems (which correspond to 24 and 25 PHR
of DMA respectively) Bell theoretically calculated, with his
stoichiometric model, the number average molecular weight between
crosslinks to be $\bar{M}_c$=331 for the 24 PHR MDA formulation while for
the 25 PHR MDA formulation $\bar{M}_c$=348. However, when $\bar{M}_c$ values of
samples made with these compositions were experimentally determined
from dynamic mechanical and glass transition temperature measurements, they showed a decrease with increasing PHR and MDA. Bell's
$\bar{M}_c$ values from the stoichiometric model and experimental measurements are shown in Table VII. For comparison, for this system the
$\bar{M}_c$ values at 24 and 25 PHR MDA obtained with the probabilistic model
are also shown in Table VII at three values of γ which bound the
experimentally reported $\bar{M}_c$'s of Bell. Clearly, since for these
samples the curing reaction was reported to be nearly complete (7)
these high values of γ (.8 $\leq \gamma \leq$ 1.0) required to obtain agreement
with the experimental results can easily be justified. Furthermore,
the fact that the probabilistic model predicts a decrease in $\bar{M}_c$ with
increasing amine content in harmony with the experimental results
further demonstrates the general applicability of our approach to
other epoxy systems.

NETWORK DESCRIPTION IN VIEW OF EXPERIMENTAL RESULTS

     The description of the network structure provided in the pre-
ceding sections by the probabilistic model, both in its basic and
extended forms has gone a long way in uncovering some fundamental
characteristics imbedded in the formation and subsequent arrangement
of the molecular species that make up a particular epoxy-diamine
network. The input parameters required to provide a quantitative
description of the resulting network structure are: a) the relative
amounts and molecular weights of the components used; b) the func-

TABLE VII

Comparison of $\bar{M}$ Values from the Probabilistic Model, Bell's (7) Stoichiometric Model and Experiments for the Diglycidyl Ether of Bisphenol A m-Phenylenediamine System (DEBA-MDS)

PHR of MDA	Calculated	$\bar{M}_c$ Results from Bell (7)		$\bar{M}_c$ Results from Prob. Model		
		From modulus $\bar{M}_c = \dfrac{\rho RT}{G}$	From Tg Shift $\bar{M}_c = \dfrac{3.9 \times 10^4}{Tg-Tg_o}$	$\gamma = 1.0$	$\gamma = 0.9$	$\gamma = 0.8$
24	331	393	420	339	379	428
25	348	372	402	329	368	416

tionality of the epoxy molecules; and c) the relative reactivity of the epoxide groups associated with each type of epoxy molecule with respect to the amine concentration employed and the temperature at which the crosslinking reactions have taken place. All these parameters can in principle be determined experimentally or controlled when making up a particular epoxy-amine formulation. If the fractions of secondary amino hydrogens that have reacted with the epoxide groups in forming the network can also be determined (usually by DSC or Titration measurements (27)) then the probabilistic model can provide a detailed description of the network structure. The reality and validity of this description will be as good as the extent at which the basic assumptions, on which the probabilistic model was based, have been clearly observed by the system under consideration. Thus the basic assumptions inherent in the probabilistic description must be critically reviewed and their effects on the network parameters examined, in light of the experimental and physical realities encountered in the formation of amine cured epoxy matrices.

One of the first assumptions that must be critically considered is the fact that only the reactions between the amino hydrogens and the epoxide groups have been presumed to be of significance. As it was mentioned initially, possible homopolymerization of the epoxies and reactions between hydroxyl and epoxide groups present in the network was considered insignificant in the curing mechanism. Unless, under special conditions, such as in the presence of a strong bases or acids (e.g. a Lewis base or acid), or high temperatures, this assumption has been found to be valid (7,15). In the experimental considerations for the constituents that made up our high performance epoxy diamine network, no strong bases or acids were present. Furthermore, for the curing conditions employed and the compositions considered, the degree of homopolymerization resulting from these considerations may also be assumed small. However, the network description may be easily extended to account for the homopolymerization of epoxide groups by appropriately modifying the expressions for the reaction states of the epoxy molecules. This added capability of the model is being explored in continuing studies of these high performance epoxy systems to describe high temperature postcuring treatments and/or further reactions in the presence of catalysts that promote homopolymerization.

The assumption of the stepwise reaction of the amino hydrogens, i.e., that the secondary amine reaction would start only after completion of the primary amine reaction, has been discussed in length (7,8,17). Bell (7) found that the primary amine reaction for MDA (similar in structure to DDS) was approximately seven to twelve times as fast as the secondary amine reaction. Studies have also appeared recently with claims of random reactions of the primary and secondary amino hydrogens with the epoxide groups (30). However, the scatter for the reported rate constants in reaching this conclusion was large enough to also support the stepwise reaction mechanism.

Thus, in our view, the stepwise reaction of amino hydrogens remains valid especially for these high performance amine cured epoxies. Although for the network description it is important to be able to distinguish between stepwise or random amino hydrogen reactions for derivation of the basic probabilities employed in the description of the total reaction states of the molecules, this distinction alone cannot conclusively demonstrate the existence of a homogeneous or heterogeneous epoxy-amine network. In fact, one needs to consider the reactivities of all unreacted groups as well as any steric and diffusional limitations that may be imposed by the network formation on the reactions past the gel point.

The assumption that the reactivities of all the unreacted epoxide groups of the same molecule are equal has been extensively discussed and found to be reasonable (8,25,29,30). Thus, the controlling feature for the formation of a heterogeneous network after gelation is the relative mobility of the system. As the crosslinked network is built up, the mobility of the chain segments would be decreased, thus lowering the collision frequency of the potential reactant groups. For further reaction to occur, increased curing temperature is often necessary (5).

As a general description of post gel properties of amine-cured epoxy matrices, our description accounts for the complete size distribution and different types of crosslinks on various molecular species that make up the network. Although in its present form only stepwise amine-epoxy reactions have been considered for systems made up of excess epoxy compositions, it meets basic requirements for describing high performance systems in terms of number average molecular weight between crosslinks, size, and makeup of dangling chains, etc. Comparison of our description with the work of Miller and Macosko (29), who have developed a generalized probabilistic treatment of network polymers based directly on weight average properties and random reactions of all functional groups, may lead to a generalized treatment for further experimental examination of the basic assumptions employed in these studies.

For our current description, a number of assumption concerning specifically the components of the TGDDM-Novalac-DDS system have been used and must be discussed. The assumption concerning the structure of the Novalac epoxy has been discussed in the preceding section. It is also assumed that the components of the system are perfectly mixed, and that they are pure. For a real system, neither of these two assumptions will be closely observed. Perfect mixing is highly idealized and represents an upper limit for the model. Some investigators have found localized micelles or small particles with sizes ranging from 80 to 90 $\overset{\circ}{A}$ (18), and regions with high crosslinking density are believed to be imbedded in regions with low crosslinking density (10). In the probabilistic model, the $\bar{M}_c$ values represent the statistical number averaged molecular weight between the cross-

linked chains. However, the inhomogeneity introduced by imperfect
mixing would affect the extent of reaction which would in turn be
reflected in the γ values that are imputed in the model. So, imperfect mixing which may be observed experimentally in forming the network, can be to a first approximation accounted for by the probabilistic model. As was observed in the experimental considerations
(Table I), the components supplied by the manufacturers are usually
impure. This is particularly true for the commercial epoxy resins.
Molecular entities of lower functionalities are usually found. For
example, difunctional and trifunctional epoxies are not uncommon
in TGDDM resins. This implies that not every chain end of the
TGDDM molecule will contain an epoxide group. Some chains will
be 'dead' chains. The same is true of the Novalac epoxy. As a
result, the calculated equivalent weight may be considerably smaller
than the actual epoxy equivalent weight. If the theoretical molecular weight of the epoxy is used, the calculated concentration of
epoxide groups would be too high and significantly different
results would be obtained for the network parameters. If all the
'impurities' present in the resin are assumed to have structures
similar to the 'pure' compound, except that they have lower epoxide
functionalities due to the presence of 'dead' chains, the use of
the equivalent molecular weight in the calculations can provide a
reasonable alternative. To illustrate this point, values for
$\bar{M}_c$, $\bar{M}_c^B$ and $\bar{M}_c^L$ for the TGDDM-Novalac-DDS network system were
calculated with the equivalent molecular weight of the components
and the results are tabulated in Table VIII for a relative reactivity of R=20. By comparison to the corresponding values of
Table VI where theoretical molecular weights of the components
were used, the calculated network parameters with equivalent molecular weights of the components were lower for all the γ and PHR DDS
values shown in the tables. The reduced epoxide concentration
resulting from the use of equivalent molecular weight provides a
higher probability for reaction. Consequently, a larger fraction
of the epoxide groups would be in higher reaction states, leading
to a network with high crosslinking density and lower $\bar{M}_c$ values.
Since the exact percent purity and the list of impure components
present in commercial resins are not easily available, the use of
the most reasonable molecular weight and the optimal functionality
that would best describe the system should yield the best results
in the probabilistic model.

The last assumption that should be critically examined concerns
the linearity that has been assumed between the probability of having
an epoxide group reacting with a primary (P') or secondary (P") amino
hydrogen and the relative reactivity (R) of the different epoxide
groups. It is clear that kinetic considerations for the resulting
species should be incorporated to account for an exact
relation which may be quite different from the linear one presently
assumed. Although such a modification to the model is planned in
future studies, a kinetic treatment will only modify the probabili-

TABLE VIII

Numerical values of $\overline{M}_c$ for the TGDDM-Novalac-DDS system using equivalent molecular weights for the components. Relative reactivity of Novalac to TGDDM epoxide groups was set at R=20.

Composition of DDS	Reaction Parameter γ	TGDDM-Novalac-DDS System $\overline{M}_c$ R = 20	TGDDM-DDS Comp. of TGDDM-Novalac System $\overline{M}_c^T$ R = 20	Novalac-DDS Comp. of TGDDM-Novalac-DDS system $\overline{M}_c^B$ R = 20
19PHR	.1	2741	3959	8909
	.5	664	786	4282
	1.0	292	328	2753
25PHR	.1	2200	2999	8254
	.5	485	558	3750
	1.0	219	240	2546
31PHR	.1	1641	2125	7196
	.5	365	409	3370
	1.0	177	191	2439
35PHR	.1	1318	1656	6455
	.5	311	344	3198
	1.0	158	169	2371
40PHR	.1	998	1214	5626
	.5	262	287	3044
	1.0	138	147	2263

ties obtained at each input value of R to the model. Thus, although
at present with the linear assumption the effect of R on the model
network parameters can only be examined on an approximate basis, the
potential effect that R may have on an experimental system can still
be elucidated.

Although a detailed comparison of the network descriptors as
provided by the probabilistic model and experimentally deduced by
dynamic mechanical measurements can be found elsewhere (27,28), it
is worthwhile to highlight some of these results also in this work
for completeness. In Figures 12 through 14, dynamic mechanical properties of film samples made with the TGDDM-Novalac-DDS system are
plotted in the traditional form of storage modulus and tan δ as a
function of temperature. All data were obtained with the Rheovibron
Viscoelastometer DDVII at a frequency of 11 Hz and a heating rate
of 1°C/minute. Of interest here is to focus on the α_1 and α_2 transitions exhibited by this system in the tan δ and modulus plots. As we
have previously reported, the α_1 transition for the TGDDM-Novalac-DDS
system is a composite transition that consists of the glass transition of the cured sample whose properties are being evaluated, with
a super-imposed effect of additional curing (5,8). As the sample
undergoes additional curing during the dynamic mechanical experiment,
it is expected that the crosslinking density is increasing ($\bar{M}_c$ is
decreasing) and thus an increase in the modulus and a decrease in
the tan δ should occur. Thus, a new network structure has been
formed for this sample as a result of the additional curing. Upon
further heating during the dynamic mechanical experiment, the sample
with a decreased molecular weight between crosslinks undergoes
through its glass transition. This is exhibited by the α_2 transition in the tan δ vs. temperature plots and a corresponding decrease
by the modulus as a function of temperature.

In Figure 12, the effects of increasing the DDS content from
19 to 40 PHR in the standard 88.5% TGDDM-11.5% Novalac epoxy mixture
is elucidated. All these samples were made with the standard curing
cycle outlined in the experimental section. Thus, we may expect
that the extent of secondary amine reaction will be approximately
the same for all samples or as has been observed experimentally,
γ increases slightly with DDS content (27,28). Consequently, in
reference to Figure 9, we expect a sample with 19 PHR DDS content
to have the highest $\bar{M}_c$ while a sample with 40 PHR DDS content the
lowest $\bar{M}_c$. These should correspond to the highest and lowest α_1
transitions in tan δ, respectively. Furthermore, as a dynamic
mechanical run is performed and the temperature of the experiment
exceed the sample curing temperature, γ should be continuously
increased toward one. For the range of DDS content of interest
(19 to 40 PHR), the $\bar{M}_c$ values obtained should not be substantially
different for $\gamma=1$. Thus, the α_2 transition exhibited in the tan δ
and modulus plots should be approximately equivalent for all samples.
Indeed, Figure 12 provides remarkable experimental support for these

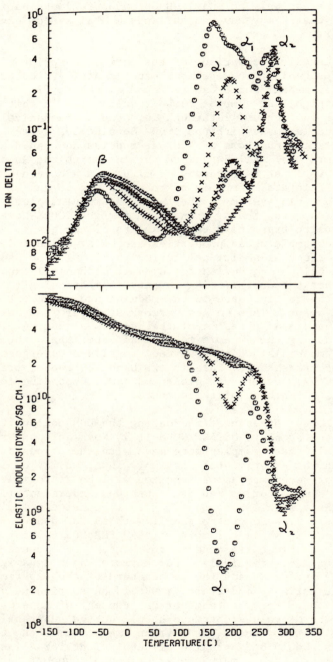

Fig. 12. Influence of amine content on dynamic mechanical properties (tan δ and storage modulus E') as a function of temperature. Film samples were made with 88.5% TGDDM, 11.5% Novalac epoxies and cured with different PHR amounts of DDS: (⊙) 19 PHR; (X) 25 PHR; (⊗) 31 PHR; (◇) 35 PHR; (⊠) 40 PHR.

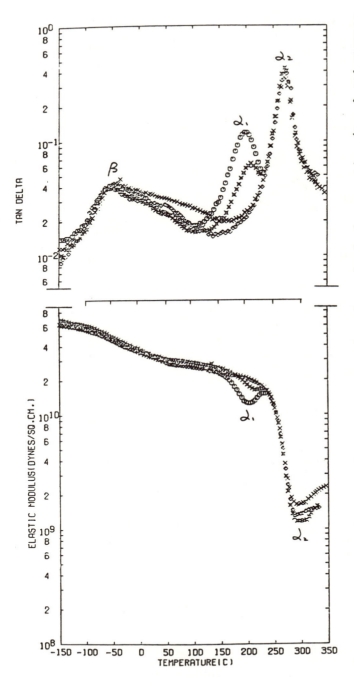

Fig. 13. Influence of postcuring treatment on dynamic mechanical properties (tan δ and storage modulus E') as a function of temperature. All film samples were made with the standard composition of 88.5% TGDDM, 11.5% Novalac epoxies and cured with 25 PHR of DDS. Data shown are for the following treatments: (○) original material cured at 177°C for two hours with no postcure; (X) sample had been previously subjected to a dynamic mechanical run up to a temperature at which the postcuring indicated by the α_1 transition was complete (~230°C); (⊠) sample had been postcured at 177°C for an additional hour than the original material; (◇) sample had been postcured at 230°C for an additional hour than the original material.

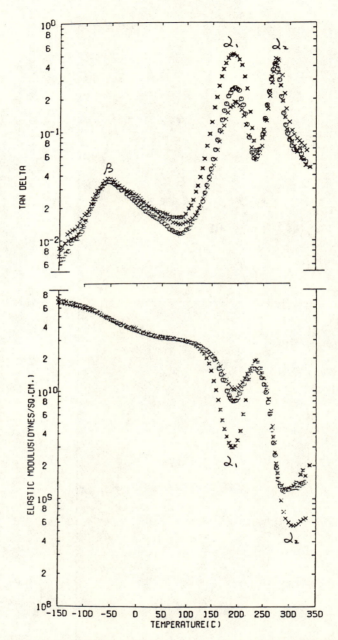

Fig. 14. Influence of the Novalac epoxy on dynamic mechanical properties (tan δ and storage modulus E') as a function of temperature.
Film samples were made with 25 PHR DDS with different epoxy ratios:
()() 0% Novalac and 100% TGDDM; (◦) Normal formulation with 11.5%
Novalac and 88.5% TGDDM; (X) 23% Novalac and 77% TGDDM.

features in the network description. Furthermore, as expected from Figure 9, the $\bar{M}_c$ of a sample with a given PHR DDS can decrease by increasing the γ parameter (i.e., by postcuring the samples). Again, this effect on the network structure is supported with the results of Figure 13 where a sample made with the standard TGDDM-Novalac epoxy proportions, 25 PHR DDS and cured under the normal curing cycle has been postcured in order to increase γ. As expected, the α_1 transition decreases with postcuring in a manner that is consistent with the corresponding increases in γ and from Figure 9 decreases in $\bar{M}_c$. As shown in Figure 13, a sample postcured at 230°C for one hour shows a reduced α_1 transition in comparison to a postcured sample at 177°C for one hour. In addition, a rerun of a sample that has been heated at 1°C/minute during a first dynamic mechanical experiment also shows a reduced α_1 transition as a result of expected γ increases. Thus, the dynamic data of Figures 12 and 13 experimentally provide support for the probabilistic description of a network structure which may be obtained by either postcuring a low DDS content sample or by curing with the standard curing cycle a sample with an increased DDS content. Finally, dynamic mechanical results of Figure 14 for TGDDM-Novalac-DDS samples prepared under the normal curing procedure but with different amounts of Novalac than the 11.5% by weight of total epoxies that is usually contained in the mixture, have explicitly demonstrated the chain concentration and high reactivity effects on the network structure that were also predicted by the probabilistic model. As can be seen, a sample containing twice the normal amount of Novalac (23% by weight), when cured under the standard curing procedure and with 25 PHR DDS exhibits an α_1 transition that is equivalent to a sample made under the standard curing procedure and epoxy content (88.5% TGDDM and 11.5% Novalac) but a much higher DDS content (See Figure 12). In general, all equivalent dynamic mechanical properties obtained for different samples of the TGDDM-Novalac-DDS system made by a variety of processing conditions and relative amounts of the constituent species can now be rationally explained in terms of the network structure that is predicted to be formed by the probabilistic model.

CONCLUSION

A network description and analysis for relatively complex epoxy-diamine systems of commercial significance as matrix materials for high performance composites has been provided in this work. The method of analysis was based on the probability of the epoxide groups to react with an amino hydrogen as a function of reactant composition and extent of secondary amine reaction. Application of this analysis was focused to epoxy-diamine systems approximating commercial compositions with stoichiometric or excess amounts of epoxies reacting with a tetrafunctional diamine. Systems made up with a single epoxy or two distinct epoxies with functionalities of two or larger were described. Network descriptors like molecular weight between cross-

links, crosslink density, dangling chains, etc., were specifically
calculated for the commercial TGDDM-Novalac-DDS system elucidating
network structures that could be formed by different compositions
and/or curing conditions. Experimental support of the elucidated
network structures was provided with dynamic mechanical measurements
on film samples made with this epoxy-diamine system. This coupling
of experimental and theoretical results identified the influence of
compositional and processing variables on the network structure
descriptors which in turn were used to analyze the exhibited dynamic
mechanical properties. Thus, this study may be viewed as a first
significant step in developing quantitative processing-structure-
property relationships for epoxy-diamine matrix materials used in
high performance composites.

ACKNOWLEDGMENTS

The authors express their appreciation to the Boeing Commercial
Airplane Company for providing continuing financial assistance for
this work. We also thank Judy Chen, A. G. Miller, and J. D. Keenan of
Boeing for many helpful discussions and continued interest in our
work.

REFERENCES

1. O. Delatycki, J. C. Shaw and J. C. Williams, J. Polym. Sci.,
   A-2, 7, 753 (1969).

2. J. Heijboer, Intern. J. Poly. Mater., 6, 11, (1977).

3. D. H. Kaelble, SPE Journal, 1071 (1959).

4. A. S. Kenyon and L. E. Nielsen, J. Macromol. Sci. Chem.,
   A3(2), 275, (1969).

5. J. D. Keenan, J. C. Seferis and J. T. Quinlivan, J. Appl.
   Polym. Sci., 24, 2375 (1979).

6. T. Murayama and J. P. Bell, J. Polym. Sci., A-2, 8, 437 (1970).

7. J. P. Bell, J. Polym. Sci., A-2, 8, 417 (1970).

8. J. D. Keenan, Master's Thesis, Department of Chemical Engineering, University of Washington (1979).

9. N. B. Chapman, N. S. Isaacs and R. E. Parker, J. Chem. Soc.,
   1925 (1959).

10. R. J. Morgan and J. E. O'Neal, Polym.-Plast. Technol. Eng.,
     10(1), 49 (1978).

11. C. E. Browning, Polym. Eng. Sci., 18, 16 (1978).

12. C. E. Browning, AFML Tech. Report, 76 (1977).

13. E. L. McKague, Jr., J. D. Reynolds, and J. E. Halkias,
     J. Appl. Polym. Sci., 22, 1643 (1978).

14. C. A. May, J. S. Fritzen and D. K. Whearty, AFML-TR-76-112,
     (1976).

15. L.Schechter and J. Wynstra, Ind. Eng. Chem., 48, 86 (1956).

16. H. Dannenberg and W. R. Harp, Analyt. Chem., 28, 86 (1956).

17. J. Dusek, S. Lunak and M. Ilavesky, J. Polym. Sci., Symp.,
     53, 45 (1975).

18. E. H. Erath and M. Robinson, J. Polym. Sci., C-3, 65 (1975).

19. J. L. Papalardo, Plastics Eng., 31, 52 (1975).

20. P. Peyser and W. D. Bascom, J. Appl. Polym. Sci., 21, 2359
     (1977).

21. T. B. Schatzki, J. Polym. Sci., 57, 496 (1962).

22. G. A. Pogany, Polymer, 11(2), 66, (1970).

23. J. Chen, private communication.

24. K. E. J. Barrett, J. Appl. Polym. Sci., 11, 1617 (1967).

25. P. J. Flory, "Principles of Polymer Chemistry," Cornell
     University Press (1953).

26. R. L. McCullough, "Concepts of Fiber Reinforced Resin Com-
     posites," Dekker (1971).

27. H. S. Chu, Master's Thesis, Department of Chemical Engineer-
     ing, University of Washington (1980).

28. H. S. Chu and J. C. Seferis, in preparation.

29. D. R. Miller and C. W. Macosko, Macromolecules, 9, 206 (1976).

30. J. M. Charlesworth, J. Polym. Sci. A-1, 18, 621 (1980).

THE TIME–TEMPERATURE–TRANSFORMATION (TTT) STATE DIAGRAM AND CURE

John K. Gillham

Polymer Materials Program
Department of Chemical Engineering
Princeton University, Princeton, NJ 08544

ABSTRACT

A generalized time-temperature-transformation (TTT) state diagram for the thermosetting process is presented in which the four physical states encountered (i.e., liquid, rubber, ungelled glass and gelled glass) are related to the time and temperature of cure. Gelation and vitrification, as a consequence of quenching morphological development and chemical conversion, respectively, are discussed with respect to control of material properties. A generalized stress response model for rubber-modified thermoset materials is presented which relates morphology to low and high strain rate behavior.

INTRODUCTION (1)

Composite materials involving an organic matrix reinforced with continuous filaments having a high tensile modulus and strength are important in applications requiring light-but-strong structures. In such materials the organic matrix is generally formed by the chemical conversion of a reactive fluid to a solid in the thermosetting process. Although thermoplastic materials can also be used, they are of limited application because of the high viscosity of their melts, their relative dimensional instability under load, and their unsuitable composite performance above the load-limiting transitions of the organic matrix (i.e. the glass transition temperature, T_g, for amorphous and the melting temperature, T_m, for semicrystalline polymers).

The most important thermosetting matrices involve network systems such as the epoxies, and semi-ladder polymers such as the

polyimides. The proper exploitation of these materials is currently
restricted because of the unsatisfactory state of the scientific and
technical information available concerning the interdependence of
their chemistry and their mechanical properties. Fundamental reasons for this include a lack of understanding of the cure process
and of the nature of the glassy state. However, from the experimental point of view, they are also inherently difficult materials
to study. They are infusible and insoluble and are therefore syn-
thesized and fabricated in one operation: because of this, their
chemistry and physics are strongly coupled. The amorphous nature
of the materials also restricts the applicability of diffraction
and morphological techniques that can be used with crystalline and
oriented samples.

The very intractability which makes the characterization of
thermosetting materials difficult is associated with the reasons for
their superior engineering behavior. A material property of particular importance that is related to the nature of the molecular networks is their dimensional stability under mechanical stress. However, in the unreinforced state the materials are often brittle and
they must therefore then be used in structural applications in the
form of fiber-reinforced composites or chemically produced two-phase
rubber-modified materials. The current interest in composites makes
it essential to understand the physical properties of these organic
matrices in relation to their chemistry. Again, from the practical
point of view, it is to be noted that homogeneous unreinforced specimens are often difficult to prepare in a defect-free state for
testing because of residual curing and thermal shrinkage stresses,
bubble inclusions introduced during cure, and surface defects introduced during test specimen preparation. In addition, the chemical
approach to the study of molecular structure-bulk property relations
has been made difficult because of the ubiquitous use of impure
reactants, proprietary formulations, and arbitrary curing conditions.
Each of these factors becomes of greater importance as the performance expected from the composite is increased.

Even with pure reactants the complexity and competing nature
of the chemical reactions involved in synthesizing the network
materials would make molecular structure-bulk property correlations
difficult to obtain. What is required is a more general understanding of the key relationships between the process of cure and the
properties of the cured state. It is to this point that the present
article is directed.

Recent research (1-9) has indicated that a Time-Temperature-
Transformation diagram (analogous to the TTT diagrams that have been
employed for many years in metallurgical processing) may be used to
provide an intellectual framework within which an understanding of
the physical properties of thermosetting systems may be achieved.
In the discussion presented below, the significance of this diagram
is discussed and it is used to explain practices and phenomena

prevalent in the technology of thermosets. This technology involves organic adhesives and protective coatings as well as structural matrix/fiber composites.

TIME-TEMPERATURE-TRANSFORMATION (TTT) STATE DIAGRAMS

Time-Temperature-Transformation (TTT) diagrams have played an important role in the control of the properties of metals by permitting thermal history paths to be chosen so that a desired microstructure can be obtained. The diagrams are specific to a particular material composition and considerable insight into the design of alloys can be achieved once the effects of additions of alloying elements on the TTT diagrams have been explored. Since thermosetting polymeric systems are prepared *in situ*, the availability of an equivalent diagram for either the pure matrix material, or a matrix containing impurities such as a dispersed rubber phase, would be of considerable technological importance. Such a diagram would permit time-temperature paths for cure to be chosen so that gelation, vitrification and phase separation occurred in a controlled manner and consequently gave rise to predictable properties of the thermosetting matrix.

Gelation and vitrification are two macroscopic phenomena which are encountered as a consequence of chemical reactions which convert a fluid to a solid in the thermosetting process. On the molecular level, gelation corresponds to the incipient formation of branched molecules of very high molecular weight. Macroscopically this process is accompanied by a dramatic increase in viscosity, the onset of elasticity, and a corresponding decrease in the condensed phase diffusional processes and in material processibility. In principle, molecular gelation starts at a fixed chemical conversion that can be predicted from the functionality of the reactants (10). (The time to molecular gelation will therefore vary exponentially with the temperature of isothermal cure.) Eventually the total mass of material can be regarded as one molecule. This network structure will be an elastomer at a given temperature if the segments between junction points of the network are flexible. If these segments are immobilized by further chemical reaction, or by cooling, the structure will change to the glassy (vitrified) state.

Vitrification, which usually follows gelation, occurs as a consequence of increasing molecular weight and further crosslinking causing a reduction in the degrees of freedom of the network. It occurs during isothermal cure when the glass transition of the reactants rises to the temperature of cure. Vitrification can further retard (or quench) chemical reactions in the matrix.

The overall transformation from liquid to gel to rubber to glass that occurs as a result of chemical reactions in the thermosetting process is termed "cure". The properties of the final material are intimately related to the details of the curing process.

In particular they depend upon the interplay between such factors as the chemical reactants involved, their mutual solubility, their viscosity prior to gelation, the volatility of the reactants and byproducts, gelation, phase separation, vitrification, overall chemical conversion, the details of the time-temperature path of the curing reaction, and the limits of the thermal stability of the materials involved.

    Figure 1A (5) shows a generalized TTT diagram obtained from isothermal experiments for a typical thermosetting process that does not involve phase separation. It displays the four distinct material states (liquid, elastomer, ungelled glass and gelled glass) that are encountered during cure. Three critical temperatures are also displayed on the diagram. These are: $T_{g\infty}$, the maximum glass transition temperature of the fully cured system; $_{gel}T_g$, the isothermal temperature at which gelation and vitrification occur simultaneously; and $_{resin}T_g$, the glass transition temperature of the reactants.

    When a thermosetting material is cured isothermally above $T_{g\infty}$, the liquid gels to form an elastomer but it will not vitrify in the absence of degradation. (Vitrification due to degradation is shown in Fig. 1.) An isothermal cure at an intermediate temperature between $_{gel}T_g$ and $T_{g\infty}$ will cause the material first to gel and then

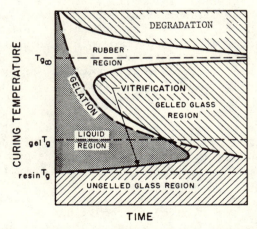

Fig. 1A.   Schematic Isothermal Time-Temperature-Transformation (TTT) diagram for the curing process: A thermosetting system not involving phase separation.

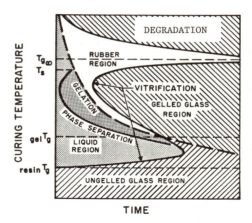

Fig. 1B.   Schematic Isothermal Time-Temperature-Transformation (TTT) diagram for the curing process: A thermosetting system in which a rubber phase may separate during cure. (T_S is the maximum temperature for phase separation to occur prior to gelation.)

to vitrify. If chemical reactions are quenched by vitrification it follows for this case that the glass transition temperature in principle will equal the temperature of cure and that such a material will not be fully cured. It also follows that temperature $_{gel}T_g$ is the glass transition temperature of the material at its point of gelation since cure at temperature $_{gel}T_g$ will lead to a vitrified state as the material gels. At temperatures below $_{gel}T_g$ but above $_{resin}T_g$ the viscous curing liquid can vitrify simply by an increase of molecular weight and, if chemical reactions are quenched by vitrification, the material will not gel.

It is immediately apparent that molecular structure-macroscopic property relationships will be meaningful only if the material is fully reacted. This is generally only possible by curing above $T_{g\infty}$.

As indicated in Fig. 1 the time to vitrify passes through a minimum between $_{gel}T_g$ and $T_{g\infty}$. This behavior reflects the competition between the increasing rate constants for reaction and the increasing chemical conversion required to achieve vitrification as the temperature is increased. Also, as indicated in Fig. 1, the time to vitrify passes through a maximum between $_{resin}T_g$ and $_{gel}T_g$. This behavior reflects the competition between the temperature-and time-dependences of viscosity of the reacting system.

The cure TTT diagram of Fig. 1A can be extended to include two phase systems with rubber dispersed in an inherently brittle

polymeric material in order to increase toughness and impact resistance. The curing of such rubber-modified systems may involve a
change from an initially homogeneous solution to a heterogeneous
multiphase morphology, the visual onset of which is shown schematically in Fig. 1B (8). Growth of the dispersed phase is considered
to continue until gelation: therefore, procedures which alter the
time and temperature to gelation can be used to control the material properties. Since the nucleation and growth of the rubber phase
involve a balance between nucleus formation and matter transport,
the degree of phase separation achieved in an isothermal process
would be expected to show a maximum at a temperature between that
for which thermodynamics favors the solubility of the rubber in the
matrix and the $_{resin}T_g$ of the matrix. (Analogously, the rate of
isothermal crystallization of a thermoplastic melt passes through a
maximum between the glass transition and melting transition temperatures.) Careful control of the cure temperature will also permit
the size and number of particles per unit volume of the dispersed
rubber phase to be modified. Relationships between the detailed
morphology and isothermal temperature of gelation are shown schematically in Fig. 2. Control of the time-temperature history of
the material during cure is a method for achieving the desired character of phase separation, but a knowledge of the TTT diagram is a
useful prerequisite. Evidence has been presented (11, 12) to show
that improved material toughness and impact resistance arise in
rubber-modified systems in which part of the rubber is phase-
separated and part is trapped in the matrix. The path taken on the
TTT diagram must therefore be chosen so as to balance the distribution of the rubber between the two phases. Cure of two-phase systems will in general involve two sequences; a first to develop a
controlled morphology by gelling at one temperature, and a second to
complete the chemical reactions by curing above $T_{g\infty}$.

A Continuous Heating Transformation (CHT) State Diagram which
is analogous to the isothermally obtained TTT state diagram can be
obtained experimentally from a series of temperature scans at different rates from below the glass transition temperature of the
reactants ($_{Resin}T_g$) to above $T_{g\infty}$ (3). A typical scan for a homogeneous reactive system will reveal in sequence: relaxations in
the glassy state, $_{Resin}T_g$, gelation, vitrification and (in the
presence of some types of degradation) revitrification. After
vitrification on cure, in these scans, the glass transition temperature will in principle equal the instantaneous scanning temperature until the rate of chemical reaction is not sufficient to
overcome the increased segmental mobility, at which temperature the
material will devitrify.

The above discussion has been intended to introduce the concept of TTT diagrams for thermosetting materials and to indicate
their utility in the control of processing that will influence the
mechanical properties of these materials. To further illustrate

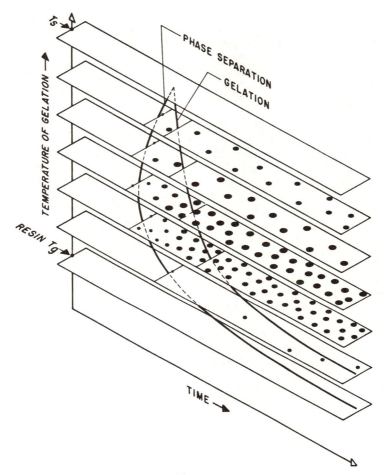

Fig. 2. Morphology Map (Schematic). Representation of the morphology developed on isothermal cure of a rubber-modified
thermoset versus temperature.

this point, the immediate discussion uses the TTT cure diagrams to
explain a number of practices current in the field of thermosets.
Later sections will include several specific examples of the effect
of undercure versus more fully developed cure on material behavior,
and the influence of gel time (and hence morphology) on material
behavior of rubber-modified systems.

If the storage temperature is below $_{gel}T_g$, a reactive fluid
material will convert to a vitrified solid of low molecular weight

which is stable and can be later liquified by heat and processed. Above $_{gel}T_g$ the stored material will have a finite shelf-life for subsequent processing since gelation will occur before vitrification. (A gelled material does not flow in the usual sense.) This concept lies at the basis of a widespread technology which includes thermosetting molding compounds (including reaction injection molding materials) and "prepregs" with latent reactivity.

In general, if $T_{cure} < T_{g\infty}$, a reactive material will vitrify and full chemical conversion will be prevented. The material will then usually need to be post-cured above $T_{g\infty}$ for development of optimum properties. For highly crosslinkable or rigid-chain polymeric materials $T_{g\infty}$ can be above the limits of thermal stability, in which case full chemical conversion of the original network-forming reactions would not usually be attainable. For composite materials in which a component other than the cured resin is thermally sensitive, $T_{g\infty}$ for the thermosetting resin should be below temperatures which would lead to damage of any part of the assembly.

Time-Temperature-Transformation diagrams have been essential in the exploitation of metallic systems and particularly in the control of their mechanical properties in the alloyed state. It is highly desirable to be able to exert equivalent control over the properties of thermosetting polymeric systems. In order to obtain such control it will be necessary to develop TTT diagrams for specific systems using pure reactants and to extend these studies to include the addition of deliberate impurities, such as rubbers, to these materials. On the basis of these diagrams, thermal history paths for the cure process may be chosen and desired final morphologies achieved. The final step in this process is to relate the materials' morphologies to their mechanical properties such as toughness, impact and fatigue resistance.

TORSIONAL BRAID ANALYSIS (TBA) - A METHOD FOR CONSTRUCTING TTT & CHT STATE DIAGRAMS

An automated, free-hanging, freely decaying torsion pendulum has been developed (2,13,14) which permits monitoring of the changes which occur throughout cure by using a substrate (e.g. braid) impregnated with the reactive system (TBA). The caption of the schematic diagram of the pendulum shown in Fig. 3 includes a brief *modus operandi* and definitions of the deduced mechanical properties of the specimen. Isothermal experiments (e.g., Fig. 4) monitoring cure logically precede temperature scans (e.g., Fig. 5) of the same specimen so as to relate the process of cure to properties of the cured state.

The TTT diagram (Fig. 1A) can be generated by measuring times to gel and to vitrify at a series of isothermal temperatures. These transformation times have been obtained from TBA plots versus time (e.g., Fig. 4), using peaks in the mechanical damping (e.g.,

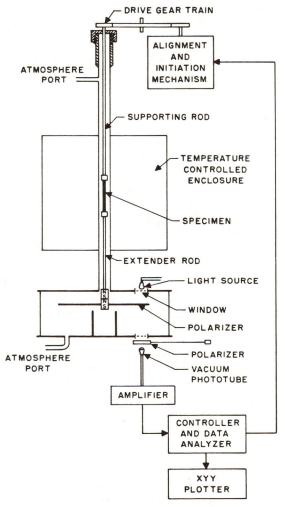

Fig. 3. Automated Torsion Pendulum (Schematic). An analog elec-
         trical signal results from using a light beam passing
         through a pair of polarizers, one of which oscillates
         with the pendulum. The pendulum is aligned for linear
         response and initiated by a computer that also processes
         the damped waves to provide the elastic modulus and
         mechanical damping data, which are plotted on an XY
         plotter versus temperature or time. The elastic modulus
         is conveniently expressed as the relative rigidity = $1/P^2$
         where P is the period of oscillation. The mechanical
         damping is conveniently expressed as the logarithmic
         decrement, $\Delta = \ln A_i/A_{i+1}$ where A_i and A_{i+1} are peak
         amplitudes of consecutive oscillations.

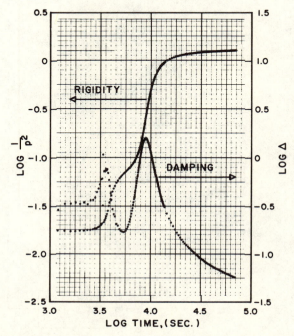

Fig. 4.   Isothermal Cure of a Thermosetting Liquid (TBA).
Two events are discerned, the first being attributed to gelation and the second to vitrification (2).
Times to each of these events are measured for different isothermal temperatures as a basis for
generating an isothermal Time-Temperature-
Transformation State Diagram.

logarithmic decrement, Δ) which correspond approximately to points
of inflection in the elastic modulus (e.g., relative rigidity,
$1/P^2$). Complementary CHT diagrams have been generated by scanning
the temperature range from below the glass transition temperature
of the reactant mixture, $_{resin}T_g$, to above $T_{g\infty}$ at a series of constant heating rates.

Details of procedures and results on structure-property relationships of thermosetting systems are provided in recent publications (2-9, 11-20). The concept of the TTT and CHT diagrams for
thermosetting systems has developed from application of the TBA
technique.

The automated TBA system is manufactured by Plastics Analysis
Instruments, Inc., P.O. Box 408, Princeton, New Jersey, USA.

EFFECT OF EXTENT OF CURE ON MATERIAL BEHAVIOR

In principle, cure at a temperature T_{cure} which is below $T_{g\infty}$
will lead to the glass transition temperature $T_g = T_{cure}$. Post-
cure above $T_{g\infty}$ will lead to $T_g = T_{g\infty}$. An epoxy was cured according
to the manufacturer's specifications and yielded the thermomechani-
cal TBA behavior "before post-cure" shown in Fig. 5 (which lists
experimental details and a summary of transitions). It is apparent
that the cure cycle was not sufficient to have $T_g = T_{cure}$. Post-
cure resulted in significant increase in T_g as well as change in
the viscoelastic behavior (e.g. the damping behavior) below the
glass transition. In particular, a small but significant decrease
in the rigidity (i.e. modulus if no dimensional changes occur on
post-cure) occurred in the glassy state at, for example 0°C, in
consequence of the post-cure.

Consider (Fig. 6) two specimens (1 and 2), one cured above
$T_{g\infty}$ at temperature $_2T_{cure}$, the other cured below $T_{g\infty}$ at temperature $_1T_{cure}$. Specimen 1 vitrified on cure to give a glass transition temperature equal to the temperature of cure. Specimen 2
reacted completely to give the maximum glass transition temperature
($T_{g\infty}$). In the absence of further reaction the specific volume of
the specimen cured at the lower temperature will be higher at
$_2T_{cure}$ than that cured at $_2T_{cure}$ (due to lower crosslink density and
more segmental free ends). The diagram, Fig. 6 (19, 20, 21), shows
that cooling of the more completely reacted material (at equal
rates) results in a higher T_g and indicates how a higher specific
volume results in the glassy state.

The higher specific volume at RT of the more highly cross-
linked material is held responsible for its lower density and lower
modulus at RT and greater water adsorption on immersion at RT (7,
19,20).

It might be noted that, as a consequence of the operational

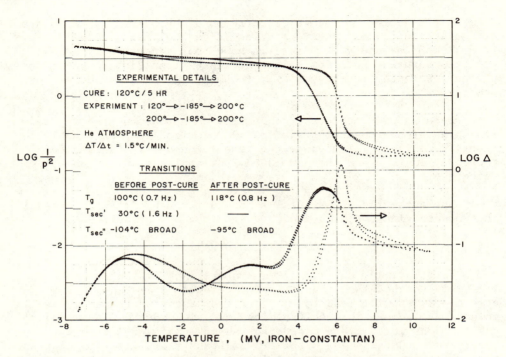

Fig. 5. Thermomechanical behavior (TBA) of an epoxy after
recommended cure and after post-cure. Note the
decreased rigidity at 0°C (OMV) after post-cure.

procedure of locating the vitrification time [and the glass transition temperature (T_g)] in dynamic mechanical analysis by the position of the maximum in mechanical damping, in practice the glass
transition temperature T_g obtained after prolonged cure at temperature $T_{cure} < T_{g\infty}$ will lead to $T_g > T_{cure}$. This is shown (4) in
Fig. 7 where $T_g \simeq (T_{cure} + 40°C)$ for $T_{cure} < $ ca. $(T_{g\infty} - 50°C)$.

It should be feasible to molecularly design systems such that
T_g is well above T_{cure}. Suggestions for reactions in the glassy
state include cure by small molecules able to diffuse in the glassy
state (as in chain reactions), incorporation of sterically well-
situated reactive groups in the polymerizing system, and employment
of systems with prominent glassy-state relaxations.

Some thermosetting curing systems are considered to give
materials with a glass transition temperature significantly higher
than the maximum temperature reached during cure (e.g. see ref. 22).
These are likely to involve rapid free radical polymerization in

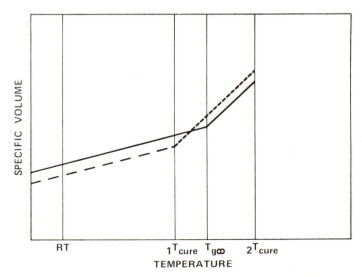

Fig. 6. Schematic: Specific volume versus temperature in the
         absence of chemical reaction. Cured at temperature
         $1^{T}cure$: dashed line. Cured at temperature $2^{T}cure$:
         solid line. Note the higher specific volume at room
         temperature of the more highly reacted system.

which even although radical–monomer and radical–radical reactions
are quenched by vitrification, subsequent heating results in vir-
tual immediate reaction as the materials begin to soften.

EFFECT OF GELATION ON MATERIAL PROPERTIES IN TWO–PHASE SYSTEMS

     The curing of rubber–modified epoxy systems can involve
change from an initially homogeneous solution of reactants containing epoxy resin, curing agent and reactive liquid rubber (e.g.
carboxy–terminated acrylonitrile butadiene copolymer) to a two–
phase system having rubber particles dispersed in an epoxy matrix.
As discussed, different morphologies arise from cure at different
temperatures due to the influence of temperature on the competition
of thermodynamic and kinetic factors. For example, cure above temperature T_S (Fig. 1B), the temperature above which phase separation
does not occur prior to gelation, leads to an optically transparent
material in which rubber is dispersed in the matrix. Cure at lower
temperatures than T_S leads to visual phase separation, the number

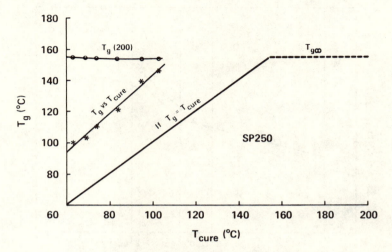

Fig. 7. T_g versus T_{cure}. Specimens were cured isothermally at
T_{cure} for prolonged times before measuring T_g (4). The
maximum glass transition temperature ($T_{g\infty}$ = 156°C) was
measured on cooling after heating to 200°C (at 1.5°C/min).
Note $T_g \simeq$ (T_{cure} + 40°C) for T_{cure} < $\sim$ ($T_{g\infty}$ – 50°C).

and size of the rubbery domains depending on the temperature. The
morphology developed controls the material properties.

An example of the influence of gelation time on morphology is
made evident by comparison of the thermomechanical TBA behavior
after the same cure of a rubber-modified epoxy cured without (Fig. 8
top) and with (Fig. 8 bot.) catalyst (18). The glass transition temperature of the rubber is much more prominent in the sample cured
without catalyst. This suggests that the extent of phase separation depends on the time available for phase separation which is
limited by the process of gelation. The higher glass transition
temperature of the epoxy matrix for the sample cured with the
longer gelation time also suggests more complete separation of the
two phases.

The profile of the relative amounts of phase-separated rubber
versus temperature (see Fig. 2) developed in a rubber-modified epoxy
system in principle can be determined conveniently by TBA temperature scans after isothermal cures. In the scans the intensity of

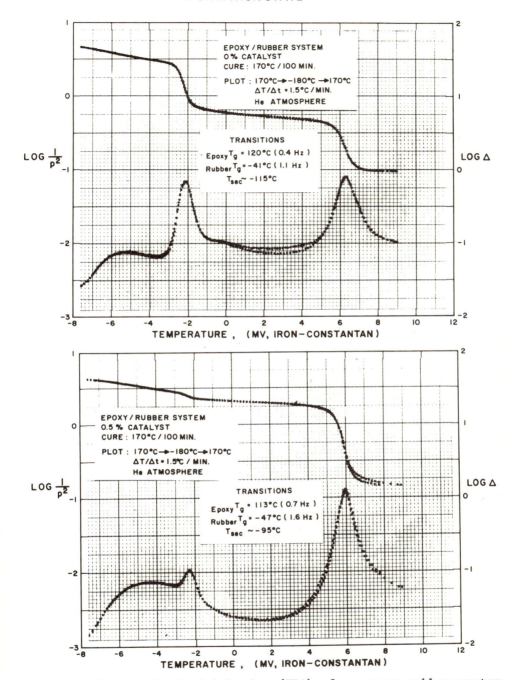

Fig. 8. Thermomechanical behavior (TBA) of an epoxy-rubber system following identical time-temperature cure paths. Top: Zero parts per hundred of catalyst. Bottom: 0.5 parts per hundred of catalyst.

the glass transition of the rubbery phase relates to the degree of phase separation of the rubber. Microscopy in its various forms is required for determining the details of the morphology.

STRESS RESPONSE MODEL FOR RUBBER-MODIFIED THERMOSETS (12)

Stress-strain behavior to failure of rubber-modified thermoset materials at low and high rates of strain can be fitted to a concise model (12) which is summarized schematically in Figure 9. The model is an extension of the Ludwik-Davidenkow-Orawan hypothesis (23). Brittle fracture stress, and plastic deformation as measured by the yield stress, are considered to be independent processes that can be represented by separate characteristic curves versus temperature. Intersection of the curves is the brittle-ductile transition. The yield stress involves a viscous dissipation mechanism and is assumed to show a higher dependence on temperature. The response mechanism is decided by arguing that whichever process will occur at lower stress will be the operative one. This will be either brittle failure or yield for a single phase material (see Fig. 9B). Other response mechanisms may operate in a two-phase system. As an example (12), the scanning electron micrograph of the failure surface of a rubber-modified epoxy with rubbery domain inclusions (see Figure 10) shows the result of extensive debonding and cavitation which occurred on fracture. The presence of domain inclusions in a rubber-modified epoxy material introduces debonding and cavitation as a stress response mechanism. This introduces a third curve in the stress response diagram (see Figures 9A and 9C).

Changing the rate of strain produces a shift in the curves. The viscous element is the most rate dependent and the yield stress curve will be shifted to higher temperatures at higher strain rates.

The mechanical response of a phase-separated, modified epoxy that contains little dissolved rubber is explained in the following manner (Figure 9A). During tensile testing at low strain rates, the stress builds up until it reaches a curve that represents a failure response mechanism. The first one reached is the debonding/cavitation curve. An apparent yield stress is noticed as the specimen stress-whitens. It fails soon afterwards since the cavities coalesce quickly to provide catastrophic failure. At higher rates of strain, the first curve reached is again the debonding/cavitation stress. The impact energy can be increased because of the presence of rubbery domains in the fracture plane. High levels of separated rubber do not usually improve the toughness or the impact energy since the domains decrease the effective area of the hard-phase epoxy in the fracture plane.

In a single phase, rubber-modified epoxy (Fig. 9B), formed by curing isothermally above temperature T_S (Fig. 1B, and Fig. 2), the yield stress curve has been shifted to lower temperatures as a result of the rubber dissolved in the epoxy matrix. The debonding/

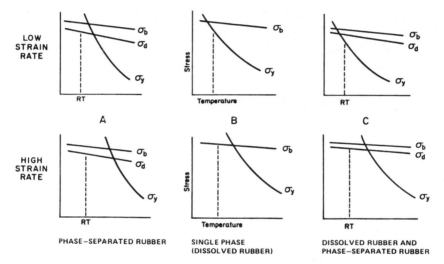

Fig. 9. Stress response model (12) for rubber-modified epoxies
with different amounts of dissolved and phase-separated
rubber: (A) two-phase system with a large volume fraction of phase-separated rubber; (B) single phase system
with all rubber dissolved; (C) two-phase system with
both phase-separated and dissolved rubber.

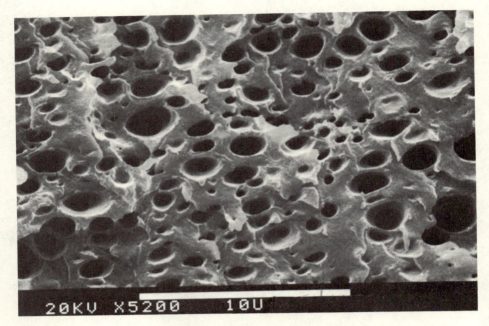

Fig. 10. Scanning electron micrograph (SEM) of tensile fracture
          surface of rubber-modified epoxy after fracturing at
          −195°C (12).

cavitation curve is absent, since the cured epoxy material is
single phase. The first curve reached at low rates of strain is
the yield stress. The material can exhibit high elongation to
break and good toughness through plastic deformation. At higher
rates of strain, the yield stress curve has shifted to higher temperatures. The first curve to be reached at room temperature will
then be the brittle failure stress since there is no intercepting
debonding/cavitation curve. The material fails in a brittle manner
and the impact energy is poor.

     Optimum properties result from the correct combination of
phase-separated and dissolved rubber (Fig. 9C). When the rubber
and epoxy are relatively compatible, the interfacial bonding is
good and the debonding stress is high. The yield stress curve is
again shifted to lower temperatures since there is a large amount
of dissolved rubber in the epoxy matrix. The system is able to
reach its yield stress at room temperature at low rates of strain.
The yield stress is unattainable at impact rates but the debonding/
cavitation curve is available. The material can then exhibit improved impact properties since the brittle failure curve is again
blocked out. This material has both improved low strain rate toughness and improved impact properties.

ACKNOWLEDGMENT

Partial support has been provided by the Chemistry Branch of the office of Naval Research and the B. F. Goodrich Company.

REFERENCES

1. J. K. Gillham, Matrix Properties, in "Organic Matrix Composites", National Academy of Sciences, National Materials Advisory Board, Report NMAB-365, 1981. Ch. 3.
2. J. K. Gillham, Polym. Eng. Sci. 19, 676 (1979).
3. A. F. Lewis, M. J. Doyle, and J. K. Gillham, Polym. Eng. Sci., 19, 683 (1979).
4. N. S. Schneider and J. Gillham, Polym. Composites 1, 97 (1980).
5. J. K. Gillham, Society of Plastics Engineers, Proceedings, Annual Technical Meeting, N.Y., 268 (1980).
6. J. B. Enns, J. K. Gillham, and M. J. Doyle, Amer. Chem. Soc., Prepr., Div. Org. Coat. Plast. Chem., 43, 669 (1980).
7. M. J. Doyle, J. K. Gillham, S. J. Washburn, and C. A. McPherson, Amer. Chem. Soc., Prepr., Div. Org. Coat. Plast. Chem., 43, 677 (1980).
8. J. K. Gillham, Amer. Chem. Soc., Prepr., Div. Org. Coat. Plast. Chem., 44, 185 (1981).
9. J. B. Enns and J. K. Gillham, Amer. Chem. Soc., Appl. Poly. Sci., Div. Org. Coat. Plast. Chem., 46, 592 (1982)
10. P. J. Flory, "Principles of Polymer Chemistry", Cornell University Press, Ithaca, N.Y. (1953).
11. L. T. Manzione, J. K. Gillham, and C. A. McPherson, J. Appl. Polym. Sci., 26, 889 (1981).
12. L. T. Manzione, J. K. Gillham, and C. A. McPherson, J. Appl. Polym. Sci., 26, 907 (1981).
13. J. K. Gillham, Amer. Inst. Chem. Eng. Journal, 20, 1066 (1974).
14. J. B. Enns and J. K. Gillham, Amer. Chem. Soc., Prepr., Div. Org. Coat. Plast. Chem., 45, 492 (1981).
15. N. S. Schneider, J. F. Sprouse, G. L. Hagnauer, and J. K. Gillham, Polym. Eng. Sci., 19, 304 (1979).
16. J. K. Gillham, Polym. Eng. Sci., 19, 319 (1979).
17. J. K. Gillham, Amer. Chem. Soc. Symp. Series, "Resins for Aerospace", 132, Ch. 26 (1980).
18. J. K. Gillham, C. A. Glandt, and C. A. McPherson, in "Chemistry and Properties of Crosslinked Polymers", Acad. Press, N.Y. (1977), p. 491.
19. S. J. Washburn, Senior Thesis, Dept. Chem. Eng., Princeton Univ., May 1980.
20. J. P. Aherne, J. B. Enns, M. J. Doyle and J. K. Gillham, Amer. Chem. Soc., Appl. Poly. Sci., Div. Org. Coat. Plast. Chem., 46, 574 (1982).
21. A. Shimazaki, J. Polym. Sci., C, 23, 555 (1968).
22. J. K. Gillham and J. C. Petropoulos, J. Appl. Polym. Sci., 9, 2189 (1965).
23. I. M. Ward, "Mechanical Properties of Polymers", Wiley, London (1971), p. 332.

CONTROL OF COMPOSITE CURE PROCESSES

Richard J. Hinrichs, Ph.D. and J. M. Thuen

Applied Polymer Technology

Tustin, California

ABSTRACT

The control of a composite cure process is an attempt to minimize thermal and pressure gradients within a laminate. The sources of these gradients (fabrication support materials, tooling, part geometry) were investigated for process significance. In addition, the concept of using engineering cure behavior transformation diagrams (heat rate and cure temperature affects on the reaction dynamics of a material) was used to evaluate the effects of gradients on the consistency of laminate processability. Control of these gradients was achieved by the use of an automated composite control system developed by Applied Polymer Technology, known as CAPS.

INTRODUCTION

The goal in laminate fabrication is to achieve manufacturing consistency and high quality (void free, uniform compaction, appropriate fiber volume) parts. To achieve this goal, a basic understanding of the principle factors related to the cure process is necessary. These factors allow us to determine the areas for investigation during engineering process development and how to examine laminate defects for the source producing the imperfections.

The dynamics of composite cure behavior are a result primarily of two integrally associated processes. These are the chemical reaction-polymerization characteristics and the physical flow-compaction properties of the material. The chemical reaction dynamics is a time-temperature pathway dependent function. This means variable reaction kinetics, polymer structure and morphology can result as a

147

function of the thermal cure history.[1,2,3,&4] These affects are
translated into variable physical flow-compaction properties of a
laminate. In addition, the physical properties are also a function
of the mechanical stress (pressure) and relaxation pathways (tooling
cavities, bleeder support materials, etc.) available during part cure.
The greater and less resistant the flow pathways available to the
resin at any particular viscosity, the greater the flow for an applied
stress. This view of composite cure reduces the considerations to
those factors which affect the time-temperature thermal history and
effective hydrodynamic (fluid) pressure within the laminate.

This study is an attempt to evaluate common daily fabrication
techniques and determine their significance of the laminate thermal
history. In addition, to introduce an efficient economical approach
to generate process cure specifications based on actual material
behavior properties. These engineering cure behavior transformation
diagrams can be used to project the affects of thermal variations,
fabrication out time, and volatile outgasing on a proposed cure
process cycle.

EXPERIMENTAL WORK AND RESULTS:

EFFECTS OF FABRICATION SUPPORT MATERIALS

The affects of fabrication support materials on thermal and
pressure gradients were evaluated in two steps. The first variable
was to examine the affects of variable bleeder plies on producing
thermal gradients within a laminate. Four 20 ply quasi-isotropic
T-300 carbon fiber laminates were prepared. They were placed on a
one-eighth inch thick steel tool. Thermocouples were placed on the
tool-laminate surface, laminate-bleeder ply surface, and in the
ambient autoclave air environment above the tool. Variable (one to
eight) plies of 120 style glass bleeder was placed on top of each
laminate. A standard vacuum bag enclosure was used to seal the
laminates for autoclave cure. The autoclave was then ramped at 5°F
per minute (air temperature) and the corresponding temperature
measurements were taken. The objective was to evaluate the laminate
temperature differential as a function of the bleeder ply insulation
capability. The results are illustrated in Figure 1. You can readily
see that the laminate surface temperature, with a single bleed ply,
tracks almost directly with the ambient autoclave air temperature.
However, with two plies of bleed material, there is a 10°F temperature
differential between autoclave and laminate surface. After four plies
of bleed material, it is apparent that the laminate is tracking the
tool surface temperature irrespective of the autoclave ambient temperature. This resulted in a 25 to 30°F temperature differential withi:
this test.

The next concern to further complement these results, was to
examine the effect of laminate thickness on creating thermal

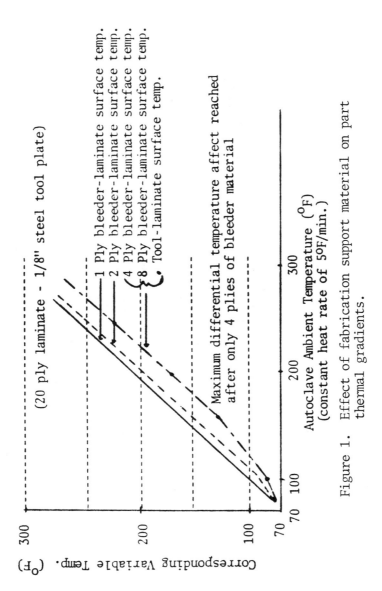

Figure 1. Effect of fabrication support material on part thermal gradients.

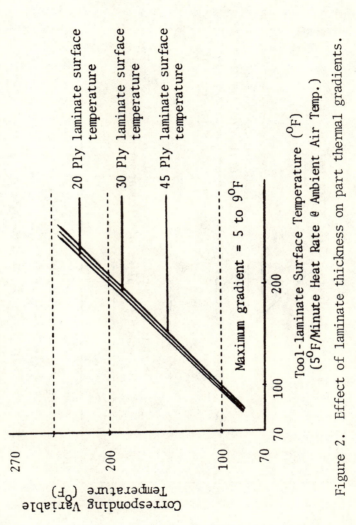

Figure 2. Effect of laminate thickness on part thermal gradients.

differentials under the same test conditions as above. In this experiment, three laminate thicknesses (20, 30, and 45 plies) were layed up, each with 8 plies of 120 glass bleeder. The bleeder, chosen as a result of the previous experiment, was to achieve a thermal barrier so that only the gradient effects due to laminate thermal conductivity would be observed. Again, thermocouples were placed on the laminate-bleed and laminate-tool surfaces. The results are plotted so that the laminate surface temperature versus tool surface (heat source due to the bleeder insulation effect) can be directly compared (Figure 2). It is evident that only a $5^{\circ}F$ to $9^{\circ}F$ gradient was observed within a 20 to 45 ply laminate. This illustrates a rapid ability of thermal conductivity (under normal $5^{\circ}F$/minute heat rates) to minimize laminate thickness affects. This clearly does not include conditions where exothermic, potential localized adiabatic heating or high heat rates are encountered.

Combined, these two experiments indicate a process fabrication concept slightly more defined than encountered in previous experiences. These experiments indicate that the bleeder-support materials (and thus fabrication techniques) can be the primary causes of thermal gradients within a laminate. If a condition can exist where the laminate can have one surface exposed to a higher heat source (less than 4 bleed plies or variable tooling surface temperatures) while the other is thermally coupled to a heat sink, then major thermal gradients can occur. (Note that both conditions must apply.) If no heat sink is present, then thermal conductivity is sufficient to allow the particular laminate section, even of considerable thickness variations, to maintain the temperature of the primary heat source (generally the tool). In situations where fabrication techniques (greater than 4 bleed plies) are used, the tooling surface controls the heat distribution and rate during a cure process.

MANUFACTURING MODEL OF LAMINATE PROCESS BEHAVIOR

The primary objective in developing a thermal cure cycle is to routinely produce a high quality laminate. The achievement of this goal can be viewed as a competitive interraction between three principle players. Through examining the relationship of these players and understanding their individual components, a direct procedure to obtain efficient process cure cycle is obtained. This particular relationship, illustrated in Figure 3, allows us to investigate distinct aspects of laminate process behavior and directs our attentions to logical sources of difficulties by which to view particular problems.

The diagram of the model is oriented towards the production world of physical/chemical laminate fabrication relationships. Process behavior is thus defined in terms of the competition between volatile gas pressure (player 3), fluid hydrostatic pressure (player 2), and

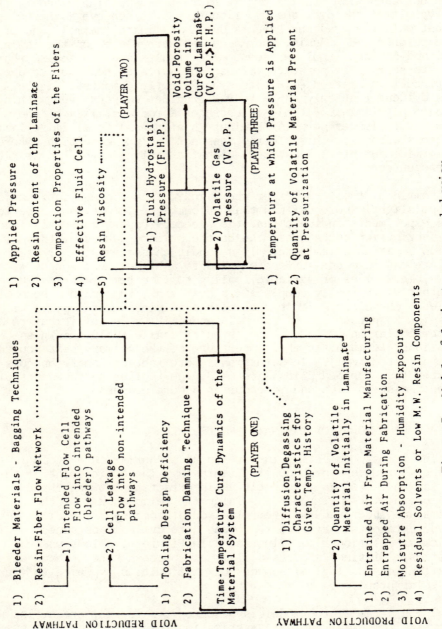

Figure 3.   Model of laminate process behavior.

chemical polymerization kinetics (player 1). The diagram is read
from right to left, thus orienting the laminate result or problem
as a function of the principle process variables which contributed
to its origin.

DEVELOPMENT OF PROCESS TRANSFORMATION DIAGRAMS

The model presented describes essentially a balance. One side
contains the fluid hydrostatic pressure (effective resin liquid pressure)
responsible for flow. The other side contains the volatile gas pressure (gas expansion force) attempting to form bubbles and migrate
through the laminate. If the effective liquid cohesive pressure is
greater than the gas expansion pressure, then no bubble formation can
occur. The major variables which interract with these two processes
are the chemical kinetic (through viscoelastic cure behavior) and the
physical tooling-fabrication techniques employed.

The objective of these experiments was then to define analytical
techniques by which to develop effective cure parameters. The procedure
had to allow for evaluation of both temperature and rate variable
affects so that tooling or manufacturing vessel criteria could be
imposed. It had to account for variable gas diffusion considerations,
yet be in such a form as to act as a tool for real time engineering
process decisions.

The result was the development of a specific series of dynamic
rheological and thermalgravimetric tests. When combined, these tests
meet the experimental objectives and are used to specify the process
cure requirements. The rheological test pattern consists of two
distinct procedures. Each material is initially characterized
(strain and frequency sweeps) to define the test parameters. The
first test is a series of isothermal temperature hold experiments
measuring the viscoelastic kinetic cure properties (Figure 4). The
temperature values selected bracket every $10^{\circ}F$, the range in which
processing is to occur. The four properties measured are the loss
modulus (viscous modulus G''), storage modulus (elastic modulus G'),
complex viscosity (n*), and the tan delta (G''/G'). However, when
mainly newtonian liquids or monomers are present (G'' $\gg$ G' & tan
delta $>$ 10), viscosity is sufficient to use for the evaluation
criteria. Figure 4 is a composite engineering diagram indicating
the rate of viscosity change versus time for a family of isothermal
temperatures. Gelation time was defined as the point where tan
delta = 1.0. This definition is very reproducible due to the technique employed and has been correlated with physical gelation by
other workers in this field.[5] This point then identifies, for a
given temperature-time history, the outer boundary where laminate
consolidation (pressurization) would become inconsistent at best.
The onset of gelation is defined as that point where a measureable
network structure is starting to form (i.e. G' begins to increase
relative to G'' and thus tan delta begins a significant decrease).

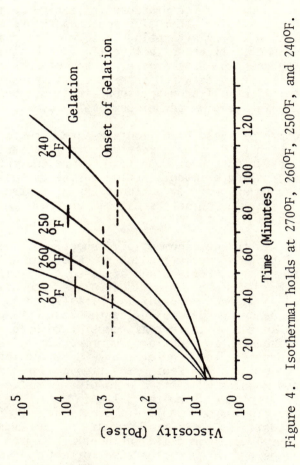

Figure 4. Isothermal holds at 270°F, 260°F, 250°F, and 240°F.

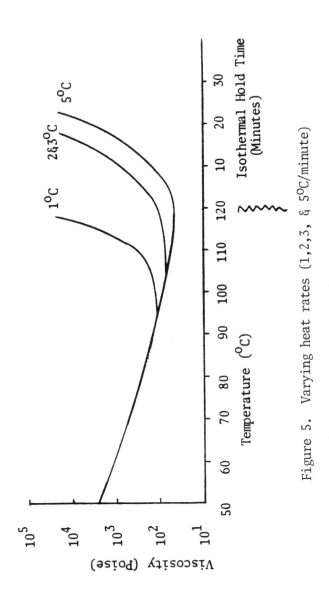

Figure 5. Varying heat rates (1,2,3, & 5°C/minute)

This point is also very reproducible and easy to define from the
rheological measurements. It represents the start of a cautionary
zone (before gelation point) where laminate consolidation should
commence. This data is therefore able to identify a materials time-
temperature process thermal sensitivity. This type of cure transformation diagrams allows the engineer to define temperature boundary
tolerances (optimum hold and differential allowable temperatures) and
process time functions (i.e. pressurization). It is thus used to
develop specific M & P laminate cure experiments based on actual
material behavior properties. The second rheological test is a
dynamic temperature ramp where variable heat rates are examined.
Heat rate interracts with the polymer chemistry to cause variations
in morphology, chemical reaction pathways and rheological process
parameters (viscosity, gelation, vitrification, etc.)6&7 Heat
rate variables are chosen which represent the possible manufacturing
dynamic range. Figure 5 illustrates this concept. Here, rates from
one to five degrees Centigrade per minute were used to heat the
specimen to the desired hold temperatures. The data was then compiled
on a single temperature axis base. It is readily apparent that
significant alterations in the viscoelastic profile is observed. Slow
heat rates tend to allow more time for polymerization processes to
occur. This affects process behavior through higher viscosities and
shorter gelation times. Fast heat rates minimize the polymerization
effects. This results in lower viscosity profiles and longer
gelation times. (Note morphology and component phase seperation
precipitation affects can also occur and would be observed in the
other rheological properties not shown in this illustration.) These
rheological time-transformation diagrams then yield the process outer
boundary limits for laminate consolidation, temperature sensitivity
for part thermal differential requirements, material kinetic behavior,
and heat sensitivities. This forms the base set for thermal process
dynamics.

The remaining parameter is the lower or minimum boundary requirements to optimize volatile gas evolution and identify affects of
fabrication out time. The gas evolution concept is based on the
thermalgravimetric properties (weight loss versus time) of a laminate
specimen at the same isothermal temperatures used in the first
rheological experiment. Clearly, the diffusion can be altered by
laminate thickness and geometry. These tests then reflect a minimum
time necessary for volatile evolution. This does not impair its use,
as only a minimum time (for a 1% moisture saturated prepreg specimen
weight loss to stabilize) is required. The maximum time is predefined
by the gelation data of Figure 4. This data forms the minimum time
boundary marker for the isothermal engineering transformation diagrams
The final affect of fabrication out time is determined by repeating
the rheological experiments on material aged as desired in controlled
environments. Compiling the aged data with the original will immediately identify out time effects and process requirements.

SUMMARY AND CONCLUSIONS

This paper has dealt with examining affects of fabrication techniques and laminate thicknesses on thermal process behavior. Here we found that if both a source for a heat sink (tool) and limited fabrication support materials (less than 4 to 8 plies of bleeder) are contacting the opposing laminate surfaces, significant thermal gradients can result. Further, that thermal conductivity is sufficient (under laminate surface insulation situations) to minimize thermal differentials and cause the laminate to track the tooling surface. This reinforces the concept of uniform thermal distribution tooling design (i.e. there is no substitute for proper fabrication and tooling techniques).

A manufacturing process model was developed describing the interractions associated with production of composite materials. Through this model, it is apparent that three major pathways contribute to quality laminate production. The first is the thermal cure profile chosen through which the materials chemistry (reaction polymerization, process morphology, viscoelastic properties) then affects the remaining two competive pathways. These are the concepts of fluid hydrostatic pressure (resulting from the physical tooling, fabrication design and laminate geometries) and entrapped volatile gas pressure (results of the diffusion characteristics and quantity of volatile material present at time of cure).

Finally, the development of a combined series of experiments which result in manufacturing cure transformation diagrams. These diagrams define process parameters based on actual material cure dynamics. These tests are more quantitative and less expensive than the current trial and error laminate test programs. They present data in graphic process engineering format for specification development and real time process control. The parameters describe time-temperature boundary volatile release limits, thermal sensitivity/tolerance, heat rate sensitivity, and fabrication out time effects.

REFERENCES

1. Lewis, Doyle, Gilham, "Effect of Cure History on Dynamic
        Mechanical Properties of an Epoxy Resin", Polymer Engineering
        and Science, Vol. 19, No. 10, 1979.

2. Schneider, Gilham, "TBA Studies of Prepreg Curing Behavior".
        Polymer Composites, Vol. 1, No. 2, 1980.

3. Hinrichs, Thuen, "Environmental Effects on the Control of Advanced
        Composite Materials Processing", SAMPE Journal, Vol. 15, No.6,
        1979.

4. Thuen, Hinrichs, "Structural Adhesives Rheological Behavior
        Response to Process-Environmental Variations", SAMPE Journal
        Vol. 16, No. 5, 1980.

5. Tung, Dynes, "Chemo-Rheological Characterization of B-staged
        Printed Wiring Board Resin", ASTM Conference, Oct. 1981.

6. Gilham, "Award Address, Formation and Properties of Network
        Polymeric Materials", Polymer Engineering and Science, Vol. 19
        No. 10, 1979.

7. Roller, "Characterization of the Time-Temperature-Viscosity
        Behavior of Curing B-Staged Epoxy Resins", Polymer Engineering
        and Science, Vol. 15, No. 6, 1975.

A STUDY OF THE THERMO-OXIDATIVE PROCESS AND STABILITY

OF GRAPHITE AND GLASS/PMF POLYIMIDE COMPOSITES

Daniel A. Scola

United Technologies Research Center

East Hartford, CT 06108

INTRODUCTION

Composite materials consisting of graphite and glass fibers in polyimide matrices are being investigated for use in aerospace and other commercial applications at temperatures of 600°F and above. Results of a continuing investigation to determine factors involved in the thermo-oxidative stability of unidirectional fiber/PMR polyimide composites will be presented at this symposium. Composite materials consisting of HTS, HMS, T-300, Celion 600 graphite fibers and S-glass fibers in two polyimide matrices, PMR-11 and PMR-15, were subjected to isothermal aging at 316°C in flowing air at two flow rates (100 cc/min and 900 cc/min) for time periods up to 600 hours. The weight losses, shear and flexural properties were determined at room temperature and 316°C. One system, Celion 6000/PMR-15, was selected for study at 335°C in flowing air at 100 cc/min over a 1015 hour period. The oxidation process of this isothermally exposed composite system was followed by optical microscopy and mechanical property measurements.

RESULTS AND DISCUSSION

Materials

PMR-11 and PMR-15 refer to addition-type polyimides in which the prepolymer molecular weights are nominally 1100 and 1500 respectively (Ref. 1). Fiber materials were obtained commercially, HTS and HMS from Hercules, Inc., Thornel 300 from Union Carbide Corporation, Celion 6000 from Celanese Corp., and S-glass from Owens-Corning Fiber Glass Corp.

159

Weight Loss of Graphite Fibers and Composites at 316°C

   The weight losses of several fibers at the two air flows are
listed in Table 1. At 316°C in air flows of 100 cc/min or 900 cc/min,
fiber stability is excellent. However, composite weight loss
data, listed in Table 2, show that composites containing the HTS
fiber showed the largest weight loss at both air flow rates.

SIMS Analysis of Graphite Surfaces

   The surfaces of HTS, HMS, Celion 6000 and Thornel 300 graphite
fibers were analyzed by secondary ion mass spectroscopy (SIMS).
The fibers were sputtered by the argon ion beam to a depth of 15
Angstroms, and then subjected to SIMS analysis. These results
in Table 3 show that HTS fibers contain a high concentration of
sodium and potassium ions relative to Celion 6000 and Thornel
300 graphite fibers. The fibers were further subjected to
sputtering by a neon ion beam to a depth of 60 Angstroms and then
subjected to SIMS analysis. The results in Table 4 also show that
HTS graphite fibers have a much higher concentration of sodium
ions than HMS, Celion 6000 and Thornel 300 fibers. Both HTS and
HMS fiber show significantly higher potassium ion levels than the
Celion 6000 and Thornel 300 fibers.

Shear Properties of Thermo-Oxidatively Exposed Composites

   Composites of these materials were subjected to isothermal
conditions at 316°C, in flowing air at 900 cc/min. The RT shear
properties of PMR-11 and PMR-15 composites are listed in Tables
5 and 6 respectively. The results clearly show that after 596
hours in flowing air at 900 cc/min, composites containing the
HTS fiber were degraded completely, while composites containing
HMS, Celion 6000 and Thornel 300 graphite fibers retained a
fraction of the original strengths varying from 23 to 99%.

   The 316°C shear strengths of the fiber/polyimide composites
after exposure at 316°C for 596 hours are listed in Tables 7 and
8. Comparisons of the 316°C shear strengths (Tables 7 and 8)
and weight losses (Table 6) of each composite system after
isothermal aging suggest that the interface in the PMR polyimide
composite system containing HTS graphite fiber is less thermo-
oxidatively stable at the higher air flows than PMR polyimide
composites containing HMS, Celion 6000, T300 or S-glass.

   The excellent thermo-oxidative stability of the HTS fibers
in flowing air (900 cc/min) relative to the other fibers and the
poor performance in the composite system suggests that the fiber-
resin interface is contributing to the instability of the composite systems. It is postulated that the high concentration
of sodium on the HTS fiber surface, and near the fiber surface

contributes to the poor thermo-oxidative stability of HTS/PMR
polyimide composites.

Thermo-oxidative Behavior of Celion 6000/PMR-15 Composites at 335°C

One composite system, Celion 6000/PMR-15, was selected for
thermo-oxidative studies at 335°C in flowing air (100 cc/min)
over a 1015 hour period. The thermo-oxidative process over this
time period was followed by optical microscopy. Micrographs of
the composite cross section perpendicular to the fiber ends and
directly on the composite top section were taken. Micrographs
of the end surfaces of the composite systems after exposures
over these time periods are shown in Figure 1. These end surfaces
were polished, and the distance required to reach a crack-free
surface was determined. Micrographs of the crack-free surfaces
of the ends of polished composites are shown in Figure 2. The
depth of the "crack" or thermo-oxidative cavity into the 2 inch
length of composite is shown under each micrograph, and is also
listed in Table 9. Figure 1a shows that no thermo-oxidative
cracks or cavities appear after 70 hours at 335°C. However,
cracks or cavities appear in 100 hours and get progressively
larger as the exposure time increases to 1015 hours. The actual
crack depth in the fiber direction over a 2 inch long composite
specimen after 1015 hours of exposure was found to be 0.177
inches (Table 9). The polished cross section shows that the
cracks extend a distance of 1.5 plies into the top and bottom
surfaces of what was a 9 ply composite. Two plies are lost
during the oxidation process, and cracks penetrate the remainder
of the composite to a depth of about 0.020 inches.

The weight losses of shear and flexure specimens after aging
at 335°C are listed in Table 10. As expected, the greatest weight
loss (11%) is experienced after 1015 hours. The weight loss takes
place mostly from the outer surface of the composite, as is
indicated by the micrographs in Figures 1 and 2 and also by the
micrograph in Figure 3. This is a top view of a composite
specimen after aging 1015 hours at 335°C. The light area of
the edges of the composite is caused by the loss of the polyimide
matrix as a result of the oxidation process. The dark areas show
the unaffected composite surface.

The shear and flexural properties at room temperature and
316°C after isothermal aging at 335°C in flowing air (100 cc/min)
over these time periods are listed in Tables 11 and 12 respectively.
It should be noted that the room temperature shear and flexural
strengths and flexural moduli are virtually unaffected up to 500
hours, and then show approximately 40 to 50% decrease in strengths
and moduli after 1015 hours. The 316°C shear and flexural properties, however, after 1015 hours isothermal aging at 335°C are
actually increased as a result of this exposure.

Table 1. Fiber weight loss at 316°C.

Fiber	After 570 hrs Air Flow 100 cc/min % Wt Loss	After 400 hrs Air Flow 900 cc/min % Wt Loss
HTS	0.20	0.25
Celion 6000	0.54	0.77
T-300	1.23	--
HMS	0.42	0.50

Table 2. Composite weight loss; 316°C, 400 hrs.

Composite	100 cc/min Air Flow % Wt Loss	900 cc/min Air Flow % Wt Loss
HTS/PMR-15	22.6	28.2
Celion 6000/PMR-15	6.0	6.5
HMS/PMR-15	4.0	15.2
HTS/PMR-11	6.5	27.5
HMS/PMR-11	5.0	14.8
S-Glass/PMR-11	4.0	15.6
T-300/PMR-11	4.0	13.0

Table 3. SIMS analysis of graphite fibers.
^{3}He (2500V), 0.17μA (Depth ~ 50Å)

Fiber	Relative Concentrations		
	Na	K	F
HTS	15.3	2.0	–
Celion 6000	0.05	0.04	82.3
T300	0.6	0.6	–

Table 4.   SIMS analysis of graphite fibers.
^{20}Ne (2500V), 0.17μA (Depth ~ 60Å)

Fiber	Relative Concentrations	
	Na	K
HTS	10.5	1.8
HMS	1.7	1.2
Celion 6000	0.2	0.1
T300	1.4	0.8

Table 5.   Shear properties of fiber/PMR-11 composites.

	RT Strength			
	Unaged		After 596 hrs at 316°C Flowing air (900 cc/min)	
	Psi	MPa	Psi	MPa
HTS/PMR-11	13,150	90.7	degraded	–
HMS/PMR-11	11,800	81.3	3060	21.1
S-Glass/PMR-11	12,500	82.8	5240	36.1
T300/PMR-11	14,900	96.5	3480	24.0

Table 6.   Shear properties of fiber/PMR-15 composites.

	RT Strength			
	Unaged		After 596 hrs at 316°C Flowing air (900 cc/min)	
	Psi	MPa	Psi	MPa
HTS/PMR-15	13,850	95.5	degraded	
Celion 6000/ PMR-15	14,800	102.3	14,700	104
HMS/PMR-15	7,250	50.0	2,600	17.9

Table 7. Shear properties of fiber/PMR-11 composites.

	316°C Shear Strength			
	Unaged		After 596 hrs at 316°C Flowing air (900 cc/min)	
	Psi	MPa	Psi	MPa
HTS/PMR-11	3940	27.2	degraded	
HMS/PMR-11	7250	50.0	1430	9.9
S-glass/PMR-11	4900	33.8	2530	17.5
T300/PMR-11	6020	41.5	1740	12.0

Table 8. Shear strength of fiber/PMR-15 composites.

	316°C Shear Strength					
	Unaged		After 200 hrs at 316°C in Flowing Air 100 cc/min		After 596 hrs at 316°C in Flowing Air 900 cc/min	
	Psi	MPa	Psi	MPa	Psi	MPa
HTS/PMR-15	8100	55.9	7850	54.1	Degraded Completely	
Celion 6000/ PMR-15	8260	57.0	8590	59.3	3210	22.1
HMR/PMR-15	4850	33.5	5880	40.6	1610	11.1

Table 9. Thermo-oxidative behavior of Celion 6000/PMR-15 composites at 335°C in flowing air, 100 cc/min.

Aging Time hrs	Depth of Thermo-oxidative "crack" into composite end, in.
0	–
70	0.000
100	0.027
200	0.029
500	0.069
1015	0.177

Table 10.   Isothermal aging of Celion 6000/PMR-15 polyimide composites.

Aging Time hrs at 335°C (635°F)	Weight % Loss	
	Shear Specimens	Flexure Specimens
70	0.77	0.59
100	1.21	1.02
200	1.89	1.49
500	4.45	3.66
1015	9.90	10.9

Table 11.   Effect of isothermal aging at 335°C (635°F) of Celion 6000/PMR-15 composites in flowing air, 100 cc/min.

Isothermal Aging at 335°C (635°F) hours	Short Beam Shear Strength			
	RT		(316°C (600°F)	
	psi	MPa	psi	MPa
0	17,850	123.3	5120	35.3
70	16,700	115.3	6130	42.3
100	16,400	113.3	6270	43.2
200	15,600	107.7	6640	45.8
500	14,450	99.8	6330	43.6
1015	10,850	79.9	5750	39.6

Table 12.   Effect of isothermal aging at 335°C (635°F) of Celion 6000/PMR-15 composites in flowing air, 100 cc/min.

Isothermal Aging at 335°C(635°F) hours	Flexural Properties							
	RT				316°C (600°F)			
	Strength		Modulus		Strength		Modulus	
	ksi	MPa	ksi	MPa	ksi	MPa	ksi	MPa
0	242.5	1675	16.4	113	48.0	331	8.30	57.4
70	234.5	1608	16.8	116	53.4	369	9.35	64.5
100	234.5	1608	16.2	111	59.2	408	8.95	61.6
200	204.0	1406	16.2	111	65.8	454	10.92	75.0
500	196.5	1356	15.0	103	78.2	540	14.1	97.5
1015	112.0	769	12.5	86.3	73.6	507.5	10.7	74.1

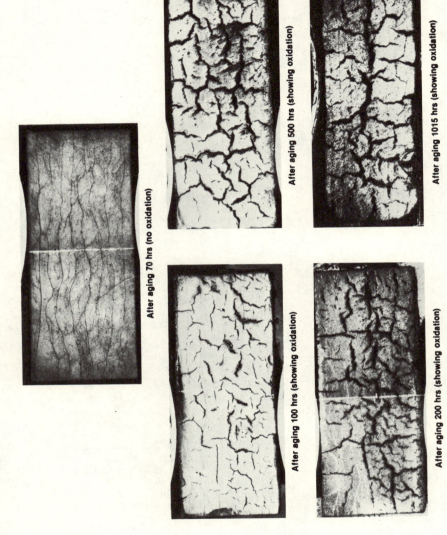

After aging 70 hrs (no oxidation)

After aging 100 hrs (showing oxidation)

After aging 200 hrs (showing oxidation)

After aging 500 hrs (showing oxidation)

After aging 1015 hrs (showing oxidation)

Fig. 1. Thermo-oxidative studies at 335°C 100cc/min flowing air.

Aged samples ground and polished — 70 hrs

Depth of oxidation, 0.0 inches over 2 inch length

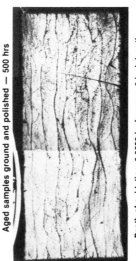

Aged samples ground and polished — 500 hrs

Depth of oxidation, 0.0695 inches over 2 inch length

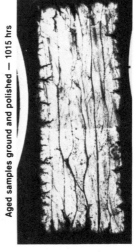

Aged samples ground and polished — 1015 hrs

Depth of oxidation, 0.177 inches over 2 inch length

Aged samples ground and polished — 100 hrs

Depth of oxidation, 0.029 inches over 2 inch length

Aged samples ground and polished — 200 hrs

Depth of oxidation, 0.027 inches over 2 inch length

Fig. 2. Celion 6000/PMR-15 thermo-oxidative studies at 335°C, 100cc/min flowing air.

Fig. 3.   Top view of Celion 6000/PMR-15 polymide composite
          (after aging 1015 hrs at 335°C, in flowing air,
          100 cc/min).

CONCLUSIONS

Thermo-oxidative stability of graphite fiber/PMR-11 and PMR-15 composite materials at 316°C and higher appear to be affected by fiber surface impurities such as sodium ions. The composite materials which exhibited the greatest thermo-oxidative stability contained low levels of sodium and potassium ions. These materials were S-glass, T-300, HMS and Celion 6000. Optical microscopy studies of the thermo-oxidative process of Celion 6000/PMR-15 composites at 335°C over a 1015 hour period in flowing air (100 cc/min) reveals that oxidation occurs at free surfaces of the composite, particularly the end surfaces in the fiber direction, and proceeds slowly in the resin matrix near the fiber/resin interface leaving fibers free of resin. There is no evidence for thermo-oxidative degradation in the matrix phase or matrix-fiber interface region in the bulk (interior) of the composite. Flexural strength and modulus and shear strength data, as well as weight loss data over the 1015 hour period provide additional evidence for the course of the thermo-oxidative process. The Celion 6000/PMR-15 composite system retains and, in fact, shows improvement in the 316°C shear and flexural properties after isothermal aging at 335°C in flowing air (100 cc/min) after 1015 hours.

REFERENCE

1. Serafini, T. T., Delvigs, P. and Lightsey, G. R.: J. Appl. Poly. Sci., Vol. 16, No. 4, pp 905-915 (1972).

RUBBER MODIFIED MATRICES

Robert Y. Ting

Code 5975
Naval Research Laboratory
P.O. Box 8337
Orlando, Florida 32856

INTRODUCTION

Fiber-reinforced organic composites are increasingly being considered for structural applications in aerospace and advanced marine systems. This emphasis is basically due to the very high specific modulus and tensile strength the composite materials have to offer. The thrust behind this trend is the potential for replacing as much metallic components as possible with composite parts such that weight reduction may be achieved resulting in energy saving. In current organic composite systems, thermosetting resins such as epoxy and polyimide are widely used as matrix materials. These resins offer the combined properties of high modulus and creep resistance because of their highly cross-linked structures. However, high cross-linking generally leads to brittle materials that are subject to failure by the growth of internal flaws and progressive crack propagation. The flaws that may be anticipated include internal cracks, surface cuts and microvoids inherently present as a result of the current processing methods employed in fabricating composite parts. The growth of these flaws and the subsequent crack propagation usually cause catastrophic breakdown in structures. The importance of the fracture properties of matrix resins in the performance of composite materials has therefore been realized and emphasized in recent years[1-4]. In the future, both next generation commercial and military aircraft will demand high strain-to-failure in composites for increasing structural applications such that the projected weight reduction and fuel efficiency may be met. For a reliable design of structures against impact damage and fatigue cracking, a minimum of 2% strain in the composites will be required to increase the current design allowables of 0.003 inch/inch. Such a requirement

171

therefore demands improved fracture toughness in matrix resins in addition to high modulus and low creep.

On a scale of increasing fracture toughness, Fig. 1 shows that thermosetting resins are only slightly tougher than inorganic glass, having fracture energies ranging from 80 to 200 J/m^2.[2] On the other hand, thermoplastic resins such as polymethyl methacrylate are tougher than the thermosets, having fracture energies of the order of 1 kJ/m^2. New high performance sulfone polymers[5] exhibit even higher toughness with fracture energies of several kilo-Joules per square meter, which represent values not far from that of, say, the 7075-T6 aluminum. The high fracture toughness of the thermoplastics is not unexpected because of the large free volume available in these amorphous materials to absorb the energies associated with crack growth and propagation. However, it is surprising to note that there is a unique class of thermosetting materials having fracture energies matching that of thermoplastic polysulfones, namely the elastomer-modified epoxy resins.

In order to mitigate the brittleness of epoxy materials, many industrial formulators use elastomeric additives in their epoxy formulations in an attempt to create a multi-phase system. This results in a soft rubbery inclusion dispersed in the brittle epoxy matrix, which is widely known by now to enhance the toughness of epoxies. The effect of the particulate elastomeric modifiers on the fracture behavior of epoxy polymers was studied by Sulton and McGarry[6], Riew et al.[1] and Bascom and Cottington[7], among others. These investigators have shown that, when an elastomer was copoly-merized with the epoxy resin prior to gelation, a uniformly dispersed phase of small rubber particles was formed in situ. These particles have diameters of a few microns or less, and they enhance the toughness of the unmodified epoxy by several orders of magnitude. The toughening mechanisms were only recently understood to involve triaxial dilatation of the rubber particles at the crack tip[6], particle elongation[8] and plastic flow of the epoxy matrix[9]. However, when such a rubber-modified resin system is used in fiber-reinforced composites, the effects of the added elastomer particles on the mechanical properties of the composite are not clear. In this paper, the balance between matrix fracture behavior and composite mechanical property will be discussed based on the result of an extensive experimental study. Epoxy polymers modified by using carboxy-terminated butadiene acrylonitriles (CTBN) from the Hexcel Corporation (Dublin, CA) were used. These epoxy formulations utilize both a liquid CTBN and a solid rubber in order to achieve a bimodal distribution of particle sizes, which was shown to give even greater fracture toughness in neat resin form than liquid CTBN additives alone[10]. This enhanced resin toughness has also been shown to be transferrable to increase the interlaminar fracture energy of composite laminates[3]. The results that will be reported here include those from shear, tensile, flexural and fatigue tests

of composite samples containing the toughened resin as organic
matrices. The role of rubber-modified matrices on the structural
properties of composite materials will then be discussed, and the
trend for future resin development briefly addressed.

EXPERIMENTAL

    The base epoxy resin is a DGEBA (diglycidyl ether of bisphenol
A) type. · The resin system composition is given in Table I. The
modified formulation contains additional 8.1% liquid CTBN (Hycar
1300, B. F. Goodrich Chemical Company) and 5.4% solid CTBN (Hycar
1472). The solid rubber was pre-cooked with the epoxy resin components of Table I by heating in methylethyl keytone solution. By
using Thornel 300/3K 24 x 23 woven graphite fabric or 7781 fiberglass as reinforcements, laminate samples were prepared by autoclave
cure with normal bagging techniques. The cure was carried out at
0.41 MPa (60 psi), 2°C/min heat-up rate and then one hour at 120°C.

    The laminates were tested at room temperature and at 93°C
(200°F) in tension, flexure and shear for screening sample mechanical
strength and modulus. Short-beam shear test was chosen for interlaminar shear strength determination because it represented a totally
matrix dominated property. The test was carried out according to
the ASTM Standard D2344. The dimensions of shear specimens were
1.27^{cm} x 0.25^{cm} x 0.71^{cm}, and 20 specimens were tested for each
laminate sample. The flexural specimens were 5.1^{cm} x 2.5^{cm} x 0.25^{cm},
satisfying the requirement of ASTM Standard D790. Six specimens
were tested in this case for each laminate. The ASTM Standard
method[11] was adapted for the tensile test, and 5 specimens, each
12.7^{cm} x 1.27^{cm} x 0.25^{cm} in size, were tested for each laminate
plate.

    Flexural fatigue studies were also performed on model uni-
directional glass laminates. Flexural specimens were clamped on
both ends of a fixture, and the necessary load was applied to produce
a 1.27^{cm} deflection amplitude at a given frequency. A cold air jet
was applied at all times to minimize hysteresis heating effects.
The specimens were tested until failure.

    A TGMDA (tetraglycidyl methylene dianiline) type epoxy cured
with DDS (diaminodiphenyl sulfone) was also included in this study
for comparison. This resin formulation is currently being used as
the matrix material in many commercial and military composite systems.
TGMDA offers higher temperature capability than DGEBA, and gives
excellent mechanical properties when used with DDS because of very
stable links produced between the epoxy and the aromatic amine. The
cure for the TGMDA/DDS system was at 177°C and 0.55 MPa (80 psi)
for two hours.

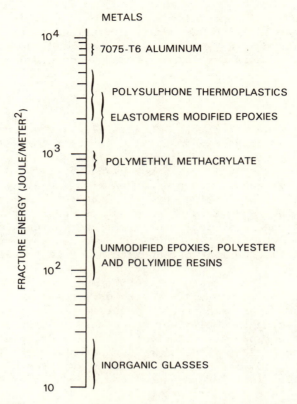

Fig. 1. Fracture energies of various structural materials.

Table I. Composition of the Base Epoxy Resin

Component	Approximate Weight %
Epoxides	73
(diglycidyl ether of bisphenol-A) (epoxidized novolac, epox. eq. wt. 165)	
Diphenols	20
(bisphenol-A) (tetrabromobisphenol-A)	
Catalysts	7
(dicyadiamide) (substituted urea)	

RESULT AND DISCUSSION

The fracture toughness, defined as the critical strain energy
release rate[4], of Hexcel model epoxy systems has been evaluated[10].
The results are given in Table II. Whether the liquid CTBN and the
solid rubber modifier were used alone or in combination, they
increased the toughness of the base epoxy resin by orders of magnitude. The combination of liquid and solid rubber increased the
toughness even more than when either modifier was used alone. The
compact tension specimen data in Table II were obtained by fracturing
the specimens in an INSTRON with a crosshead speed of 0.125 ^{cm}/sec.
The Izod impact test data represented the case of a much higher
loading rate, since the fracture time was only about 1 msec. It
may be noted that the fracture property of the base epoxy resin was
independent of the loading rate. For the rubber-modified resins,
there was a general decrease in fracture toughness with increasing
loading rate[10,12]. However, even during the impact testing, the
modified epoxies generally exhibited 10 to 20 times higher toughness
values than the base resin.

It has also been demonstrated that the enhanced resin toughness
can be transferred into an improved interlaminar fracture energy for
a composite sample. By using a width-tapered double cantilever
beam specimen, Bascom and Bitner showed that the interlaminar fracture energy was increased by a factor of ca. 4 by using the
rubber-modified epoxies as matrix resins[3]. Some of their results
are given in Table III. Post-failure fractography study also
revealed that gross plastic deformation took place in the resin
around the reinforcement fibers, and the development of a porous
structure in the resin was evident. These observations suggested
that the basic toughening mechanisms involved in the case of composite interlaminar fracture are similar to those in the fracture of
neat resin samples[8]. A typical SEM photograph of the fracture
surface of composite samples is shown in Fig. 2.

The increased toughness in either the neat resin or the interlaminar fracture of composites, however, can not come without some
sacrifice of the mechanical properties, specially the initial
matrix-dominated properties in a composite such as interlaminar
shear strength, high temperature capability and wet strength retention. As the matrix resin is toughened by elastomer modification,
these properties will be adversely affected. In general, the
modulus of the modified resin is lowered by a proportional volumetric
amount of the "soft" second phase, which essentially reduces the
effective load bearing area[13]. When applied as a matrix material
for fiber-reinforced composites, this reduction in matrix modulus
leads to reductions in initial matrix dominated mechanical properties
because polymer modulus is the most dominant resin property that
controls the composite performance. The results shown in Table IV
clearly demonstrate these points.

Table II: Neat-resin fracture toughness
            of rubber-modified epoxies (kJ/m²)

Sample Number	Liquid CTBN (wt. %)[a]	Solid CTBN (wt. %)[b]	Compact Tension Specimen[c]	Izod Impact Specimen
x205[d]	None	None	0.23 ± 0.04 (6)	0.29
x206	8.1	None	2.50 ± 0.3 (8)	1.10
x207	8.1	1.0	4.10 ± 0.2 (8)	1.38
x210	None	8.1	3.20 ± 0.2 (8)	1.34
F185	8.1	5.4	5.83 ± 0.9 (8)	2.50

a) Average molecular weight of 3,500.

b) Average molecular weight of 260,000.

c) INSTRON cross-head speed at 0.125 cm/sec.

d) Base epoxy formulation of Table 1.

Table III: Effect of CTBN additives on
              the interlaminar fracture
              energy of composites

Sample	Fiber Volume (%)	Fracture Energy (kJ/m^2)
x205 resin[a]	0	0.23
F185 resin[b]	0	5.83
x205/glass	59.9	1.0
F185/glass	59.7	4.4
x205/graphite	61.0	0.6
F185/graphite	57.9	4.6

a)   Base epoxy of Table I.

b)   Modified epoxy shown in Table II.

Table IV: Mechanical Properties of Epoxy
           Polymers and Composites

NEAT RESINS (DRT (1))

Property	x205 (2)	F185 (3)	TGMDA/DDS
Tensile Strength (psi)	10.0 x 10³	7.7 x 10³	-----(4)
Tensile Modulus (psi)	0.4 x 10⁶	0.32 x 10⁶	0.58 x 10⁶(4)
Tensile Strain (%)	2.5	7.0	-----(4)
Fracture Energy (kJ/m²)	0.23	5.83	0.08

COMPOSITES

Test	Property	#7781 Glass (5)		T300/3K Graphite (6)		
		x205	F185	x205	F185	TGMDA/DDS
Tensile (DRT)	Strength(7)	56.0	52.0	103.0	103.0	93.0
	Modulus (8)	3.6	3.5	10.9	10.2	11.3
Tensile (200°F)	Strength	28.0	26.5	67.4	51.6	98.4
	Modulus	2.8	2.5	9.0	6.0	10.7
Flexure (DRT)	Strength	83.0	79.0	118.0	99.6	135.0
	Modulus	3.1	2.8	7.2	7.4	7.8
Flexure (200°F)	Strength	33.0	16.0	54.0	14.7	123.0
	Modulus	2.0	1.1	6.0	1.2	7.1
(DRT)	SBSS (9)	9.5	6.5	9.8	7.8	10.7
(200°F)	SBSS	3.0	0.6	4.1	1.1	8.7

(1) Dry room temperature value.
(2) Base epoxy of Table I.
(3) Modified epoxy shown in Table II.
(4) Premature failure due to extreme brittleness
     and flaw sensitivity.
(5) 52x45 E-glass 8 H.S., all laminates at 65%
     fiber volume.
(6) 24x23 HMG 8 H.S., all laminates at 63% fiber
     volume.
(7) Strength values in psi x 10³.
(8) Modulus values in psi x 10⁶.
(9) Short beam shear strength in psi x 10³.

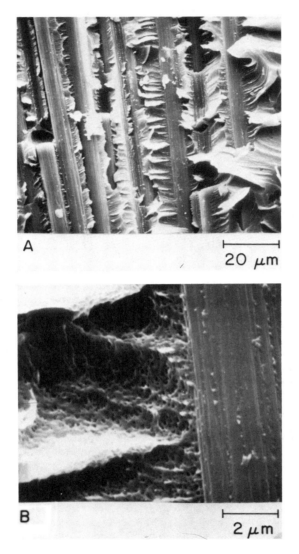

Fig. 2:   SEM photographs of the fracture surface
          of the F185/graphite composite.

It is noted that by increasing the resin fracture toughness from 0.23 kJ/m^2 in the x205 sample to 5.83 kJ/m^2 in the rubber-modified F185 resin, tensile strain-to-failure is also increased by ca. three-folds. On the other hand, both the tensile strength and modulus are reduced by ca. 20%. The TGMDA/DDS system, being a more highly cross-linked, higher temperature resin, offers a higher tensile modulus than the DGEBA epoxy. But its extremely brittle nature gives a fracture toughness value of only 0.08 kJ/m^2, whereas the tensile strength and strain can not be measured since the specimens always failed in premature failure due to flaw sensitivity.

For composite samples, both the tensile and the flexural test results show that initially the effect of elastomeric additives in matrix resins only represents a very small decrease in either the strength or the modulus of the composite laminates. But at higher temperatures these reductions become greater. For instance, the flexural strength of the T300 graphite laminate reduces by about two-thirds at 200°F if the rubber modifiers are present in the matrix. Similarly the flexural modulus is decreased by approximately 80%. High temperature testing was performed at 200°F because this represents the current upper limit design temperature for commercial aircraft. TGMDA/DDS epoxy clearly shows its superior properties in this temperature range since its ambient strength and modulus remain virtually unchanged when the samples are dry. The most pronounced effect of rubber-modified matrices in composites is manifested in the lowering of the interlaminar shear strength as determined by the short-beam shear test. Even at room temperature a large reduction in shear strength is clear. At 200°F, this shear strength is further decreased to ca. 1 ksi or less, becoming totally unacceptable for any structural application. This strong effect perhaps is not at all surprising, since the short-beam shear strength is a totally matrix dominated property for a constant fiber geometry.

In Fig. 3, the high-rate fracture toughness of matrix resins is plotted against the short-beam shear strength. The rapid drop-off of shear strength with increasing resin toughness clearly demonstrates the trade-off between matrix-dominated mechanical property and the improved fracture property of a rubber-modified epoxy. This quantitative inverse relationship may be applicable to any of the current available composites. Other mechanical properties tested at different temperatures and under different environmental conditions could also be substituted for the vertical ordinate axis to give a similar curve. The 200°F flexural modulus data are shown as an example. Furthermore, Fig. 3 clarifies a long-standing supposition that short-beam shear test provides a measure for the toughness of the matrix material. It can be seen that short-beam shear strength only varies inversely with the matrix toughness, and a new test such as interlaminar fracture[3] is necessary to determine the resistance of a composite laminate to interlaminar crack growth and propagation.

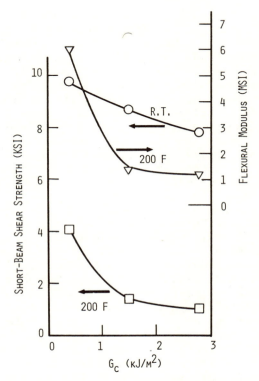

Fig. 3: Short-beam shear strength and flexural modulus
         as a function of high-rate fracture energy
         of the neat resins.

Figure 4 shows how high-temperature strength retention of a graphite composite is affected by using rubber-modified matrices. The high-temperature TGMDA epoxy gives the highest initial ambient flexural strength, which maintains its high level as temperature increases. For the DGEBA resin, its temperature capability is intermediate, whereas the CTBN modified F185 resin shows the most rapid drop-off, and hence the poorest high temperature strength retention. A similar trend for composite flexural modulus is also observed as shown in Fig. 5.

The results in Table IV seems to indicate that the rubber-modified matrix has a greater effect on the composite mechanical properties in flexure than in tension. Flexural fatigue tests were therefore performed by using unidirectional S-glass laminate samples. Figure 6 shows the fatigue test result of the F185/S-glass composite in comparison with that obtained by using the TGMDA/S-glass system. Elastomer modification of the matrix resin is found to increase the laminate fatigue life by a factor of ca. 10, even though the static interlaminar shear strength of the F185/glass system is ca. 30% lower. Furthermore, fatigue data for the modified system seem to show considerably less scattering. With both of these effects combined, the composite systems with rubber-modified matrices would offer a much higher design limit for fatigue.

In addition to the improved flexural fatigue property, the rubber-modified matrices provide two other advantages in composite applications. First, by virtue of its extremely high fracture toughness in neat resin form, these polymers offer very good adhesive peel strength and are ideal for use as structural adhesives. In bonding a thin composite laminate skin to a honeycomb structure, these rubber-modified polymers may be used as both the adhesive and the composite matrix material. The advantage then is that the structural component may be co-cured all together in a simple "one-shot" operation. Second, the rubber-modified resins offer rheological properties that are extremely advantageous for improving the laminate processability. Figure 7 illustrates this concept. The two dotted horizontal lines indicate the desired viscosity range for one-shot honeycomb bonding. Curve A represents the viscosity of the current unmodified TGMDA/DDS epoxy system as the resin is heated toward complete cure. The low viscosity means too high a flow during the cure, resulting in a resin-starving laminate-honeycomb structure, an end product that is unacceptable. On the other hand, the rubber-modified resin gives much higher viscosity, which enables the processing engineer to precisely define the process "windows" for curing operation. This controlled flow characteristic makes the desired one-shot bonding of composite laminate to honeycomb structure possible.

In summary, rubber-modified polymers exhibit improved fracture toughness over that of the unmodified resin. Whether one uses the

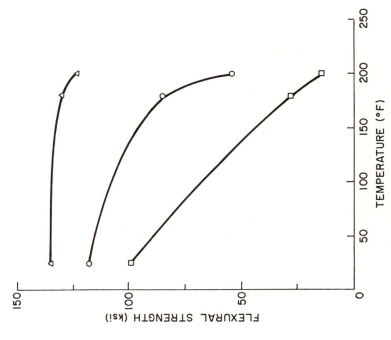

Fig. 4: Temperature dependence of the flexural strength of graphite laminates, (□ F185, ○ x205, △ TGMDA/DDS).

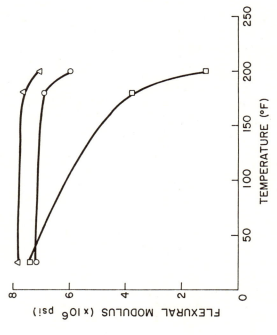

Fig. 5: Temperature dependence of the flexural modulus of graphite laminates, (□ F185, ○ x205, △ TGMDA/DDS).

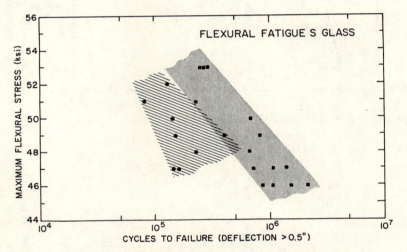

Fig. 6: Flexural fatigue test results of S-glass
         laminates. (● TGMDA/DDS, ■ F185)

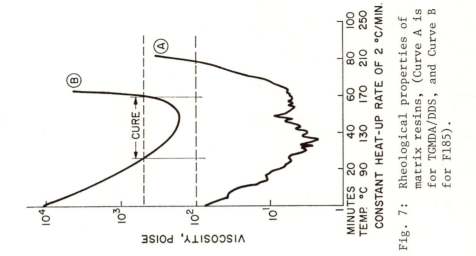

Fig. 7: Rheological properties of matrix resins, (Curve A is for TGMDA/DDS, and Curve B for F185).

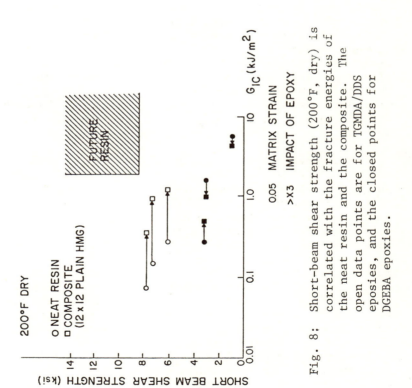

Fig. 8: Short-beam shear strength (200°F, dry) is correlated with the fracture energies of the neat resin and the composite. The open data points are for TGMDA/DDS eposies, and the closed points for DGEBA epoxies.

liquid and the solid rubber alone or in combination, the fracture
energy may be increased by orders of magnitude. When these materials
are used as matrix resins in composites, the "toughening" effect is
transferred into the composites to enhance their interlaminar fracture energy, flexural fatigue and other off-axis mechanical properties. The rubber-modified resins also offer the additional advantages of high peel strength and controlled flow such that they are
ideal for the "one-shot" bonding process between composite and
honeycomb. But, unfortunately, the reduced resin modulus of the
rubber-modified matrices will adversely affect composite properties
to reduce the interlaminar shear strength, to cause a fast drop-off
of strength and modulus at elevated temperatures, and to lead to a
poor wet strength retention.

FUTURE TREND

     It has been shown that the fracture toughness and matrix strain
properties can be significantly improved by using rubber-modified
resins. However, this improvement presently can only come at a
prohibitive sacrifice of the required mechanical strength and modulus
of the composites. In recent years, the fiber industry has made
significant progress in producing high strain fibers for composite
application. According to micromechanics analysis such as that of
Chamis[14] or Christensen[15], a magnification factor of 5 to 10 is
needed for current matrix materials in a (0,90) composite laminate
over the strain of the fiber. When combining this desired matrix
strain property with the good mechanical properties needed in composites for engineering structural design, the overall requirement
indeed becomes a very severe one. This may be further explained
with the aid of Fig. 8, which also indicates the necessary directions
for future resin development.

     Short-beam shear strength at the elevated temperature is plotted
in Fig. 8 against the fracture energies of both the neat resin and
the woven graphite composite. The shear strength is selected for
discussion because it is totally matrix dominant and is perhaps the
parameter **most** sensitive to **any** change in resin properties. It is
noted that for brittle unmodified epoxy resins (i.e. those having
fracture energy less than 0.3 kJ/m^2) their composite fracture energy
is greater than that of the neat resin. Bascom et al.[3] suggested
that this difference is due to the contributions to toughness from
the inelastic deformation processes involving fiber breakage and
pull-out, and to the tortuous path provided by the weave of the
graphite fabric for interply load transfer. On the other hand, the
rubber-modified resins exhibiting toughness values in excess of
1 kJ/m^2 show a reversed trend, namely, their composite toughness is
reduced from the neat resin value. This is likely caused by the
restrain the fiber layers imposed on the development of the crack-tip
deformation zone, a situation similar to that found in thin adhesive

bond lines[7]. The threshold for the occurrence of this toughness suppression from resin to composite is approximately 1 kJ/m^2. Therefore, this establishes the lower bound of resin fracture energy for future matrix resins. Such a polymer toughness would remove the threat of crack propagation resulting from flaw sensitivity due to fiber type and lay-up geometry, resin enrichment near fiber crossing and the deposit of excessive curative particles. This toughness value of 1 kJ/m^2 may also provide a matrix strain up to 5%, which will alleviate premature failure at matrix/fiber interface due to uneven stress concentration factors between fibers under transverse stress conditions. In order to meet high impact requirement, however, the first phase of a rubber-modified resin also needs to be significantly improved. The soft, second phase (elastomeric domain) generally is rate-dependent, as shown by Ting and Cottington[12], and is less effective for toughening under a high-rate loading. One expects such a rate effect because the elastomers require time to allow the various mechanisms to become operative for energy dissipation. The first phase may be improved by choosing thermoplastics or thermoset resins containing linear blocks. Of course, these new resins, in addition to being compatible with the elastomer phase, must also be able to deliver sufficiently high shear strength, say 7×10^4 Pa (ca. 10 ksi) at the operating temperature. Thermoplastic polymers, having an amorphous structure, would cause concern because of their tendency to creep, their poor solvent resistance and the high processing temperature they require. One potential solution may be the introduction of a small percentage of crosslinking in the polymer structure[16]. In any event, whether the future resin be thermoplastics, multi-phase hybrids or thermoplastic/thermoset combinations, its development represents a great challenge to both polymer chemists and material scientists.

ACKNOWLEDGEMENT

The author gratefully acknowledges the efforts of J. Brown and R. L. Cottington in taking the experimental data. Dr. W. D. Bascom performed the fractographic study of Fig. 2. Dr. R. J. Moulton provided the samples and gave much insight into the discussion presented here. Ms. D. A. Beard typed the manuscript. Their kind assistance is greatly appreciated.

REFERENCES

1. C. K. Riew, E. H. Rowe and A. R. Siebert, in Toughness and Brittleness of Plastics, Amer. Chem. Soc. Adv. in Chem. Series, 154, 326 (1976).
2. W. D. Bascom, J. L. Bitner and R. L. Cottington, Amer. Chem. Soc. Org. Coat. Plast. Preprint, 38, 477 (1978).

3. W. D. Bascom, J. L. Bitner, R. J. Moulton and A. R. Siebert, Composites, $\underline{11}$, 9 (1980).

4. R. Y. Ting and R. L. Cottington, J. Appl. Polym. Sci., $\underline{25}$, 1815 (1980).

5. R. Y. Ting and R. L. Cottington, in Rheology ed. G. Astarita, G. Marrucci and L. Nicolais, Plenum Press. NY, Vol. 3, p. 349 (1980).

6. J. N. Sulton and F. J. McGarry, Polym. Eng. Sci., $\underline{13}$, 29 (1973).

7. W. D. Bascom and R. L. Cottington, J. Adhesion, $\underline{7}$, 333 (1976).

8. W. D. Bascom, R. L. Cottington and C. O. Timmons, J. Appl. Polym. Sci., Symp. Ser. $\underline{32}$, 165 (1977).

9. C. B. Bucknall, Toughened Plastics, Appl. Sci. Publ., London, p. 221, (1977).

10. W. D. Bascom, R. Y. Ting, R. J. Moulton, C. K. Riew and A. R. Siebert, J. Materials Sci., in press (1981).

11. E. A. Rothman and G. E. Molter, ASTM STP 460, p. 72 (1969).

12. R. Y. Ting and R. L. Cottington, Polymer Bulletin, $\underline{2}$, 211 (1980).

13. Z. H. Ophir, J. A. Emerson and G. L. Wilkes, J. Appl. Phys., $\underline{49}$, 5032 (1978).

14. C. C. Chamis, M. P. Hanson and T. T. Serafini, Modern Plastics, $\underline{5}$, 90 (1973).

15. R. M. Christensen and E. M. Wu, Optimum Design of Anisotropic Fiber-Reinforced Flywheels, Lawrence Livermore Laboratory, Univ. Calif., Nov. 1976.

16. J. Jaquish, G. H. Shepherd, S. G. Hill, E. E. House and W. Symonds, Graphite Reinforced Thermoplastic Composites, Rept. D180-26067-1 for Naval Air Systems Command, Boeing Aerospace Company, Seattle, WA, Aug. 1980.

A PRELIMINARY STUDY OF COMPOSITE REACTION INJECTION MOLDING*

V. M. González, J. M. Castro** and C. W. Macosko

Department of Chemical Engineering and Materials Science
University of Minnesota
Minneapolis, MN 55455

ABSTRACT

A promising method for producing a composite reaction injection
molded material, is to place a fiberglass mat into the mold before
injection. In this work we examine the behavior of two typical RIM
systems, one mixing (polyurethane) and the other heat activated
(styrene-based). The effect of a fiberglass mat during filling and
curing is analyzed. Theoretical predictions are compared to experimental results for the two RIM systems.

INTRODUCTION

Reaction Injection Molding can be defined as the filling of a
mold with an initially low viscosity polymerizing mixture, which
solidifies in the mold by means other than cooling [1]. In RIM processes, polymerization can be initiated by two different methods:
mixing and heat transfer. For the mixing activated process, two
highly reactive monomers or prepolymers are brought into intimate
molecular contact by impingement mixing. From the mixhead they flow
into the mold and react rapidly to form a solid part. A sketch of
the process is shown in figure 1. The mold wall temperature (T_w)
is not much different than the starting material temperature (T_0),
since the monomers are highly reactive at T_0. Polyurethane and
nylon-6 are examples of mixing activated polymerizations.

---

* To be presented at NSF-US Italy Composites Conference, Capri,
  June 16, 1981.
** Present address: Planta Piloto de Ingeniería Química, Universidad
   Nacional del Sur, Bahía Blanca, Argentina.

189

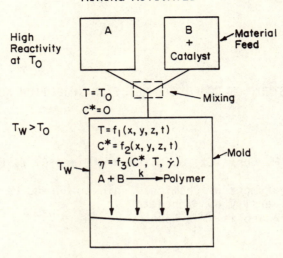

Figure 1.   Schematic diagram of a mixing activated RIM system
            (ref. [16]).

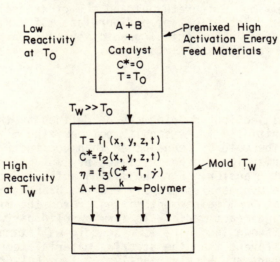

Figure 2.   Schematic diagram of a heat activated RIM system
            (ref [16]).

In the thermally activated RIM process, the monomers do not react appreciably at T_0, but are highly reactive at T_w. The speed of the mixing step is therefore not as critical; it may be carried out in a separate operation or equipment. The reaction starts after the material comes in contact with the hot walls. The mold wall temperature is much higher than the starting material temperature. A schematic diagram of the process is shown in figure 2. Chemical systems that are being studied for thermally activated RIM include epoxies, silicones, unsaturated polyesters and other vinyl systems [2].

Not every polymerizing system can be RIM processed. Polymers that are formed rapidly from addition reaction (with no byproducts) such as those mentioned above are good candidates. Some consideration is also being given to Interpenetrating Polymer Networks (IPNs) [3].

During the past few years reaction injection molding has become a well established polymer processing method. Simultaneously, the use of composite materials has increased significantly. This has lead to reinforced reaction injection molding, where the benefits of RIM, namely low filling pressure, large part capability, energy efficiency, and fast cycle times, are augmented by the advantages of reinforced materials: enhancement of structural properties and low thermal expansion coefficient.

Most of the developments on reinforced RIM have been on milled glass fibers in polyurethanes. This brings significant physical property improvement [4,5] but has some limitations, principally flow orientation of the fibers and large viscosity increase of the resins with fiber concentration. The latter limits the maximum fiberglass content that can be processed [5,6]. Because of these problems, RIM using resin injection through a fiber mat preplaced in the mold looks attractive. The main goals of the mat are to obtain isotropic products, and to increase the reinforcement content in the RIM part which will reduce the linear coefficient of thermal expansion and enhance the structural properties.

In the present work, a general model including both the filling and curing steps for RIM with fiberglass mats is formulated. The model is solved for a particular case in which the filling step can be assumed isothermal and the reaction during filling neglected. The predictions of the model are compared to experiments using a polyurethane (mixing activated) and a styrene-based copolymer (heat activated) system.

MATHEMATICAL MODEL

To apply modelling equations it is necessary that the boundaries of the system in question be mathematically described. In the case of flow through a fiberglass mat, the complexity of the structure makes such description extremely difficult. Thus, a mathematical theory should take the form of either a statistical theory, or a theory based upon laws describing only the macroscopic features of the flow [8]. If we assume that there is a region (in the mold cavity) large enough compared with the pore size and small enough compared with the mold dimensions, it is possible to use Darcy's law instead of the linear momentum balance. This law states that the flow through a porous medium (the fiberglass mat in our case) is directly proportional to the gradient of the driving forces and inversely proportional to the fluid viscosity. The constant of proportionality is called permeability. Physically it measures the conductivity of the medium to the flow. Mathematically this is expressed as

$$\underset{\sim}{v} = - \frac{\rho}{\eta} \underset{\sim}{K} \cdot \nabla\Psi \qquad\qquad (1)$$

where $\underset{\sim}{K}$ is the permeability tensor and $\nabla\Psi$ the driving gradient, which includes pressure and gravity.

$$\nabla\Psi = \frac{\nabla p}{\rho} + \underset{\sim}{g} \qquad\qquad (2)$$

Further assumptions of the model are:
1.   Radial laminar flow into a disk-shaped mold (see fig. 3).
2.   Constant flowrate, Q, (this is the operating mode of RIM equipment).
3.   Fixed, isotropic mat, thus, the permeability tensor has constant and only diagonal components, $K = KI$.
4.   Constant density (ρ), and thermal properties $\sim$(C_p, ΔH_r and k).
5.   Newtonian fluid, that is the viscosity is only an explicit function of the extent of reaction and temperature.
        $\eta = \eta (C^*, T)$
     No shear thinning has been observed for the polyurethane system [9]. For the styrene-based copolymer system preliminary experimental results lead to the same conclusion [10].
6.   Neglect foaming.
7.   Fibers are inert to reactants.
8.   Neglect wall effects on the flow field.
9.   Neglect molecular diffusion [11].
10. Neglect viscous dissipation.
11. Unidirectional heat conduction in the direction perpendicular to the flow.
12. Local thermal equilibrium.

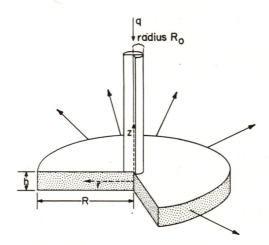

Figure 3.   Disk shaped mold diagram.

The last assumption implies that at a point, the temperature of the fiber surface and the fluid that is in contact with it, are the same at any time. More research is currently in progress to understand the transient heat transfer problem.

The material properties of the system (resin plus fiberglass) are to be evaluated as a function of volume fraction of resin, ϕ, which we shall call porosity. The effective density ($\hat{\rho}$) and specific heat ($\hat{C}_p$) will be given by a volume and weight average respectively.

$$\hat{\rho} = \phi \rho_f + (1-\phi)\rho_g \tag{3}$$

and

$$\hat{C}_p = w_f C_{pf} + w_g C_{pg} \tag{4}$$

where

$$w_f = \frac{\phi \rho_f}{\hat{\rho}} \quad ; \quad w_g = \frac{(1-\phi)\rho_g}{\hat{\rho}}$$

The subscripts "f" and "g" denote fluid (resin) and fiberglass respectively.

The effect of the fiberglass in the thermal conductivity is not simple [12]. For our case, we will take a volume average of the thermal resistivity (the reciprocal of the conductivity). That is,

$$\frac{1}{\hat{k}} = \frac{\phi}{k_f} + \frac{(1-\phi)}{k_g} \tag{5}$$

Notice that equation (5) represents the conductivity assuming a series arrangement of the components with respect to the direction of heat flow.

With the above assumptions and definitions the balance equations become

Darcy's law $$v_r = -\frac{K}{\eta}\frac{\partial P}{\partial r} \tag{6}$$

Continuity $$Q = 4\pi r \phi \int_0^{h/2} v_r dz \tag{7}$$

Reactive group balance $$\frac{\partial C^*}{\partial t} + v_r \frac{\partial C^*}{\partial r} = R_M \tag{8}$$

Energy balance $$\frac{\partial T}{\partial t} + v_r \frac{\partial T}{\partial r} = \hat{\alpha}\frac{\partial^2 T}{\partial z^2} + \phi \frac{(-\Delta H_R)C_o}{\hat{\rho}\,\hat{C}_p} R_M \tag{9}$$

where $\hat{\alpha} = \dfrac{\hat{k}}{\hat{\rho}\,\hat{C}_p}$

To solve the above equations we also need a viscosity relation of the form $\eta=\eta(C^*,T)$ [11], and kinetic information $R_M = f(C^*,T)$. Since viscosity is a function of the state of reaction (reactive group conversion) and temperature, the four balance equations are coupled and simultaneous numerical solution is needed. Research is in progress for the solution of this model as well as for more complex geometries. In this work we take a more simplified approach below.

A SIMPLIFIED APPROACH

If we neglect reaction and heat transfer during filling, we can decouple filling and curing as explained by Castro and Macosko [7]. This was found to be applicable for polyurethanes when the gelling potential is small [1,7]. It should also be true for free radical copolymerization when the filling time is smaller than the time at which sufficient initiator radicals have been generated to consume completely the inhibitor. Under these assumptions (6) and (7) are decoupled from (8) and (9) with η approximately constant. Using equations (6) and (7) we find that the pressure rise as the mold is being filled is given by

$$\Delta P = \frac{Q\eta}{2\pi\phi Kh} \ln \frac{r}{R_0} \tag{10}$$

where $r = \left(\frac{Qt}{\pi h\phi} + R_0^2\right)^{1/2}$

Defining the dimensionless variables

$$t^* = \frac{Qt}{\pi h\phi R_0^2} \quad \text{and} \quad p^* = -\left(\frac{4\pi\phi Kh}{Q\eta_0}\right)\Delta P$$

Equation (10) becomes

$$p^* = \text{Ln}(t^* + 1) \tag{11}$$

which is plotted in figure 4.

For the curing step the energy and reactive group balance (equations 8 and 9) with the terms containing v_r set equal to zero were integrated numerically using explicit finite differences [10]. The proper initial and boundary conditions are

$$t = 0 \qquad C^* = 0 \qquad T = T_0$$

$$z = \pm \frac{h}{2} \qquad\qquad T = T_w$$

$$z = 0 \qquad\qquad \frac{\partial T}{\partial z} = 0$$

Some results of the numerical solution of this model are compared below with experiments.

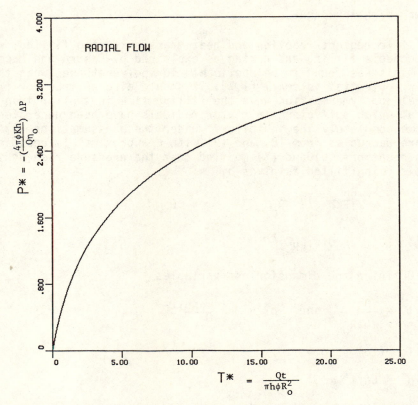

Figure 4. Dimensionless pressure drop versus dimensionless time
          for isothermal radial flow through a fiberglass mat.

EXPERIMENTAL

Permeability Determination

The purpose of this experiment was to measure the permeability
K. Once K is known, equation (10) can be used to predict the pressure needed to fill the mold. A center gated parallel plate mold
(see fig. 3) and a chopped strand fiberglass mat (type AKM from PPG
Industries, Pittsburgh, PA) were used. The mat is composed of
chopped fiberglass strands, laid down uniformly and bonded together
with a resin soluble adhesive.

A pressure transducer was placed 3 cm from the center on the
upper mold wall and the gate in the center of the bottom wall. The
experiments were carried out isothermally using water. The pressure
drop as a function of flow rate and porosity was measured using a
0-3 atm. pressure transducer (Dynisco PT311JA Waltham, MA). The

mold thickness was varied between 0.4 to 1 cm. The permeability obtained in this way is applicable to the resin flow since, it does not depend on the fluid but only of the topology of the porous medium. We should note here that during actual filling there is a moving front through the mat. This flow and capillary effects may contribute to the pressure but the effect is believed to be very small.

Curing

A 1.5 cm aluminum mold wall packed with fiberglass was pre-heated in an oven. Fast response copper constantan thermocouples were placed in the center and in the wall. When the system reached the desired temperature, the mold was taken out of the oven and the resin injected. The response of the center line thermocouple was recorded.

Two different RIM materials were used. The mixing activated was the experimental RIM type polyurethane system described by Castro and Macosko [9]. This system has been found to follow a second order kinetics. The heat activated was a styrene-based copolymer system for which phenomenological kinetics were obtained by fitting the following expression to DSC data.

$$\frac{\partial C^*}{\partial t} = (1-C^*) (k_0 + k_1 C^*) \tag{12}$$

The values of $k_0 = k_1 \times 10^{-4}$ and $k_1 = 7 \times 10^{12}$ EXP($-25800/R_c T$) were obtained. A rather complete kinetic model for free radical copolymerization with multiple initiators and inhibition has been developed by Stevenson [15]. The differences in process behavior between the two systems have been studied by Castro, González and Macosko [14].

RESULTS AND DISCUSSION

Filling

Figure 5 shows a plot of pressure drop versus flow rate for several porosities (ϕ). As expected, if Darcy's law is applicable, the pressure drop varies linearly with flow rate for constant porosity. From the slope of these lines the permeability was calculated and is plotted versus volume fraction of resin in figure 6. Note that at $\phi \approx 0.36$ the permeability decreases sharply. There appears to be a critical value ($\phi \approx 0.32$) beyond which flow stops. This graph and eq. 10 can then be used to predict filling pressure in any radial mold where there is not appreciable reaction or heat transfer during filling (as long as the topology of the fiberglass is similar).

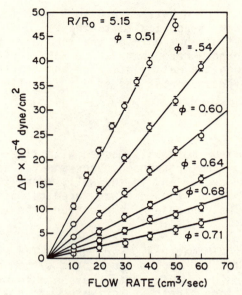

Figure 5. Effect of porosity and flow rate on the pressure drop
for isothermal filling through a fiberglass mat in
radial geometry.

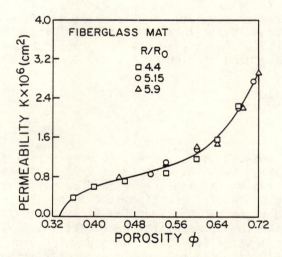

Figure 6. Permeability of the fiberglass mat as a function of
porosity.

Curing

To have a picture of the differences between both systems in the curing step we have reproduced three figures from Reference [16]. Figure 7 shows a comparison of the center line exotherm for both cases without fiberglass mat. Notice that the maximum temperature occurs earlier for the polyurethane (u) than for the styrene-based (s) system and the temperature rise is very steep for the styrene-based system. Figures 8 and 9 show the extent of reaction versus axial position at different times for the polyurethane and styrene-based systems respectively. The difference between them are not only quantitative but also qualitative. The thermally activated system starts reacting first near the wall. As time proceeds, the reaction front moves in. The mixing activated system is just the opposite. Even though the maximum temperature occurs first for the polyurethane system, the styrene-based system reaches total conversion first.

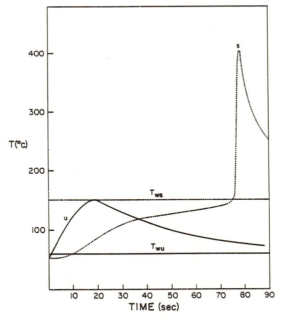

Figure 7.   Comparison of the center-line temperature rise for the polyurethane (u) and styrene-based (s) systems, (ref. [16]).

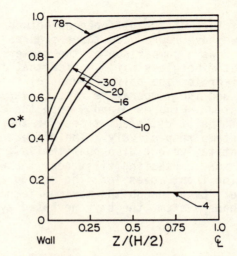

Figure 8.   Extent of reaction versus axial position at several
times for the polyurethane (mixing activated) system.
(ref.   16)

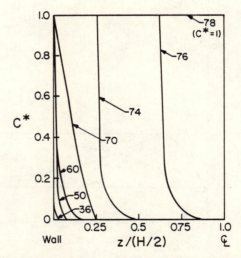

Figure 9.   Extent of reaction versus axial position at several
times for the styrene-based (thermally activated)
system. (ref. [16])

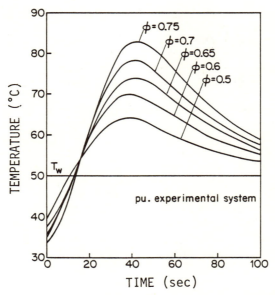

Figure 10. Center line temperature at different porosities
for the mixing activated system (h = 0.5cm)

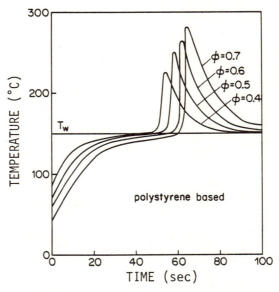

Figure 11. Center line temperature at different porosities
for the heat activated system (h = 0.5cm).

The effect of the amount of fiberglass in the centerline
temperature for both polyurethane and styrene-based systems as
predicted by our model are shown in figures 10 and 11, respectively.
As expected, the maximum temperature decreases as ϕ drops, and is
delayed as ϕ increases, since the thermal conductivity of the
polymer is smaller than the one of the fiberglass.

Experimental results are compared with the model predictions
in figures 12 and 13. The dotted lines are experimental results
and the solid lines model predictions. The initial temperature
for the numerical solution was taken to be the resin-mat adiabatic
mix temperature (T_0). For both cases, this T_0 was larger than the
experimental value. The agreement between the model predictions
and experimental results is better for the mixing than for the heat
activated system. This reflects the fact that we have better kinetic
information for that system. Also note that for the heat-activated
system, the theoretical temperature rise is very sharp. It may be
difficult for the thermocouple to follow this rise.

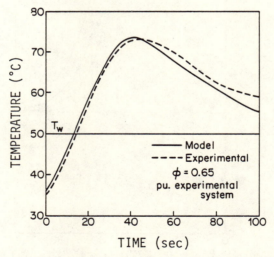

Figure 12.   Center line temperature for the mixing activated
             system, sample thickness 0.5cm.

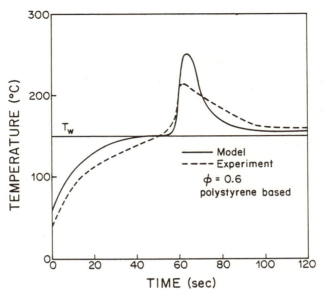

Figure 13. Center line temperature for the heat activated system, sample thickness 0.4 cm.

CONCLUSION

A simple case of RIM with fiberglass mat is analyzed. Darcy's law seems to be valid for constant viscosity filling in common mold cavities. The curing step is qualitatively well predicted by the model (figures 12 and 13). The temperature rises up to a maximum value and then decreases due to heat conduction to the mold walls. The contrast between the two systems results from the different nature of their reaction kinetics, in the mixing activated the reaction starts from the beginning while in the heat activated the resin must first reach a temperature where the initiator radicals are generated and start consuming the inhibitor. After this, the reaction is very fast giving rise to a sharp temperature peak.

The major difference between both systems occurs in the conversion field. The conversion and temperature history of the different material particles are critical to the development of the final part properties. With care, a thermocouple can provide reliable temperature data but conversion can not be obtained readily by experimental measurements. However, with models like the one presented here, conversion can be accurately predicted and conditions for a particular RIM process can be quickly studied and optimized.

More research is in progress to understand the transient filling step. This will give a better estimate for the temperature distribution to be used in the curing step. Experimental results for
the effect of fiberglass in the thermal conductivity of composite
RIM materials is also necessary.

ACKNOWLEDGMENT

This work was supported by the Union Carbide Corporation and
the U.S. National Science Foundation, Division of Material Research,
Polymer Program. A fellowship to V.M.G. from The Mexican National
Council on Science and Technology (CONACYT) is gratefully acknowledged.

REFERENCES

[1]   J. M. Castro, "Studies of Mold Filling and Curing in Polyure-
       thane Reaction Injection Molding", Ph.D. Thesis, University
       of Minnesota, 1980.
[2]   R. S. Kubiak, "Taking RIM Beyond the Urethanes", Plast. Eng.,
       36, 55 (1980).
[3]   G. Ferber, "New Materials for RIM", Soc. of Plast. Eng. Tech.
       Papers, 26, 56 (1980).
[4]   M. W. Liedtke, "Reinforced Reaction Injection Molding", Journal
       of Cellular Plastics, March/Apr., 102 (1978).
[5]   B. C. Chisnall and D. Thorpe, "RRIM - A Novel Approach Using
       Chopped Fiberglass", 35th Annual Conference of the Reinforced
       Plastics/Composites Institute of the Society of the Plastics
       Industry, Feb. 1980, New Orleans, LA.
[6]   Donald McNally, "Short Fiber Orientation and Its Effects on
       the Properties of Thermoplastic Composite Materials", Polym.
       Plast. Tech. Eng. 8 (2), 101 (1977).
[7]   J. M. Castro and C. W. Macosko, "Studies of Mold Filling and
       Curing for the Reaction Injection Molding (RIM) Process",
       AIChE J. to appear (1981).
[8]   R. E. Collins, "Flow of Fluids Through Porous Materials",
       Reinhold, New York, 1961.
[9]   J. M. Castro and C. W. Macosko, "Kinetics and Rheology of
       Typical Polyurethane Reaction Injection Molding Systems",
       Soc. Plast. Eng. Tech. Papers, 38, 434 (1980).
[10]   V. M. Gonzalez, Ph.D. Dossier, University of Minnesota (1981).
[11]   E. Broyer and C. W. Macosko, "Heat Transfer and Curing in
       Polymer Reaction Molding," AIChE J., 22, 268 (1976).
[12]   D. Hands, "The Thermal Transport Properites of Polymers",
       Rubber Chem. and Tech., 50, 481 (1976).
[13]   R. P. Tye, "Thermal Conductivity", Vol. 1, Academic Press,
       New York, (1969).

[14] J. M. Castro, V. M. Gonzalez and C. W. Macosko, "Process Behavior Differences Between Mixing and Thermally Activated RIM Type Chemical Systems", Soc. Plast. Eng. Tech. Papers 27, 363 (1981).

[15] J. F. Stevenson, "Processing of Reactive Materials: a Kinetic Model for Free Radical Copolymerization," Soc. of Plast. Eng. Tech. Papers, 28, 452 (1980).

CHARACTERIZATION OF HIGH PERFORMANCE COMPOSITE MATRICES[*]

Roger J. Morgan

Lawrence Livermore National Laboratory, L-388
University of California, P.O. Box 808
Livermore, California 94550

INTRODUCTION

The increasing use of high performance, fibrous composites in critical structural applications has led to a need to predict the lifetimes of these materials in service environments. The durability of the epoxy matrices (primarily amine-cured epoxies) used in these high performance composites has been a cause for concern. A number of studies have indicated that the combined effects stress, sorbed moisture and thermal exposure can cause significant changes in the mechanical response of the composite.[1,2] The structural and mechanical integrity of both the epoxy matrix and the fiber-matrix interfacial region can be modified by these environmental factors.

To predict the durability of a composite in service environment requires a basic understanding of (1) the role the matrix and its interface with the fibers play in the composite deformation and failure processes, (2) the relations between the structure, deformation and failure processes, and mechanical response of the matrix and (3) how such phenomena are modified by environmental factors. At LLNL we are actively studying these areas. In this paper we review the fundamental areas of the structure-property relations of composite matrices that are necessary for making meaningful durability predictions. These areas include the chemical and physical structure of epoxies and

---

[*]This work was performed under the auspices of the U.S. Department of Energy by the Lawrence Livermore National Laboratory under Contract No. W-7405-ENG-48.

their modes of deformation and failure. We will consider those
structural parameters that control the matrix properties that play
a role in the composite mechanical performance.

CHEMICAL STRUCTURE

     It is important to know the chemical cure reactions and
resultant chemical structure of epoxies for the following reasons.
(1) A knowledge of the cure reactions is necessary to ensure that
the epoxide and amine monomers are mixed in the correct quantities
such that no unreacted groups or low molecular weight material
remains in the cured resin. Such groups could subsequently react
in service environment and/or any low molecular weight material
could diffuse out of the resin. Both of these phenomena could
modify the resin and composite performance. (2) The chemical
structure determines the strength, flexibility and extensibility of
the cross-linked network. Furthermore, this structure determines
the degree of molecular interaction between (1) the epoxy and the
chemical groups on the fiber surface and (2) environmentally sorbed
low molecular weight material such as water.

     In amine-cured epoxies, networks generally result from
addition reactions of epoxide groups with primary and secondary
amines.[3] However, additional reactions can occur as a result of
the epoxide groups undergoing a ring-opening homopolymerization to
form polyether linkages.[4-13]

     The chemical reactions that occur in amine-cured epoxies have
only been directly determined on model compounds by chromatographic
studies.[10-13] Infrared spectroscopy has been used to monitor the
disappearance of the epoxide band and appearance of the hydroxyl
band during the cure of epoxies.[13-15] However, to determine more
definitely the chemical structure of amine-cured epoxy networks
requires monitoring the disappearance of all the reactants and the
appearance of all of the products. Fourier-transform infrared
spectroscopy (FTIR) studies using spectral stripping, which reveals
differences in the spectra recorded at different stages of cure,
has the potential to monitor such changes.

     Diaminodiphenyl sulfone (DDS) cured tetraglycidyl
4,4'-diaminodiphenyl methane (TGDDM) epoxies are currently the most
commonly used matrices for high performance fibrous composites.
The cure reactions of the TGDDM-DDS epoxy system are complex. The
commercially available prepreg resins such as Narmco 5208,[*]
Fiberite 934 and Herculese 3501 all primarily consist of the
TGDDM-DDS epoxy system; the latter two systems also contain a

$BF_3:NH_2-C_2H_5$ catalyst. Initial IR studies of the cure
reactions of Narmco 5208 and Fiberite 934 indicate different
degrees of cure occur in these systems for similar cure conditions
and, hence, glasses of different physical properties are
produced.[16]

More recently we have studied the chemical structure of
TGDDM-DDS epoxy systems from systematic FTIR studies as a function
of DDS concentration and cure conditions.[17,18] To consume all
epoxide groups in the TGDDM-DDS epoxy system exclusively by
epoxide-primary and secondary amine addition reactions would
require 37 wt% DDS. However, commercial TGDDM-DDS epoxy systems
contain only 20-25 wt% DDS which indicates other reactions in
addition to epoxide-amine addition reactions must occur to consume
all the epoxide groups in these systems.

From our FTIR studies to date we have arrived at the following
conclusions.[17,18] (1) The TGDDM epoxide groups below $200^{\circ}C$
undergo homopolymerization to form linear ether linkages and also
isomerize to form $- C(OH) = CH_2$ groups. Such reactions are
accelerated by epoxide impurities and $BF_3:NH_2-C_2H_5$ catalysts.
(2) Studies on TGDDM-DDS systems as a function of cure conditions and
DDS concentrations (5-35 wt% DDS) reveal epoxide-amine addition
reactions occur and all primary amine groups and 80% of epoxide groups
are consumed at $177^{\circ}C$ after 2.5 hours. The secondary amine groups,
however, are never all consumed even at $300^{\circ}C$. Also, both inter-
and intramolecular TGDDM epoxide homopolymerizations can occur to form
linear ether linkages and six-membered morpholine rings respectively,
with the latter ether mechanism predominating. At the higher cure
temperatures, $\geq 225^{\circ}C$, decreases in the hydroxyl and secondary
amine band intensities indicate further complex reactions occur.
(3) In the TGDDM-DDS-BF_3 systems all the epoxide groups are consumed
at $177^{\circ}C$ after 2.5 hours. The relative contributions of the
epoxide-amine addition reactions and the epoxide homopolymerization
and isomerization reactions to the resultant epoxy structure are
different than those that occur in the absence of the catalyst.

These FTIR studies to date indicate the chemical structures of
commercial TGDDM-DDS resins will vary considerably with cure

---

*Reference to a company or product name does not imply approval or
recommendation of the product by the University of California or the
U.S. Department of Energy to the exclusion of others that may be
suitable.

conditions and the presence of catalysts. Indeed, our studies on Narmco 5208 and Fiberite 934 reveal different T_g's, densities, network morphologies and long-term moisture sorption characteristics.

PHYSICAL STRUCTURE

The primary physical structural parameters that control the durability of epoxy composite matrices are the microvoid characteristics and the cross-linked network structure.[11,12,19,20]

Microvoids can result when air, moisture or other low molecular weight material is trapped in the system during cure and subsequently vaporized and possibly eliminated from the glass during postcure. The low molecular weight material results from either inhomogeneous mixing of epoxide and curing agent or from the aggregation of unreacted constituents. For example, in TGDDM-DDS epoxy systems the melting and volatilization of unreacted DDS crystallites during cure produces microvoids.[11,21]

Epoxies can form networks with heterogeneous cross-link density distributions primarily as a result of phase separation in the form of microgelation during cure or from heterogeneous polymerization conditions.[21] The network morphology of cross-linked glasses is, however, difficult to determine experimentally. The majority of the evidence for heterogeneous regions of cross-link density in epoxies is derived from electron microscope investigations. These microscopy studies involve carbon-platinum replication of etched and nonetched free surfaces and fracture surfaces. In our experience, however, artifacts can often result from replication techniques, and we question evidence of nodular morphology based on these techniques for the following reasons. (1) similar nodular structures that are observed by replication of epoxy surfaces can also be produced by replication of inorganic glass slides; (2) blisters that are produced as a result of the etching of epoxy surfaces can be misinterpreted as nodular regions of high cross-link density; (3) the fracture topographies can exhibit a nodular appearance in the initiation region as a result of fractured craze fibrils; and (4) the nodular surface structures observed in carbon-platinum replicas are rarely observed by scanning electron microscopy studies of gold-coated surfaces.

More confidence can be placed in bright-field transmission electron microscopy studies of the morphology of thin epoxy films strained directly in the electron microscope.[19,22] The morphology of such thin films, however, may not be representative of the bulk. However, if the heterogeneous regions are sufficiently large as in the case of Fiberite 934 resin such regions can be detected directly by optical microscopy.[16,23]

Small angle x-ray scattering studies have the potential to monitor cross-linked network morphologies.[24] However, to interpret such results it is necessary to know the relationship between the density and degree of cross-linking of the network. This knowledge requires an understanding of the network structure on the molecular level.

We are presently studying epoxy network structure in terms of the extensibility of molecular segments between cross-links, bond strengths, and local free volume.[25] These studies include computer modeling the epoxy networks in terms of the molecular geometries of the epoxide and amine monomers. The extensibility of segments between cross-links in these networks is a critical parameter that determines the amount of local deformation that can occur prior to network bond rupture. Bond rupture is the initial stage of failure in these glasses. These studies have revealed that molecular segments in cross-linked networks may not exist in their minimum energy rotational isomeric configurations because of molecular geometrical constraints imposed on such segments when the 3D network is formed. Furthermore, the cross-links can inhibit close packing of molecular segments which results in an increase in glassy-state free volume with cross-link density.

DEFORMATION AND FAILURE PROCESSES AND MECHANICAL RESPONSE

Although epoxies are generally considered relatively brittle polymeric glasses they can deform in tension in the glassy state by >15%.[11,12,19] Upon such deformations our recent photoelastic studies have revealed that these ductile systems initially undergo homogeneous plastic deformation followed by the development of inhomogeneous plastic deformation in the form of shear bands.[25] Failure occurs in these glasses in the region of a highly strained shear band. Our studies of these deformation processes as a function of network structure reveals the fully reacted most highly cross-linked networks exhibit the highest ductility. Lower cross-linked, not fully reacted epoxies inherently contain molecular defects in the form of unreacted groups whereas bond rupture is necessary to form a similar concentration of defects in the fully cross-linked system. Hence, a flaw of critical dimensions will form via chain scission at lower elongations in the lower cross-linked not fully reacted systems.

Failure generally occurs in tension in unnotched epoxies by growth of a single crude craze.[11,19,22,26] In notched specimens that fail under controlled crack propagation, however, only a small plastic zone is formed ahead of the crack tip.[26-33] Whether this plastic flow at the crack tip occurs by shear yielding or yielding under normal stresses is difficult to ascertain experimentally for small plastic zones and will depend on the stress fields imposed on the epoxy immediately ahead of the crack tip.

Failure on the molecular level is initiated by bond rupture within the network. This rupture will depend on the extensibility of the segments between cross-links and/or the steric distortions imposed on the cross-link sites. The extensibility of the segments between cross-links will depend on the minimum energy rotational isomeric configurational states within these segments and how such configurations are modified by the geometrical constraints of the cross-linked network. This segmental extensibility will exhibit a time dependence and will therefore depend on the strain rate of the deformation imposed on the glassy network.

To correlate network molecular parameters with the role the resin plays in composite performance we presently are evaluating the additional strains imposed on the epoxy matrix as a result of resin shrinkage, the mismatch in the coefficients of expansion between the resin and fiber and strains produced in the resin as a result of fabrication induced plastic flow. Together with our studies on the microscopic deformation and failure processes of composites by photoelastic studies we plan to correlate the composite matrix performance with measurable, molecular structural parameters of the resin.

ACKNOWLEDGMENTS

I wish to acknowledge the contributions of my colleagues, Dick Christensen, Fung Ming Kong, Jay Lepper, Eleno Mones, Larry Nielsen and Connie Walkup, to these studies.

REFERENCES

1. Air Force Durability Workshop, Battelle Columbus Laboratories, Columbus, Ohio, September 1975.
2. Air Force Conference on the Effects of Relative Humidity and Temperature on Composite Structures, March 1976, University of Delaware, AFOSR-TR-77-0030 (1977).
3. H. Lee and K. Neville, "Handbook of Epoxy Resins," McGraw-Hill, New York, 1967, Ch. 5 and 7.
4. H. C. Anderson, SPE Journal, 16:1241 (1960).
5. T. Kakurai and T. Noguchi, J. Soc. Org. Syn. Chem. Japan, 18:485 (1960).
6. T. K. Kwei, J. Polym. Sci. A-1:2985 (1963).
7. P. V. Sidyakin, Vysokomol. Soyed. A14:979 (1972).

8.   Y. Tanaka and T. F. Mika, "Epoxy Resins," C. A. May and
     Y. Tanaka, eds. Marcel Dekker, Inc., New York (1973), Ch. 3.

9.   D. A. Whiting and D. E. Kline, J. Appl. Polym. Sci.
     18:1043 (1974).

10. K. Dusek, M. Bleha, and S. Lunak, J. Polym. Sci., (Polym. Chem.
     Ed.) 15:2393 (1977).

11. R. J. Morgan, J. E. O'Neal, and D. B. Miller, J. Mater. Sci.
     14:109 (1979).

12. R. J. Morgan, J. Appl. Polym. Sci. 23:2711 (1979).

13. N. S. Schneider, J. F. Sprouse, G. L. Hagnauer, and
     J. K. Gillham, Polym. Eng. and Sci. 19:304 (1979).

14. E. Narracott, Brit. Plast. 120:26 (1953).

15. H. Dannenberg, SPE Trans. 3:78 (1963).

16. R. J. Morgan and E. T. Mones, Composites Tech. Rev. 1, No. 4:17
     (1979).

17. E. T. Mones and R. J. Morgan, "FITR studies of the cure
     reactions of amine-cured epoxies," Paper presented at
     American Physical Society, Phoenix, 1981; Bull. of APS 26,
     No. 3:366 (1981).

18. E. T. Mones and R. J. Morgan, Polymer Preprints (ACS) 22,
     No. 2 (in press).

19. R. J. Morgan and J. E. O'Neal, Polym. Plast. Technol. Eng.
     10:49 (1978).

20. R. J. Morgan and J. E. O'Neal, J. Macromol. Sci. Phys.
     B15(1):139 (1978).

21. R. J. Morgan, "Developments in reinforced plastics-1,"
     Applied Science Publishers Ltd., G. Pritchard, ed. (1980),
     Ch. 7.

22. R. J. Morgan and J. E. O'Neal, J. Mater. Sci. 12:1966 (1977).

23. R. J. Morgan and E. T. Mones, "Resins for aerospace," ACS
     Symposium Series 132, C. A. May, ed., ACS, Washington, D.C.
     (1980), Ch. 18.

24. F. M. Kong, E. T. Mones, and R. J. Morgan, "The relationship
     between the structure of epoxies and their mechanical
     properties," Paper presented at American Physical Society,
     Phoenix, 1981; Bull. of APS 26, No. 3:429 (1981).

25. R. J. Matyi, D. R. Uhlmann, and J. A. Koutsky, J. Polym. Sci.
     Polymer Physics Edition, 18:1053 (1980).

26. R. J. Morgan, E. T. Mones, and W. J. Steele, Polymer (in press).

27. S. Yamini and R. J. Young, Polymer 18, 1075 (1977).

28. R. A. Gledhill, A. J. Kinloch, S. Yamini, and R. J. Young,
     Polymer, 18:574 (1978).

29. D. C. Phillips, J. M. Scott, and M. Jones, J. Mater. Sci.
     13:311 (1978).

30. R. A. Gledhill and A. J. Kinloch, Polym. Eng. Sci. 19:82
     (1979).

31. J. Mijovic and J. A. Koutsky, Polymer, 20:1095 (1979).

32. A. J. Kinloch and J. G. Williams, J. Mater. Sci. 15:987 (1980).

33. S. Yamini and R. J. Young, J. Mater. Sci. 15:1823 (1980).

DISCLAIMER

ENVIRONMENTAL AGING OF EPOXY COMPOSITES

A. Apicella, L. Nicolais, and C. Carfagna

Istituto di Principi di Ingegneria Chimica

Universita' degli Studi di Napoli,
Piazzale Tecchio, 80125, Napoli,Italy

INTRODUCTION

Fiber reinforced plastics (FRP) are increasingly utilized for structural application where their long-term properties are of primary importance. The strength enhancement and resistence to fracture which occurs when one combines high-strength fiber with low strength, low modulus polymer matrix, depends to a great extent on the polymer-filler interfaces. However, FRP undergo a slow degradation process when subjected to the physical and chemical aging in aggressive environments (1-5). The synergistic action of the moisture penetration, high temperatures, frozen-in residual stresses and other aggressive agents (i.e. ozone or gasoline), in fact, strongly influence the morphology of the matrix and interfaces (6-8). The polymeric matrices do not present an effective barrier to the diffusion of the aggressive agents through the fiber and then the degradation mechanism acts on the fiber, the interface and the matrix. In particular, sorbed moisture, acting both as a plasticizer and a crazing agent, differently deteriorates the mechanical integrity of the epoxy matrix. However, while the plasticization is reversible since it disappears upon drying, the microcavitation (localized crazing in regions of different crosslinking densities or around the inclusions) is permanent.

EXPERIMENTAL

Materials

Specimens were prepared from Epikote 828 (Shell Italiana), using commercial triethylene-tetramine (TETA), (Montedison S.p.A.) as curing agent. Distilled water was used in sorption experiments.

Composites were prepared using Vacu-Blast glass microspheres with
particle sizes in the range of 32 to 40 microns. A 7.5% by volume
composite was analyzed. Test procedures are descibed in (6).

RESULTS AND DISCUSSION

Water sorption mechanism in epoxy resin: Influence of the
previous hygrothermal conditioning.

    The diffusion of small molecules in glassy polymers is often
associated to morphological modifications of the bulk material.
This fact involves sorption kinetics which are not described by
ordinary diffusion Fick's law and which are reported as "anomalous"
(9-12). The physical modification occurring in epoxies exposed to
liquid and vapour phase water has been associated with micromechanical cavitation (crazing) occurring in a region of different crosslinking densities (6-9). Figure 1 shows the liquid water sorption
in amino-hardened epoxy films at 70°C (left) and 20°C (right).
The equilibrium water uptake at 70°C was clearly higher than the
value at 20°C, 4.00% and 2.96% respectively, however, when the
sample conditioned at 70°C was brought down to 20°C additional
water was absorbed and finally reached the value of 4.60%. The
excess of water present in the sample previously conditioned at
70°C, 1.64%, was recognized as trapped in craze microcavities formed
during the thermal aging. The increase of the apparent solubility
at the lower temperature for the same sample should reflect

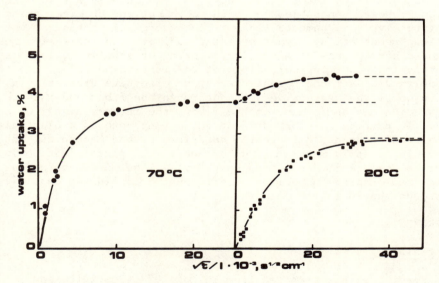

Fig. 1 - Liquid water sorption at 70°C (left) and
        20°C (right) in DGEBA amine-cured resin.

the slight exothermic water sorption process, although the equilibrium water uptakes of samples conditioned at increasingly higher temperatures, first column of Table 1, seem to indicate the opposite behavior.

Table 1: Mechanical properties at 20°C and 50°C and apparent solubilities of unfilled epoxy resins equilibrated in liquid water at increasing temperatures.

WATER CONDITIONING TEMPERATURE, °C	S_{app}, %	E, Kg/mm^2 20°	E, Kg/mm^2 50°C	$_b$, Kg/mm^2 20°C	$_b$, Kg/mm^2 50°C	ε_b, % 20°C	ε_b, % 50°C
2	2.58	139	76	3.40	2.40	4.9	5.4
20	2.96	149	92	3.58	2.70	4.5	5.5
50	3.22	156	105	3.80	2.92	4.4	5.4
70	4.00	179	112	4.05	3.10	4.5	5.3
DRY	--	216	129	4.40	3.40	4.0	4.2

Tensile tests performed at 20°C and 50°C on wet samples, however, showed that the higher degree of plasticization was reached by the epoxy conditioned at lower temperatures. The elastic moduli and the stresses at break, compared in Table 1 for the dry and water saturated epoxy, are, in fact, progressively lowered by the lower temperature conditioning, irrespective of the corresponding higher water uptakes.

More evidence of the micromechanical nature of the damage induced at high temperatures in presence of sorbed water has been obtained from the analysis of the vapor sorption kinetics (5,6). The Dual Mode Sorption Theory has been successfully developed to correlate the presence of hypothesized "pre-existing holes" or "free volume elements" frozen in the out-of-equilibrium glassy state to characteristic anomalous sorption behavior of gasses and vapours. If the changes of the apparent water solubilities after the aging are attributed to a microcavitational damage (of the size of nucleating crazes (13), the increase of the number of high energy sorbing sites in which the diffusing molecules may be trapped should be evident in a reduction of the effective diffusion coefficient, following the increase of the Languimirian capacity (9).

The analysis based on the cited model gave the following relationship (9)

$$K' / K'' = D'' / D' \qquad (1)$$

between the apparent solubilities K' and K", expressed as Henry's
law constants, and the effective diffusion coefficients D' and D"
obtained in tests performed in the same conditions of temperature
and Relative Humidity but on samples with a higher (') and lower (")
degree of damage. As previously pointed out, lower degrees of damage characterize samples conditioned in water at low temperature,
however, epoxies have been found to be differently deteriorated also
when equilibrated in environments of different relative humidities.
The equilibrium sorption histeresis at 60°C of an "as cast" resin
progressively equilibrated at higher R.H. is shown in Figure 2.

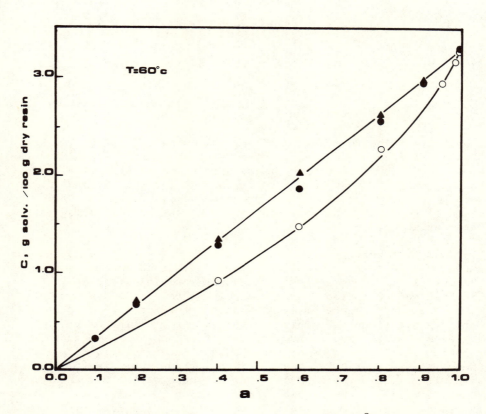

Fig. 2 - Equilibrium sorption hysteresis at 60°C vs external
         water vapor activity: (○) first sorption, (●)
         second sorption and (▲) third sorption.

The isotherm appears upward when the sample experiences progressively higher R.H. (open circles) while it remains linear once the maximum R.H. is reached (full circles) also in a third set of sorptions (full triangles).   The upward shape of the curve in the first
set of sorptions has been related to a progressive damage of the
resin since the histeresis disappeared for any subsequent sorption.
The initial slope of the upward isotherm at low R.H. gives a parameter to estimate the Henry's law constant of the water in the undamaged resin.   The sorption isotherms solubility constants from
$75^{\circ}C$ to $20^{\circ}C$ referred both to the undamaged state (open circles)
and to samples conditioned at 100% R.H. (full circles) are reported in the vant'Hoff diagram of Figure 2.

   The data relative to the undamaged state lie on a straight line
while, accordingly with the previous analysis, positive deviations
at temperatures higher than $30^{\circ}C$ have been found for the 100% R.H.
damaged samples.

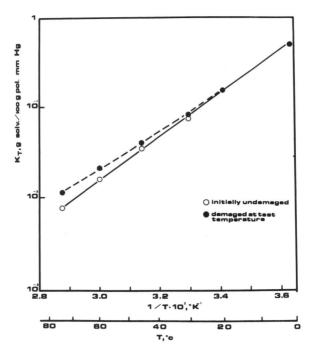

Fig. 3 – Solubility constants versus the reciprocal of the tempera-
         ture for undamaged (o) and aged at the test temperature
         and 100% R.H. epoxies (●).

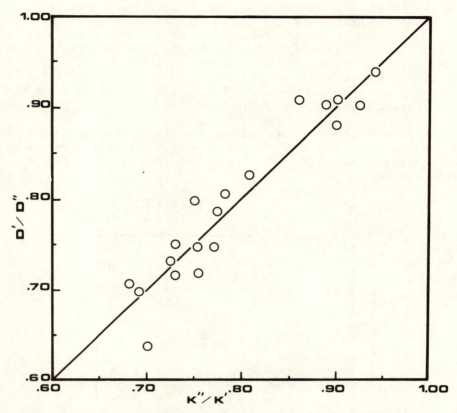

Fig. 4 - Comparison between the experimental (O) and theoretically predicted (line) diffusion coefficient depressions as a function of the solubility increases.

Water sorption solubility and kinetic parameters obtained by testing samples damaged in environments of different temperature and R.H. were used to test the validity of the equation 1. Figure 4 compares the diffusion coefficient depressions, D'/D", and the corresponding solubility increases, K" /K' , experimentally found (open circles) and theoretically predicted by equation 1 (full line).

The good agreement between the experimental data and theoretical
predictions strongly supports both the hypothesis of the microcavitational damage and its dependence on temperature and relative humidity.

Epoxy-water interactions

It has been widely described how sorbed moisture plasticizes
the epoxy resins depressing its glass transition temperature, however, the nature of the association process between the water and
the polymer is not clearly understood. The free volume theory (14),
in fact, satisfactorily applies only to polymer-diluent couples
which do not present strong mutual molecular interactions. An alternative description of the wet glass transition, which designates
configurational entropy rather than free volume as the temprature-
dependent function, has been proposed (15) for epoxies and other
polymers in which water may be absorbed by localization at strongly
polar molecular groups (hydrogen bonding). Evidence has been given,
in fact, by broad line Nuclear Magnetic Resonance analysis (16) that
the plasticization effect of the water on a crosslinked epoxy may
be related to the strong interactions between the dissolved water
and segments or groups of the polymer, although the exact sorption
sites to which water may be bonded are still uncertain.

The dependence of the water sorption as a function of the
prepolymer composition of amino hardened epoxy will be discussed
further on to resolve the sorption process into effects arising
from morphological (microcavitation) as well as chemical (hydrogen
bonding) origins. Moreover, free volume and entropy models will be
used to evaluate the nature of the epoxy-water interactions by
comparing the experimental and theoretically predicted values of
the glass transition temperature depressions of epoxies crosslinked
with different amounts of hardener and subsequently saturated at
high temperature in liquid water. Liquid water sorption and calori-
metric behavior for three types of amino hardened epoxies have been
extensively discussed in a previous work (17). The densities, the
dry and wet glass transition temperatures, the equilibrium water
uptakes at 70°C and 20°C for the "as cast" resins and at 20°C for
previously equilibrated at 70°C samples and the difference in the
specific heat at the glass transition temperature are summarized
in Table 2. The resin crosslinked with a large excess of hardener
(25 PHR), when conditioned in liquid water at 20°C and 70°C, sorbs
10.80% and 8.66% of water which is much higher than for the other
two compositions. The water uptakes for the 5 PHR are, in fact,
of only 1.50% and 1.92% at 20°C and 70°C while they are of 2.20%
and 3.30%, respectively, for the stoichiometric composition.

The strong influence of the prepolymer composition on the
equilibrium water uptakes may be related to the increased number
of hydrogen bonding sites due to the presence of the amines where
water may be absorbed.

Table 2: Experimental values of some physical properties of the DGEBA
epoxy resin cured with different amounts of TETA.

PREPOLYMER COMPOSITION	$\rho_{20°C}$, g/cm^3	T_{gd}, °C	T_{gw}, °C	S^{70}, %	S^{20}, %	S^{20}_{70}, %	ΔS^{20}, %	Δc_p, cal/g°C
5 PHR	1.174	109	105	1.92	1.50	2.42	0.92	0.11
14 PHR	1.219	142	109	3.30	2.70	3.90	1.20	0.11
25 PHR	1.107	95	59	8.66	10.80	10.80	0.00	0.13

The actual water solubilities, calculated as discussed in the
previous section ($S^{70}_{act} = S^{70} - S^{20}$), have been used to predict
the wet glass transition temperatures of the plasticized resins by
means of the theoretical free volume (15) and entropy models (16).
The values of the physical parameters to be used in the models are
reported in Table 2 and 3.

Table 3: Apparent and actual water solubilities in the DGEBA
epoxy resin cured with different amounts of TETA. Comparison between the experimental ($T_g$wet) and calculated
(T_{gw}'s) glass transition temperature of water-saturated
samples.

PREPOLYMER COMPOSITION	S_{app}, %	S_{act}, %	Entropy Model T_{gw} °C	Free Volume Model T_{gw} °C	T_{gw}, °C
5 PHR	1.92	1.00	92	100	105
14 PHR	3.30	2.10	110	115	109
25 PHR	8.66	8.66	57	52	59

Free Volume Model:

The expression used to correlate the variation of the glass
transition temperature of a plasticized polymer is

$$T_{gw} = (\alpha_p V_p T_{gp} + \alpha_d(1-V_p) T_{gd}) / (\alpha_p V_p + \alpha_d(1-V_p)) \qquad (2)$$

where V_p is the volume fraction of the resin in the resin-water mixture, T_{gp} and α_p are the dry glass transition temperature and the difference between the volume expansion coefficients of the polymer in the rubbery and glassy state, while T_{gd} and α_d are the characteristic transition temperature and the volume expansion coefficient of the water. Previous authors (18,19) identified T_{gd} and $_d$ with a temperature near the freezing point of the water and the corresponding expansion coefficient: $4^{o}C$ and $3.66 \times 10^{-3o}C^{-1}$.

Entropy Model:

The expression derived assuming the configurational entropy as the temperature dependent function is

$$T_{gw} = T_{gp}(1-R \; y(r) \; /(M_s \triangle c_p))\qquad(3)$$

with $y(r) = r \; \ln(1/r) + (1-r) \; \ln 1/(1-r)$ and $r = (M_s/M_w)f$

where R is the universal gas constant, M_w is the molecular weight of the water, $M_s = N_A / N_s$ is the effective formula weight of the hydrogen bond sites with N_A the Avogadro's number (mole $^{-1}$) and N_s the number of hydrogen bond sites for grams of resin and, finally, f are the grams of moisture for gram of dry resin. $\triangle c_p$ is the change of the specific heat at the glass transition.

The values of the wet glass transition temperatures obtained from the free volume model, T_{gw} (S_{act}), better match the effective wet glass transition temperature, T_{gw}, of the resin made with a defect of hardener. Conversely, the entropy model adequately describes both the plasticizations observed for the resin made with a large excess of hardener and of stoichiometric composition, when the actual values of the solubilities are used.

The effectiveness of two distinct models in describing the plasticization due to the water sorption indicates that different types of polymer-diluent interactions may be present: strong polymer-diluent interactions lead to a mechanism driven by hydrogen bond formation while, in the opposite case, the free volume changes become effective.

Furthermore, the main role of the amino-hydrogen bonds compared to the hydroxyl-hydrogen bonds in the plasticization of moisture penetrated epoxy resins is confirmed both by the analysis based on the entropy model and by the strong sensitivity on the amino content of the water solubilities.

Water sorption in epoxy composites: Influence of the frozen-in stresses
───

The internal tensile stresses, like those arising from the
shrinkage of the glassy matrix around the fillers in a composite,
further increase the polymer tendency to craze. In fact, recongni-
tion that crazing requires void formation leads to the suggestion
that the phenomenon should be favored in presence of a dilatational
component of the stress.

An epoxy composite containing 7.5 percent by volume of 30-40
microns glass beads has been used to evaluate the influence on the
mechanical performance of the thermal aging in presence of sorbed
moisture. Tensile tests were performed at $20^{\circ}C$ and $50^{\circ}C$ on samples
previously equilibrated in liquid water at $50^{\circ}C$ and $70^{\circ}C$.

Liquid water sorptions for the unfilled resin and the composite are compared in Figure 5 for samples first equilibrated at $70^{\circ}C$
(left) and then brought down to $20^{\circ}C$ (right). The water sorption
behavior of an "as cast" sample conditioned directly at $20^{\circ}C$ is
also reported in Figure 5. The water uptakes for the composite
are referred to the dry weight of the polymeric component.

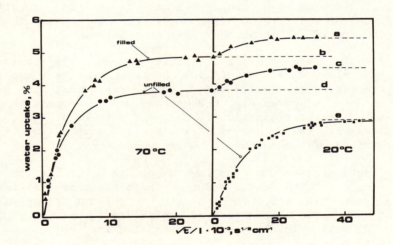

Fig. 5 - Liquid water sorptions at $70^{\circ}C$ (left) and $20^{\circ}C$ (right)
          for an epoxy composite and unfilled resin.

The two sorption curves show a difference of about 1% (b > d) in the
equilibrium water uptakes at $70^{\circ}C$ (4% for the resin and 5% for the
composite). Such difference was maintained when the same samples
were subsequently conditioned at $20^{\circ}C$ (a > c, 4.60 for the resin and
5.60 for the composite). The water uptakes for the resin condition-

ed directly at 20°C and for the resin and the composite previously
aged at 70°C were, respectively, 2.96%, 4.60% and 5.60%. The large
excess of water found for the thermally aged composite, 90% more
than the resin conditioned directly at 20°C, could be partially
attributed to the presence of the fillers. The Electron Scanning
Micrographs of the fracture surfaces of undamaged and water conditioned at 70°C composites are reported in Figure 6a and 6b, respectively.

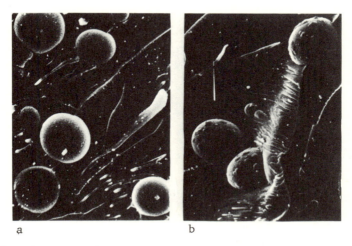

Fig. 6 - SEM of the fracture surfaces of an epoxy
composite aged at 2° (a) and 70°C (b).

The unconditioned composite shows "clean" bead surfaces while the
thermally aged composite has fragments of the bulk resin adherent
to the fillers. The adhesion between the matrix and the fillers
is weaker than the bulk strength for the unconditioned samples
while it becomes relatively stronger when the matrix was locally
deteriorated (crazed). The residual stresses induced around the
inclusions by the shrinkage of the resin during the curing could
favor a localized crazing when the composite is exposed to a
severe environment. However, the mechanical strength reduction
observed for a composite exposed to liquid water is due to both
the contributions of the water plasticization and the polymer-
filler interface failure. The role of the water plasticization
and crazing on the mechanical properties of the matrix and the
composite aged at 50°C and 70°C is summarized in Table 4.

The presence of the inclusions increases the elastic modulus
of the dry samples, i.e. stress-strain curves "a" and "b" in Figure
7 for tests at 20°C, but strongly decreases and embrittles the
sample after the aging, i.e. s-s curves "c" and "d" in Figure 7.

Table 4: Mechanical properties at 20% and 50% of the unfilled resin and the glass bead composite equilibrated in liquid water at 50° and 70°C. The values of the "as cast" dry samples are also reported.

Temperature	20°C						50°C					
Sample	Dry		Sat.@ 50°C		Sat.@ 70°C		Dry		Sat.@ 50°C		Sat.@ 70°C	
Vol.Frac.,%	0	7.5	0	7.5	0	7.5	0	7.5	0	7.5	0	7.5
E, Kg/mm^2	129	175	105	115	112	127	216	250	156	143	179	164
σ_b, Kg/mm^2	3.4	3.2	2.9	2.2	3.1	2.3	4.4	4.3	3.8	2.8	4.1	2.8
ε_b, %	4.2	3.0	5.4	2.6	5.3	2.4	4.0	3.3	4.4	2.9	4.5	2.7

The degradation process is particularly evident for the composites conditioned at 70°C. In fact, while only the plasticization due to the sorbed water lowers the elastic modulus of the matrix (reduction of 17%), the wet composite is subjected to the influence of both plasticization and the strong localized crazing (reduction of 34%). The increased strength losses observed for the high temperature aged composites should be attributed to the lower number of beads acting as effective reinforcements once the polymer-filler interfaces failed.

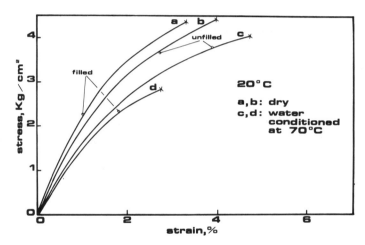

Fig. 7 - Stress-strain curves for the composite and the matrix at 20°C.

The chemical nature of the epoxy-water interactions leads to a particularly strong degradation process in presence of ozone. The DSC traces and infrared spectra of samples exposed for 3 days at 70° to dry and 100% R.H. environments containing 3% of ozone are compared in Figure 8 to those of an "as cast" epoxy resin.

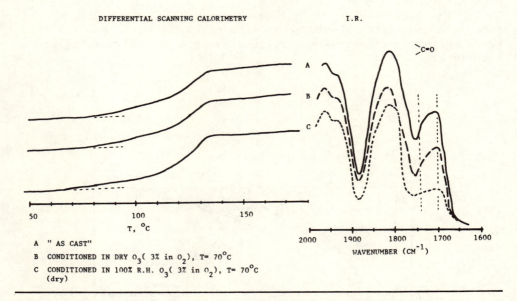

Fig. 8 - DSC thermograms and Infra Red spectra of "as cast" (A)
and aged at 70°C in a dry (B) and humid (C) atmosphere containing O_3.

The synergistic action of the ozone and the moisture is evident in the I.R. spectra of the aged samples where the presence of
carbossilic groups is noted only for the sample conditioned with
ozone in the humid atmosphere. The nucleophilic character of the
nitrogen present, which is responsible for the epoxide ring opening reaction during cure, may be also responsible for the chain
breaking of the amino hardened DGEBA epoxy resins. The calorime-
tric analysis, however, did not clearly indicate a reduction of
the glass transition temperature although the phenomenon could be
attributed to the considerable crosslinking which accompanies chain
breaking during degradation (18).

The low stability of this amine-cured epoxy is a result of the
incorporation in these structures of nucleophilic centers in the
form of hardener nitrogen atoms.

REFERENCES

1. T.K.Kwei and H.M.Zupko,J.Polym.Sci.,A2,7,876 (1969)
2. L.Nicolais,E.Drioli,H.B.Hopfenberg and A.Apicella,Polymer,20, 459 (1979)
3. U.T.Kreibich and R.Schmid,J.Polym.Sci.,A2,7,876 (1969)
4. A.S.Kenyon and L.F.Nielsen,J.Macromol.Sci-Chem.,A3,275 (1969)
5. R.J.Morgan and J. O'Neal,J.Mater.Sci.,12,1966 (1977)
6. A.Apicella,L.Nicolais,Ind.Eng.Chem.Prod.Res.Dev.,20,138 (1981)
7. C.E.Browning,Polym.Eng.Sci.,18,16 (1978)
8. E.L. McKague,Jr.,J.E.Halkias, and J.D.Reynolds,J.Compos.Mater., 9,2 (1975)
9. A.Apicella,L.Nicolais,G.Astarita,E.Drioli,Polym.Eng.and Sci, 21,18 (1981)
10. H.B.Hopfenberg,R.H.Holley and V.T.Stannet,Polym.Eng.Sci.,9,242 (1969)
11. N.Thomas and A.H.Windle,Polymer,19,255 (1978)
12. L.Nicolais,E.Dioli,H.B.Hopfenberg and G.Caricati,J.Membr.Sci., 3,231 (1978)
13. R.R.Kambour,E.E.Ramagosa,and C.L.Grummer,Marcomolecules,5,335 (1972)
14. M.L.William,R.F.Landel,J.D.Ferry,J.Am.Chem.Soc.77,3701 (1955)
15. H.G.Carter,K.G.Kiblet,J.Comp.Mat.,11,265 (1977)
16. P.Moy,F.E.Karasz,Polym.Eng.and Sci.,20,315 (1980)
17. C.Carfagna,A.Apicella,L.Nicolais,"The Effect of the Prepolymer Composition of Amino Hardened Epoxy Resins on the Water Sorption Behaviour and Plasticization",J.Appl,Polym.Sci.,in press
18. E.C.Leisegang,A.M.Stephen and J.C.Paterson-Jones,J.Appl.Polym. Sci.,14,1961 (1970)

FACTORS AFFECTING THE DEVELOPMENT OF

NEW MATRIX RESINS FOR ADVANCED COMPOSITES

C. E. Browning

Air Force Wright Aeronautical Laboratories
Materials Laboratory
Wright-Patterson Air Force Base, Ohio 45433

INTRODUCTION

Historically, advanced composites utilizing high performance reinforcements such as graphite fibers evolved using matrix resins (principally epoxies) having a pre-existent data base. In most cases a substantial amount of processing and performance experience with these matrix resins had already been achieved in other materials systems and applications such as those involving glass reinforced systems. Therefore, it was natural that the use of these resins was extended to advanced composites. In the early stages of evolution of advanced composites, most requirements for improved matrix resins were satisfied by reformulating the epoxies. However, as advanced composites enter the era where they are in routine production of primary aircraft structure with further applications on the horizon, it is reasonable to ask what factors will determine the directions for the types of new matrix resins to be developed for use in advanced composite materials.

In general, there are two broad categories from which these factors are drawn: (1) properties, and (2) processing, with the most important specific factors to be dealt with being: (1) moisture resistance, (2) toughness, (3) improved processability, and (4) elevated temperature performance. These factors invariably are the result of a particular deficiency in current resins relative to specific requirements.

The effect of these factors on the development of new candidate matrix resins is discussed in terms of replacing an existing resin system having specific deficiencies. This replacement mode is extremely difficult because of the added requirement that the

newly developing material must retain all of the attributes of the
old material while removing its deficiencies.

DISCUSSION OF FACTORS

Moisture Resistance

     One of the most important factors affecting the development
of new matrix resins is the environmental durability of current
epoxy resin systems.

     Moisture absorption from high humidity environmental exposures
has been shown to have a deleterious effect on the elevated temperature properties of current epoxy matrix resins.[1,2] The absorption of water by epoxies can be attributed to the affinity of functional groups of a highly polar nature in the epoxy for water
molecules as follows:

$$\sim CH_2 - \underset{\underset{H}{|}}{\overset{\overset{:\ddot{O}:}{|}}{C}} - CH_2 \sim \qquad \sim R - \ddot{O} - \langle \phi \rangle \sim \qquad \sim R - \underset{\underset{H}{|}}{\overset{\overset{|}{}}{\ddot{N}}} - R \sim$$

Physically, the absorption and diffusion process produces a weight
gain behavior typified by that shown in Fig. 1.

     The mechanism of moisture-induced elevated temperature prop-
erty losses in epoxies has been attributed to a classical plasticization process wherein water plasticizes the epoxy resin, producing the classical characteristics of swelling of the resin,
lowering its T_g, and shifting its modulus-temperature curve to
lower temperatures as shown in Fig. 1.

     This mechanical/physical effect, in turn, results in a
reduction of elevated temperature properties of matrix-dominated
composites as shown in Fig. 1.

     Clearly, an improved, moisture resistant resin is required
to raise the use temperatures of composites above those of current
epoxy systems. Approaches to a moisture resistant, improved durability resin include:

     (1) Utilization of higher T_g matrices having acceptable
         moisture-induced property reductions.

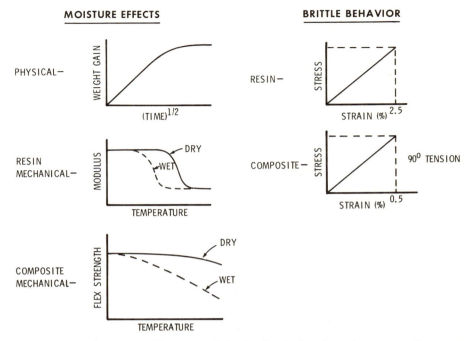

Fig. 1. Moisture effects and brittle behavior in current epoxy
         resins and derived graphite fiber composites.

(2) Development of new chemistry matrices having moisture
    insensitive chemistry with concomitant reductions in
    moisture weight gain and enhanced chemical or hydrolytic
    stability.

An approach to new resins that falls under category (1) is
the development of modified bis-maleimide resins having the following type of chemistry:

These resins may be modified or formulated with other ingredients
to achieve processability characteristics possessed by epoxies,
such as tack and drape, vacuum bag/autoclave cures, etc., with the
major non-epoxy trait being a high temperature postcure of 225°C.
The cure and resulting chemistry produces a higher as-cured, dry
T_g than epoxies. This T_g can be affected by moisture, but the
relative reduction still provides acceptable levels of properties
at temperatures higher than epoxies as shown in Table 1.

Table 1.   Properties of Modified Bis-Maleimide Resin/Graphite
           Fiber Composites

	0° Flexural Strength, N/m^2 (psi)	
T(°C)	Dry	Wet (1.8% W.G.)
RT	1725 (250)	1695 (246)
177	1380 (200)	965 (140)

A promising approach under category (2) involves the development of a totally new chemistry matrix resin, acetylene terminated
sulfone (ATS):

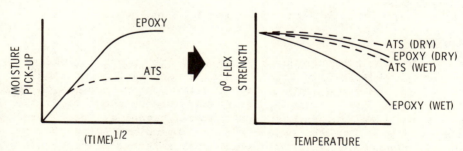

that undergoes cure through reaction of the acetylenic end groups.
These resins possess the important processing attributes of
epoxies: hot-melt prepregging, vacuum bag/autoclave cures, addition cure reactions, etc., in addition to their chemically stable,
moisture insensitive chemistry which produces a relative low moisture weight gain compared to epoxies (Fig. 2).   This new chemistry
and low weight gains result in substantially greater retention of
wet, elevated temperature composite properties versus epoxies as
illustrated in Fig. 2.

Fig. 2.   Moisture effects in acetylene terminated sulfone (ATS)
          composites.

Toughness

Another factor having a major impact on the development of advanced composite matrix resins is toughness.

Many of the current epoxy resin formulations have in common the following two ingredients:

$$\left(H_2C \overset{O}{\diagdown} CH - CH_2 \right)_2 N - \langle \bigcirc \rangle - CH_2 - \langle \bigcirc \rangle - N - \left(CH_2 - CH \overset{O}{\diagdown} CH_2 \right)_2$$

tetraglycidylmethylenedianiline (TGMDA)

and

$$H_2N - \langle \bigcirc \rangle - \overset{O}{\underset{O}{\overset{||}{S}}} - \langle \bigcirc \rangle - NH_2$$

diaminodiphenylsulfone (DDS)

The combination of these tetrafunctional ingredients gives rise to a cured resin having a high crosslink density and, in turn, a relatively high degree of brittle character as typified by the stress-strain curve in Fig. 1. This brittle character of the resin translates into brittle behavior in the composite as typified by the low transverse strain-to-failure as shown in Fig. 1. Additional brittle composite behavior would include low impact strength, low shear strength, etc.

In order to improve on these characteristics of current epoxies, a more ductile or tougher matrix resin is warranted. As was previously noted, in achieving these improvements it is necessary to maintain all of the property and processing attributes of epoxies.

Approaches to achieving improved toughness include:

(1) Utilizing existing methodology which principally includes two areas - elastomeric, secondary phase toughening of thermosets and thermoplastics.

(2) Developing totally new chemistry matrices having "tough" structures.

Elastomeric, secondary phase toughening has been shown to be effective with epoxy adhesives[3,4] but has not proven to be as successful in advanced composites such as graphite composites where high fiber volumes with close packing of the filaments tend to interfere with the mechanism of rubber particle toughening. Nevertheless, this is an approach being actively pursued for the development of tougher advanced composite matrix resins. As an example, recent work by St. Clair and St. Clair[5] has extended this toughening approach to high temperature polyimide resins. As shown in Table 2, this mechanism of toughening has potential for high temperature resin systems as evidenced by substantial increases in the peel strength and fracture toughness.

Thermoplastic resins have always been recognized as very ductile, tough polymer systems compared to thermosets but have only recently been utilized with graphite fibers for advanced composite applications.[6] The characteristics of a "tough", high performance thermoplastic resin (e.g., polysulfone, polyphenylenesulfide, etc.) are shown in Fig. 3 - a stress-strain curve with nonlinear behavior and very high strain-to-failure. This characteristic translates into "tough" composite properties such as higher transverse strain-to-failure (Fig. 3), improved impact strength, improved delamination resistance, etc.

The second area of developing totally new chemistry matrices having "tough" structures, while maintaining epoxy-like attributes, is typified by the development of systems such as the acetylene terminated quinoxaline (ATQ)[6] shown in Fig. 4. This resin combines the addition cure chemistry of acetylenes with the toughness associated with quinoxalines to give the type of stress-strain curve shown in Fig. 4. Compared to the epoxy (Fig. 1), this type of matrix can be expected to provide composite properties associated with improved toughness matrices - higher transverse strain-to-failure, improved impact strength, etc.

This last example typifies an ideal case where the desired properties of a resin were tailored by design prior to the start of synthesis.

Table 2. Properties of LARC-13 Polyimide Adhesive Modified by Addition of a Silicone Elastomer (SI) and a Butadiene-Nitrile Elastomer (B-N)[5]

Resin	T-Peel Strength N/m (lbs/in)	G_{I_C} (J/m^2)
LARC-13	230 (1.3)	117
LARC-13/SI	540 (3.1)	146
LARC-13/B-N	960 (5.5)	393

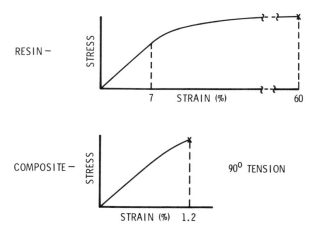

RESIN –

COMPOSITE –   90° TENSION

Fig. 3.   Mechanical behavior of a thermoplastic matrix
          resin and derived graphite fiber composites.

CHEMISTRY –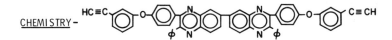

RESIN –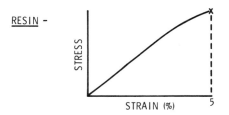

Fig. 4.   Mechanical behavior of an acetylene terminated
          quinoxaline matrix resin.

Improved Processability

A third major factor affecting the development of new matrix
resins is improved processability.

Processing relative to current epoxy matrix technology consists
of the major steps outlined in Fig. 5. The related important characteristics are shown below each step. Current epoxy matrix technology involves two participants - the vendor (prepregger) and the
user. The vendor buys raw materials (resins, hardeners, etc.),
prepares a formulated resin product by mixing and cooking these
constituents, coats carbon fibers with this resin to give the
prepreg product that is shipped to a user. Each of the vendor's
steps requires various degrees of control from simple weight
measurements to sophisticated physiochemical inspection techniques
such as chromatography. These controls are required to ensure
materials consistency because epoxy resins constantly change with
time.

The user receives the prepreg material, stores it in a
freezer, subsequently prepares the prepreg lay-up, then prepares
the vacuum bag lay-up and inserts it into the autoclave for cure
according to a given time, temperature, pressure profile. Important characteristics of this sequence are that the user must also
perform physiochemical inspections and must cure the material to
its final state by carrying out large-scale chemical reactions in
the autoclave. A poor part either goes through a costly repair
procedure or is scrapped.

A significant improvement in this process can be realized by
developing thermoplastic resins for use in advanced composites.

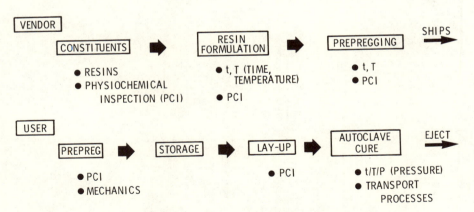

Fig. 5. Processing steps for advanced composites based on
         current epoxy matrix technology.

A typical composite processing sequence based on thermoplastic
technology is shown in Fig. 6. The vendor supplies a sheetstock
which was prepared by laminating thermoplastic film with graphite
fabric. Since the thermoplastic matrix is already a high molecular
weight polymer, little physiochemical control is required by the
vendor. The sheetstock shipped to the user is not a changing
material like the epoxy and, therefore, requires little physiochemical inspection by the user. The user utilizes a thermoforming
process to fabricate the part and, therefore, does not have to contend with the complexities associated with prepreg lay-ups and
vacuum bag/autoclave cures. Overall then, the thermoplastic process is less complex and less variable than the epoxy process,
resulting in reduced manufacturing costs. Further, the potential
exists to reform a thermoplastic matrix part that may initially be
unsatisfactory.

    There are at least two developmental issues with thermoplas-
tics. The principal one is solvent sensitivity, and the second
one is the cost of the tooling required for the thermoforming
operation. These issues are currently being addressed with favorable results. Many thermoplastic resin candidates have been developed with the required solvent insensitivity.

    Another processing requirement affecting resin development is
the desire to achieve cure of thermosetting resin systems in a more
rapid time than the several hours required by current epoxies. One
approach to rapid cure (30 min. or less) resins is unsaturated
resins (high vinyl butadienes, unsaturated polyesters, etc.) which
cure by free radical reactions with peroxide initiators (Fig. 7).
These resins have many of the processing attributes of epoxies yet
can be cured in 30 minutes or less. These reduced cure times
translate into substantially reduced costs for those applications
requiring high-rate production unattainable with epoxy technology.

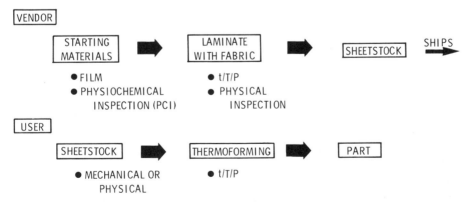

Fig. 6. Processing steps for advanced composites based on
         thermoplastic matrix technology.

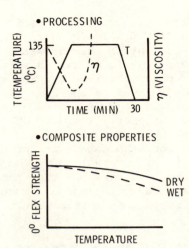

Fig. 7. Processing and properties of rapid cure, butadiene
matrix resin composites.

The high-vinyl butadienes also have a reduced moisture pick-up
versus epoxies and, in turn, improved retention of "wet", elevated
temperature properties (Fig. 7).

Elevated Temperature Performance

The final major factor affecting the development of new matrix
resins for advanced composites is elevated temperature performance.
The technology in this area (temperature on the order of 250-325°C)
has generally been based on polyimide (PI) resins (Fig. 8). The
original PI technology consisted of amic-acid varnishes in high
boiling solvents such as N-methyl pyrrolidone. The processing of
these systems was extremely difficult because of the difficulty
of removing the volatile condensation products from the chemical
reactions and the volatile solvent. These volatiles would routinely produce high degrees of porosity in cured laminates.

Subsequent developments in this area were directed at removing
these volatile-induced deficiencies, resulting in resins with
volatileless, addition-type cure reactions and/or the use of easily
removed, low-boiling solvents such as alcohols (Fig. 8), while
retaining properties at elevated temperatures.

A new chemistry approach to easy processing, elevated temperature performance matrix resins is based on a concept known as

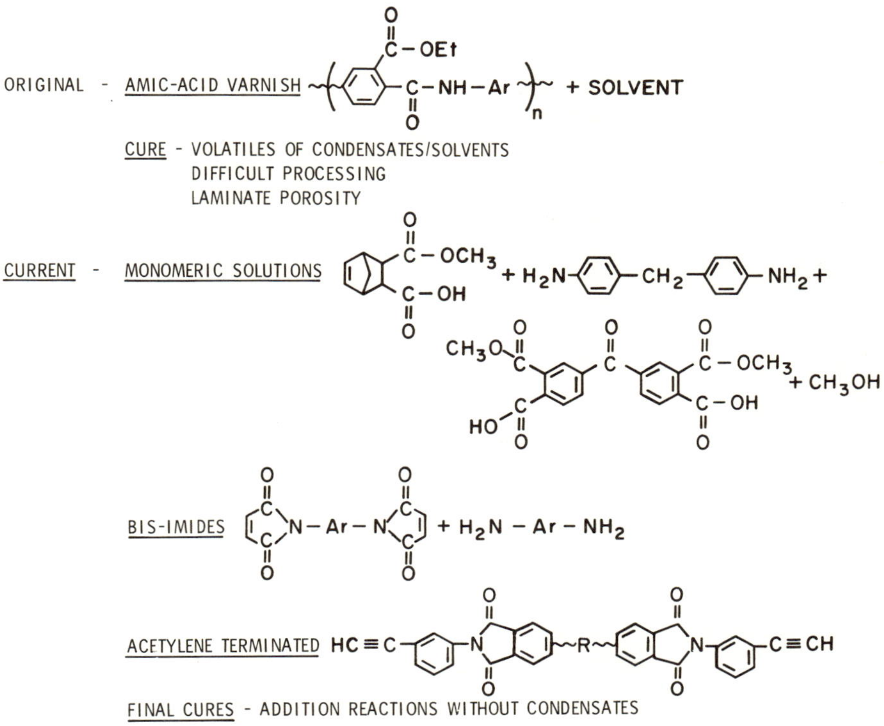

Fig. 8.   Characteristics of original and current polyimide
          technology matrix resin.

intra-molecular cyclization (IMC)[7] as outlined in Fig. 9.   These
resins would first be processed to an uncyclized stage.   In this
processing step, the resin is more flexible, has a lower T_g, and
is more easily processed.   In the next processing step, at a temperature higher than that of the first step, the resin is converted
to the cyclized structure via the solid-state, IMC reaction.   This
results in a resin with a higher T_g and use temperature.   Therefore,
this new chemistry approach offers the potential of simplified
processing with retention of elevated temperature properties.

SUMMARY

     The major factors affecting the development of new matrix
resins for advanced composites include: (1) moisture resistance,
(2) toughness, (3) improved processability, and (4) elevated temperature performance.

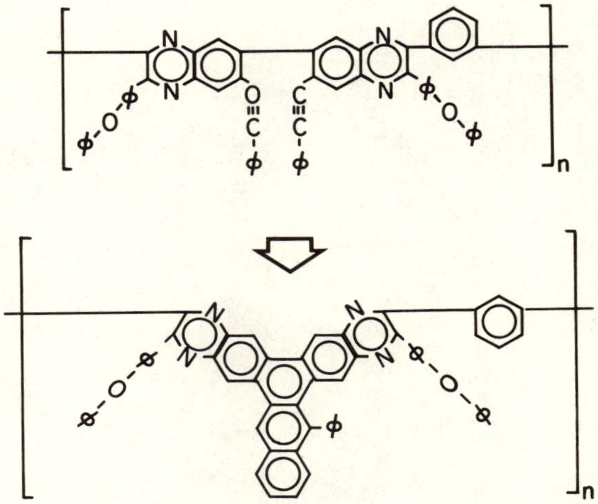

Fig. 9.   Processing characteristics of matrix resins based on
          intramolecular cyclization (IMC) technology.

These factors dictate the directions of matrix materials
development activities, with the two principal directions being:
(1) adapting existing chemistry to new matrix requirements, and
(2) developing new chemistry matrices. New matrix resin systems
which have emerged from these directions include:

- modified bis-maleimides

- acetylene terminated sulfone

- elastomeric, secondary phase toughened thermosets

- thermoplastics

- acetylene terminated quinoxalines

- unsaturated resins

- intra-molecular cyclization systems

REFERENCES

1. Browning, C. E., *Polymer Engineering and Science*, Vol. 18,
    No. 1, p. 16, January 1978.

2. Browning, C. E., Husman, G. E., and Whitney, J. M., *Composite
    Materials: Testing and Design (Fourth Conference)*, *ASTM STP
    617*, American Society for Testing and Materials, p. 481, 1977.

3. Rowe, E. H., Siebert, A. R., and Drake, R. S., *Modern Plastics*,
    Vol. 49, p. 110, 1970.

4. Sultan, J. N. and McGarry, F. J., *Polymer Engineering and
    Science*, Vol. 13, p. 29, 1973.

5. St. Clair, A. K. and St. Clair, T. L., *Proceedings of 12th
    National SAMPE Technical Conference*, Vol. 12, p. 729, 1980.

6. Eddy, S. R., et al, *Adhesives Age*, p. 18, February 1980.

7. Hedberg, F. L. and Arnold, F. E., *Journal of Polymer Science*,
    Vol. 14, p. 2607, 1976.

A STRESS TRANSFER MODEL FOR THE DEFORMATION AND

FAILURE OF POLYMERIC MATRICES UNDER SWELLING CONDITIONS

Daniel Cohn* and Gad Marom

Casali Institute of Applied Chemistry
School of Applied Science and Technology
The Hebrew University of Jerusalem
Jerusalem 91904   ISRAEL

*Presently at: Dept. of Chemical Engineering
Benson Hall, University of Washington
Seattle, WA 98195   U.S.A.

ABSTRACT

    Polymers in which the diffusion mechanism was characterized by
a sharp advancing boundary between the swollen shell and the core,
showed a highly anisotropic swelling response. The anisotropy of the
swelling strains was caused by the mechanical constraints exerted
mutually by the two regions of the specimen. The swelling stresses
developing during the process eventually led to fracture of the
polymer specimen. An analytical model which explained the modes of
failure of the polymer under the swelling stresses, was developed.
The proposed approach was based on the general analogy existing
between the studied swollen specimens and composite materials. A
model for the prediction of the anisotropic hygroelastic response
of the swollen systems was also proposed.

INTRODUCTION

    In the last years polymers and composite materials are being
used more and more as structural materials in various fields, and are
a preferred substitute for conventional materials. Since their use
almost invariably entails contact with liquids (either aqueous or
organic) their behavior under such conditions must be of prime con-

245

sideration as a limiting factor. The reasons are that these materials may undergo various chemical degradation processes, may change
their dimensions significantly, or change their mechanical properties
by loss of strength and rigidity.

The interactions between polymeric materials and liquid and
vapor environments have been widely studied over the past twenty
years. Despite the broadness of the subject and the numerous possible
variations, most of the studies can be classified into two conceptual
approaches[1]. The first focuses on the mechanism of liquid penetration, where emphasis is put on questions related to the transport
kinetics and diffusion mechanism[2]. The second approach is concerned
with the interaction existing between the solvent and the polymer,
and its results, such as chemical degradation, polymer leaching,
cracking[3] and crazing[4-6].

The present study focuses on an additional approach, based on
the dimensional behavior of the polymeric material exposed to liquid
penetration. Although this issue is closely related to the other
two, different questions now receive priority. The guiding concept
employed in our studies of dimensional changes due to swelling is to
develop analogy between this phenomenon and thermoelasticity. In
thermoelasticity, deformation (strain) is determined by a temperature
change, ΔT. Correspondingly, hygroelasticity, i.e., the dimensional
response (mostly elastic) of polymer matrices as a result of liquid
penetration, is treated by examining the relative dimensional change
in relation to the relative volume of penetrated liquid. This
approach results in the following definition[7] of the coefficient of
hygroelasticity:

$$\mu = \frac{\Delta L/L_o}{\Delta V^*/V_o} \tag{1}$$

where $\Delta L/L_o$ is the relative length change, ΔV^* is the volume of the
diffused liquid and V_o is the initial volume of the polymer.

In general, diffusion behavior has been categorized according
to the relative rates of mobilities of the liquid and of the polymer
segments[8]. A liquid mobility, either smaller than the rate of segmental relaxation or comparable to it, results in normal Fickian and
anomalous behavior, respectively. A liquid mobility greater than
segmental relaxation rates, is distinguished by a Case II mechanism,
characterized by a sharp advancing boundary between the swollen shell
and the unswollen core.

This study examines the behavior of polymers in which the diffusion mechanism creates a sharp penetrating front, this feature being
of primary importance since it dictates the dimensional response and

modes of failure of the swollen polymer. The road has been paved
for this study by previous ones dealing with the significance of
the coefficient of hygroelasticity and its physical implications[7],
the effect of the polymer morphology[9] and the effect of reinforcement[10-11].

The two fundamental issues which determine the performance of
a polymer in service conditions which expose it to a liquid environment, are its dimensional stability and mechanical resistance. In
view of this, and in conjunction with our experimental findings,
analytical tools were developed, which enabled us to predict both
the hygroelastic response of the polymer and its modes of failure.
The analysis is based on a shell-core stress transfer model.

EXPERIMENTAL

The experimental work was carried out with the following resin-
liquid system: Epoxy resin (Araldite MY 750/HT 972) methylene
chloride. Twelve specimens, having six different geometries (see
Table I) were taken out periodically for weighing and dilatation
measurements. The recordings were carried out at 23°C and each
specimen was cooled down in methylene chloride before readings were
taken. ΔL was measured by placing each specimen in a stainless steel
fixture attached to a dial gauge capable of measuring to half a
micron. The maximum temperature variation allowed during ΔL measurements was ± 0.5°C, and this resulted in ΔL change of less than about
± 1 micron.

The dimensional changes due to the solvent penetration were
measured for each specimen along the three axes, x, y, z (width,

Table I. Theoretical and Experimental Directional Coefficients
of Hygroelasticity for Different Epoxy Specimens Swollen
by Methylene Chloride at 35°C

μ_z		μ_x		μ_y		Specimens geometry (cm^3)
theor.	exp.	theor.	exp.	theor.	exp.	
0,665	0,670	0,299	0,290	0,060	0,064	10,0x1,9x0,9
0,798	0,782	0,161	0,170	0,032	0,040	10,0x1,9x0,4
0,861	0,845	0,086	0,097	0,017	0,022	10,0x1,9x0,2
0,600	0,583	0,268	0,266	0,134	0,148	4,0x1,9x0,9
0,769	0,762	0,154	0,147	0,077	0,091	4,0x1,9x0,4
0,834	0,820	0,083	0,089	0,042	0,048	4,0x1,9x0,2

length, and thickness, respectively). The three relative dimensional
changes were then plotted against $\Delta V^*/V_O$ and the corresponding coefficients of hygroelasticity, μ_x, μ_y, and μ_z, were worked out from the
initial slope of the resulting curves.

RESULTS AND DISCUSSION

     Fig. 1 presents the swelling data of an epoxy specimen (10.0 x
1.9 x 0.4 cm^3) by methylene chloride, as expressed by the relative
dimensional changes along the x, y, and z directions versus the
liquid immersion period. By plotting the swelling strains versus
the solvent take-up as expressed by $\Delta V^*/V_O$, the directional coefficients of hygroelasticity are derived. Fig. 2 contains the experi-
mental results of $\Delta L/L_O$ along the three axes versus $\Delta V^*/V_O$ for another
epoxy specimen (10.0 x 1.9 x 0.9 cm^3) immersed in methylene chloride
for more than 1200 hours. The response pattern of the different
specimens is generally similar indicating that the specimen geometry

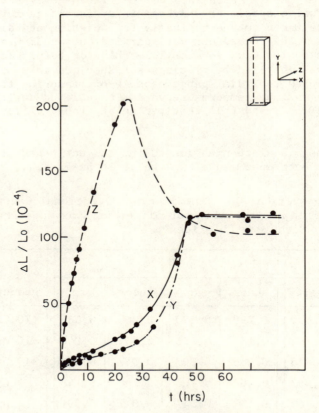

Fig. 1. Swelling data of an epoxy specimen (10.0 x 1.9 x 0.4 cm^3)
         by methylene chloride at 35°C

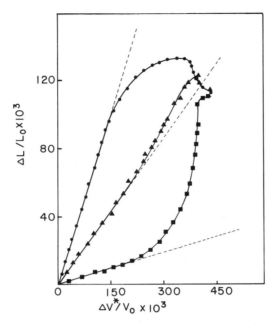

Fig. 2. Plots of the swelling strains versus $\Delta V^*/V_0$ for an epoxy
resin ($10.0 \times 1.9 \times 0.9$ cm^3) immersed in methylene chloride
at $35°C$

■ = y axis   ▲ = x axis   ● = z axis

affects the behavior only quantitatively. Thus, the data presented
in Figs. 1 and 2 are viewed as exemplifying the essential features
of the hygroelastic response of the polymer under the swelling conditions investigated.

The complicated hygroelastic behavior obtained was characterized
by the following elements: (a) The system showed a highly anisotropic
dimensional response during the first stages of the absorption process, with very small swelling strains along the long axes of the
specimen, and very large strains along the specimen's thickness; (b)
With the continuation of the process a drastic change in the hygroelastic behavior of the specimen occurred resulting in a rearrangement of the stresses; and (c) An approximately isotropic behavior,

with equal directional strains along the three axes, was observed
after the sharp change.

A previous work[7] showed that the upper bound for the coefficient
of hygroelasticity in an isotropic system is 0.33. However, the data
plotted in Fig. 2 yields the three directional coefficients of
$\mu_x = 0.290$, $\mu_y = 0.064$, and $\mu_z = 0.670$. The anisotropy of the system
is obvious and it is important to note that μ_z is significantly
larger than the upper bound found for μ in isotropic systems.

The solvent penetration into the polymer is characterized by a
sharp advancing front which separates the inner glassy core, essentially free of penetrant, from the swollen rubbery outer shell. Such
a system is viewed in this study, as analogous to a composite material
comprising a stiff "fibre" embedded in a soft matrix. The size and
relative dimensions of the core are determined by the initial geometry
of the specimen, and will vary with the absorption process. It is
maintained that the anisotropy of the swelling strains is caused by
the existence of the two regions in the specimen, the shell and the
core, and the mechanical constraints imposed by the core on the shell.
These mechanical constraints are transferred across the surface of the
core, defined here, as the interfacial surface. Due to this, it is
clear that the size and dimensions of the core are of primary importance in determining the anisotropic response of the specimen. The
fundamental phenomenon causing the special behavior of the swollen
system is the appearance of tensile and shear stresses at the interfacial surface, between the swollen expanding shell and the glassy
stiff core. Since these stresses depend on the geometry of the core
they are different in the three directions and they vary with the
swelling period. In an attempt to establish the mechanism, the geometrical factors affecting this behavior were investigated. In light
of our findings it is clear that the extent of the restraint along a
certain axis will increase, implying smaller swelling strains, as
the size of the core in that direction increases. This is so due to
the fact that the mechanical restraint imposed by the core on the
swollen shell along a certain axis, is a function of both the absolute
and the relative dimensions of the core along that specific direction.
Thus, the strains along the y axis are expected to be the smallest.
This significant longitudinal restraint of the expansion of the
swollen network, "squeezes out" the swollen polymer matrix along the
z axis, producing especially large strains along this axis.

As the diffusion process continues the geometry of the core
varies, it becomes smaller and the ratios of its dimensions change.
The stresses applied on the core by the swollen shell increase
steadily, and concomitantly, the extent of the mechanical restraint
imposed by the core on the strains of the swollen shell, decreases.
Eventually, a stage will be reached at which the core will attain a
critical geometry and will be too small to affect the dimensional
response of the specimen significantly. At this point the strains

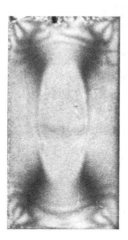

Fig. 3.   The advancing boundary between the shell and the core and
          the stresses developed

of the specimen rearrange, and a significant relaxation of its
stresses is observed. The swelling process will continue in a
mostly anisotropic way until its completion.

    As the swelling proceeds, the sharp boundary advances steadily
and the size of the core reduces. Consequently, the surface to
volume ratio of the core increases or, in a simpler two-dimensional
approach, the perimeter to cross-section ratio, P/A, increases.

    The tensile and shear stresses imposed by the swollen shell on
the core through the interfacial surface may, eventually, lead to
specimen failure. Fig. 3 shows the advancing boundary between the
shell and the core and the stresses developed. Figs. 4 and 5 show
two different specimens after prolonged swelling periods where two
types of cracks are evident: (a) boundary cracks, formed at the

Fig. 4.   An epoxy specimen (10.0 x 1.9 x 0.9 cm^3) after a prolonged
          exposure to methylene chloride

Fig. 5.   An epoxy specimen (5.0 x 3.0 x 0.5 cm^3) after a prolonged
          exposure to methylene chloride

advancing boundary, and (b) core cracks, formed within the brittle
core. It is important to note that all the cracks are in the x-z
plane, resulting in a symmetrical array along the y axis. Evidently,
the dimensional constraint imposed by the core on the shell is relaxed
when a crack is formed (see Fig. 6).

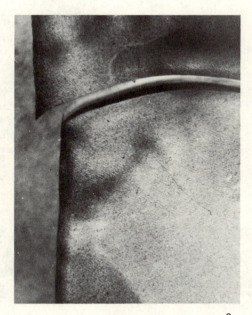

Fig. 6.   An epoxy specimen (10.0 x 1.9 x 0.4 cm^3) showing a crack
          developed as a result of the immersion in methylene chloride

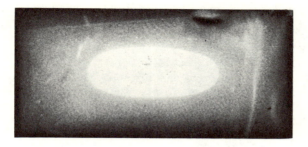

Fig. 7. The cross-section of a partially swollen epoxy specimen,
along the x-z plane

A SHELL-CORE STRESS TRANSFER MODEL

An analytical model which explains the modes of failure and the
symmetrical arrangement of the cracks in the specimen, is proposed.
Our appro-ch is based on the general analogy existing between a
swollen specimen and a composite material reinforced by a single
fiber. Only two assumptions were made: (a) the core has a truncated ellipsoid shape (see Fig. 3), having an elliptical cross-
section (see Fig. 7), and (b) the maximum shear stress is obtained
at the edge of the core along the y axis and then decreases towards
the center (see Fig. 8). The model refers to a sample having at a
given time t the following dimensions: X_o, Y_o, and Z_o, X, Y, and Z

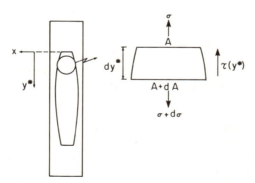

Fig. 8. A two-dimensional schematic description of a partially
swollen specimen

are the three axes of the ellipsoid from which the truncated one is derived. Since a 20% truncation is assumed, the actual dimensions of the core are 0.8Y, X and Z. The shear stress at the boundary is given by Eq. 2:

$$\tau(y*) = \tau_{max} - By* \tag{2}$$

where y* is the distance from the edge of the core along the y axis and $\tau(y*)$ is the shear stress at the interfacial surface. The stress distribution along the length of the core is understood by considering the equilibrium of a small element of the core, as shown in Eq. 3:

$$A(y*)\sigma(y*) + \tau(y*)P(y*)dy* = [A(y*) + dA][\sigma(y*) + d\sigma] \tag{3}$$

where $A(y*)$ is the cross-sectional area of the core, $P(y*)$ is the perimeter of the core and $\sigma(y*)$ is the tensile stress in the core. Note the analogy to the stress transfer mechanism considered in the analysis of the fibre critical length in fibre-reinforced composites[12].

The above differential equation (Eq. 3) was solved and an expression (Eq. 4) based on the varying geometrical parameters of the core was obtained. It describes the variation of the tensile stress along the core, as a function of the distance y* from its edge:

$$\sigma(y*) = \frac{\sqrt{2}\, Y(X^2 + Z^2)^{1/2}}{XZW}\left[(W)^{1/2}\left[(0,5y* - 0,2Y)\tau_{max} - \right.\right.$$

$$\left. - B[0,1Y(2y* - 0,8Y) - 0,33W]\right] + Y^2\left[\arcsin\left[\frac{-2y* + 0,8Y}{Y}\right]\right.$$

$$\left.\left.(0,05YB - 0,13\,\tau_{max}) + C(o)\right.\right] \tag{4}$$

where $W = -y*^2 + 0.8Yy* + 0.09Y^2$.

This model was examined with six different epoxy specimens swollen by methylene chloride at 35°C, and the experimental results were in close agreement with our theoretical predictions. Fig. 9 is a graphical presentation of the tensile stress in the core, as a function of the distance from its edge, for an epoxy specimen (10.0 x 1.9 x 0.9 cm^3), as obtained by Eq. 4. Experimentally, the first core crack appeared after 970 hours of swelling, at a distance of 1.2 cm from the edge. The maximum core stress and the coordinate of its location as a function of the swelling time, are presented graphically in Fig. 10. Cracking will occur whenever $(\sigma_{core})_{max} = \sigma_F$, where σ_F is the fracture stress of the dry polymer. Since the tensile strength of dry epoxy resin is in the range of 80–85 MPa, it is seen that our experimental findings are in very good agreement with our theoretical predictions.

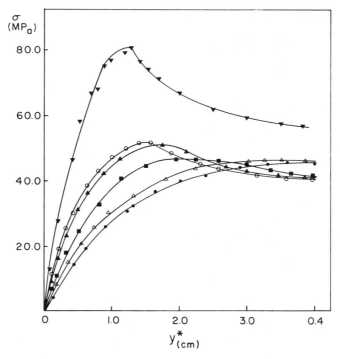

Fig. 9. The tensile stress in the core as a function of the distance
         of its edge for an epoxy specimen (10.0 x 1.9 x 0.9 cm^3)
         as obtained by Eq. 4

●	△	■	▲	0	▼	
100	250	550	750	850	960	hours

Fig. 11 shows a plot of the perimeter to cross-sectional area
ratio (P/A) versus the swelling time. It can be seen that the proportions of the core undergo, at a particular moment (for this specimen after 930 hours), a sharp transition. The P/A ratio is directly
related to the efficiency of the stress transfer mechanism acting
between the swollen shell and the core. The shear stresses at the
interface are transferred to the core, building up tensile stresses
in it. As a result, the bigger the P/A ratio is, the higher is the
tensile stress at the core at a given distance from its edge. It is
important to note that the drastic increase in the tensile stresses
at the core seen in Figs. 9 and 10, leading ultimately to its fracture, occurs simultaneously with the sharp growth in the P/A ratio
(see Fig. 11). This emphasizes, once again, the primary importance
of the geometrical parameters in determining whether or not the
specimen is going to undergo cracking.

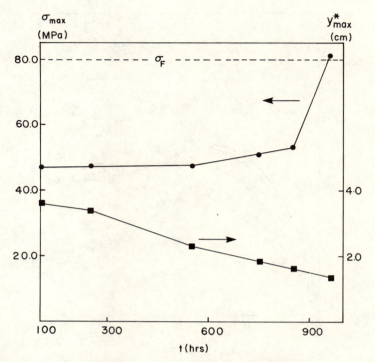

Fig. 10. The maximum core stress and the coordinate of its location
          as a function of the swelling time for an epoxy specimen
          (10.0 x 1.9 x 0.9 cm³) immersed in methylene chloride
          ● – σ_{max} ■ – y^*_{max}

    The formation of the cracks in the x-z plane along the long
axis, y, and their symmetrical arrangement are also explained by the
proposed model. The first is a direct consequence of the fact that
higher shear stresses develop along the longest axis, and the second
is a result of the symmetry of the problem whereby σ_{max} are located
symmetrically at the same distance from both core edges. Thus, it
is possible to predict if and when a specimen will fail, based on
the theoretical model proposed and the kinetics of the diffusion
process.

ANISOTROPIC HYGROELASTICITY

    In the second phase of this study a model for the prediction of
the anisotropic hygroelastic response of swollen specimens, based on
the directional coefficients of hygroelasticity was developed. The
analytical tools proposed were based on the relation existing between

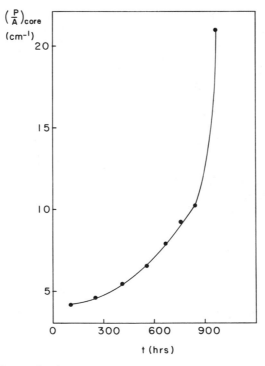

Fig. 11.   A plot of the perimeter to cross-sectional area ratio
           (P/A) versus the swelling time for an epoxy specimen
           (10.0 x 1.9 x 0.9 cm^3) immersed in methylene chloride

the anisotropy of the hygroelastic behavior of the polymer and the
deviation of the specimen geometry from a cube.

    Since the anisotropy of the swelling strains is caused by the
mechanical constraints exerted mutually by the two regions of the
specimen, the plane across which these constraints are transferred
is of primary importance. Hence, it is clear that the size and
dimensions of the interface are essential in determining the anisotropic response of the specimen.

    The proposed model regards the geometry of the specimen as being
that of the core at time zero. In addition, it assumes that as the
swelling proceeds the core will maintain the original relative dimensions of the specimen. This will be so until the core geometry, and
more specifically, its P/A ratio, will undergo a sharp change, coin-
ciding with the limit of the applicability of the proposed model.

In light of our experimental findings, it is clear that a cubic specimen will show an isotropic hygroelastic behavior. Because the anisotropy of the hygroelastic response of the polymer is directly linked to its geometry, the proposed analysis considers the ratios of the initial dimensions of the specimen and those of a cube with an identical surface area, regarded as an "equivalent cube."

The anisotropic swelling is studied in terms of the directional coefficients of hygroelasticity, μ_i ($i = x, y, z$). The bulk coefficient of hygroelasticity, μ_v, is defined as $\mu_v = \Sigma \mu_i$ ($i = x, z, y$). The specimen's dimensions are taken as X, Y and Z. The deviation ratios, Q_J, were defined as the ratios between the side lengths of the equivalent cube and the length of each of the specimen axes. On this basis, a simple expression for the directional coefficients of hygroelasticity was developed:

$$\mu_i = \frac{\mu_v}{J\Sigma(1/J)} \qquad \begin{array}{l} (i = x, y, z) \\ (J = X, Y, Z) \end{array} \qquad (5)$$

where J are the initial dimensions of the specimen.

This equation was examined on six different epoxy specimens and the experimental results were found to be in excellent agreement with theoretical predictions (see Table I).

CONCLUSIONS

The main conclusions of the present study are summarized as follows:

1) Swelling systems in which the diffusion mechanism is characterized by an advancing front show an anisotropic hygroelastic behavior;

2) The size of the core and its relative dimensions determine the directional swelling strains of the polymer specimen;

3) Rearrangement of the specimen strains and a significant relaxation of its stresses are observed when a critical core geometry is reached;

4) The cracking under swelling stresses is explained in terms of the general analogy between the swollen systems and fibre-reinforced composite materials. On this basis, a stress transfer model is developed and successfully tested; and

5)   A simple expression for the directional coefficients of hygro-
      elasticity, based on the initial geometry of the specimen, is
      proposed, and is in very good agreement with the experimental
      results.

REFERENCES

1.   D. Cohn and G. Marom: Plastics and Rubber: Materials and
     Applications, 5(4), 165 (1980).
2.   H. B. Hopfenberg, L. Nicolais, and E. Drioli: Polymer, 17,
     195 (1976).
3.   D. J. Steel: Trans. J. Plast. Inst., 35, 429 (1967).
4.   E. H. Andrews and L. Bevan: Polymer, 13, 337 (1972).
5.   E. J. Kramer, E. G. Krenz and D. G. Ast: J. Mater. Sci., 13,
     1093 (1978).
6.   K. Iisaka, Y. Nishimoto and K. Shibayama: J. Polym. Sci. Polym.
     Phys. Ed., 17, 791 (1979).
7.   D. Cohn and G. Marom: Polym. Eng. Sci., 18, 1001 (1978).
8.   N. L. Thomas and A. H. Windle: Polymer, 17, 195 (1976).
9.   D. Cohn and G. Marom: Polymer, 20, 501 (1979).
10. G. Marom: Polym. Eng. Sci., 17, 799 (1977).
11. G. Marom and D. Cohn: J. Mater. Sci., 15, 631 (1980).
12. B. D. Agarwall and L. J. Broutman: "Analysis and Performance
     of Fibre Composites," John Wiley and Sons, New York (1980).

CONSTITUTIVE RELATIONSHIPS FOR SHEET MOLDING MATERIALS

R. L. McCullough, G. J. Jarzebski, and S. H. McGee

Department of Chemical Engineering and
Center for Composite Materials
University of Delaware
Newark, Delaware 19711

ABSTRACT

Constitutive relationships (or "Combining Rules") are described to predict the Young's moduli (E_1 , E_2 , E_3), shear moduli (G_{12} , G_{13} , G_{23}), Poisson's ratios (ν_{12} , ν_{13} , ν_{23}), and coefficients of thermal expansion (α_1 , α_2 , α_3) for two-component (fiber/resin) and three-component (fiber/filler/resin) sheet molding materials. Two important features of these relationships are the incorporation of (i) a quantitative orientation parameter to characterize states of fiber orientation intermediate between random and collimated and (ii) an effective aspect ratio to account for the affect of filament aggregation and "swirling." The model predictions are shown to agree, within experimental error, with characterizations for SMC-65 (47%, by volume, E-glass fiber; 53% Polyester resin); SMC-25 (18%, by volume, E-glass; fiber, 32% Calcium Carbonate filler; 50% Polyester resin) and SMC-R50 (35%, by volume, E-glass fiber; 11% Calcium Carbonate filler; 54% Polyester resin).

INTRODUCTION

Demands for weight saving have focused attention on reinforced polymeric materials which can be molded into rigid structural elements. Unlike the traditional materials, these heterogeneous materials can exhibit a wide range of properties which are dependent on composition as well as the internal microstructure developed during fabrication. Sensitivity to processing is manifest in both the magnitude and directional dependence of the mechanical, thermal and transport properties.

The wide range of possible compositions, fiber length distributions, and fiber orientations that may be utilized, as well as generated during the manufacture of the composite, precludes a total reliance on direct laboratory characterizations of these anisotropic materials. Consequently, constitutive relationships, or "combining rules," which relate the composition and processing dependent micro-structure to thermomechanical performance are important components of the technology for these materials.

Numerical methods could be used to predict the behavior of heterogeneous materials if a complete point-by-point description of the microstructure was available. It is unrealistic to expect that such detailed knowledge will be available. Indeed, the labor and expense involved in generating detailed structural information for a given specimen could exceed the effort to obtain a direct characterization of the thermoelastic properties. Consequently, an essential component for the development of models for these materials is the identification of appropriate structural descriptors which are susceptible to experimental evaluation. It can be anticipated that such structural descriptors can serve as design variables which may be subject to control by the proper selection of processing conditions.

In the following sections, relationships are described which are capable of predicting the small deformation elastic response behavior of composite materials as characterized by

- The longitudinal ("1"), transverse ("2"), and through-the-thickness ("3") Young's moduli: E_1 , E_2 , E_3 .
- The shear moduli: G_{12} , G_{13} , G_{23} .
- The Poisson's ratios: ν_{12} , ν_{13} , ν_{23} .
- The coefficients of Thermal Expansion: α_1 , α_2 , α_3 .

in terms of

- The properties of the resin and reinforcing agents;
- The volume fraction composition of the components;
- An orientation parameter, "f," which characterizes the average state of orientation of the reinforcing agents;
- An effective aspect ratio, "a_e ," which reflects the role of the average shape of the effective reinforcing agent in determining load transfer from the compliant matrix.

These structural parameters are described in the following section and subsequently incorporated into model relationships to predict thermoelastic properties.

STRUCTURAL DESCRIPTORS

The orientation parameter, "f," and the effective aspect ratio,

"a_e ," have emerged as convenient quantitative descriptors of the microstructure of composite materials. Both of these parameters may be evaluated from observations of micrographs. The orientation parameter, "f," may be obtained from a knowledge of the orientation distribution and is scaled such that f = 0 represents a random orientation distribution, while f = 1 corresponds to perfect alignment. Intermediate values of f are associated with varying degrees of alignment. The effective aspect ratio, "a_e , " can be viewed as the ratio of the length to diameter of an effective cylindrical (or ellipsoidal) reinforcing agent. Accordingly, a continuous filament corresponds to $a_e \to \infty$; spherical particles correspond to $a_e = 1$; short-fiber reinforcements are represented by intermediate values of a_e .

As illustrated in Figure 1, various types of composite materials may be distinguished as special cases of the structural features described by these parameters. This comparison points out the simple microstructure of continuous fiber laminates. For this special material system, the continuity ($a_e \to \infty$) of the perfectly collimated (f = 1) fibers assures the simplifying condition that the strain field parallel to the aligned fibers is essentially uniform. As a consequence of this special condition, the longitudinal properties may be predicted from the simple "rule of mixtures"; however, the variation of the stress and strain fields transverse to the fiber direction must be taken into account in order to obtain relationshisp to predict the transverse properties and shear moduli.

Particulate filled systems represent the next level of complexity. In these systems, the reinforcing agents are spherical (or near spherical) so that the effective aspect ratio is unity. The marked discontinuity of particulate filled systems introduces significant fluctuations in the internal stress-strain fields which complicate the analysis of thermomechanical behavior. On the other hand, the simple spherical reinforcing geometry precludes a dependence on the orientation of the reinforcing agent so that consideration of this structural feature is not required in the analysis of these systems.

Short-fiber reinforced materials represent yet a higher level of complexity since variations in both the aspect ratio and fiber orientation must be taken into consideration. As shown in Figure 1, the short-fiber systems can be further distinguished according to the microstructural features induced by fabrication procedures. Sheet molding materials are produced by dropping chopped "rovings" (of approximately 1 to 3 inches in length) on a moving belt with the subsequent application of the resin (or resin/filler) component. Under these manufacturing conditions, the rovings are confined to lie in the plane of the moving belt. Although the intent of this process is to produce a random distribution of fibers within the plane of the sheet material, the motion of the belt with respect

TYPE	SYMMETRY	MICROSTRUCTURE

Continuous Fibers

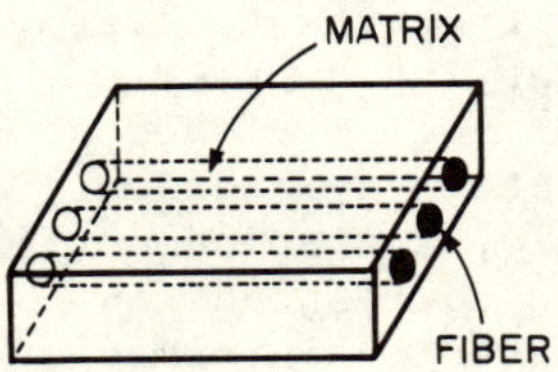

Orthotropic

Transversely
  Isotropic

Aspect Ratio: $a_e \to \infty$

Orientation: Aligned
               (f=1)

Particles

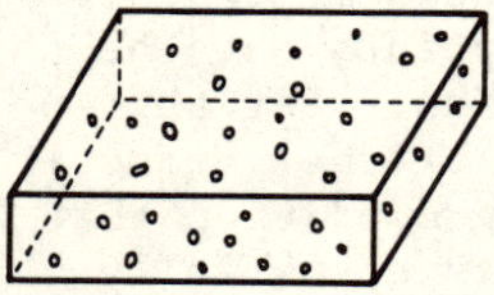

Isotropic

Aspect Ratio: $a_e=1$

Orientation:
   Independent

Short Fibers

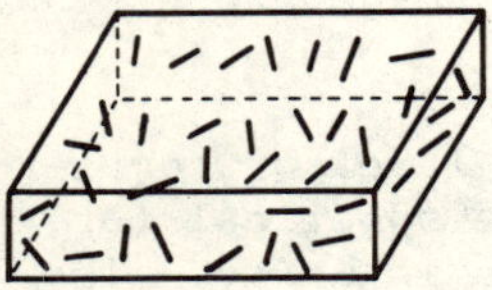

Orthotropic

Transversely
  Isotropic

$1 <$ Aspect Ratio $< \infty$
        (a_e)

Planar Random (f=0)
   $\leq$ Orientation (f)
   $\leq$ Aligned (f=1)

**Injection/Transfer
Molded Materials**

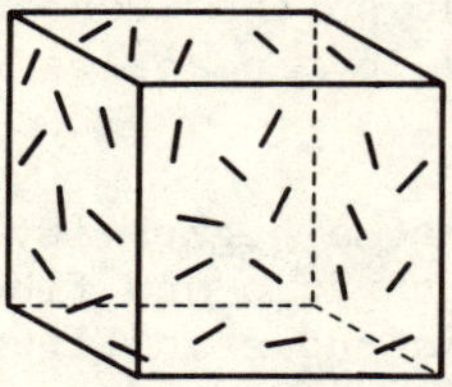

Orthotropic

Transversely
  Isotropic

Isotropic

$1 <$ Aspect Ratio $< \infty$
        (a_e)

Random (f=0)
   $\leq$ Orientation (f)
   $\leq$ Aligned (f=1)

Fig. 1.   Schematic definition of the structural features that
distinguish various classes of composite materials.

to the falling strands may induce partial orientation. Since the
fibers are confined to lie within a plane, the state of orientation
must be described by a special two-dimensional, planar distribution.

An additional structural feature of sheet molding materials is
the tendency for the integrity of the rovings to be preserved so
that aggregate bundles of filaments may be distinguished, as shown
in the micrograph of Figure 2. Such aggregates of close-packed
filaments tend to behave as an integral unit so that the aggregate
bundle diameter is several orders of magnitude greater than the
filament diameter, the effective aspect ratio is considerably
smaller than the "ideal" aspect ratio obtained as the ratio of the
filament length to filament diameter.

Filament aggregation is reduced in materials produced by injection and transfer molding procedures. The milling actions associated with these processing methods tend to produce highly dispersed filaments. However, these processing methods require fibers
of substantially shorter initial lengths (1/8 to 1/4 inch) which may
become fragmented during processing to produce even shorter filaments with significantly reduced aspect ratios. A further important distinction between sheet molding materials and injection/
transfer molded materials is due to the nature of the orientation
distribution of the fibers. The flow fields within the mold cavity
may carry the short fibers into a three-dimensional orientation.
This three-dimensional character of the fiber orientation causes
injection/transfer molded materials to exhibit properties which
differ from those of sheet molding materials with restrcited two-
dimensional fiber orientation. It can be anticipated that the
"out-of-plane"fibers of injection/transfer molded materials will
enhance the "through-the-thickness" properties as compared to sheet
molded materials. However, the utilization of some of the fibers
for out-of-plane reinforcement reduces the number of fibers contributing to the in-plane reinforcement so that the improvements in
through-the-thickness properties are achieved at the expense of the
longitudinal and transverse properties.

These complex and variable features of the microstructure point
out the futility of attempting detailed numerical analyses. Similarly, a rigorous theoretical treatment of the classical boundary
value problems of elasticity theory appears to be hopeless. Bounding approaches present the advantage that the bounds are always
valid; unfortunately, even improved bounds are too far apart to
serve as meaningful brackets on the behavior of these materials.
As a consequence, resort must be made to model formulations, guided
by structural observations, which attempt to capture the important

Fig. 2.   Photomicrograph of a polished surface of SMC–25 illustra-
ting filament aggregation. The composition of SMC–25 is
given in Table 5.

aspects of the role of the microstructure. An "Aggregate Model"
is described in the following section which serves to incorporate
the structural descriptors f and a_e into model relationships
capable of predicting thermoelastic behavior of sheet molding
materials.

AGGREGATE MODEL

The central feature of this model is the partitioning of complex materials into subregions, or grains, of characteristic material. Under this approach, a composite can be viewed as an aggregate of these regions, as illustrated in Figure 3. As in the case
of grainy metals, each equivalent grain is treated as an apparent
homogeneous (but anisotropic) material which may be described by
an array of elastic constants, [C] (or compliance constants, [S])
and a coefficient of thermal expansion vector $\underline{\alpha}$ (e.g., α_1 , α_2 , α_3).
The relationship of the elastic constants, $\overline{[C]}$, and compliance
constants, [S] (with $[C] = [S]^{-1}$) to the traditional Engineering
constants (Young's moduli, E; shear moduli, G; and Poisson's ratios,
ν) are summarized in Table 1.

The observation that filaments agglomerate into coherent bundles, as illustrated in Figure 2, provides guidance for the specification of the typical grain. These bundles can be identified as
the domain in which the filaments are all aligned parallel to an
axis that makes some angle, ϕ, with respect to an external body
axis (e.g., the machine direction of the moving belt). These
agglomerates tend to behave as highly correlated, integral units.

The material shown in Figure 2 contains low concentrations of
fibers which agglomerate into relatively long bundles. Similar
agglomerates can be identified in material with higher concentrations of fibers as shown by the micrograph in Figure 4 for SMC-65.
In this case, the contact and compaction of the rovings produce a
marked "swirling." As a consequence of this swirling, the dimensions of the domains over which a local collection of parallel
filaments can be identified are substantially reduced. Representative domains characterized by local collections of parallel filaments are indicated in Figure 4. The general shape of the domain
may be approximated as ellipsoidal. In this case, the aspect
ratio of some of the ellipsoidal domains has been reduced to near
spherical geometry ($a_e \sim 2$). It should be noted that the apparent
discontinuity of the domains of Figure 4 is an artifact resulting
from polishing the specimen. The domains are connected by the
filaments of the initial roving that weave through each of the
domains. In this case, the partitioning into domains of parallel
filaments reflects tangential loading acting on a section of the
swirled roving.

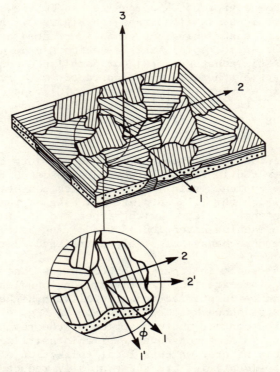

Fig. 3. Schematic definition of the aggregate model
         for sheet molding materials

Table 1. Summary of Relationships Between Various
Forms of Materials Descriptors

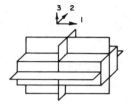

Elastic Constants

$$\begin{pmatrix} \sigma_1 \\ \sigma_2 \\ \sigma_3 \\ \tau_{23} \\ \tau_{13} \\ \tau_{12} \end{pmatrix} = \begin{pmatrix} C_{11} & C_{12} & C_{13} & 0 & 0 & 0 \\ C_{12} & C_{22} & C_{23} & 0 & 0 & 0 \\ C_{13} & C_{23} & C_{33} & 0 & 0 & 0 \\ 0 & 0 & 0 & C_{44} & 0 & 0 \\ 0 & 0 & 0 & 0 & C_{55} & 0 \\ 0 & 0 & 0 & 0 & 0 & C_{66} \end{pmatrix} \begin{pmatrix} \epsilon_1 \\ \epsilon_2 \\ \epsilon_3 \\ \gamma_{23} \\ \gamma_{13} \\ \gamma_{12} \end{pmatrix}$$

Compliance Constants

$$\begin{pmatrix} \epsilon_1 \\ \epsilon_2 \\ \epsilon_3 \\ \gamma_{23} \\ \gamma_{13} \\ \gamma_{12} \end{pmatrix} = \begin{pmatrix} S_{11} & S_{12} & S_{13} & 0 & 0 & 0 \\ S_{12} & S_{22} & S_{23} & 0 & 0 & 0 \\ S_{13} & S_{23} & S_{33} & 0 & 0 & 0 \\ 0 & 0 & 0 & S_{44} & 0 & 0 \\ 0 & 0 & 0 & 0 & S_{55} & 0 \\ 0 & 0 & 0 & 0 & 0 & S_{66} \end{pmatrix} \begin{pmatrix} \sigma_1 \\ \sigma_2 \\ \sigma_3 \\ \tau_{23} \\ \tau_{13} \\ \tau_{12} \end{pmatrix}$$

Engineering Constants

$$S_{11} = 1/E_1 \qquad S_{22} = 1/E_2 \qquad S_{33} = 1/E_3$$
$$S_{12} = -\nu_{12}/E_1 \qquad S_{13} = -\nu_{13}/E_1 \qquad S_{23} = -\nu_{23}/E_2$$
$$S_{44} = 1/G_{23} \qquad S_{55} = 1/G_{13} \qquad S_{66} = 1/G_{12}$$

Fig. 4.   Photomicrograph of a polished surface of SMC-65 illustra-
ting the effects of filament swirling in reducing the
effective aspect ratio. Domains of locally parallel filaments are indicated as ellipsoidal regions. The composition is given in Table 5.

Under this abstraction, sheet molding materials may be viewed as an assemblage of ellipsoidal inclusions. The aspect ratio of the inclusion, a_e , corresponds to the dimensions of the domain of locally parallel filaments rather than the ideal aspect ratio of the filament. These effective aspect ratios are determined by the extent of agglomeration and swirling of the filaments.

The utilization of the Aggregate Model reduces the analysis to that of a one-component grainy material in which the grains assume a variety of orientations with respect to the external axes of the specimen. The notion of a characteristic grain conveniently separates orientation effects from the influence of composition and reinforcing geometry as reflected in the specification of the typical grain.

Orientation Dependence

The load-deformation characteristics of the bulk specimen are dependent upon the relative orientation of the material axes of the grains with respect to those of the specimen. The external "1" axis of the specimen (shown in Figure 3) is taken as the machine direction along which the filaments in each grain are aligned. As illustrated in Figure 3, the angle ϕ is the angle within the "1-2" plane between the "1" and "1'" axes. The contribution of an individual grain (oriented at angle ϕ) to the behavior of the specimen is obtained by projecting the appropriate set of properties through the angle ϕ onto the external specimen axes. The net contribution from all grains is obtained by averaging these projections over the orientation distribution of the grains.

The orientation distribution, $n(\phi)$, is given as the fraction of grains whose unique (fiber) axis, "1'," make a common angle, ϕ, with the external axis, "1." Since the grains were selected to contain locally parallel filaments, the orientation distribution of the grains is equivalent to the orientation distribution of the filaments. Thus, if all the filaments are aligned parallel to the "1" direction, $n(\phi)$ assumes the nature of a delta function; viz., $n(\phi = 0) = 1$ and $n(\phi \neq 0) = 0$. Alternately, if the filament sections are uniformly (randomly) distributed in the "1-2" plane, then $n(\phi) = n$, a constant for all values of ϕ.

As a consequence of symmetry in the distribution function (i.e., $n(\phi) = n(-\phi) = n(\pi + \phi)$), only those projections involving the quantities $\cos^2\phi$ and $\cos^4\phi$ survive the averaging process. It is useful to convert these surviving trigonometric averages into the orientation parameters f_p and g_p defined in the following fashion:

$$f_p = 2<\cos^2\phi> - 1 \tag{1a}$$

$$g_p = (8<\cos^4\phi> - 3)/5$$

with

$$<\cos^m\phi> = \int_0^{\pi/2} n(\phi)\cos^m\phi d\phi \qquad (m = 2,4) \tag{1b}$$

and

$$\int_0^{\pi/2} n(\phi)d\phi = 1$$

The subscript "p" on the orientation parameters is introduced to emphasize that the definitions of Equations (1) are specific for a two-dimensional (planar) distribution. Alternate definitions are required for three-dimensional distributions.

These orientation parameters are constructed to provide a convenient scale for characterizing the state of orientation. For $f_p = g_p = 1$, the filaments are aligned along the "1" axis. For $f_p = g_p = 0$, the filament sections are uniformly (randomly) distributed within the "1-2" plane. Values of f_p and g_p between 0 and 1 represent intermediate states of orientation.

It is possible to obtain values for these parameters from micrographs, such as those shown in Figures 2 and 4. From these direct observations, the number of filament segments, $N(\phi)$, which are oriented between $\phi - \Delta\phi$ and $\phi + \Delta\phi$ can be counted and a histogram of the orientation distribution constructed. The quantities $<\cos^2\phi>$ and $<\cos^4\phi>$ can be obtained from numerical integration of the data supplied by the histograms; viz.,

$$<\cos^m\phi> = N^{-1} \sum_j^N N(\phi_j)\cos^m\phi_j \qquad (m = 2,4) \tag{2}$$

where N is the total number of filaments so that $N(\phi)/N = n(\phi)$.

Recently, laser scattering techniques (McGee and McCullough, 1981a) have been developed which permit a less tedious evaluation of the orientation parameters.

For reasonable forms of the planar distribution (e.g., $n(\phi) = k\cos^b\phi$), the orientation parameters are related; viz,

$$g_p = 2f_p(7 - 2f_p)/[5(4 - 2f_p)] \tag{3}$$

so that only one parameter is required to describe the state of planar orientation.

The resulting averages of the projected properties can be expressed in terms of these orientation parameters and certain combinations of the properties of the grain which group to form orientation invariants. Since it is not evident at this stage of the analysis whether the compliance arrays, [S], elastic constant arrays, [C], or some intermediate combination should be projected, attention will be directed to the transformation behavior of general contracted tensorial quantities, B_j and A_{jk} so that the resulting expressions will have wider applicability.

The second rank tensors, B_{ij} , of particular interest correspond to expansional strains which exhibit the property $B_{ij} = 0$ for $i \neq j$. Contraction of the B_{jj} to B_j (j = 1,2,3) and averaging yields:

$$< B_1 > = \beta_o + \beta_f f_p$$

$$< B_2 > = \beta_o - \beta_f f_p \qquad (4a)$$

$$< B_3 > = \dot{\beta}_n$$

and

$$\beta_o = \frac{1}{2}(B_1 + B_2)$$

$$\beta_f = \frac{1}{2}(B_1 - B_2) \qquad (4b)$$

$$\beta_n = B_3$$

These relationships further illustrate the utility of the orientation parameter, f_p , in isolating the rotational invariants: e.g., "β."

The corresponding averages of the contracted second rank tensor [A] are summarized in Table 2. For [A] → [C], the relationships of Table 2 yield the orientation average of the elastic constant array; for [A] → [S], the orientation average of the compliance array is obtained. The factor B is introduced to account for the contraction of the compliance constants and elastic constants to second order tensors; for [A] → [C], B = 1; for [A] → [S], B = 4.

The remaining task is the specification of the properties B_j and A_{jk} of the typical grain to serve as input to Equation (4) and Table 2 for the computation of the corresponding averaged properties $< B_j >$ and $< A_{jk} >$.

Table 2.   Orientational Averages of Contracted Tensors for
           Planar Orientation Distributions

$$\langle A_{11} \rangle = A_{11}^{0} - \left[a_{11} + \frac{5}{8} a_{66} \right] f + \frac{5}{8} a_{66} g$$

$$\langle A_{12} \rangle = A_{12}^{0} + 4 a_{12} f - 5 a_{12} g$$

$$\langle A_{13} \rangle = A_{13}^{0} - a_{13} f$$

$$\langle A_{22} \rangle = A_{22}^{0} - \left[a_{22} + \frac{5}{8} a_{66} \right] f + \frac{5}{9} a_{66} g$$

$$\langle A_{23} \rangle = A_{23}^{0} - a_{23} f$$

$$\langle A_{33} \rangle = A_{33}^{0}$$

$$\langle A_{44} \rangle = A_{44}^{0} - a_{44} f$$

$$\langle A_{55} \rangle = A_{55}^{0} - a_{55} f$$

$$\langle A_{66} \rangle = A_{66}^{0} + 4 a_{66} f - 5 a_{66} g$$

$$a_{ij} = A_{ij}^{0} - A_{ij}$$

$A_{11}^{0} = A_{22}^{0} = k^{A} + \mu^{A}$ $\qquad\qquad k^{A} = \frac{1}{4} \left[A_{11} + A_{22} + 2 A_{12} \right]$

$A_{13}^{0} = A_{23}^{0} = \ell^{A}$ $\qquad\qquad \mu^{A} = \frac{1}{8} \left[A_{11} + A_{22} - 2 A_{12} + \frac{4}{8} A_{66} \right]$

$A_{44}^{0} = A_{55}^{0} = 8 \gamma^{A}$ $\qquad\qquad \ell^{A} = \frac{1}{2} \left[A_{13} + A_{23} \right]$

$A_{12}^{0} = k^{A} - \mu^{A}$ $\qquad\qquad \gamma^{A} = \frac{1}{28} \left[A_{44} + A_{55} \right]$

$A_{33}^{0} = n^{A}$ $\qquad\qquad n^{A} = A_{33}$

$A_{66} = 8 \mu^{A}$

Grain Models

The orientation relationships developed in the preceding treatment were obtained by projecting the appropriate response characteristic of the grains. Alternately, the determination of the effective averaged properties can be cast into an equivalent problem which corresponds to taking various directions of imposed loads (or deformations) acting on a collection of aligned grains. In this form, the primary structural element can be viewed as a micro-laminate of aligned elliptical inclusions. Since the ellipsoidal domains were identified as regions of locally aligned filaments, the micro-laminate assumes the character of an aligned collection of fibers. However, the effective aspect ratio of the fiber is given by the dimensions of the local domain of parallel filaments rather than the actual filament aspect ratio.

The behavior of an assembly of aligned inclusions (of arbitrary aspect ratio) can be developed from the general relationships, described by Wu and McCullough (1977) and summarized in Table 3 along with the extension to thermal stress coefficients and coefficients of thermal expansion (Rosen and Hashin, 1970; Mukhopadhyay, 1980). The quantities in brackets denote 6×6 matrix arrays; underlined quantities denote vectors; starred quantities refer to the predicted effective properties for the micro-laminate. The elastic constant arrays $[C_i]$, $[C_m]$, and $[C^\circ]$ refer to the inclusion, "i," the resin phase, "m," and an arbitrary reference material, "o," respectively. The term $[E^\circ]$ is similar to the "Eshelby Tensor" (Eshelby, 1957) and takes into account correlation effects. The elements of $[E^\circ]$ are dependent upon the aspect ratio of the inclusion, a_e, as well as the particular choice for the reference material represented by $[C^\circ]$. Relationships for the elements of $[E^\circ]$ in terms of these quantities are given by Wu and McCullough (1977).

The versatility of the relationships summarized in Table 3 is manifest through the quantity $[C^\circ]$. Assigning a reference material zero rigidity yields the primitive Reuss (1929) uniform stress model; assigning the reference material an infinite rigidity yields the Voigt (1910) uniform strain model. If the reference material is taken as the matrix phase ($[C^\circ] = [C_m]$), or the fiber phase ($[C^\circ] = [C_f]$), improved upper and lower bounds are obtained, respectively. Unfortunately, these improved bounds are too far apart to serve as useful brackets of material behavior. Equating the reference material to the unknown effective properties ($[C^\circ] = [C*]$) yields a "self-consistent field model."

The notion of a reference elasticity, $[C^\circ]$, was introduced to incorporate all structural information not accounted for by the parameters in the relationships of Table 3. Consequently, it is reasonable to treat $[C^\circ]$ as a lumped parameter that takes into account the unknown (or experimentally inaccessible) structural

Table 3.   Summary of Relationships for Transversely
           Isotropic Ellipsoidal Inclusions Embedded
           in an Isotropic Matrix Phase

v_j = volume fraction of component j

P_j = property of component j

j = "i" for inclusion

j = "m" for matrix

EFFECTIVE ELASTIC CONSTANTS:

$$[C*] = [C^\circ] + \{[M^\circ] + [E^\circ]\}^{-1}$$

$$[M^\circ] = \Sigma_j v_j [m_j]$$

$$[m_j] = \{[H_j^\circ] - [E^\circ]\}^{-1}$$

$$H_j^\circ = \{[C^\circ] = [C^\circ]\}^{-1}$$

THERMAL STRESS COEFFICIENT:

$$\underline{\Gamma}* = \Sigma v_j \underline{\Gamma}_j + \frac{1}{2}\{[1] + \Sigma v_j [m_j][E^\circ]\}^{-1}\underline{Q}$$

$$\underline{\Gamma}_j = [C_j]\underline{\alpha}_j$$

$$\underline{Q} = \{\Sigma v_j [m_j]\}[E^\circ]\{\Sigma v_j \underline{\Gamma}_j\} - \Sigma v_j [M_j][E^\circ]\underline{\Gamma}_j$$

COEFFICIENT OF THERMAL EXPANSION:

$$\underline{\alpha}* = [C*]^{-1}\underline{\Gamma}*$$

influences. A pragmatic view would further suggest that $[C^\circ]$ be
determined from an experimental characterization of some "standard
reference material," The standard reference material should be
selected so that its behavior mimics the influence of the higher
order correlations between reinforcing agents and thereby captures
the influence of structural features not accounted for by the
parameters of the relationships given in Table 3. A reasonable
choice for such a "standard reference material" would be a composite
of spherical "beads" of the same material as the inclusion and at
the same volume fraction concentration as the inclusion.

Recently (McGee and McCullough, 1981b), a model for particulate systems has been developed which accurately predicts the
behavior of a wide variety of particulate filled systems; viz.

$$[S*] = v_r[S_{Lo}] + v_p[S_{Hi}] + \frac{1}{2} v_r v_p ([S_{Lo}] - [S_{Hi}]) \qquad (5)$$

where v_r and v_p are the respective volume fractions of resin and filler particles. The quantity $[S_{Lo}] = [C_{Lo}]^{-1}$ where $[C_{Lo}]$ is obtained from the relationships in Table 3 with $[C°] = [C_{resin}]$ and the aspect ratio appearing in the $[E°]$ terms is taken as a = 1. Similarly, the quantity $[S_{Hi}] = [C_{Hi}]^{-1}$ is obtained with $[C°] = [C_{filler}]$ and a = 1. The success of this semi-empirical "S-Combining Rule" in correlating the behavior of particulate filled systems disposes of the necessity for the experimental determination of the behavior of the reference system.

Assigning $[C°]$ the properties of an equivalent "bead" (or particulate) filled composite permits the use of the relationships of Table 3 to estimate the properties of the typical micro-laminate in terms of the effective aspect ratio, a_e , and the composition and properties of the matrix and inclusion.

A bead reference material with the properties for $[C°]$ computed by the "S-Combining Rule" can be used in conjunction with the relationships of Table 3 to predict the properties of continuous fiber laminates (with $a_e \to \infty$). The values obtained for the thermoelastic properties were in good agreement with experimental data and only marginally different from the values predicted by the Halpin-Tsai relationships (Ashton, et al., 1969). However, at lower aspect ratios ($a_e \sim 1-50$) the results obtained utilizing the bead reference depart from the Halpin-Kardos (1976) modification for finite aspect ratios.

Effective Aggregate Properties

The application of the orientation averaging relationships of Table 2 to the properties predicted utilizing a bead reference model requires the further specification of the properties to be averaged; e.g., [A] = [S] or [A] = [C]. Since the micro-laminate is aniso-tropic, an average conducted on the compliance array, [S], will differ from an average conducted on the elastic constants, [C]. An average conducted on the assignment [A] = [S] implies a condition approaching uniform stress throughout the material while an average conducted under the assignment [A] = [C] implies a condition approach-ing uniform strain. If the stress/strain fields fluctuate between grains, the actual behavior of the aggregate will lie somewhere between these extremes so that a combination of $<[S]>^{-1}$ and $<[C]>$ could be required to describe the aggregate behavior. The resolu-tion of this problem requires additional information concerning the stress/strain distributions between the grains. The nature of these distributions can be estimated by considering yet another structural feature of sheet molding materials; the connectivity of the domains.

The filaments within the roving of a typical sheet molding material are from 1 to 2 inches in length. However, as shown in Figure 4, agglomeration and/or swirling give rise to domains of locally parallel filaments which act as the principal reinforcing agent. Nonetheless, the filaments are considerably longer than the dimensions of these domains so that contiguous domains are connected by rigid filaments. This sharing of common filaments between domains promotes a tendency toward uniform strain throughout the aggregate. Accordingly, the average behavior of the aggregate can be approximated by averages conducted on the elastic constant array; viz, [A] = [C].

In contrast, Figure 2 shows that for relatively low concentrations of fibers the swirling effect is mitigated. However, the filaments are agglomerated in bundles to form isolated domains of locally parallel filaments. In this case, the effective aspect ratios of these domains are relatively large (~200). At these high aspect ratios, the relationships of Table 3 reduce to the case for $a_e \to \infty$, so that the micro-laminate tends toward the behavior of an aligned continuous fiber laminate. Under these conditions, the compatible strain arguments of Christensen (1979) are appropriate and the laminate analogy of Halpin and Pagano (1969) is applicable. As before, for a condition approaching uniform strain, the average behavior of the aggregate can be approximated by averages conducted on the elastic constant array.

It should be cautioned that the use of filaments of substantially shorter initial lengths would mitigate these arguments. In this case, the behavior could tend toward an average conducted on the compliance array; viz. [A] = [S].

In summary, the effective thermomechanical properties of a two-component fiber/resin sheet molding material may be predicted by the following procedure:

- Convert the Engineering constants (E, G, ν) into an elastic constant array for each component via the relationships given in Table 1.

- Compute the reference elasticity, [C°], utilizing Eq. (5) for a system of equivalent volume fraction of resin and particles ($a_e = 1$) with the particles assigned the properties of the fiber.

- Use the values for [C°] and the observed effective aspect ratio, a_e, in conjunction with the relationships of Table 3 to predict the properties of the grain; viz. $\underline{\Gamma}*$ and [C*].

- Use the resulting values in conjunction with Equations (4) ($B_j = \Gamma_j$) and Table 2 ($A_{jk} = C_{jk}^*$) to obtain the effective (average) properties for a specified state of orientation, f_p. Values of g_p for Table 2 can be estimated from Eq. (3).

- Convert to familiar material descriptors: the effective Young's moduli, shear moduli, Poisson's ratio, and thermal expansion coefficients of the composite using Tables 1 and 2.

These computational procedures can be extended to three-component resin/fiber/filler systems by introducing the notion of a "surrogate matrix." In this approach, the resin (r) and particulate filler (p) are viewed as an isolated matrix phase. The properties of the surrogate matrix, $[C_m]$, can be predicted by the application of the "S-Combining Rule," viz. Eq. (5), for a particulate system with the apparent volume fractions v_r' and v_p' for the resin and filler. These apparent volume fractions are related to the true volume fraction, v_r and v_p, through the relationships

$$v_r' = v_r/(v_r + v_p)$$

$$v_p' = v_p/(v_r + v_p) = 1 - v_r' \tag{6}$$

The composite is now viewed as a two-component composite of fibers with properties $[C_f]$, embedded in the surrogate matrix, with properties $[C_m]$. The volume fraction of the surrogate matrix phase is given by $v_m = v_r + v_p = 1 - v_f$, where v_f is the volume fraction of the fiber component.

COMPARISONS OF PREDICTIONS WITH EXPERIMENTAL VALUES

The resort to an approximate modeling approach demands that the validity of the simplifying abstractions of the Aggregate Model be tested by comparisons between predicted and experimentally determined thermoelastic properties. Furthermore, the preceding analyses emphasize the importance of conducting structural characterizations, in addition to mechanical and thermal characterizations, in order to determine the orientation parameter, f_p, and the effective aspect ratio, a_e. A knowledge of the properties of the components of a composite material is essential to the application of the model relationships. Typical properties of common, isotropic, components of short-fiber composites are summarized in Table 4.

The thermoelastic properties of the two-component sheet molding material, SMC-65 (shown in Figure 4) have been reported by Sanders, et al. (1978). The composition of SMC-65 is given in Table 5. The glass rovings are 1 inch (2.54 cm) in length; however, Figure 4 shows that filament agglomeration and swirling reduce the effective aspect ratio to $a_e \simeq 2$.

Table 4. Typical Properties of the Isotropic Components
          of Short-Fiber Composites

Property	Resin (Polyester)	Fiber (E-Glass)	Filler (CaCO$_3$)
Specific Gravity	1.20	2.55	2.40
Young's Modulus E, GPa (10^6 psi)	3.25 (0.51)	72.5 (10.5)	47.8 (6.93)
Shear Modulus G, GPa (10^6 psi)	1.38 (0.20)	27.8 (4.0)	18.1 (2.62)
Poisson's Ratio, ν	0.3	0.3	0.323
CTE α, mm/mm°C 10^{-6} (in/in°F)	100 (56)	5.0 (2.8)	5.0 (2.8)

Table 5. Compositions of Three Sheet Molding Compounds

Designation	E-glass fiber (vol %)	Polyester resin (vol %)	Calcium Carbonate Filler
SMC-65	47	53	0
SMC-25	18	50	32
SMC-R50	35	54	11

The characterization program for SMC-65 (Sanders, et al., 1978) was conducted under the assumption that the material was planar random. Consequently, test specimens were samples from the sheet without regard for the relative orientation of the sample with respect to the machine direction. The results from several specimens were pooled to obtain average values and standard deviations. Accordingly, the predicted values were obtained under the assumption that $f_p = 0$.

A comparison of predicted and observed values is given in Table 6. Experimental values for the "through-the-thickness" properties (E_3, ν_{23}, G_{23}, and α_3) were not reported due to the difficulty involved in making these measurements; nonetheless, predicted values for these quantities are included in Table 6. These comparisons show good agreement (well within the standard deviations) between the predicted and experimental values.

The standard deviations of the experimental data given in Table 6 are relatively large and may be due to anisotropic behavior resulting from partial orientation of the filaments in the machine direction. Since the random sampling of test specimens did not take into account directional dependence, anisotropic material behavior was superimposed on the material variability indicated by the standard deviation.

Specimens of the SMC-65 material were subsequently characterized along the machine (i.e., longitudinal) and transverse directions of the specimens to test for anisotropy (Taggart, et al., 1979). Analysis of the micrograph of SMC-65 in Figure 4 indicates a slight orientation with $f_p = 0.3$. A comparison of these experimental values with the model predictions with $f_p = 0.3$ is given in Table 7. Again the predictions from the model fall well within the standard deviations of the experimental values.

These experimental results show that the distinction between different directions can reduce the variability of the measured properties as indicated by the decrease in the standard deviations. As indicated in Table 7, pooling of longitudinal and transverse data gives a larger standard deviation. Large standard deviations in characterization data have been attributed to the "inherent" variability of sheet molding materials. These results show that part of the variability can be attributed to the anisotropic character of the material resulting from partial orientation of the filaments. Consequently, the direction of testing should be taken into account in characterization programs.

The thermoelastic properties of a three component sheet molding material, SMC-25 (composition given in Table 5) were also reported by Sanders, et al. (1978). As shown in Figure 2, the filaments agglomerate into long straight bundles with an effective aspect ratio of $a_e \simeq 200$. The specimens were sampled without regard to

Table 6.   Comparisons of Experimental with Predicted Values
           for a Two-component Sheet Molding Material, SMC-65

State of Orientation, f_p = 0 (assumed)

Effective Aspect Ratio, a_e = 2

Property	No. of Samples	Experimental*	Predicted
Specific Gravity	6	1.82 ± 0.01	1.83
Young's Modulus, $E(10^6$ psi)			
In-Plane ($E_1 = E_2$)	17	2.1 ± 0.7	2.1
Out-of-Plane (E_3)	--	--	1.7
Poisson's Ratio			
In-Plane (ν_{12})	6	0.26 ± 0.03	0.26
Out-of-Plane, ($\nu_{13} = \nu_{23}$)	--	--	0.29
Shear Modulus, $G(10^6$ psi)			
G_{12}	(computed)	0.9 ± 0.3	0.8
$G_{13} = G_{23}$	--	--	0.7
CTE (in/in°F)10^{-6}			
In-Plane ($\alpha_1 = \alpha_2$)	3	(13.7 ± 0.8)	15.8
Out-of-Plane (α_3)	--	--	22.6

*Sanders, et al. (1978)

direction so that the pooled data correspond to a planar random
orientation with f_p = 0.   Predicted and experimental values are
compared in Table 8.   Again, the predicted values are in good
agreement with experimental values.   The results reported by
Taggart, et al. (1979) for SMC-25 which were tested for anisotropy
are in good agreement with predictions based on f_p = 0.2 - 0.3.

Table 7. Comparison of Experimental with Predicted Values of SMC-65 Tested for Anisotropy

Property	No. of Samples	Experimental*	Predicted ($f_p = 0.3$, $a_e = 2$)
Young's Modulus			
$E_L (10^6$ psi)	3	2.3 ± 0.3	2.2
$E_T (10^6$ psi)	3	2.0 ± 0.2	1.9
$E_{avg} (10^6$ psi)	6	2.15 ± 0.5	--
Shear Modulus			
$G_{12} (10^6$ psi)	3	0.78 ± 0.05	0.81
Poisson's Ratio			
ν_{12}	2	0.32 ± 0.03	0.29

*Taggart, et al. (1979)

Table 8. Comparisons of Experimental with Predicted Values for a Three-component Sheet Molding Material, SMC-25

Property	No. of Samples	Experimental*	Predicted
Specific Gravity	6	1.85 ± 0.01	1.82
Young's Modulus, $E(10^6$ psi)			
In-Plane ($E_1 = E_2$)	69	1.9 ± 0.5	2.0
Out-of-Plane (E_3)	--	--	1.6
Poisson's Ratio			
In-Plane (ν_{12})	13	0.25 ± 0.03	0.28
Out-of-Plane, ($\nu_{13} = \nu_{23}$)	--	--	0.28
Shear Modulus, $G(10^6$ psi)			
G_{12}	(computed)	0.8 ± 0.2	0.8
$G_{13} = G_{23}$	--	--	0.6
CTE (in/in°F)10^{-6}			
In-Plane ($\alpha_1 = \alpha_2$)	3	13 ± 2	13
Out-of-Plane (α_3)	--	--	21

*Sanders, et al. (1978)

A further test of the model can be made by comparisons with the experimental values reported for a three-component sheet molding material, SMC-R50, with a larger content of glass fibers (composition given in Table 5). As before, the specimens were not tested for anisotropy so that the pooled data generate an apparent random planar orientation ($f_p = 0$); the effective aspect ratio is large, $a_e \simeq 200$. The results of the model predictions based on these structural parameters are compared with reported thermomechanical properties (Denton, 1979) in Table 9. As was the case with the previous comparisons, the predicted and observed values are in good agreement.

Table 9. Comparisons of Experimental with Predicted Values
for a Three-Component Sheet Molding Material, SMC-R50

State of Orientation: $f_p = 0$ (assumed)

Effective Aspect Ratio: $a_e = 200$ (assumed)

Property	Experimental*	Predicted
Young's Modulus, $E(10^6$ psi)		
In-Plane ($E_1 = E_2$)	2.27 ± 0.04	2.4
Out-of-Plane (E_3)	1.2 ± 0.1	1.5
Poisson's Ratio		
In-Plane (ν_{12})	0.31 ± 0.02	0.28
Out-of-Plane, ($\nu_{13} = \nu_{23}$)	0.22 ± 0.04	0.29
Shear Modulus, $G(10^6$ psi)		
G_{12}	0.9 ± 0.2	0.9
$G_{13} = G_{23}$	$0.4 \pm ?$	0.6
CTE (in/in°F)10^{-6}		
In-Plane ($\alpha_1 = \alpha_2$)	$9.4 \pm ?$	12.5
Out-of-Plane (α_3)	--	29

*Data summarized by Denton (1979)

SUMMARY

An essential step toward the understanding of the behavior of
composite materials is the identification of the appropriate
descriptors of the system which are susceptible to experimental
characterization. It is apparent that the following information
is essential to the development of any model to describe the
behavior of composite materials:

- the thermoelastic properties of each of the components and

- the volume fraction concentration of each component.

In order to distinguish between the behavior of widely different
classes of composite materials (e.g., continuous fiber composites,
particulate reinforced composites, and short-fiber composites)
additional information is required; viz,

- a measure of the load transfer efficiency associated with
  various reinforcing geometries (e.g., the effective aspect
  ratio, a_e) and

- a measure of the state of fiber orientation (e.g., the
  orientation parameter, f_p).

In view of the complex and variable features of the internal
microstructure, a rigorous treatment was abandoned in favor of the
development of a model, guided by structural observations, which
attempts to capture the dominant features of the reinforcing geometry and the state of orientation. The resort to a modeling
approach requires that the model predictions be tested against
experimental observations. The current verification of the Aggregate Model focused on the technologically significant glass fiber/
polyester resin systems. It would be useful to test the model for
other fibers (e.g., graphite, high-performance aramids, etc.);
unfortunately, a data base covering both the thermomechanical properties and micro-structure, to the extent of the glass fiber system
is not currently available. It is anticipated that the Aggregate
Model will serve to predict the thermoelastic behavior of these
alternate fiber systems.

An important feature of the model relationships is the incorporation of quantitative structural parameters associated with the
state of orientation, f_p , and the internal load tranfer characteristics as reflected by the effective aspect ratio, a_e . The sensitivity of the properties of composite materials to these structural parameters emphasized the futility of attempting to compare
the properties of short-fiber composite specimens of unknown
microstructure.

The state of orientation as well as the effective aspect ratio can be altered (and in some instances controlled) by variations in processing conditions. The verification of the model relationships provides justification for using the models to explore the influence of processing induced changes in the state of orientation and effective aspect ratios.

ACKNOWLEDGMENTS

This work was supported through the University-Industry Research Program of the Center for Composite Materials and by a grant from the Rogers Corporation, Rogers, Connecticut.

REFERENCES

Ashton, J. W., Halpin, J. C., and Petit, P. H., 1969, "Primer on Composite Materials: Analysis," Technomic, Stamford, Conn.

Christensen, R. M., 1979, "Mechanics of Composite Materials," John Wiley and Sons, New York.

Denton, D. L., 1979, "Mechanical Properties Characterization of an SMC-R50 Composite," Society of Automotive Engineers, Dearborn, Michigan, June 11-15.

Halpin, J. C., and Kardos, J. L., 1976, Polym. Eng. Sci., 16:344.

Halpin, J. C., and Pagano, N. J., 1969, "The Laminate Approximation for Randomly Oriented Fibrous Composites," J. of Comp. Mat., 3:732.

McGee, S. H., and McCullough, R. L., 1981a, An Optical Technique for Measuring Fiber Orientation in Short Fiber Composites, in Proceedings: "The Role of the Polymeric Matrix on the Processing and Structural Properties of Composite Materials," Plenum, New York.

McGee, S. H., and McCullough, R. L., 1981b, Combining Rules for Predicting the Thermoelastic Properties of Particulate Filled Polymers, Polyblends and Foams, Polymer Composites, in press.

Mukhopadhyay, A. K., 1980, "On the Thermoelastic Response of Composite Materials," Ph.D. dissertation, Department of Mechanical and Aerospace Engineering, University of Delaware, Newark, Delaware; Center for Composite Materials Report CCM-80-10.

Ruess, A., 1929, Z. Angew. Math. Mech., 9:49.

Rosen, B. W. and Hashin, Z., 1970, Int. J. Engg. Sci., 8:157.

Sanders, B. A., Warner, G. G., and Heimbuch, R. A., 1978,
    "Mechanical Properties of Three Glass Fiber Reinforced
    Polyester Materials, An Interim Report," Report No. MD76-016,
    General Motors Manufacturing Development, G. M. Technical
    Center, Warren, Michigan.

Taggart, D. G., Pipes, R. B., Blake, R. A., Gillespie, Jr., J. W.,
    Prabhakaran, R., and Whitney, J. M., 1979, "Properties of SMC
    Composites," Center for Composite Materials, University of
    Delaware, Report CCM-79-01, Newark, Delaware.

Voigt, W., 1910, "Lehrbuch der Kristallphysik," Teubner, Leipzig.

Wu, C. D., and McCullough, R. L., 1979, Constitutive Relationships
    for Heterogeneous Materials, in: "Developments in Composite
    Materials," (G. S. Holister, ed.), Applied Science Publishers,
    London.

EFFECTS OF MATRIX CHARACTERISTICS IN THE

PROCESSING OF SHORT FIBER COMPOSITES

Lloyd A. Goettler

Rubber Chemicals Division
Monsanto Company
260 Springside Drive
Akron, OH 44313

INTRODUCTION

The widely differing consistencies of composite matrices, that can range from carrier vehicles for laying out preforms to highly viscoelastic melts, dictate which processing methods are feasible and impose limitations on their operation and output. The following comparison of techniques employed for fabricating short fiber composites in the thermoset, thermoplastic, and rubber industries will illustrate these differences. Details of matrix effects are given within each field.

PROCESSING OF REACTIVE SYSTEMS

This classification includes thermosets and those thermoplastics formed by reaction molding. Since the polymerization is completed subsequent to the fabrication operation, the materials during forming are of low molecular weight and exhibit the low viscosity and elasticity characteristic of monomers or pre-polymers. These properties facilitate the production of large or complex parts, but pose some restraints upon the inclusion of short reinforcing fibers.

The following phenomena occurring in reinforced reaction injection molding (RRIM) and thermoset premix molding become more severe as the matrix viscosity drops or the fiber length (or aspect ratio) and concentration increase:

1. Sedimentation of fibers during mixing, as in the cylinders of a RRIM machine.

2. Balling of fibers during mixing in monomers, such as the molten caprolactam of the nylon 6 RIM system.

3. Fiber plugging at small gates or orifices, which causes segregation from the matrix.

4. Low shear stress development that retards dispersive mixing of fiber clumps.

5. Highly flattened velocity profiles that impede distributive mixing.

With suspensions of chopped fiberglass in glycerine, Folgar and Tucker[1] found the establishment of a central solid fiber network during shear flow to depend on fiber length. With 1/8" fibers, it occurs only at concentrations above 20 volume percent, whereas if the fiber length is increased to 1/2", the unsheared core will form at concentrations as low as 5% by volume.

In a similar work using A-staged epoxy resins, Takano[2] has found a dependency of this boundary effect upon the viscosity of the suspending fluid. At a 40 volume percent concentration of 1/8" fibers, an increase in shear orientation can be observed when the Newtonian carrier viscosity exceeds 1,000 Pas. On the other hand, if its viscosity drops too low (to 5-6 Pas), a substantial velocity differential occurs between the fibers and the fluid, even in a uniform channel. Under this condition, the 1/8" fibers can also readily plug a 1/8" x 1/8" square orifice.

The plugging effect was also described in an early Russian work by Stankoi et al[3]. More recently, Murty and Modlen[4] have quantified the jamming effect by defining an efficiency factor relating the fiber volume fraction in the effluent suspension to that in the original liquid. For the low concentration range in which they worked, the efficiency was constant above a viscosity of 50 Pas, but decreased linearly with the logarithm of viscosity at lower levels.

Low viscosity carrier vehicles can also be used to prepare highly aligned preform mats that can later be impregnated and molded into directional composites or composite laminates. The best known is the use of a reciprocating trough for producing aligned sheets, which was developed by the Explosives Research and Development Establishment in England[5] and extensively studied by Kacir and Narkis[6]. The glycerine carrier fluid used here has also been employed by Takano[7] to produce elongated grains of parallel fibers that can then be used as a molding compound. The exceptional parallelism, end overlap, dispersion, and length integrity of the fibers preserved through the gentle Couette flow action in his preparation results in especially high tensile strength as well as modulus in compression molded composites. A comparison in Table I with similar epoxy compositions shows the superiority of these techniques over high production fabrication methods. Their success in perfecting the composite structure derives from the low fiber concentrations employed as well as the low viscosity of the carrier liquid which generates only low levels of shear and normal forces and excludes viscoelastic effects.

TABLE I

Comparison of Preform and Melt Flow Fabrications of Fibers in Epoxy

Process	Fiber Content, v/o	Avg. Fiber Angle, Degrees	E GPa	Tensile Strength MPa
Conventional Injection Molding	40	40-60	10.3	48.3
Injection Molding with Controlled Fiber Orientation	40	19	26.9	186.
Hand lay-up of Extruded Prepreg.	40 57	20 20	24.1 34.5	228. 241.
Hand lay-up of Encapsulated Grain Preforms	57	7	39.3	310.
E.R.D.E. Process	57*	–	96.6	393.

*Graphite fiber, Modmor Type II

Exact alignment techniques are difficult to apply for thermoplastic composites due to the high viscosity of the melt. Solution coating is preferred, but precautions must be taken to insure solvent removal. The best resin-reinforcement contact in a thermoplastic system can be obtained when the two are combined before polymerizing the resin[8]. Then viscosity of the monomer is in the centipoise range. In addition, void elimination is enhanced, and high fiber loadings with less fiber damage are possible.

For example, composites prepared by polymerizing ε-caprolactam (nylon 6) around a highly aligned array of glass fibers display high tensile strength and elongation to failure in the transverse direction. The tensile strength of 83 MPa, equal to that of the matrix, and an elongation of 2.0%, achievable at 30 v/o fiber content, are unusually high for non-elastomeric polymer systems. In contrast, a typical brittle epoxy reinforced with fibers transverse to the stress direction breaks at .5% strain and about 55 MPa tensile stress. Such a material could surpass classical design limits.

THERMOSET MOLDING

An early investigation of the effects of mold flows on fiber alignment in short fiberglass reinforced epoxy compounds was included in the joint ONR/ARPA contract by Washington University and Monsanto Company around 1970. Chopped 3 mm and 6 mm E-glass was incorporated into an epichlorohydrin/bisphenol A type epoxy resin cured with 30 phr of methylenedianiline. After blending the fibers and MDA into the resin at room temperature using either a sigma blade mixer or an aqueous slurry encapsulation process developed by Andersen and Morris[9], the compound was allowed to harden to a non-tacky B-stage at room temperature. The degree of reaction determined the viscosity of the matrix during the molding process, as shown in Figure 1. The extent of epoxide reaction prior to fabrication typically varied from 43 to 56%. Above 50% reaction a network structure begins to form and above 56% the compound becomes intractable and cannot be fabricated.

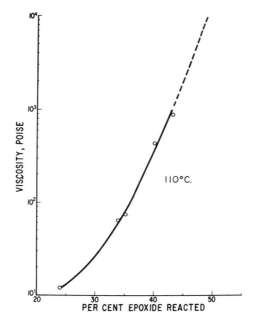

Fig. 1. Viscosity of epoxy matrix.

One geometry studied was a 6 mm thick x 25 mm wide x 150 mm
long bar molded through an assortment of interchangeable gate geometries at one end. A plunger injected the preheated stock from a
holding pot to the cavity under 35-100 MPa pressure. The expansion
in the flow emerging from the gate into the bar cavity produced a
central transverse orientation of the reinforcing fibers, as shown
in Figure 2, which severely reduces the useful tensile properties
of the molded part[10].

It was found that softer molding compounds (lower degree of
epoxide reaction or higher mold temperature) produce a larger shear
layer of parallel orientation at the walls of the cavity and consequently a smaller core of transverse fibers. This is similar to the
viscosity effect found with elastomeric composites in expanding hose
dies[11], but contrary to the results of Takano[2] working with strictly
parallel orientation in a converging flow.

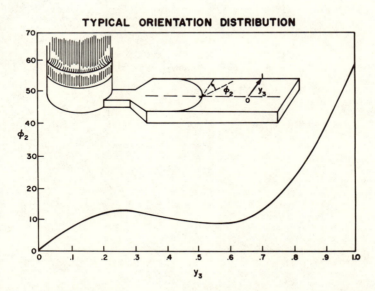

Fig. 2. Bar cavity geometry and resulting transverse fiber
orientation.

The wetting of the fibers by the resin and subsequent adhesion in mechanical testing of the molded bar were found to depend on the degree of B-stage during processing. If its cure is too far advanced, the stresses generated during flow appear to be capable of pulling the resin away from the fiber surfaces[12]. Void content also more than doubles as the resin B-stage is increased from 43 to 57%.

Mechanical properties similarly suffer. Over the same increase in reacted epoxide groups in the B-stage, the fully cured tensile strength reduced by 60% from the usual 35 MPa representative of transverse fiber orientation for this 40 v/o fiberglass composite[12]. Since similar reductions were observed under reduced temperature and pressure conditions, when little flash formed, it can be concluded that strength is enhanced by a sufficiently soft matrix that properly wets the fibers and eliminates molding defects.

In some moldings through a converging flow into one end of a 12 mm rod, it was found that the resin viscosity exerts no effect on the degree of fiber orientation over a 24-fold range. This result pertains only to the parallel orientation produced by the converging flow in which extensional and viscous forces act in tandem. In diverging flows, the shearing action operative with a low viscosity matrix, as described earlier, would destroy the transverse orientation established by the extensional field.

These same molding compounds containing 40 v/o of 3 mm fiberglass were also extruded into a 6 mm diameter rod as aligned prepreg for subsequent compression molding[9]. The surfaces of these extrudates were quite rough, owing to the low strength of the epoxy resin in its B-staged condition. They could not consequently be used directly, but served a a means for obtaining fiber alignment, as shown in Table I.

THERMOPLASTIC COMPOSITES

The orientation and flow behavior of thermoplastic and thermosetting types of fiber reinforced molding compounds are very similar. Crowson, Folkes and Bright have recently reported studies on fiber orientation in short (.5 and 10 mm) glass fiber reinforced polypropylene undergoing converging, diverging, and shear flows[13]. By using a contact radiography technique, they found that the flow enters a capillary die through a very small cone angle from the reservoir; a high degree of fiber orientation parallel to the flow obtains as the entrance is approached, but much of this is lost during shear flow through the tube. This behavior, along with the highly transverse orientation produced by a divergent flow, exactly parallels that found by Goettler[14] in epoxy BMC materials.

However, the higher viscosity levels of thermoplastic melts in comparison to thermoset pre-polymers will cause different levels of behavior. For example, shear development in mold flow would be expected to be different, affecting the velocity profile and yielding a different degree of fiber orientation. A reduction in fiber effects would follow from the fiber degradation that might occur from those higher stresses during mixing and processing. Chan et al[15] have recently found a standardized compounding procedure to be 20% less damaging to the chopped fiberglass reinforcements with a polystyrene matrix than with a polyethylene matrix. They conclude that the nature of the resin matrix is more influential in this regard than the fiberglass concentration, although they do not discuss which resin characteristics are meant; the measured viscosities are identical. The degradation sustained by brittle fibers during compound preparation would subsequently reduce the level of fiber orientation generated by the flow.

The expanding flow hose dies described in[11] have also been found applicable to orienting fiberglass contained in thermoplastic resins into the circumferential direction. However, due to attrition, long glass fiber is required for anisotropy development, and even then the level is below that generated in cellulose fiber/rubber composites. A die with 4:1 radius expansion imparts only a 1.6:1 ratio of hoop to axial Young's modulus in polypropylene reinforced with 20 wt. % 9.5 mm glass fiber. Composites containing the shorter .6 mm length remain isotropic. A major drawback -- the extrudates have rough surfaces due to low melt strength. A suitable hose can, however, be successfully extruded from cellulose reinforced plasticized PVC, which has better melt characteristics and fibers that maintain a high aspect ratio.

REINFORCED ELASTOMERS

These materials are characterized by both a higher viscosity and melt elasticity in processing compared to either thermosetting or thermoplastic resins. Reinforcement with chopped textile fibers

or short natural wood cellulose[16] is common in the rubber industry.
In the latter case, the fibers of 100:1 aspect ratio may be incorporated at loadings as high as 20 volume percent in typical compounded rubber stocks to produce a tensile strength of 25 MPa and
Young's modulus of 275 MPa[17]. Such material can be considered as
high performance composites in view of the high 40:1 enhancement
over matrix stiffness parallel to the direction of alignment.

Molding

Rubber composites can be injection molded as readily as reinforced plastics. Since these materials usually require a crosslinking vulcanization to produce useful properties at and above
room temperature, a heated mold and water-cooled injection barrel
are used, as in BMC molding.

The anisotropy developed in the moldings by preferred orientation of the reinforcing fibers follows the same principles as in
reinforced plastic molding. (Converging flows produce parallel
alignment and diverging flows a transverse alignment.) The viscoelastic character of the polymer appears to have little effect,
and the fiber orientations are governed principally by the flow
kinematics as defined by the mold geometry.

As an example, end-gated plaques measuring 76 mm in width x
152 mm in length x 1.9 mm thickness were molded from the rubber
formulations given in Table II consisting of 22 or 30% cellulose
fiber by weight in a natural rubber compound. The fiber has been
treated for easy dispersion in the rubber compound as it is mixed
in an intensive internal mixer, and for subsequent bonding to the
polymeric matrix during the vulcanization process.

The narrow 3.2 mm x 1.6 mm deep gate produces a predominantly
transverse fiber orientation pattern as a result of the divergent
flow, which is practically a pure shear field, in the vicinity of
the gate. The fiber pattern is qualitatively the same as that
obtained with the fiberglass reinforced epoxy molded through a
similar geometry, despite considerable differences in the molding
compound. (See Table III). However, in consideration of the
differences in fiber level shown in the table, the more viscous
rubber stock develops a somewhat higher degree of anisotropy.

TABLE II
Cellulose Fiber Reinforced Natural Rubber

Component	Stock A	Stock B
Natural Rubber	100	100
Silica	50	50
Cellulose Fiber	50	75
Zinc Oxide	5	5
Stearic Acid	2	2
Polymerized 2,2,4-trimethyl-1,2-dihydroquinoline	2	2
N,N'-m-phenylenebismaleimide	1.6	1.6
N-t-butyl-2-benzothiazylsulfenamide	2	2
Sulfur	0.6	0.6
Resorcinol Resin	5	5
Hexamethylemetetramine	$\underline{1}$	$\underline{1}$
	219	244
Vol. % Fiber	20	28

Young's Modulus, MPa[a].

Longitudinal Direction (E_L)	17.2	17.2
Transverse Direction (E_T)	31.0	58.6
E_T/E_L	1.8	3.4

[a].Measured at center of plaque.

TABLE III
Comparative Fiber Orientations in Epoxy and Rubber Composites

	Chopped Fiberglass in B-Staged Epoxy	Wood Cellulose in Natural Rubber	
Fiber Length, mm	3.2	1.5	
Matrix Shear Viscosity, Pas [a.]	100	5,000	
Matrix Elasticity	0	20% die swell	
Fiber Content, vol. %	51	20	28
E_T/E_L	2-3	1.8	3.4

[a.] At low shear rate ($.1s^{-1}$), $110^{\circ}C$

Since stiffness relates directly to fiber directionality, the Young's modulus ratio may be used to indicate the degree of preferred orientation. The increased transverse orientation obtained at higher fiber level is commonly observed and is attributed to a reduction in the shear component of the flow.

Two additional moldings of fiber reinforced rubber, utilizing a stagnation flow geometry, are shown in Fig. 3. In both cases an EPDM rubber was reinforced with 17 wt. % cellulose fiber. Tensile properties are given in Table IV. The very high directional ratio of Young's modulus in the disk derives from the high expansion in radius from 2.4 mm at the gate to 70 mm at the edge of the part.

On the other hand, the modest 4:1 expansion in the cup mold is comparable to the radial expansion built into the special extrusion dies used to orient fibers circumferentially in a hose[11] and the resulting anisotropy is also the same.

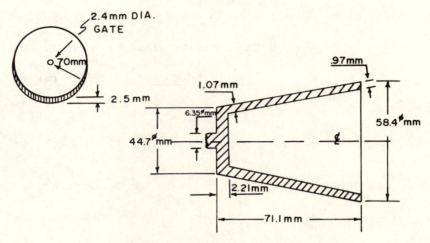

Fig. 3.   Rotationally symmetric mold geometries.

TABLE IV
Properties of Center Gated Injection Moldings[a]

Mold Type	Expansion[b]	Tensile Property, M	Direction		M_C/M_R
			Circum.(C)	Radial (R)	
Disk	58	Modulus, MPa	116	16.	7.2
		Strength, MPa	12.3	4.0	3.1
Cup	4	Modulus, MPa	75.2	36.6	2.1
		Strength, MPa	9.7	6.9	1.4

[a]17 wt. % cellulose fiber in EPDM rubber.
[b]Cross sectional area to flow at outer cavity wall/area at gate.

Extrusion

The greater viscoelasticity of the elastomeric matrix phase
in rubber composites also allows a considerably wider latitude in
free surface forming than is possible with plastic melts. All such
processes common to the rubber industry -- milling, calendering, and
extrusion -- are feasible with short fiber reinforcement.

Rubber extrusion is aided by the lower rigidity of the organic
fiber reinforcement commonly used, so that the protruding fibers
obtained in epoxy prepreg extrusions or the foams generated in
thermoplastics extruded through short land dies[13,18] are rare. When
brittle reinforcements such as fiberglass are employed, the large
stresses developed by the high viscoelasticity in rubber processing
reduce them to particulate dimensions. Glass fiber chopped to 12 mm
length and incorporated into a NR/SBR rubber blend at 125 phr produced a directional Young's modulus of only 35 MPa when extruded
through an orienting die[11], whereas cellulose fiber at only 60 phr
gives 94 MPa. The 11 MPa tensile strength of the cellulose fiber
composite was also 45% above that of the fiberglass.

Several phenomena are common to all polymers reinforced with
discontinuous fibers when they are extruded through converging dies.

1. The die swell is practically nil, despite the high
    normal stresses generated in these composites.

2. The entrance pressure drop is high, due to the
    high elongational viscosity.

3. The surfaces tend to be rough, tear, or branch.

The last effect derives from the weakness of the aligned composite
melt in shear. It becomes worse as the melt strength of the polymer
matrix diminishes. Thus, highly viscoelastic rubber compounds yield
smoother extrudates than reinforced thermoplastics, which in turn are
superior to reinforced thermosets (epoxy, polyester BMC) that are
extruded prior to gelation. It is the rough surface condition,
which may even penetrate into the interior of the extrudate in the
form of clevage cracks and voids, that imposes the severest constraint on extrusion of short fiber composites.

Even among rubbers, the surface becomes smoother the harder the
stock. Not only is the melt shear strength then higher, but if the

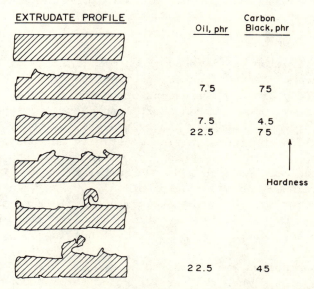

Fig. 4. Effect of rubber stock hardness on appearance of NR/SBR
extrudates containing 8 wt. % cellulose fiber.

fiber mass flows with a flatter velocity profile, shearing action
concentrates at a thin region near the die wall where the fiber
concentration is lower. The photographs of a Garvey die extrusion
in Fig. 4 show the improved surface as the hardnes of the stock is
increased by increasing carbon black or decreasing oil level.
Reducing the overall fiber concentration is likewise beneficial.

Without resorting to change in the rubber composition, there are some physical changes that help to promote a smoother extrudate.

1. Increase the land area (length at constant cross section) of the die. A length of 8 times the channel thickness is recommended.

2. Heat the die to produce a lower viscosity in the slip region against the die wall. A ΔT of $15^{\circ}C$ is generally sufficient.

3. Taper the entrance region.

4. Change the predominant fiber orientation out of the axial direction in order to reduce the incidence of weak shear planes. This can be accomplished by incorporating a special geometry into the die, as will be explained below.

An example of a die with long land length and tapered entrance is designated die "C" and shown in Fig. 5. An extrudate produced from this die with and without heating is compared with one from the unheated, shorter die designated "A" of Fig. 6 as samples E,F, and A of Fig. 7. Note the improvement in perfection of the 90° corners as well as in the smoothness of the surface. Both dies are superior to a simple plate die, which produces considerable tearing and sinuous flow, as in the first sample of the figure. The extrudates are consistently smoother at the lower extrusion rates. Entries B and C show the further improvement that would be possible by eliminating the high degree of axial orientation imparted by the converging flow. Following the principles developed[19] and demonstrated in molding and extrusion[10,11] for fiber rotations in extensional flows, the channel constriction upstream of the land in Fig. 8 will tend to redirect the fibers into the transverse direction. The extent of blockage, indicated by the expansion ratio in area or linear dimension following the constriction, is kept small in this die to minimize the degree of rotation of the fibers from their desired predominant axial orientation. The stock for these samples comprises 8.5 volume percent cellulose fiber in a highly extended EPDM rubber (105 phr process oil, 200 phr carbon black) and has a ML-4 ($100^{\circ}C$) Mooney viscosity of 66.

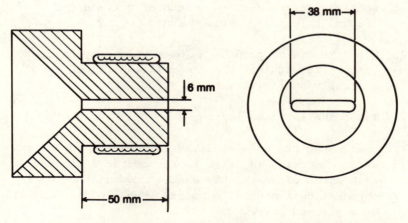

Fig. 5.   Heated long land die for axial strip extrusion (Die C).

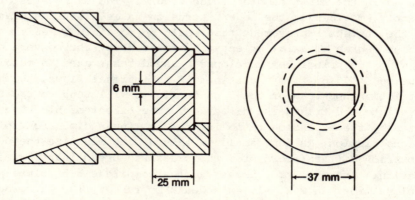

Fig. 6.   Unheated short die for axial strip extrusion (Die A).

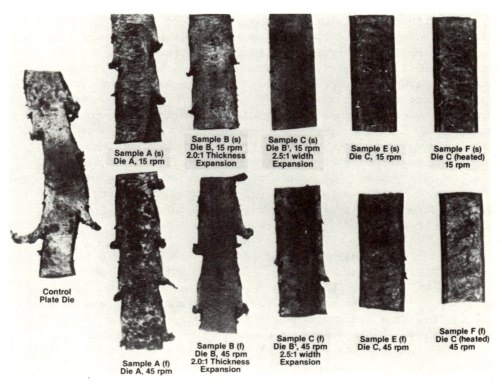

Control
Plate Die

Sample A (s)
Die A, 15 rpm

Sample B (s)
Die B, 15 rpm
2.0:1 Thickness
Expansion

Sample C (s)
Die B¹, 15 rpm
2.5:1 width
Expansion

Sample E (s)
Die C, 15 rpm

Sample F (s)
Die C (heated)
15 rpm

Sample A (f)
Die A, 45 rpm

Sample B (f)
Die B, 45 rpm
2.0:1 Thickness
Expansion

Sample C (f)
Die B¹, 45 rpm
2.5:1 width
Expansion

Sample E (f)
Die C, 45 rpm

Sample F (f)
Die C (heated)
45 rpm

Fig. 7. A comparison of surface appearances on extruded strips.

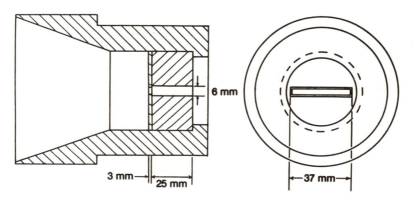

6 mm

3 mm

25 mm

37 mm

Fig. 8. Constricted flow strip die (Die B).

A further exhibit of the smoothing effect caused by re-orienting
the fibers is afforded by a triangular shaped die, as might be used
in the extrusion of certain vehicular tire components. If the triangular profile is nearly equilateral (symmetric), the extrudate will
emerge straight from the extrusion die. However, a non-uniform axial
velocity profile will establish across the face of an asymmetric
profile and either cause the extrudate to bend into a curvature directed away from the side of higher velocity, produce localized
ripples in the extrudate, or result in tearing. It is especially
important in the case of extrusion of fiber filled stocks that a
uniform exit velocity be maintained because of the low melt extensibility of those stocks (which prevents them from being straightened
by stretching) and also the low resistance to tear propagation parallel to the localized fiber direction.

Fig. 9 shows the effects of adding 4 and 8 wt. % cellulose fiber
to a NR/SBR compounded rubber blend extruded through the plate die
shown in the figure. Relieving the back face of the die in the
vicinity of the small angle to reduce the land length and consequently the flow resistance where the clearance is tight results in
an insufficient improvement in the curvature of the extrudate and
its consequent tearing when straightened. The equation for pressure
flow of a power law fluid through slit dies suggests that the land
length should vary with the (n + 1) power of the local channel
thickness to maintain a uniform velocity. This would be very difficult to obtain in practice if the channel thickness approaches zero
at one end of the profile.

A further refinement involves the partial blockage of the wide
end of the profile with an obstruction attached to or in back of the
die plate, as shown in Fig. 10. The following expansion required
to fill out the die channel will reduce the velocity in proportion
to the ratio of dimensions. The resulting uniform velocity produces
a straight extrusion with well defined edges, even at 8% fiber content, as shown in the figure. The obstructing plate can be reduced
to 1.6 mm thickness and structured so as to avoid blockage of the
back-relieving at the tip of the die orifice.

The expansion taken in thickness to slow the velocity through
the wide part of the die may be insufficient to substantially alter
the axial positioning of the reinforcing fibers. That will be satisfactory for many applications, where axial strength and stiffness
is required in the extrudate. However, cases do arise in belting
and tire components where reinforcement is needed in the transverse

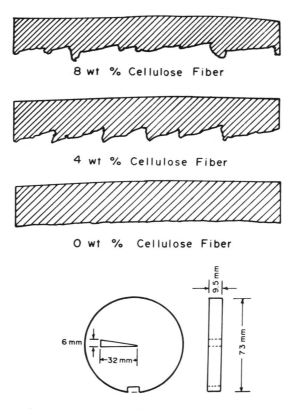

8 wt % Cellulose Fiber

4 wt % Cellulose Fiber

0 wt % Cellulose Fiber

Fig. 9.   Triangular plate die and resulting extrusions.

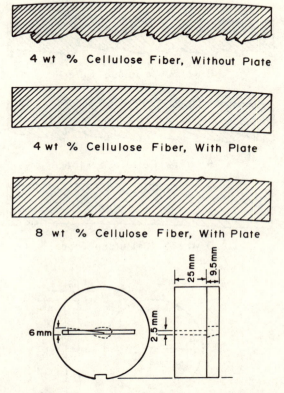

Fig. 10.   Partial die blockage to produce uniform
velocity profile in asymmetric die.

direction of an oblong profile. The large (>4:1) channel area increase required to transpose the fibers from their axial disposition
will lead to a jetting phenomenon at the center of the profile. The
usual die modifications employed to channel the flow will unfortunately alter the desired fiber orientation pattern.

The proper die design can be constructed by application of the
principle that changes to the shape of the channel do not alter the
axial component of fiber orientation as long as a constant cross
sectional area is maintained[11]. Thus, to avoid jetting, the expansion can be initially taken uniformly in both transverse directions,
allowing smaller linear changes to produce the same area expansion.
The shape can subsequently be changed through a mated die piece to
produce the desired profile while keeping the area constant. The
net result is as if a direct expansion were taken to the final shape.

To further reduce the tendency to flow faster in the center
of the channel, which results from the expansion, a partial blockage
can be introduced into the center and then removed during the subsequent step of channel shape alteration described above.

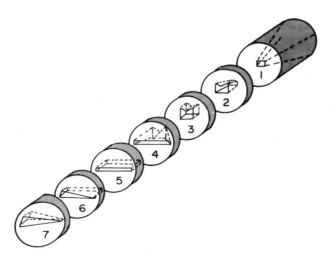

Fig. 11. Steps in constructing a die to produce an oblong profile
          with transverse fiber orientation.

An example of such a die used to produce a tire bead filler is shown in Fig. 11. The die consists of as many as 7 components stacked back-to-back in a holding ring. They are, in the order seen by the flow:

1. A flow constrictor.

2. A fish-tail nozzle to expand the flow in both transverse directions and simultaneously introduce the center con- striction.

3. A spacer of constant cross section to equalize the flow.

4. A transition to a desired (slit) cross section while simul- taneously removing the center constriction.

5. A land length for this cross section.

6. A further transition to a different cross sectional shape.

7. A final land, which may be of non-uniform thickness to produce an even flow if the shape is asymmetric.

In transition pieces 4 and 6, the cross sectional area is kept constant. Pieces 6 and 7 may be omitted if the die is dedicated to a single shape produced directly in piece 4. The overall result of incorporating all of these elements is to produce a smooth straight extrudate which contains a high degree of fiber orientation into one transverse direction. If die piece 2 is omitted, the flow is too fast in the center, relative to either edge, by a factor of 2 or 3.

The anisotropy generated in such an extrusion die for producing a 58 mm wide x 2.7 mm thick strip or a triangular profile with an altitude of 4.7 mm to an offset apex is given in Table V and Fig. 12. The area expansion ratio of 5:1 in this die is comprised mainly of a linear expansion in width (the direction of preferred orientation) of 4.5:1.

Despite differences in polymer type, viscosity, and fiber level, all stocks of Table V display the same flow behavior in the die and produced a uniform extrudate. The degree of orientation into the

TABLE V

Comparative Properties of Transversely Oriented Profile Extrusions

Stock Type	A EPDM	B NR/SBR	C NR/SBR
Wt. % Cellulose Fibers	10.	28.	17
ML - 4 (100°C) Mooney Viscosity	67	106	66
Transverse-to-Axial Young's Modulus Ratio (E_T/E_A) in Strip	1.6	2.8	1.9
Axial-to-Transverse Solvent Swell Ratio ($\varepsilon_A/\varepsilon_T$) in Strip	–	24.	3.6
$\varepsilon_A/\varepsilon_T$ in Triangular Profile	–	3.5	–

transverse (width) direction is best measured by the ratio of the Young's modulus in that direction (E_T) to the modulus in the axial (flow) direction (E_A). Another, but less exact, estimate is the inverse swell ratio $\varepsilon_A/\varepsilon_T$ measured after soaking in toluene. Although the results of this test are influenced by the out-of-plane deformation under the induced multiaxial stress field, it is suitable for non-uniform sections.

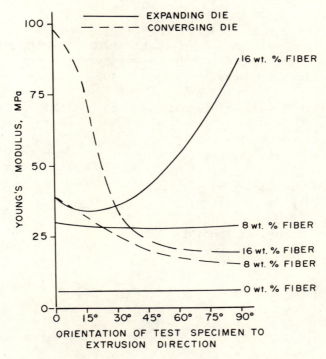

Fig. 12.   Orientation distributions in expanding
            and converging strip dies.

The data of Fig. 12 pertain to extrusion of the rectangular strip from a NR/SBR tire stock containing 8 or 16 wt. % (6.6 or 13.2 vol. %) cellulose fiber, 65 phr of HAF carbon black and 10 phr process oil. Here the actual variation in modulus with angle is shown in comparison to the predominant axial orientation produced when the expanding sections 2, 3, and 4 are omitted.

Both Fig. 12 and Table V show that the transverse reinforcement increases with increasing fiber level. The disproportionately high ratios for stock B, in spite of the higher fiber content, can be attributed to its higher viscosity. Heating of the extrusion die by up to 30°C had little effect on the anisotropy. In the triangular section the principal fiber orientation is still transverse but to a lesser extent. In particular, the fibers become re-oriented back toward the axial direction in the thin portion of the cross section due to a combination of local converging flow and high shear. In the absence of this effect, about the same degree of transverse fiber orientation is obtained in the slit dies as in the expanding hose dies described elsewhere[11].

These hose dies utilize an expansion in either the diameter or the thickness of the annular flow channel in order to turn the fibers out of the axial direction. A circumferential orientation produced by flaring the mandrel and outer die wall concentrically is preferred for strengthening the hose against a contained hydraulic pressure.

A part of that study deals with the effect of stock viscosity on the efficiency with which the fibers are re-oriented. It was previously mentioned that the fiber reinforced melt flows with a very flat velocity profile. A high matrix viscosity appears to strengthen the inner core which flows as a plug in these dies. The viscoelastic nature of the rubber compound may also aid in preserving the plug by establishing an effective yield stress. In such materials there is little influence of shear, which is restricted to a thin region near the wall.

It is interesting that the effect of viscosity here, as in the molded epoxy bars described earlier, appears to be opposite to that found by Takano[2] in low viscosity epoxy compounds. The transverse fiber orientation in these samples may make them better resistant to shear than Takano's parallel alignment as the viscosity is increased.

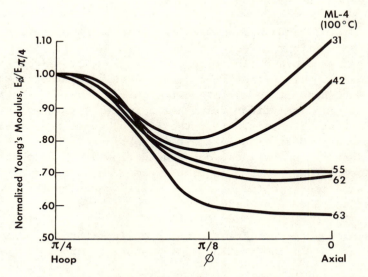

Fig. 13. Fiber orientation distribution in cellulose reinforced
          rubber hoses extruded through 4:1 expanding mandrel
          die[11].

        Figure 13 shows that when the rubber viscosity decreases, the
region of shear near the die walls enlarges and contributes a larger
component of axial orientation in the cellulose fiber reinforced
hoses. The orientation distribution is then bimodal and reflects
a sandwich structure in which a circumferentially oriented core is
surrounded by sheaths of axial orientation.

The concept of short fiber reinforced hose is extended to shaped hoses that are curved as they exit from the extrusion die[20]. The incorporation of the reinforcement, in the form of discontinuous cellulose fibers, in the rubber stock as it is mixed permits this on-line shaping in place of the conventional mandrel forming during vulcanization that is practiced with cord reinforcement. Both of the separate steps of cord application and mandrel assembly can thus be eliminated in the production of various automotive hoses.

Hose shaping occurs by moving the inner and outer hose dies out of concentricity in a controlled sequence. The non-uniform velocity profile across the offset die face causes a natural curvature in the extruded hose which is then set by the presence of the fibers. There are two ways in which the matrix viscosity can influence the performance of this process:

1. The radius of curvature of the hose bend is smaller (the bend is tighter) the lower the exponent in the power law flow model for the reinforced melt. Although the presence of the short fibers alone tends to reduce the exponent by flattening the velocity profile, the matrix exerts an additional effect.

2. High viscosity in the matrix aids in the retention of the shape produced at the extruder die as the hose is carried through post-extrusion handling and curing operations. The matrix effect is secondary to the stability imparted by the fibers.

Roll Sheeting

Calendering rolls are used to form plastic and rubber sheeting. In some modes of operation, pigs of stock are first banded around one of the rolls to initiate the feed. A similar physical situation exists on 2-roll friction mills. Frequently the plasticated stock will not band to the roll but tends to either (i) loosen at the top where the slab is returned to the bank between the counter-rotating rolls or (ii) stick to the opposite roll as it exits from the nip and so be pulled loose and sometimes off the initial roll. The addition of short reinforcing fibers to the stock generally makes the bagging problem described in (i) above more severe through the increased stiffness of the stock. The roll preference described in (ii) continues to be an issue.

In an extensive unpublished study of the milling behavior of short fiber reinforced rubbers, Goettler has identified the stock stiffness (E), viscosity (η) and adhesive bond strength to the metal rolls (A) as the critical parameters affecting banding. Thus, the mill gap, G, for which the stock will prefer the desired front roll is given by the form

$$G/R \geqslant C_1 + C_2 \frac{GA}{R^2 E} - C_3 \frac{\eta\omega}{E} \tag{1}$$

where R is the roll radius

ω is the angular velocity of the rolls

C_1, C_2, C_3 are constants that also account for the
    friction ratio of the mill.

A second limiting mill gap defines the region of top roll separation leading to bagging. The acceptable regime can be described by the dimensional equation

$$G \leqslant C_1' + C_2' A - C_3' \eta$$

Thus it is clear that matrix properties significantly affect milling performance through their influence on the composite stock's stiffness, viscosity, and adhesion. In addition, it is found that bagging is less likely to occur with polymers of broad molecular weight distribution.

SUMMARY

The role of the matrix on the processing of short fiber composites has been examined with regard to thermoset, thermoplastic, and rubber polymers. The matrix serves as more than a carrier vehicle for the fibers; it imparts its own characteristics upon the reinforced melt: viscosity level, elasticity, stiffness, melt strength, wetting or adhesion characteristics. The wide variations in some of these parameters between the different classes of polymeric matrices dictate applicable processing methods. All are amenable to closed cavity molding, but only the more coherent melts allow free surface forming, as in extrusion and calendering.

Within each class, matrix variations determine the ease of processing. Thus, higher viscosity matrices produce a smoother extrudate surface, and tackier matrices allow better mill banding.

Low viscosity is useful in prepregging, but may result in resin
segregation in molding; whereas matrix viscosity will alter the
velocity profile that determines the shearing effect on fiber orientation, a high level may cause attrition of brittle reinforcements.

Some additional details are given on the design of extrusion
dies for producing straight and smooth extrudates from fiber reinforced elastomers. These results will translate in form to other
types of polymers.

ACKNOWLEDGMENT

The collaboration of my colleague John A. Sezna on the profile
extrusions of rubber composites is greatly appreciated.

REFERENCES

1. F. Folgar and C. L. Tucker, "Properties of Short Fiber
   Reinforced Polymers I. Fiber Orientation During Flow",
   39th SPE ANTEC, Boston, May 1981, p. 35.

2. M. Takano, "Viscosity Effects on Flow Orientation of
   Short Fibers", Monsanto/Washington University ONR/ARPA
   Ass'n. Report No. HPC 73-165, U. S. Gov't. Report
   AD 772563, 1973.

3. G. G. Stankoi, E. B. Trostyanskaya, Yu. N. Kazanski,
   V. V. Okorokov, and Ya. Mikhasenok, Soviet Plastics,
   p. 47 (Sept. 1968).

4. K. A. Murty and G. F. Modlen, Poly. Eng. Sci., 17,
   848 (1977).

5. G. E. G. Bagg, M. E. N. Evans, and A. W. H. Pride,
   Composites, 97 (Dec. 1969).

6. L. Kacir, M. Narkis and O. Ishai, Poly Eng. Sci., 17,
   234 (1977).

7. M. Takano, "Rheology of Rod-like Particles in Viscous
   Media, Part I. Formation of Composites from Single
   Fibers", AD 849-285, Feb. 1969.

8. L. A. Goettler, T. B. Lewis and L. E. Nielen, ACS
   Polymer Preprints, 14(1), 436(1973).

9. H. M. Andersen and D. C. Morris, 23rd SPI Reinforced
    Plastics/Composites Conference, 1968, Section 17-E.

10. L. A. Goettler, Modern Plastics, 48, 140 (April 1970).

11. L. A. Goettler, A. J. Lambright and R. I. Leib, Rubber
    Chem. Technol., 52, 838 (1979).

12. L. A. Goettler, SPE Annual Technical Conference, Montreal,
    1973, p. 559.

13. R. J. Crowson, M. J. Folkes and P. F. Bright, Poly. Eng.
    Sci., 20, 925 (1980).

14. L. A. Goettler, SPI Reinforced Plastics/Composites Con-
    ference, 1970, Section 14-A.

15. W. W. Chan, J.-M. Charrier, S. Padmanabhan and P. Vadnais,
    39th SPE ANTEC, Boston, 1981, p. 53

16. A. Y. Coran, K. Boustany and P. Hamed, Rubber Chem.
    Technol., 47, 396 (1974).

17. A. Y. Coran, P. Hamed and L. A. Goettler, Rubber Chem.
    Technol., 49, 1167 (1976).

18. E. A. Cole, F. N. Cogswell, J. Huxtable and S. Turner,
    Poly Eng. Sci., 19, 12 (1979).

19. H. L. Goldsmith and S. G. Mason, "The Microrheology of
    Dispersions", in Eirich, F. R., ed., "Rheology",
    Vol. 4, Academic Press, N. Y. 1967, pp. 86-250.

20. L. A. Goettler, A. J. Lambright, R. I. Leib and P. J.
    DiMauro, Rubber Chem. Technol., to be published May/
    June 1981.

TIME DEPENDENT PROPERTIES OF INJECTION MOULDED COMPOSITES

M.W. Darlington   and   M.A. Christie

Department of Materials
Cranfield Institute of Technology
Cranfield, Bedford MK43 0AL, England

SUMMARY

The uniaxial tensile creep behaviour of several thermoplastics
reinforced with short glass fibres has been studied over a wide
range of temperatures using specimens machined from a variety of
injection moulded bars, discs and plaques. The samples cover a
range of fibre volume fractions from 0.10 to 0.22 and mean fibre
aspect ratios from 18.5 to 43.0. A detailed quantitative assessment of fibre orientation distribution (FOD) and fibre length
distribution has been carried out for each type of moulding used
with each material. This structural information, together with
fibre volume fraction, fibre modulus and matrix creep data, has
been used in the prediction of composite tensile modulus both
parallel and transverse to the major flow direction for each type
of moulding. For the theoretical predictions, the FOD was treated
using the 'laminate analogy' approach developed by Halpin et al;
the properties of a ply containing uniaxially aligned short fibres
being calculated using several different theoretical approaches.
At low strains, one of these approaches led to remarkably good
agreement between theoretical and experimental composite moduli,
over most of the very wide range of materials and temperatures
studied. An empirical extension of this approach is also shown
to give reasonably accurate predictions of composite creep behaviour
at finite strains, where both the matrix and the composite exhibit
non-linear viscoelastic behaviour.

INTRODUCTION

The relatively low stiffness and strength of plastics can be
significantly improved by the incorporation of stiff fibres such as

glass or carbon into the polymer matrix. For maximum benefit,
relatively long fibres are needed, together with the ability to
control fibre orientation. These requirements are easily met when
using thermosetting resins. However, the production techniques
usually employed with fibre reinforced thermosets are not in general
suitable for use with thermoplastics and most of the fibre reinforced
thermoplastics components produced to date have been made using the
conventional thermoplastics injection moulding process. This, and
other conventional thermoplastics processing methods do not, at
present, permit control of fibre orientation and severely limit the
fibre length. There is therefore scope for the development of
special processes which offer orientation control of long fibres in
thermoplastic matrices[1]. However, full exploitation of fibre
reinforced thermoplastics depends on their successful use with the
conventional mass production processes.

The usual methods of compounding the fibres and matrix, and the
processing of the resulting granules by injection moulding, leads
to a broad distribution of fibre lengths in the moulded components.
Methods are available for coating continuous fibres with a thermoplastic. However, the 'long' fibres in the resulting granules are
usually severely degraded during injection moulding, if good fibre
dispersion is required[2]. Fibre lengths in moulded components are
typically in the range 0.05 mm to 1.0 mm. The individual fibres
are usually fairly well dispersed and, with fibre diameter typically
10μ, this gives fibre aspect ratios in the range 5:1 to 100:1.
Unfortunately this is precisely the range where composite stiffness
varies significantly with aspect ratio.

Detailed structural studies on moulded components have shown
that, even in simple mouldings, the fibres are oriented in a complex
manner; the fibre orientation distribution (FOD) varying both through
the thickness and from place to place in the moulding[3-5]. The
resulting anisotropy and inhomogeneity of mechanical properties
leads to difficulties in the presentation of suitable stiffness
design data and the use of this data in design with these short
fibre reinforced thermoplastics (SFRTP)[5-7].

An extensive programme of work aimed at improving our understanding of these materials and hence simplifying both the generation
of relevant engineering data and the associated design procedures
is being carried out at Cranfield[5,8]. An important part of this
programme has been the verification of theories for the influence
of fibre aspect ratio on composite stiffness for SFRTP materials.
Ideally, samples in which the fibres were all of one length,
perfectly uniaxially aligned and uniformly dispersed in the thermoplastic matrix would have been preferred in the initial studies.
Such samples would also have been extremely useful in studies on
the influence of fibre orientation on properties. Unfortunately it
is not possible to prepare such samples in the fibre length range

of interest (0.05 mm to 1 mm for glass fibres of 10μ diameter) and
samples produced by injection moulding have had to be used throughout the programme.

   The above approach was made possible by the careful selection
of mouldings in which most of the fibres were oriented in the plane
of the moulding (planar FOD), and by the successful application of
the technique of contact-micro-radiography (CMR) to the accurate
quantitative determination of the complex, but planar, fibre
orientation distributions in each type of moulded sample used[3,9].
The resulting FOD data, together with data on fibre length, fibre
volume fraction, fibre stiffness and short term stiffness data
on the thermoplastic matrix have been used in simple theories for
the prediction of composite stiffness in uniaxial tension[10],
flexure[11] and simple shear[12]. The predictions were compared with
experimental data, obtained at low strains, for all three modes
of deformation using specimens from a range of bar and plaque
injection mouldings in a range of short glass fibre reinforced
thermoplastics. The level of agreement between theory and experiment
was considered to be very encouraging in view of the complexity of
the fibre orientation and length distributions and the relative
simplicity of the theoretical treatment.

   The above studies were restricted to low strains where the
behaviour of the materials could be regarded as linear viscoelastic.
The possibility of using the same simple theories for the prediction
of composite tensile creep behaviour at finite strains (where both
the matrix and the composite exhibit non-linear viscoelastic
behaviour) was subsequently explored (at room temperature) for a
short glass fibre reinforced polypropylene composite[8]. Again,
encouraging agreement was obtained between theory and experiment.

   The purpose of the present programme was to extend the studies
in uniaxial tension to an examination of the long term creep
behaviour, both at very low strains and at finite strains, for a
wide range of test temperatures and materials. The samples used
covered a range of fibre aspect ratios from 18.5:1 to 43:1 and
fibre volume fractions from 0.10 to 0.22. Some of the mouldings
selected enabled modulus measurements to be carried out both
parallel and perpendicular to the predominant fibre alignment
direction. Materials and test conditions were chosen such that
tests could be carried out below, in and above the glass transition
region of the various thermoplastic matrices. This also provided
varying degrees of time and stress dependence of matrix and composite
moduli. The intention was to provide a more critical assessment
than has been carried out previously of theories for composite
stiffness, by examining their utility over a very wide range of
conditions.

   The range of theories for the influence of fibre aspect ratio on composite stiffness has been extended in this paper and the treatment of fibre orientation distribution is, in all cases, based on the well knownlaminate analogy developed by Halpin and co-workers[13,14]. Much of the basic data obtained in the previous studies[8,10,15] has therefore been included in the new set of calculations in the present paper to obtain a more uniform overview.

It should be emphasised that, at the start of this programme, the main aim was to find the simplest theory for the prediction of composite stiffness in SFRTP materials which would be acceptable to engineers, i.e. would combine ease-of-use with reasonable accuracy over a wide range of materials and test conditions. Some of the theories used to date were therefore selected for their relative simplicity and as being representative of different approaches to the problem.

An outline of the theoretical approaches to fibre length distribution and fibre orientation distribution used in this paper is given in the following section. This is followed by a summary of epxerimental procedures; including details of the materials and mouldings used and the structural and mechanical characterisation of the test samples. A detailed comparison between experimental and predicted composite creep modulus for the linear viscoelastic case is made in the next two sections. Finally the utility of an empirical extension of the theoretical approaches, to permit prediction of composite creep behaviour in the non-linear viscoelastic region, is examined.

THEORETICAL CONSIDERATIONS

For continuous, uniaxially aligned fibres, a rule of mixtures based on a parallel model has been found to give predictions of composite Young's modulus that agree well with experimental values for the tensile modulus of specimens tested in the direction of fibre alignment. On this model, the Young's modulus of the composite, E_c, is given by

$$E_c = E_f V_f + E_m (1 - V_f) \tag{1}$$

where E_f and E_m are the Young's moduli of the elastic fibre and elastic matrix respectively and V_f is the fibre volume fraction.

The above simple equation is only obtained if the Poisson's ratios of the fibre and matrix are assumed to be equal.

For uniaxially aligned fibres of finite length, Equ.(1) may be modified by the inclusion of a length correction factor, η_L, such that

$$E_c = \eta_L \, E_f \, V_f + E_m \, (1 - V_f) \tag{2}$$

Various expressions have been developed for η_L. Based on a 'shear lag' analysis, Cox[16] gives

$$\eta_L = 1 - \frac{\tanh \, (\beta L/2)}{(\beta L/2)} \tag{3}$$

where L is the fibre length and β is given by

$$\beta = \left[\frac{2\pi \, G_m}{E_f A_f \, \ln \, (R/r_o)} \right]^{\frac{1}{2}} \tag{4}$$

G_m is the matrix shear modulus, A_f the cross-sectional area of the fibre, r_o the fibre radius and R the mean separation of fibres normal to their length.

To take account of the wide distribution of fibre lengths encountered in SFRTP materials, Equ.(3) may be modified to give

$$\eta_L = \frac{1}{V_f} \left\{ \sum_{i=1}^{n} \, (V_f)_i \left[1 - \frac{\tanh \, (\beta L_i/2)}{(\beta L_i/2)} \right] \right\} \tag{5}$$

where n is the number of length intervals into which the fibre length distribution is divided.

Equ.(2) can be further modified to include the case of a distribution of fibre orientations :

$$E_c = \eta_o \, \eta_L \, E_f \, V_f + E_m \, (1 - V_f) \tag{6}$$

For the general case of a planar fibre orientation distribution, Krenchel[17] proposed that η_o be given by the simple form

$$\eta_o = \sum_{k=1}^{b} \, a_k \, \cos^4 \, \Theta_k \tag{7}$$

where a_k is the fibre fraction oriented at angle Θ_k to a reference axis in the plane of the distribution. (The reference axis direction will be the direction in which it is desired to calculate E_c). b is the number of angle intervals into which the fibre orientation distribution is divided.

It should be noted that η_o does not give a unique representation of the fibre orientation distribution. Despite this, it has been found useful as an indicator of the overall degree of fibre alignment in each of the various samples used in the present study. (See following section).

The use of Equ.(6), together with Equ.(5) and Equ.(7), represents a very simple method for the prediction of the tensile stiffness of SFRTP materials. Despite its simplicity, reasonable agreement between predicted and measured Young's moduli has been obtained for specimens in which the tensile axis coincided with the direction of predominant fibre alignment or for specimens cut from samples in which the fibre orientation distribution approximated to random-in-the-plane[8,10]. However, it is apparent that these simple equations predict that fibres lying transverse to the direction of applied stress offer no reinforcement to the matrix. It has been shown elsewhere that the Krenchel treatment of fibre orientation becomes significantly inferior to more rigorous treatments when composites containing reasonably well aligned short fibres are tested (in uniaxial tension) transverse to the predominant fibre alignment direction[12].

A more rigorous treatment of the stiffness of a composite in which there was a complex fibre orientation distribution, and fibre length distribution, was developed by Halpin et al[13,14,18]. They proposed that such a composite could be modelled mathematically as a laminated system, whereupon the properties of the "laminate" could be calculated using classical laminated plate theory[19].

Consider, for example, a short fibre composite in which there is a homogeneous, planar, biased fibre orientation distribution, with the fibres all of one aspect ratio. The experimentally measured fibre orientation distribution shows the fraction of fibres oriented in each chosen angle interval in the plane of the distribution. This distribution is modelled as a complex laminate consisting of unidirectionally oriented plies combined to make a laminate which is symmetric about the mid-plane. The total thickness of the plies oriented at an angle Θ_k to a reference axis in the plane of the sheet is proportional to the fibre fraction, a_k, at the angle Θ_k in the measured distribution. The fibre volume fraction in each ply is V_f, the overall fibre volume fraction in the composite. For a ply, the tensile moduli parallel, E_{11}^p, and transverse, E_{22}^p, to the fibre alignment direction, the shear modulus G_{12}^p and the Poisson's ratio ν_{12}^p are required. The stiffness coefficients for the ply, C_{ij}^p, are calculated from these quantities. Standard transformation equations are then used to calculate the stiffness coefficients of each ply with respect to a chosen set of coordinate axes for the laminate, $\bar{C}^P_{ij}$. The stiffness coefficients for the laminate in this coordinate system C^L_{ij}, are then obtained by summing the transformed ply stiffnesses, $\bar{C}^P_{ij}$, through the total thickness of the "laminate"; the $\bar{C}^P_{ij}$ of each ply being weighted according to the fraction of the total laminate thickness it occupies. E^L_{11}, E^L_{22} etc. for the laminate in the chosen coordinate system can be readily calculated from the C^L_{ij}. Full details of this approach may be found elsewhere[14,18,20].

For the samples used in the present study, the fibre orientation distribution changes significantly through the thickness direction (but is approximately symmetrical about the mid-plane) and the experimental FOD data consists of individual FOD's measured at various through-thickness positions[3,12]. For problems in which the macroscopic stress field is uniform and all stresses are in the plane of the laminate (e.g. uniaxial tension), the individual FOD's can be added together, with an appropriate weighting to take account of the relative thickness of each layer of differing orientation. The resulting through-thickness-average FOD can then be used as the basis for the laminate analogy. The present paper is restricted to such problems (i.e. those where, in the equivalent homogeneous material, characterised by the effective moduli of the heterogeneous composite, the state of stress is uniform).

For problems such as flexure, the above simple procedure can not be used. Instead, the effective moduli of each separate layer of differing fibre orientation through the thickness must be calculated separately (using a laminate analogy based on the FOD of the individual layer). The overall flexural 'modulus' of the sample must then be calculated with the appropriate weightings both for the actual thickness and through-thickness position of each layer of fibre orientation. An example of the use of this procedure with SFRTP materials may be found elsewhere[11].

Many possibilities now exist for the calculation of the elastic properties of an individual ply containing uniaxially aligned, discontinous fibres[21,22]. Halpin and Tsai developed simple, generalised equations to approximate the results of more rigorous mathematical analyses, whilst still apparently retaining an acceptable level of accuracy[19,23]. The combination of the Halpin-Tsai equations with the laminate analogy therefore appears to offer a reasonable balance of ease-of-use and mathematical rigour. Furthermore the limited comparisons carried out to date on SFRTP materials with biased FOD's suggest that the approach is capable of reasonable accuracy[10,12,14]. It is therefore one of the approaches that has been assessed in detail in this present study and will be referred to by the abbreviation H-T/L.A. below.

The above discussion has assumed the use of elastic fibres in an elastic matrix whereas the thermoplastic matrices used in the present study exhibit non-linear viscoelastic behaviour. Fortunately, it has been shown that, for quasi-static problems, a (linear) viscoelastic solution can be approximated by an elastic solution in which all the elastic constants are replaced by the corresponding time-dependent moduli or creep compliances[24]. Thus to predict the creep moduli (i.e. the reciprocals of the creep compliances) for the composite in terms of the creep moduli of the matrix we may re-write the Halpin-Tsai equations (taken from reference 25) as :

$$E_{11}^{P}(t) = E_{m}(t) . \left[\frac{1 + 2 A V_f \eta_1 (t)}{1 - V_f \eta_1 (t)} \right] \tag{8}$$

$$\text{where } \eta_1 (t) = \frac{\left[E_f / E_m (t) \right] - 1}{\left[E_f / E_m (t) \right] + 2A} \tag{9}$$

and A = fibre aspect ratio.

$$E_{22}^{P}(t) = E_{m}(t) . \left[\frac{1 + 2 V_f \eta_2 (t)}{1 - V_f \eta_2 (t)} \right] \tag{10}$$

$$\text{where } \eta_2 (2) = \frac{\left[E_f / E_m (t) \right] - 1}{\left[E_f / E_m (t) \right] + 2} \tag{11}$$

$$G_{12}^{P}(t) = G_{m}(t) . \left[\frac{1 + V_f \eta_3 (t)}{1 - V_f \eta_3 (t)} \right] \tag{12}$$

$$\text{where } \eta_3 (t) = \frac{\left[G_f / G_m (t) \right] - 1}{\left[G_f / G_m (t) \right] + 1} \tag{13}$$

$$\nu_{12}^{P}(t) = V_f \nu_f + V_m \nu_m (t) \tag{14}$$

where $E_{11}^{P}(t)$ and $E_{22}^{P}(t)$ are the tensile creep moduli parallel and perpendicular respectively to the direction of fibre alignment and

G_{12}^p (t) and ν_{12}^p (t) are the in-plane shear creep modulus and Poisson's ratio respectively, for the uniaxially aligned ply. $E_m(t)$ is the tensile creep modulus of the matrix at time t, defined by $\sigma_0/\varepsilon(t)$, where σ_0 is the applied constant tensile stress and $\varepsilon(t)$ the tensile strain in the sample measured at time t after application of the constant stress.

Thus to calculate the value of E_{11}^p (t) corresponding to a time under constant stress of 100 seconds at temperature T, the 100 second value of E_m (t), obtained from a creep test on the matrix at temperature T, is inserted into equations (8) and (9). Data for E_{11}^p (t), E_{22}^p (t) etc. for the ply, all calculated for a creep time t, and temperature, T, are then used in the laminate analogy to calculate the creep moduli of the "laminate" for the creep time, t and temperature, T.

For the linear viscoelastic case, and the levels of time-dependence of modulus observed in the present studies, the error associated with the use of this quasi-elastic method is expected to be extremely low.

It is apparent that only E_{11}^p is a function of fibre aspect ratio in the Halpin-Tsai equations. For samples in which there was a distribution of fibre aspect ratios, Halpin and co-workers recommended that the number average mean of the distribution be used in the calculations[14,18]. However, although this can represent a useful simplification, it has been found on occasion to lead to significantly greater deviation between theory and experiment than that obtained with predictions based on the full distribution.[26] Predictions based on the use of the full fibre aspect ratio distribution are therefore used throughout this paper. To take account of the complex fibre length distributions occuring in the samples used in this study, equations (8) and (9) are modified as follows:

$$E_{11}^p \ (t) = E_m \ (t) \sum_{i=1}^{n} \left(\frac{v_i}{v_f}\right) \cdot \left[\frac{1 + 2 \ V_f \ A_i \left[\eta_1 (t)\right]_i}{1 - V_f \left[\eta_1 (t)\right]_i}\right] \qquad (15)$$

$$\left[\eta_1 (t)\right]_i = \frac{\left[E_f/E_m(t)\right] - 1}{\left[E_f/E_m(t)\right] + 2A_i} \qquad (16)$$

where v_i is the volume of fibre of aspect ratio A_i and v_f is the total volume of fibres.

It was noted above that only E_{11}^p is a function of fibre aspect ratio in the Halpin-Tsai equations. It is therefore possible to examine and compare theories for the dependence of E_{11}^p on aspect ratio by replacing the H-T equation for E_{11}^p with other relations for E_{11}^p. Three such relations have been examined in the present study.

Firstly, the relation due to Cox, given by Equ.(2) and Equ.(5) above has been modified to give

$$E_{11}^p(t) = E_f V_f \eta_L(t) + (1 - V_f) E_m(t)$$

where the time-dependence of $\eta_L(t)$ arises from the dependence of $\beta(t)$ on $G_m(t)$, the shear creep modulus of the matrix (see Equ.4). This will be referred to as the Cox/L.A. approach below.

Secondly, a relation given by Ogorkiewicz and Weidmann (Equation 3 of reference 27) for a simple geometrical 'prism-within-prism' model has been modified to allow for a fibre length distribution and time-dependent moduli to give

$$\frac{E_{11}^p(t)}{E_m(t)} = \sum_{i=1}^{n} \left(\frac{v_i}{v_f}\right) \left[\frac{E_f\left[A_i V_f + V_f^{\frac{1}{2}} - V_f\right] + E_m(t) \cdot \left[A_i - A_i V_f - V_f^{\frac{1}{2}} + V_f\right]}{E_f\left[V_f^{\frac{1}{2}} - V_f\right] + E_m(t) \cdot \left[A_i - V_f^{\frac{1}{2}} + V_f\right]} \right] \qquad (17)$$

The Ogorkiewicz and Weidmann model is based on a model due to Counto[28]. The use of the above equation in the laminate analogy will be referred to as the OWC/L.A. approach below.

Thirdly, we use the theoretical predictions of Laws and McLaughlin[29] who investigated the application of the self-consistent method to the problem of determining overall moduli for uniaxially aligned short fibre composites. The theory permits all five compliances of the transversely isotropic composite to be calculated, but the results can not be presented as simple, closed-form equations. Instead, numerical results obtained on a computer for specific values of V_f, E_f, E_m, ν_m and fibre aspect ratio, A, are presented. The computer programme has been used to determine values of E_{11}^p as a function of fibre aspect ratio, A_i, for a limited number of

combinations of V_f, E_m etc. For each such combination, the value
of E^P_{11} to be used in the laminate analogy has been calculated by
summing over the respective fibre length distribution:

$$E^P_{11} = \sum_{i=1}^{n} \left(\frac{v_i}{v_f}\right) . E^P_{11} (A_i) \qquad (18)$$

The use of the above equation in the laminate analogy will be
referred to as the Laws/L.A. approach below.

As an alternative to the above approach, the values of E^P_{11},
E^P_{22}, G^P_{12}, and ν^P_{12} for the ply can <u>all</u> be calculated using the
theory of Laws and McLaughlin. The numerical results presented by
them[29] show that E_{22}, G_{12} and ν_{12} are almost independent of fibre
aspect ratio for aspect ratios greater than 5 and fibre volume
fractions in the range 0.1 to 0.2. However, in this study values
of E^P_{22}, G^P_{12} and ν^P_{12} corresponding to the full fibre length
distribution together with E^P_{11} calculated using equ.(18) have been
used in the laminate analogy to accommodate the FOD. This
procedure will be referred to as the Full Laws/L.A. approach below.
It is apparent that this approach (and the Laws/L.A. approach)
requires extensive use of the computer and its use has therefore
been restricted to a few check calculations in the present paper.
A more detailed study of this approach will be presented elsewhere.

Finally, it must be emphasised that in all the above approaches
it is assumed that the properties of the matrix in the composite
are the same as those measured on the equivalent unfilled thermoplastic.

EXPERIMENTAL PROCEDURES

Materials and Mouldings

The materials used in the present study and their abbreviations
used in the text are as follows :

GFPP   Polypropylene containing (25.7 ± 0.05)% by weight of
            coupled glass fibres of diameter (10.2 ± 0.5) μm. (ICI
            'Propathene' grade HW70GR).

GFPA66   Nylon 66 containing (32.0 ± 0.3)% by weight of short
            glass fibres of diameter (10 ± 0.5) μm. (ICI 'Maranyl'
            grade A190).

GFPET   Poly(ethylene terephthalate) (thermoplastic polyester)
           containing (17.0 ± 0.5)% by weight of short glass fibres
           of diameter (10.3 ± 1.0) μm. (Akzo Plastics 'Arnite' grade
           AV4 331).

GFPBT/20   Poly(butylene terephthalate) containing (20.0 ± 0.05)% by
           weight of short glass fibres of diameter (13.6 ± 1.5) μm.
           (Akzo Plastics 'Arnite' grade TV6 240).

GFPBT/35   Poly(butylene terephthalate) containing (35.9 ± 0.05)% by
           weight of short glass fibres of diameter (13.6 ± 1.5) μm.
           (Akzo Plastics 'Arnite' grade TV4 270).

     The corresponding unfilled grades of PP, PA66, PET and PBT were
also studied.

     The PP, GFPP, PA66 and GFPA66 materials were supplied in the
form of injection moulded ASTM tensile bars and edge-gated discs
of diameter 100 mm and thickness 3 mm (see Fig.1). The PET and
GFPET materials were supplied as injection moulded ISO I tensile
dumb-bells.

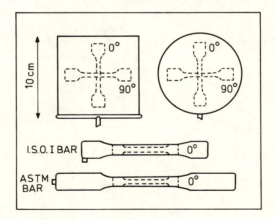

Fig. 1.   Details of the bar, disc and plaque injection mouldings
          showing the sites from which the creep specimens were
          machined in the bars and the definitions of 0° and 90°
          specimens in the discs and plaques.

The PBT and GFPBT materials were supplied in the form of injection
moulded ISO I tensile bars and 2 mm thick, 100 mm square, flash-
gated plaques.

Tensile Creep Studies

Tensile creep tests were carried out on dumb-bell shaped
specimens machined from the mouldings as shown in Fig. 1. The tests
were conducted on highly accurate, well proven apparatus described in
detail elsewhere[30]. The apparatus permits the simultaneous
measurement of axial and lateral strains. In all tests, the
temperature of the specimen was controlled to ± 1°C. For the PA66
and GFPA66 materials, the creep tests were carried out in 'dry' air
on materials in the 'dry' state, to avoid any problems associated
with the influence of moisture uptake. The 100 second tensile
creep modulus data were produced using the procedures defined in
British Standard, BS4618, for the production of 100 second
isochronous stress strain curves[31,32].

Throughout the creep programme, great care was taken to eliminate
differences in the measured moduli of the various samples which might
otherwise have occurred due to differences in storage history between
moulding and testing[33,34]. For all of the long term creep tests at
temperatures above 23°C, the specimens were held at the test
temperature for 18 hours prior to application of the creep load.
A fresh specimen was used for each creep test and the creep data were
corrected, where necessary, for the response of zero load monitor
specimens[32]. (In all cases the correction was either negligible
or fairly small, depending on material and test temperature). For
the 100 second modulus tests at low strain over a range of test
temperatures, the same specimen was tested at each temperature and
in this case, the storage time at each test temperature (prior to
test) was reduced to 2 hours. It is worth noting that misleading
comparisons between theory and experiment can result if considerable
care is not taken in the planning and execution of the mechanical
test programme, for thermoplastics materials.

A summary of the tensile moduli of all the mouldings at 23°C
is given in Table 1.

It should be noted that no direct measurements of matrix shear
modulus, G_m, were carried out for this programme. Instead, G_m
was calculated from the tensile data using the relation

$$G_m(t) = \frac{E_m(t)}{2\left[1 + \nu_m(t)\right]}$$

Table 1. Comparison of Experimental and Predicted
100 Second Tensile Creep Moduli at 23°C

Specimen	100 Sec Creep Modulus (GPa)[a]			
	(0°)		(90°)	
	Exp.	Pred.[b]	Exp.	Pred.[b]
GFPP bar	5.45	5.32		2.81
" disc	3.30	3.25	3.95	4.14
GFPET bar	7.27	7.17		5.00
GFPA66 bar	9.95	9.66		5.68
" disc	7.75	7.86	6.50	6.54
GFPBT/20 bar	7.00	6.60		4.24
" plaque	5.95	6.00	4.00	3.95
GFPBT/35 bar	11.0	10.35		5.17
" plaque	9.75	9.07	5.32	5.42

[a]All moduli measured at 100 second strains of the order of 0.1%
[b]Predictions based on the OWC/L.A. approach.

Again, the error incurred by the use of this pseudo-elastic
treatment is expected to be extremely low and this has also been
confirmed by direct experiment elsewhere[35].

Structural Characterisation

Full structural analyses of all the SFRTP mouldings listed
above were conducted in terms of the fibre orientation distribution
(FOD) in layers through the moulding thickness, the fibre length
distribution (FLD), fibre diameter and fibre volume fraction (V_f),
using techniques described elsewhere[3,9,10]. The samples for these
analyses were taken from the gauge length sections of tested creep
specimens.

The technique of contact micro-radiography (CMR) was used to
determine the FOD[9]. Thin slices were cut through the thickness
from each type of moulding in each material; the plane of the slices
being normal and transverse to the major flow direction in each
type of moulding. CMR examination of these slices showed that,
apart from the ISO I bars of GFPBT/35, the fibres lay predominantly
in planes parallel to the top and bottom surfaces of the mouldings.
Thus to a good approximation, the three-dimensional FOD could be
characterised by the projected FOD in the plane of the mouldings
(i.e. the plane of the page in Fig.1). For this purpose, thin
slices were cut, with the plane of each slice parallel to the plane
of the mouldings, at various levels through the thickness of the

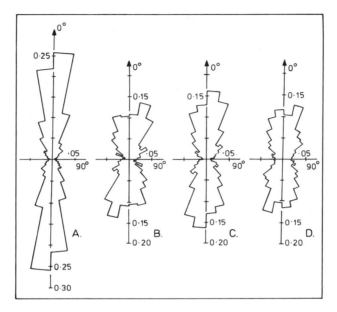

Fig. 2. Total through thickness fibre orientation distributions
         in the plane of the mouldings for (A) GFPBT/35 Bar
         (B) GFPBT/35 Plaque   (C) GFPBT/20 Bar   (D) GFPBT/20
         Plaque.

moulding. The FOD varied significantly with position through the
thickness.   However, by measuring the orientation of the fibres
relative to the same chosen axis from the micro-radiograph of
each slice, the total (through-thickness) projected FOD in the
plane of the moulding could be determined as the weighted sum of
the FOD's of the separate slices. Details of this procedure are
given elsewhere[3].

    Examples of the appearance of the FOD in the through-thickness
slices of the 3 mm disc mouldings in GFPP and GFPA66 may be found

in Fig.6 (discs D and E) of reference 3. These micro-radiographs
also show that the individual glass fibres were reasonably well
dispersed.

Full details of the FLD and the planar FOD in each layer through
the thickness of the GFPBT/20 plaque moulding are given in reference
12. Details of the FLD's and the total (through-thickness)
projected planar FOD's of the bar and disc mouldings of GFPP, GFPA66
and GFPET are given in reference 10. The total projected FOD's in
the plane of the bar and plaque mouldings of GFPBT/20 and GFPBT/35
are presented in Fig.2

For the GFPBT/35 bar mouldings, it was observed that rather more
fibres lay out of the plane of the moulding, compared to all the
other mouldings. The through-thickness slices (i.e. slices with
their plane normal to the plane of the moulding) were therefore
used to estimate the fraction of fibres which were aligned in
selected angle intervals to the plane of the mouldings. In order
to estimate the effect of this out-of-plane component of the FOD
on the predicted values of composite stiffness, the following
procedure was adopted:

The out-of-plane FOD was used in a laminate analogy procedure
to calculate the ply stiffnesses, C_{ij}^{p}. These values were then used
in the full laminate analogy treatment of the FOD in the plane of
the mouldings.

For comparison purposes the above procedure was carried out for
all the GFPBT mouldings. The predicted values of composite stiffness obtained using the projected planar FOD and the above approximation to the full three-dimensional FOD are compared in the
following section. In most cases the difference in the calculated
values is very small.

A summary of the relevant structural information on all the
mouldings is given in Table 2. (It may be of interest to note that
η_{o} would be 0.375 for a random-in-the-plane FOD).

Comparison of the experimental modulus data of Table 1 with the
η_{o} data of table 2 shows a close correlation between the modulus
and FOD data (as represented here by η_{o}). It is apparent that the
FOD in the edge-gated discs of GFPP and GFPA66 is not strongly biased
and the anisotropy of modulus is relatively low. Nevertheless, the
use of the GFPP 0° disc specimen and the GFPA66 90° disc specimen
does still permit a comparison between theory and experiment of
the moduli of these materials transverse to the direction of
predominant fibre alignment. The reversal of anisotropy apparent
between the GFPP and GFPA66 disc mouldings has been discussed in

Table 2. Fibre Volume Fraction, Mean Fibre Aspect Ratio and
Krenchel Orientation Factors for all the Mouldings.

Specimen	V_f	Mean Aspect Ratio	Krenchel Factor, η_o	
			2-D[a]	3-D[b]
GFPP bar	0.109	43.0	0.64	(0.64)
" disc	0.109	39.2	0.45[c]	(0.45)[c]
GFPET bar	0.100	32.8	0.62	(0.62)
GFPA66 bar	0.175	18.5	0.68	(0.68)
" disc	0.175	18.5	0.47	(0.47)
GFPBT/20 bar	0.115	25.1	0.67	0.64
" plaque	0.115	23.9	0.60	0.58
GFPBT/35 bar	0.223	18.8	0.76	0.66
" plaque	0.223	19.4	0.62	0.57

[a]For the projected FOD in the plane of the moulding.
[b]For the full three-dimensional FOD, obtained by allowing
for the out-of-plane component. (The numbers in brackets
indicate those samples for which no allowance was made
for the out-of-plane component in the calculations).
[c]Measured with respect to the 90° direction (In all other
cases, η_o is measured with respect to the 0° direction).

detail elsewhere[5,8].

The GFPBT plaque mouldings show a greater degree of fibre
alignment than the discs, and the modulus values in the 0° direction
of the plaques are not much below those obtained for the respective
ISO I bar specimens.

No direct measurements of the elastic properties of the glass
fibres were carried out during this study. Instead, values of
72.0 GPa and 0.20 for the fibre Young's modulus and Poisson's
ratio respectively were taken from the literature[36].

COMPARISON OF EXPERIMENTAL AND PREDICTED 100 SECOND MODULI

This section is concerned with data obtained at strains of
the order of 0.1%. In this strain region, it was found experimentally that the behaviour of both the unfilled thermoplastics and the
fibre reinforced composites could be regarded as linear viscoelastic
(i.e. the 100 second creep moduli were virtually independent of
strain, or stress, for 100 second strains <0.2%)

A comparison between predicted and measured 100 sec. moduli
at 23°C for the edge-gated discs of GFPP and GFPA66 is presented

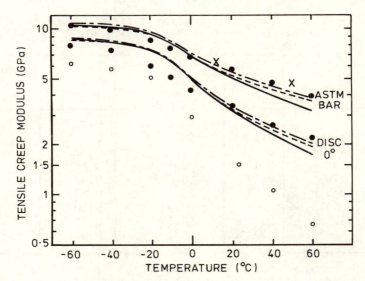

Fig. 3.   Variation of low strain, 100 second tensile creep
          modulus with temperature for PP and the bar and
          0° disc specimens of GFPP.
          Experimental data : un-reinforced PP ○ , GFPP ●
          Predictions ;   —— — ——   OWC/L.A., ─ ─ ─ Cox/L.A.
                          ——————————   H-T/L.A.,   X   Laws/L.A.

in Table 1. The disc anisotropy is correctly predicted in each
case and the OWC/L.A. modulus predictions agree closely with
experiment. (The Cox/L.A. predictions were particularly low at both
0° and 90° for the GFPA66 disc[10].)

    The variation of experimentally measured 100 second tensile creep
modulus with temperature for the tensile bar specimens of GFPP,
GFPET and GFPA66, the 0° specimens from the GFPP disc and the 0° and
90° specimens from the GFPBT/20 and GFPBT/35 plaque mouldings are
presented in Figs.3 to 7. Comparable experimental data on the
unfilled thermoplastic are included in each Figure. It is apparent
that the data span the transition region for each material.

    Also included in Figs. 3 to 7 are the predictions of composite
stiffness, determined using the modulus data shown for the unfilled
matrix material, together with the structural data for the composite,
in the four variations of the laminate analogy approach discussed
above.

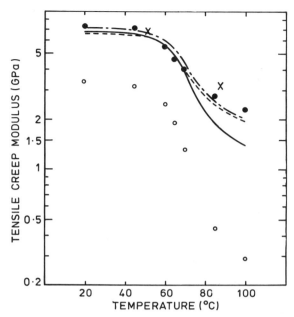

Fig. 4.   Variation of low strain, 100 second tensile creep
          modulus with temperature for PET and the bar specimen
          of GFPET.
          Experimental data : unreinforced PET ○ , GFPET ●
          Predictions : symbols as Fig. 2.

        Similar presentations for the ISO I bar data of GFPBT/20 and
GFPBT/35 may be found in Fig.4 of reference 26. Some data from
this Figure are summarised in Table 3.

        The predictions for the GFPBT/20 and GFPBT/35 samples were
based on their respective full, three-dimensional FOD's. The
effect on the predicted composite stiffness of using the projected
FOD in the plane of the moulding (2-D) instead of the 3-D FOD is
illustrated in Table 4 for calculations based on the OWC/L.A.
approach. Similar trends were noted for the predictions of the
H-T/L.A. and Cox/L.A. approaches. It is apparent that the difference
in the values of the predicted moduli,for the 2-D and 3-D FOD's,
is only significant for the ISO I bar of GFPBT/35, being 7% at 0°C
and 10% at 80°C for this specimen. In all other cases the difference
is of the order of 3% or less. The error associated with the use
of the 2-D FOD for the GFPET, GFPA66 and GFPP samples, will therefore

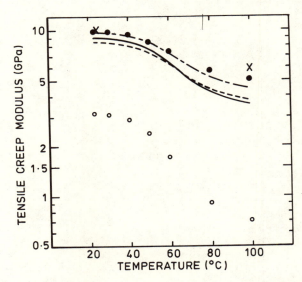

Fig. 5.   Variation of low strain, 100 second tensile creep
          modulus with temperature for PA66 and the bar
          specimen of GFPA66.
          Experimental data : unreinforced PA66 O , GFPA66 ●
          Predictions : symbols as Fig.2.

be very small.   (The FOD in these samples was at least as planar as
that for the GFPBT/20 samples).

     The following points emerge from an examination of Figs. 3 to
7 and Table 3 for temperatures below the transition region :

     There is remarkably close agreement between the experimental
values and the predictions of composite modulus, both parallel and
perpendicular to the direction of predominant fibre alignment. This
is particularly encouraging in view of the complexity of the fibre
orientation and length distributions in the test samples.

     For the predictions of composite modulus parallel to the
predominant fibre alignment direction (i.e. the bar and 0° plaque
specimens) there is a clearly discernable trend in the level of
agreement between the Cox/L.A. and H-T/L.A. predictions and
experiment as a function of fibre aspect ratio, A.   Thus for the

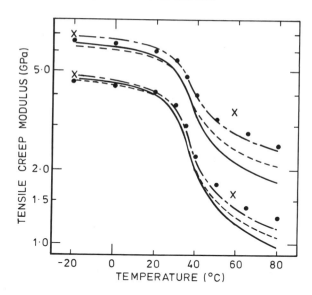

Fig. 6. Variation of low strain, 100 second tensile creep
modulus with temperature for the 0° and 90° plaque
specimens of GFPBT/20.
Experimental data : unreinforced PBT (see Fig.7),
GFPBT/20 ●
Predictions : symbols as Fig.2.

GFPP bar ($A = 43, V_f = 0.11$), the predicted values of modulus for
all four approaches show a spread of only 5% at -60°C and all lie
very close to the experimental value. For the GFPET bar ($A = 34$,
$V_f = 0.10$), the predicted values of modulus for the Cox/L.A. and
H-T/L.A. approaches fall below the experimental value at 20°C, and
all four of the predicted values show a spread of 9%. For the 0°
plaque and tensile bar specimens of GFPBT/20 ($A = 25$, $V_f = 0.115$),
the four predicted values of modulus show a spread of some 13% at
-20°C and, on average, the Cox/L.A. and H-T/L.A. predictions fall
further below experiment. For the 0° plaque and tensile bar
specimens of GFPBT/35 ($A = 19$, $V_f = 0.22$) at -20°C, and the GFPA66
bar specimen ($A = 18.5$, $V_f = 0.18$) at 20°C, the spread in the four
predicted values of modulus has increased to some 17% and, on
average, the Cox/L.A. and H-T/L.A. predictions show their greatest
deviation from experiment.

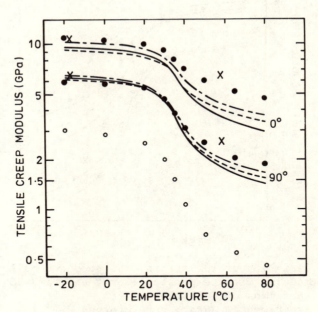

Fig. 7.   Variation of low strain, 100 second tensile creep
          modulus with temperature for PBT and the 0° and 90°
          plaque specimens of GFPBT/35.
          Experimental data : unreinforced PBT O , GFPA66 ●
          Predictions : symbols as Fig.2.

Table 3.   Comparison of Experimental and Predicted 100 Second
           Tensile Creep Moduli at -20°C and 80°C for the ISO I
           Bar Specimens of GFPBT/20 and GFPBT/35.

Material	Temp. (°C)	100 Sec Creep Modulus (GPa)[a]			
		Exp.	OWC/LA	H-T/LA	Cox/LA
GFPBT/20	-20	7.8	7.27	6.78	6.62
"	80	3.28	2.75	2.04	2.44
GFPBT/35	-20	12.0	11.17	10.44	9.88
"	80	5.6	4.27	3.40	3.90

[a]All moduli measured at 100 second strain of the order of 0.1%

Table 4.   Comparison of Experimental Tensile Creep Moduli
            at 0°C and 80°C with Predicted Values of Composite
            Moduli Obtained, Either Using the Projected FOD in
            the Plane of the Moulding (2-D), or Allowing for
            the Out-of-Plane FOD (3-D).

Specimen		100 sec. Creep Modulus (GPa)[a]					
		0°C			80°C		
		Exp.	2-D	3-D	Exp.	2-D	3-D
GFPBT/20 Bar		7.40	7.08	7.01	3.28	2.82	2.75
"	Plaque 0°	6.42	6.65	6.64	2.50	2.43	2.41
"	Plaque 90°	4.35	4.49	4.55	1.28	1.14	1.16
GFPBT/35 Bar		11.70	11.58	10.83	5.60	4.73	4.27
"	Plaque 0°	10.55	10.15	10.02	4.75	3.84	3.73
"	Plaque 90°	5.80	6.06	6.23	1.89	1.59	1.65

[a]Predicted values obtained using the OWC/L.A. approach.
 (All moduli correspond to 100 second strains of the order of
 0.1%).

     No trends with fibre aspect ratio are apparent in the predictions
of composite modulus transverse to the direction of predominant
fibre alignment (i.e. for the 0° disc and 90° plaque specimens).

     It is noticeable that the few modulus predictions made using the
Laws/L.A. approach are always higher than those of the H-T./L.A.
approach. A similar trend was noted in a recent, wholly theoretical
study by Chou et.al. in which they compared predictions for the
'longitudinal' modulus of uniaxially aligned short fibres obtained
using the Halpin-Tsai equations and the self-consistent approach[37].

     The following points emerge from an examination of Figures 3
to 7 and Table 3 for temperatures above the transition region:

     Above the transition region significant differences develop
between the experimental values and the predictions of composite
modulus, both parallel and perpendicular to the direction of
predominant fibre alignment, for all four theoretical approaches.
In all cases the Cox/L.A., OWC/L.A. and H-T/L.A. predictions fall
below experiment whilst the Laws/L.A. predictions lie above the
experimental values. The H-T/L.A. approach shows by far the biggest
discrepancy with experiment, whereas, below the transition region,
the Cox/L.A. approach gave the biggest error.

     As the modulus predictions of any one approach fall the same
side of the experimental values at 0° and 90° for the plaque

specimens, the observed discrepancies between theory and experiment
are unlikely to be due to an error in the FOD determination. Thus,
for example, in the H-T/L.A. approach, if the 'correct' FOD had more
fibres tending to lie in the 0^U direction, this would have the effect
of raising the predicted modulus in the 0° direction, but the predicted modulus in the 90° direction would then be even lower.

Other calculations concerning the effects of possible errors in
FOD also show that errors in FOD measurement alone can not account
for the discrepancies that develop between theory and experiment as
the test temperature is raised beyond the transition region[26].

The smallest discrepancy between the experimental composite
moduli and the predictions of all four theoretical approaches, above
the transition region, again occurs with the GFPP samples (A = 43,
V_f = 0.11).

Overall, the best agreement between theory and experiment is
obtained using the OWC/L.A. approach but even this approach gives
errors of 22% and 24% for the moduli of the 0° plaque and ISO I bar
specimens respectively of GFPBT/35 (A = 19, V_f = 0.22).

Apart from the above two observations, trends in the agreement
between theory and experiment as a function of aspect ratio are not
as clear as those observed at temperatures below the transition
region.

In general, comparison of the experimental modulus data, and the
predictions of all four theoretical approaches, in the transition
region itself, shows a steady trend with increasing temperature and
no surprising features. The main exception to this appears to be
the GFPP 0° disc specimen, where the predicted modulus values
deviate significantly above the experimental values in the transition
region. There is a hint of a similar trend in the GFPP bar specimen.
One possible explanation for this effect could be that the properties
of the PP matrix in the GFPP samples are not the same as those
measured on the unfilled PP specimens. However, it is apparent
that the trend could also be due to a combination of small errors
in the FOD measurement for the disc and the modulus measurements
themselves. Further work is needed to clarify this point.

Finally, it is of interest to note the effect on the predicted
composite moduli of using the Full Laws/L.A. approach, rather than
the simpler Laws/L.A. approach used above (see section on Theoretical
Considerations for an explanation of these approaches). The Full
Laws/L.A. approach has been used to calculate the tensile modulus
of the GFPA66 bar at 23^OC and 100^OC. Values of 10.26 GPa and 6.11
GPa were obtained. These values may be compared with values of

10.16 GPa and 6.06 GPa obtained, at 23°C and 100°C respectively, using the Laws/L.A. approach. It is apparent that, in this example, the modulus values obtained using the more rigorous Full Laws/L.A. approach do not differ significantly from those obtained using the simpler Laws/L.A. approach.

COMPARISON OF EXPERIMENTAL AND PREDICTED LONG TERM CREEP BEHAVIOUR

Linear Viscoelastic Region

The comparisons in the preceding section were carried out using data obtained from creep tests of 100 second duration, at very low strains, where both the matrix and the composite could be regarded as linear viscoelastic.

As a first extension of this study, long term creep data obtained for several stress levels (and hence covering a range of finite strains) have been extrapolated to very low strains and hence experimental creep moduli for the matrix and the composite determined for the 'linear viscoelastic region'. At each creep time of interest, t, the matrix modulus, E_m (t), has then been inserted into the theories to give the predicted composite modulus for the same creep time, t. Two examples of the results of such an exercise are given in Fig.8 for the 0° disc specimen and the tensile bar specimen of GFPP. The results of a similar exercise for the 0° and 90° plaque specimens of GFPBT/35 are presented in Fig.6 of reference 26. (Note in Fig.6 of ref.26, the upper set of data are for the 0° plaque specimen and the lower set for the 90° plaque specimen).

For all four examples, the shape of the creep modulus/log time curve is accurately predicted. The OWC/LA approach predicts both the shape and the absolute magnitude extremely accurately.

Non-linear Viscoelastic Region

At finite strains, both the composite and the matrix exhibit non-linear viscoelastic behaviour. It is apparent that the rigorous extension of any of the theoretical approaches used above to the non-linear viscoelastic case would lead to a level of complexity that would be difficult to justify at the present time. The approach adopted here is therefore empirical and was intended to be of assistance to those concerned with data generation for SFRTP materials.

The choice of matrix creep data on which to base the creep modulus/log time relation required for use in the prediction of the composite creep curve at a particular stress level is by no means obvious. The following approach has been adopted in this paper :

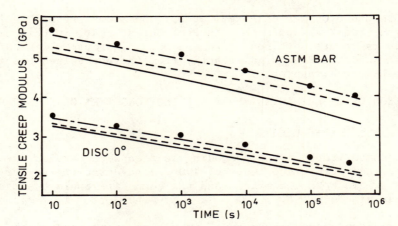

Fig. 8. Variation of experimental and predicted tensile creep
         modulus with log (time) for the bar and 0° disc specimens
         of GFPP at 23°C.
         ●   Experimental data (obtained by extrapolation to
               very low strains).
         Predicted data : —— - —— OWC/L.A., — — — Cox/L.A.
                          ——— H-T/L.A.
         (All predictions based on unreinforced PP data
         extrapolated to very low strain).

We require to predict a creep curve (i.e, tensile strain versus
log time) for the composite. The 100 second strain, x, of the
required composite creep curve is noted. A matrix curve having a
100 second strain of x is produced, either by direct experiment
(i.e. selecting the stress, σ_m, such that the 100 second strain in
the test will be x) or by interpolation between existing creep
curves[38]). From this matrix creep curve at stress σ_m the variation
of matrix creep modulus, E_m (t), with time is calculated. At each
selected value of time, the appropriate value of E_m (t) is inserted
into the theories to give the corresponding value of the predicted
composite modulus for the same creep time. Hence the creep modulus/
log time relation for the composite is established. This relation
is considered to be applicable for a composite creep test in which
the 100 second strain is x. The appropriate stress level, σ_p, for
this data is therefore obtained by multiplying the predicted 100
second creep modulus for the composite by x. The predicted

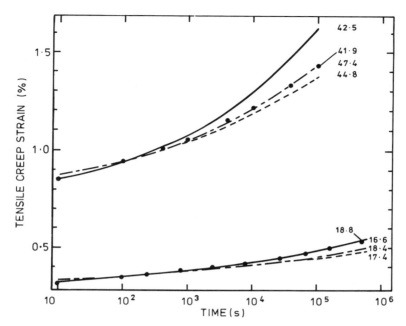

Fig. 9.   Variation of tensile creep strain with log time
for the GFPP bar specimens at 23°C.
Experimental, data: ●
Predicted data :   —— — ——— OWC/L.A.,   ———— H-T/L.A.,
— —— Cox/L.A.
(The label for each curve is the associated stress level
in MPa).

composite creep strain at each selected time is then obtained by
dividing σ_p by the appropriate value of the predicted composite
creep modulus.   In this way the variation of tensile strain with
time for the composite is predicted for a 100 second strain of x
and a stress, σ_p.

The above procedure can be repeated for other selected composite
stress (or 100 sec. strain) levels, and hence a family of creep
curves established for a range of stress levels.

This procedure is referred to as the "equal 100 second strain"
approach.   It has been found to give better predictions than the

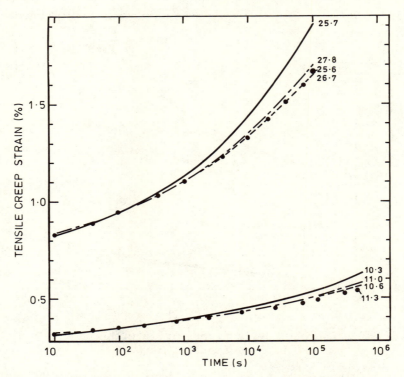

Fig. 10. Variation of tensile creep strain with log time for
          the GFPP 0° disc specimens at 23°C (symbols as for
          Fig.9).

"equal stress" approach in which matrix creep data obtained at the
same stress level as that for the composite is used[8].

    Predicted creep curves obtained using the above procedure are
compared with experimental creep data, at two stress levels, for
the GFPP bar and 0° disc specimens in Figs. 9 and 10 and at one
stress level for the GFPA66 bar and 90° disc specimens in Fig.11.
The predictions are based on the Cox/L.A., OWC/L.A. and H-T/L.A.
approaches. Each predicted creep curve is labelled with the stress
(σ_p) that is predicted to be necessary to produce it for the
particular approach. These values of σ_p can then be compared with
the stress level, σ_e, used to produce the experimental data. When

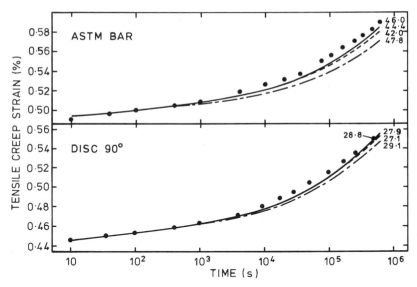

Fig. 11. Variation of tensile creep strain with log time for the
bar and 90° disc specimens of GFPA66 at 23°C. (Symbols
as for Fig.9).

examining Figs. 9 to 11 it must be remembered that the chosen
procedure matches the strains at 100 seconds. The points of
interest are therefore the difference between σ_p and σ_e and the
comparison of predicted and experimental creep curve shape.

Examination of Figs. 9 to 11 shows surprisingly good agreement
between the predicted and experimental creep curve shapes and the
associated stress levels, particularly for the OWC/L.A. approach.
The agreement at the higher stress (or strain) levels for the GFPP
specimens is particularly noteworthy since the PP and GFPP materials
are highly non-linear at strain levels above 1% and GFPP is not
normally used at such strain levels in service.

Similar detailed comparisons of experimental and predicted
composite creep behaviour, at two stress levels, for the GFPBT/20
and GFPBT/35 bar and plaque specimens may be found in Figs. 7 to 11
of reference 26. The data were obtained at three temperatures :
23°C where the materials were creeping into the transition at long
times, 30°C where the materials were creeping through the transition
and 50°C where they were creeping out of the transition. These

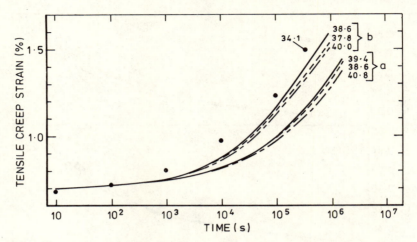

Fig. 12.   Variation of tensile creep strain with log time for
           the 90° plaque specimen of GFPBT/35 at 23°C.
           ●   Experimental data (from Fig.8 of reference 26)
           a   Predictions from the standard "equal 100 second
                    strain" approach.
           b   Predictions based on the "equal 100 second strain"
                    approach but with allowance for reduced matrix
                    volume in the composite.
           Predicted data : ── ─ ── OWC/L.A.,   ──── H-T/L.A.,
                             ─ ── Cox/L.A.
           (The label for each curve is the associated stress level
           in MPa).

data therefore provide a very critical test of the empirical approach
used in the predictions. Again, good agreement was obtained in
general between predicted and experimental creep curve shapes and
the associated stress levels, particularly for the OWC/L.A. approach.

   The main exception to the above-noted agreement was for the 90°
plaque specimen of GFPBT/35 at the higher stress level at 23°C,
although it must be emphasised that the tensile strain reached
the rather high level of 1.5% at long times (see Fig.12 of this
paper). It is also worth noting that this material contains the
highest volume fraction of fibres (V_f = 0.22). In the above "equal
100 second strain" approach, no allowance is made for the reduced
volume of thermoplastic in the composite ; it is the macroscopic
strains that are matched. This is most likely to cause errors at
the higher strain levels (since modulus changes more rapidly with

increasing strain at higher strains) and for the higher volume
fractions.

The result of allowing for the reduced matrix volume in the
composite when carrying out the "equal 100 second strain" approach
is illustrated in Fig.12 for the higher stress level of the GFPBT/35
90° plaque data. The agreement between the experimental and
predicted creep curve shapes is certainly improved by the modified
procedure. Applying this procedure to the 0° plaque specimen
similarly leads to improved agreement.

Further developments in this empirical approach to the
predictions of long term creep behaviour in the non-linear viscoelastic region, and its application to data generation for SFRTP
materials, are in progress.

It should be noted that the fibres are well coupled to the matrix
in each of the composites studied in this programme. It is well
known that, at <u>finite</u> strains, the creep behaviour of a well-coupled
SFRTP material will differ significantly from that of the equivalent
material in which there is no coupling between the fibres and
matrix[39,40]. The above empirical approach can not therefore also
be expected to give reasonable predictions for poorly coupled
systems.

CONCLUDING REMARKS

The fibre orientation and length distributions (FOD and FLD) in
injection mouldings of short fibre reinforced thermoplastics
(SFRTP) are relatively complex. Nevertheless, remarkably accurate
predictions of composite stiffness have been obtained by using the
<u>quantitative</u> FOD data as the basis for a "laminate analogy" and
calculating the stiffnesses of the basic ply of uniaxially aligned
short fibres using experimental data on fibre length distribution,
fibre volume fraction and matrix modulus. The general level of
agreement between theory and experiment reported above suggests
that this approach with injection moulded samples provides a reasonable method for the evaluation of theories for the influence of
aspect ratio on composite stiffness for SFRTP materials.

The procedure employed for the quantitative determination of
the fibre orientation distributions in the present study is somewhat laborious, but rapid, automated versions are being developed[41].
These should greatly increase the usefulness of the approach.

At very low strain, the behaviour of all the materials used in
this study approximated to linear viscoelastic. In this region,
the use of the simple Halpin-Tsai equations for the calculation of
the stiffnesses of the uniaxially aligned short fibre plies, in

conjunction with the laminate analogy, gave reasonably good predictions of composite tensile moduli at temperatures below the glass transition region for each material and hence for a range of fibre aspect ratios and volume fractions. However, above the glass transition the predicted values were significantly lower than the experimental values.

At temperatures below the glass transition region, the OWC/L.A. and the Laws/L.A. approaches yield very accurate predictions of composite tensile moduli over the entire fibre aspect ratio and volume fraction ranges encompassed by the range of materials used in this study. However, at temperatures above the glass transition, the OWC/L.A. predictions fall below experiment, whilst the Laws/L.A. predictions are higher than experiment.

Of the four theoretical approaches investigated, closest agreement between theoretical and experimental composite tensile moduli, over the widest range of materials and test temperatures, was obtained by replacing the equation for E_{11}^p in the Halpin-Tsai

equations by a simple equation due to Ogorkiewicz and Weidmann (i.e. the OWC/L.A. approach). However even this approach gave errors approaching 20% above the transition region for some samples.

All four theoretical approaches investigated yield predictions that deviate more from experiment at temperatures above the glass transition than at lower temperatures. It is apparent that, for each material, the ratio of E_f : E_m is very much higher above the

transition than below it. The trend observed could therefore reflect an increasing inadequacy of all four theoretical approaches as E_f :

E_m increases, rather than any special significance in operation above

the glass transition region.

At this point it is of interest to note the results of a preliminary study on tensile bars of a polycarbonate reinforced with 13% by volume of extremely rigid short fibres. For this material, the ratio E_f : E_m was approximately 80 at 23°C (i.e. the ratio E_f :

E_m at a temperature well below the glass transition of polycarbonate

was similar to that found at temperatures above the glass transition for the materials reported above). The results obtained to-date suggest that the OWC/L.A. approach again gives the closest agreement between the predicted and experimental tensile moduli; the Laws/L.A. prediction being some 10% higher and the Cox/L.A. and H-T/L.A. predictions some 20% lower than the OWC/L.A. value. This pattern is therefore similar to that observed at temperatures above the transition region for the four glass fibre reinforced thermoplastics described in detail in this paper.

The above remarks were limited to the deformation of the materials at very low strains, where the behaviour approximated to linear viscoelastic. In this region, and over the temperature ranges employed in this study, the composites stiffness is not expected to be dependent on the degree of bonding between matrix and fiber (provided the phases remain in good contact). Some supporting experimental evidence is available[42].

At finite strains, all the fibre reinforced and the unfilled thermoplastics used in this study exhibited non-linear viscoelastic behaviour. In this region, the creep behaviour of the composite is known to be dependent on the degree of coupling between fibre and matrix. The fibres are considered to be well-coupled to the matrix in all of the SFRTP materials used in the present study. For all of these materials it has been found that reasonably accurate predictions of composite creep behaviour at finite strains can be made over a wide temperature range using the OWC/L.A. approach and matrix creep data obtained on an "equal 100 second strain" basis.

Finally, it is apparent that the results of this study emphasise the importance of using a wide range of test conditions when attempting to assess the utility of any theory for the stiffness of short fibre reinforced thermoplastics.

ACKNOWLEDGEMENTS

Grateful acknowledgement is made to the Polymer Engineering Directorate of the Science Research Council (U.K.) and to Akzo Plastics bv for support of this work. Thanks are also due to ICI Ltd. (Plastics Division) and Akzo Plastics bv for the supply of injection moulded samples. Much of the experimental data was obtained with considerable care by Dr. G.R. Smith, Mr. P. Hopkin and Mr. J. McEvoy. Finally thanks are due to Professor N. Laws for valuable discussions and for making available to us the computer programme associated with his theory for the effect of fibre length on composite stiffness.

REFERENCES

1. L.N. Philips, Fabrication of reinforced thermoplastics by means of the film stacking technique, in ; "Fabrication Techniques for Advanced Reinforced Plastics", Proceedings of Symposium at Salford University, UK, April, 1980, IPC Science and Technology Press, Guildford, UK(1980).

2. M.W. Darlington, B.K. Gladwell and G.R. Smith, Structure and properties in injection moulded discs of glass fibre reinforced polypropylene, Polymer, 18 ; 1269 (1977).

3. M.W. Darlington, P.L. McGinley and G.R. Smith, Structure and anisotropy of stiffness in glass fibre reinforced thermoplastics, J. Mater. Sci., 11 : 877 (1976).

4. D. McNally, Short fibre orientation and its effects on the properties of thermoplastic composite materials, Polym. Plast. Technol. Eng., 8 : 101 (1977).

5. P.F. Bright and M.W. Darlington, Factors influencing fibre orientation and mechanical properties in fibre reinforced thermoplastics injection mouldings, Plastics and Rubber Processing and Applications, 1 : 139 (1981).

6. R.C. Stephenson, Design information for thermoplastic composites : coupled glass-fibre-reinforced polypropylene, Plastics and Rubber : Mats. and Applications, 4 : 45 (1979).

7. R.C. Stephenson, S. Turner and M. Whale, The mean stiffness of thermoplastic composites, Composites, 10 : 153 (1979).

8. M.W. Darlington and G.R. Smith, Design data for stiffness in short fibre reinforced thermoplastics, in "Fibre Reinforced Materials : Design and Engineering Applications", Proceedings of Conference at Inst. Civil Eng., UK, March, 1977, Inst. Civil Engineers, London (1977).

9. M.W. Darlington and P.L.McGinley, Fibre orientation distribution in short fibre reinforced thermoplastics, J. Mater. Sci., 10 : 906 (1975).

10. M.W. Darlington, P.L. McGinley and G.R. Smith, Creep anisotropy and structure in short fibre reinforced thermoplastics : Part 1, Prediction of 100 sec. creep modulus at small strains, Plastics and Rubber : Materials and Applications, 2 : 51 (1977).

11. G.R. Smith, M.W. Darlington and D. McCammond, Flexural anisotropy of glass fibre reinforced thermoplastics injection mouldings, J. Strain Analysis, 13 : 221 (1978).

12. M.A. Christie, M.W. Darlington, D. McCammond and G.R. Smith, Shear anisotropy of short glass fibre reinforced thermoplastics injection mouldings, Fibre Sci. and Tech., 12 : 167 (1979).

13. J.C. Halpin and N.J. Pagano, The laminate approximation for randomly oriented fibrous composites, J. Comp. Mat., 3 : 720 (1969).

14. J.L. Kardos, Structure property relations in short fibre reinforced plastics, CRC Critical Reviews in Solid State Sciences, 3 ; 419 (1973).

15. G.R. Smith, Creep anisotropy in short glass fibre reinforced thermoplastics, Ph.D. Thesis, Cranfield Institute of Technology (1976).

16. H.L. Cox, The elasticity and strength of paper and other fibrous materials, British J. Appl. Phys., 3 : 72 (1952).

17. H. Krenchel, "Fibre Reinforcement", Akademisk Forlag, Copenhagen (1964).

18. J.C. Halpin, K. Jerina and J.M. Whitney, The laminate analogy for 2 and 3 dimensional composite materials, J. Comp. Mat., 5 : 36 (1971).

19. J.E. Ashton, J.C. Halpin and P.H. Petit, "Primer on Composite Analysis", Technomic Publishing Co., Stamford, Conn. (1969).

20. L. Nicolais, Mechanics of composites, Poly. Eng. and Sci., 15 : 137 (1975).

21. T.W. Chou and A. Kelly, Mechanical properties of composites, Ann. Rev. Mater. Sci., 10 : 229 (1980).

22. T.S. Chow, Review - The effect of particle shape on the mechanical properties of filled polymers, J. Mater. Sci., 15 : 1873 (1980).

23. J.C. Halpin and J.L. Kardos, The Halpin-Tsai equations : a review, Poly. Eng. and Sci., 16 : 344 (1976).

24. R.A. Schapery, Viscoelastic behaviour and analysis of composite materials, in : "Composite Materials, vol.2, Mechanics of Composite Materials", G.P. Sendeckyi, ed., Academic Press, New York (1974).

25. J.C. Halpin, Stiffness and expansion estimates for oriented short fibre composites, J. Comp. Mat., 3 : 732 (1969).

26. M.A. Christie and M.W. Darlington, Creep anisotropy in short fibre reinforced thermoplastics, Proceedings of Third International Conference on Composite Materials, Paris, August 1980, published in "Advances in Composite Materials", A.R. Bunsell, C. Bathias, A. Martrenchar, D. Menkes and G. Verchery, eds., Pergammon Press, Oxford (1980).

27. R.M. Ogorkiewicz and G.W. Weidmann, Tensile stiffness of a thermoplastic reinforced with glass fibres or spheres, J. Mech. Engineering Sci., 16 : 10 (1974).

28. V.J. Counto, The effect of the elastic modulus of the aggregate on the elastic modulus, creep and creep recovery of concrete, Mag. Concrete Res., 16 : 129 (1964).

29. N. Laws and R. McLaughlin, The effect of fibre length on the overall moduli of composite materials, J. Mech. Phys. Solids, 27 : 1 (1979).

30. M.W. Darlington and D.W. Saunders, Anisotropic creep behaviour Chapter 10 in "Structure and Properties of Oriented Polymers", I.M. Ward, ed., Applied Science Publishers, London (1975).

31. British Standard, BS 4618, Recommendations for the presentation of plastics design data, Parts 1.1 and 1.1.1. (1970).

32. D.A. Thomas and S. Turner, Experimental technique in uniaxial tensile creep testing, in "Testing of Polymers", volume 4, W.E. Brown, ed., Interscience, New York (1969).

33. M.A. Christie and M.W. Darlington, Tensile creep of polyamide 6, Plastics and Polymers, 43 : 149 (1975).

34. C.M.R. Dunn and S. Turner, Thermal history and mechanical properties, Polymer, 15 : 451 (1974).

35. J.W. Ballard and M.W. Darlington, Shear and tensile creep of thermoplastics at finite strains, Brit. Polymer J., 10 : 79 (1978).

36. R.E. Lowrie, Glass fibres for high strength composites, in : "Modern Composite Materials", L.J. Broutman and R.H. Krock, eds., Addison-Wesley, Reading, Mass. (1967).

37. T.W. Chou, S. Nomura and M. Taya, A self-consistent approach to the elastic stiffness of short fibre composites, J. Comp. Mat., 14 : 178 (1980).

38. S. Turner, Deformation data for engineering design, in : "Testing of Polymers", volume 4, W.E. Brown, ed., Interscience, New York (1969).

39. L.C. Cessna, J.B. Thomson and R.D. Hanna, Chemically coupled glass-reinforced polypropylene, SPE Journal, 25 : 35 (1969).

40. L.C. Cessna, Cyclic creep test data for GR polypropylene, SPE Journal, 28 : 28 (1972).

41. F. Polato, P.Parrini and G. Gianotti, A new technique for the measurement of glass fibre orientation in composite materials,

Proceedings of Third International Conference on Composite Materials, Paris, August, 1980, published in "Advances in Composite Materials", A.R. Bunsell, C. Bathias, A. Martrenchar, D. Menkes and G. Verchery, ed., Pergammon Press, Oxford (1980).

42. W.M. Speri and C.F. Jenkins, Effect of fibre-matrix adhesion on the properties of short fibre reinforced ABS, Poly. Eng and Sci., 13 : 409 (1973).

VISCOELASTIC PROPERTIES OF PARTICULATE REINFORCED MATRICES

C. Migliaresi

Istituto di Principi di Ingegneria Chimica

University of Naples, 80125 Naples, Italy

ABSTRACT

The viscoelastic behavior of both polystyrene and epoxy resin/ glass bead composites has been studied by means of stress relaxation experiments over a wide range of temperatures in the linear visco- elastic region. The glass transition temperature of the thermoset- ting resin is increased because of the presence of the second phase while no effect has been observed for the thermoplastic matrix. Ac- cording to the time-temperature superposition principle all the stress relaxation data have been shifted to obtain single master curves at a reference filler content by taking into account the reinforcing effect of the filler and the glass transition tempera- ture of the matrix.

INTRODUCTION

During the past few years many publications have appeared in the literature on the effect of particulate fillers on the mechani- cal properties of polymeric materials[1]. The increasing use of such polymeric composite systems in many practical applications and the viscoelastic character of the matrices require an accurate predic- tion of their long term mechanical properties.

The characterization of time dependent mechanical properties of polymers and polymer composite systems has its origin in the studies of Voigt, Boltzmann and Maxwell.

Historically J.C. Maxwell (1857) has been credited with the first quantitative discussion on the viscoelasticity of the matter. His proposed model has become a basis for modeling the time dependent behavior of polymers and composites.

The linear viscoelastic behavior, mathematically formulated by Boltzmann (1876), is expressed by the equation[2]:

$$\sigma(t) = \int_0^t E_r(t-q)\,\frac{d\varepsilon}{dq}\,dq$$

where: $\sigma(t)$= stress at time t
E_r= relaxation modulus
ε = strain
q = current time

This definition requires that the stress relaxation modulus and, likewise, all other viscoelastic properties are independent of the state of extension or load at which they are measured. The results obtained by Gauchel[3] through forced dynamic oscillation tests on glass bead composites, indicate that the linear viscoelasticity limit is much lower than that of linear elastic behavior and that it decreases as the filler content and the temperature increase.

The time-temperature superposition principle[4] brings an important simplification to the study of the viscoelastic properties separating the variables, i.e. time and temperature, on which the viscoelastic properties depend and expressing them as function of each variable which can be determined experimentally.

Many equations have been proposed in order to predict the mechanical properties of a composite material by the knowledge of the properties of the matrix and the filler and of its shape and content[5]. Kerner's equation[6] and Eiler's equation[7] have been widely used for particulate composites in their glassy or rubbery state respectively.

The mechanical properties of polymer composites filled with a particulate filler or with short fibers are greatly dependent on the adhesion of phases. The discrepancy between the moduli of particulate composites and the Kerner's prediction reported by Ziegel and Romanov[8] has been attributed to strong interfacial adhesion and immobilization of the matrix layer surrounding the filler particles. Manson and Chiu[9] report for epoxy resin/glass bead systems deviations from predicted mechanical, sorption and permeation behavior which are probably due to restricted mobility of polymer segments because of the presence of the filler. A strong interfacial interaction is believed to underlie the increase in the glassy transition temperature of the matrix with increasing filler content. Ferry[10] reports that, for a filled crosslinked polymer in its transition region, the general shape of the relaxation distribution curve is

not affected by the filler, but that the modulus is increased and
the time dependence is shifted to lower frequencies. In contrast,
Schwarzl[11] reports that, for a series of concentrations of salts
in a rubbery matrix, there is no effect of filler on the glass transition temperature. Nielsen[12], in discussing the effect of filler
on the creep and dynamic properties of filled polyethylene, reports
a modulus effect similar to that reported by Ferry[10]. He also suggests a possible change in the matrix properties due to the filler-
matrix interactions. This corresponds to the idea of Rigbi[13] who
proposed a broadening of the relaxation spectrum due to filler interactions. Analogous results have been obtained by Migliaresi et al.[14]
who show an increase of T_g (glass transition temperature) in the
glass bead/epoxy resin and glass bead/poly-2-hydroxyethylmethacrylate
composites as consequence of the presence of filler.

In this paper the viscoelastic behavior of polystyrene and epoxy
resin glass bead composites has been studied through stress relaxation experiments in the linear viscoelasticity region. The time-
temperature superposition principle and a phenomenological equation
which takes into account the effect of filler on the modulus and on
the glass transition temperature of the composites permit a good
superposition of all the experimental data.

EXPERIMENTAL

The resins used were polystyrene Montedison's Edistir FA and
Shell's epoxy resin Epon 828 cured with triethylenetriamine.

Composites with the 10, 20 and 30 percent by volume of glass
beads were prepared using Vacu-Balst microspheres with a particle
size in the range of 32 to 40 microns. For more details on the
sample preparation see reference 14.

Stress relaxation experiments were performed on an Instron mod.
1112 equipped with a thermostatically controlled oven, at the temperatures of 60, 70 and 80°C for the polystyrene samples and at the
temperatures of 80, 100, 120, 130, 135 and 140°C for the epoxy resin
samples.

The imposed strain was, in all the experiments, equal to 0.0015,
which is in the region of linear viscoelasticity[3].

RESULTS AND DISCUSSION

a) Thermoplastic resin based composites

Stress relaxation moduli for the unfilled polystyrene are re-

ported in Figure 1 at the different tested temperatures. The behavior is typical of thermoplastic polymers with a modulus which quickly decreases as the temperature approaches the T_g of the system (i.e.
83°C[14,15] for both unfilled polystyrene and composites), while the
polymer will reach regions of viscous flow at higher temperatures.
The behavior of the composites, not reported, is similar.

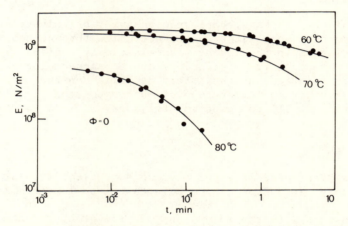

Figure 1.   Stress relaxation moduli, E, versus time t, for unfilled
            polystyrene at different temperatures.

    The time-temperature superposition principle[4], expressed by
the equation:

$$E\ (t,T) = E\ (a_T \cdot t,\ T_r)$$

has been used to obtain the master curves of Figure 2, at the
reference temperature $T_r = 60°C$, which enables us to calculate
the stress relaxation modulus of the material at the time t and
the temperature T by knowledge of the shift factor a_T.

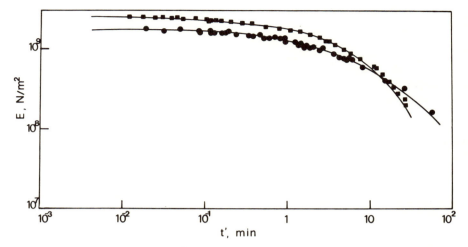

Figure 2.   Stress relaxation modulus master curves versus reduced
time, t'=t·a_T, for unfilled polystyrene (●) and poly-
styrene/glass bead composite at φ=0.3 (■).

The shift factors used are reported in Figure 3 for the different temperatures and indicate that the presence of filler does not
affect the relaxation times of the matrix.

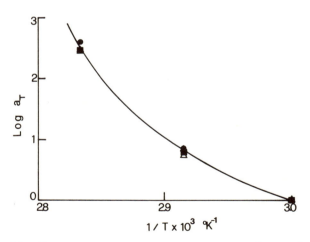

Figure 3.   Shift factors a_T versus temperature for polystyrene/glass
bead composites at φ=0 (●), φ=0.1 (△), φ=0.2 (▲) and
φ=0.3 (■).

A vertical shift, which takes into account the reinforcing effect of the filler, gives the generalized master curve of Figure 4, where in the insert the experimental vertical shift factors K (ϕ), calculated at the low values of time are reported. The procedure gives a good superposition between the data at low times while a significant scatter can be observed between the moduli of the matrix and composites in the transition zone.

The faster decline of the stress relaxation modulus of the higher concentration glass bead composites could be explained by comparison with the similar effect observed in the shear viscosity of glass bead suspensions in non-Newtonian liquids.

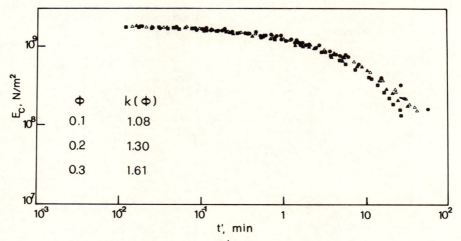

Figure 4.   Generalized stress relaxation master curve at a reference temeprature of 60°C for polystyrene/glass bead composites at ϕ=0 (●), ϕ=0.1 (△), ϕ=0.2 (▲) and ϕ=0.3 (■).

b)   Thermosetting resin based composites.

The effect of filler on the viscoelastic properties of epoxy resin is quite different because of the shift of the glass transition temperature of the matrix as described in the introduction and reported in Table 1 for the epoxy/glass bead composites.

TABLE   1

Glass transition temperature of epoxy resin/glass bead composites.

ϕ	T_g (°C)
0	125
0.1	130
0.2	135
0.3	135

The time-temperature superposition principle permits to obtain, from the stress relaxation data, the master curves reported in Figure 5 where a glassy plateau, a transition region and a rubbery plateau can be observed.

As previously reported for the thermoplastic resin based composites, the shift factors a_T (Figure 6) are not affected by the presence of the filler. Similar results have been obtained by other authors [16,17], also for long fiber composites and for the superposition of different time dependente mechanical properties, at deformation higher than the linear viscoelastic limits of the system.

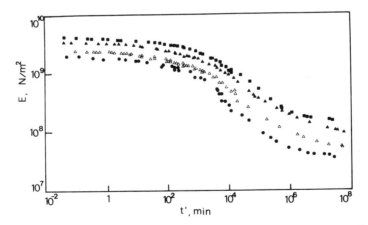

Figure 5.   Stress relaxation modulus, E, versus reduced time, t'=a_T·t, for 'epoxy resin/glass bead composites at the reference temperature of 80°C and at ϕ=0 (●), ϕ=0.1 (△),ϕ=0.2 (▲) and ϕ=0.3 (■).

From the slope of the straight line which fits the experimental points in the glassy region (i.e. 80-125ºC) an activation energy for the relaxation phenomenon can be calculated through the equation[4]:

$$\log a_T(T) = \frac{\Delta H}{2.3 \, R} \left(\frac{1}{T} - \frac{1}{T_r} \right) \qquad (3)$$

where: ΔH = Activation energy
R = Gas constant
T = Temperature of the test
T_r= Reference temperature

The calculated value of 2×10^5 joules/mole agrees quite well with the results obtained from Miyano et al.[17] (i.e. 1.76×10^5 joules/mole) for epoxy resin/long carbon fiber composites.

In order to construct a generalized master curve the shift of the glass transition temperature and the reinforcing effect due to

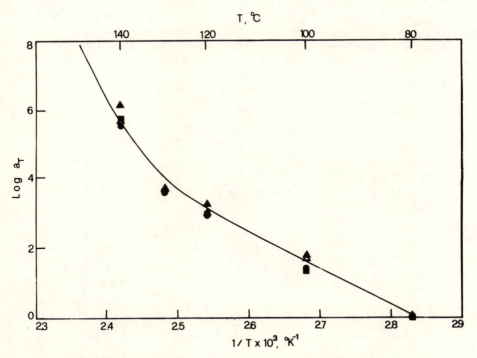

Figure 6.   Shift factors a_T versus temperature for epoxy/glass bead composites at $\phi=0$ (●), $\phi=0.1$ (△), $\phi=0.2$ (▲) and $\phi=0.3$ (■).

the presence of the filler have to be taken into account.

A horizontal shift along the time axis (i.e. the correction due to the different T_g), was performed by translating the master stress relaxation curves of the composites to lower times by a factor $a = t'_o/t'_\phi$ where t'_o and t'_ϕ represent the times corresponding to the mid-point of the master stress relaxation curves of Figure 5, for the unfilled resin and the composites respectively.

The reinforcing effect of the filler in the glassy state may be correlated using the Kerner's equation expressed by[6]:

$$E_r = E_c/E_p = \frac{1 + AB\,\phi\,(\text{eff})}{1 - B\,\phi\,(\text{eff})} \qquad (4)$$

where E_c is the elastic modulus of the composite, E_p is the elastic modulus of the matrix and A, B and $\phi(\text{eff})$ are defined by:

$$A = \frac{7 - 5\nu_p}{8 - 10\nu_p} \qquad (5)$$

$$B = \frac{(E_f/E_p) - 1}{(E_f/E_p) + A} \qquad (6)$$

$$\phi(\text{eff}) = (1 + \frac{(1 - \phi_m)}{\phi_m^2}\,\phi_f)\,\phi_f \qquad (7)$$

ν_p being the Poisson's ratio of the polymer, E_f the elastic modulus of the filler, ϕ_m the maximum packing fraction for the filler and ϕ_f the volumetric filler content in the composite.

In the rubbery state the Eilers equation[7] can be used:

$$E_r = (1 + \frac{1.25\,\phi_f}{1 - \phi_f/\phi_m})^2 \qquad (8)$$

Values of $\nu_p = 0.3$, $E_f = 20$ and $\phi_m = 0.64$ have been used.

By shifting horizontally the master stress relaxation curves by the amount a and each point vertically, at each $t^* = a_\phi \cdot t'$, by the amount $K(\phi)\,a(t^*)$, which is the Kerner's equation in the glassy state, the Eilers equation in the rubbery state and a linear combination of the two equations in the transition zone, one can obtain the stress relaxation generalized master curve reported in Figure 7.

Figure 8 reports the values of the shift factors used.

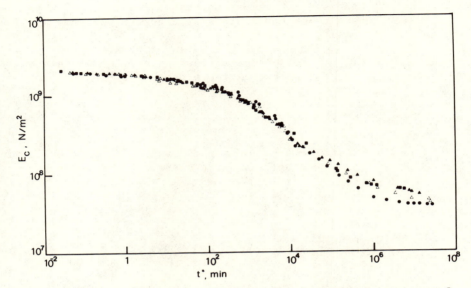

Figure 7.   Generalized stress relaxation master curve at a reference temperature of 80°C for epoxy resin/glass bead
composites at φ=0 (●), φ=0.1 (△), φ=0.2 (▲)   and
φ=0.3 (■).

From the data of Figure 7 and the knowledge of the horizon-
tal shift factor a_ϕ and the vertical shift factor $K(\phi)\,\alpha(t^*)$, it
is possible to calculate the relaxation modulus of each composite
at the time t' and the reference temperature T_r, through the
equation:

$$E(t') = E_c\,(t \cdot a_T \cdot a_\phi)/K(\phi)\,\alpha(t^*) \qquad (9)$$

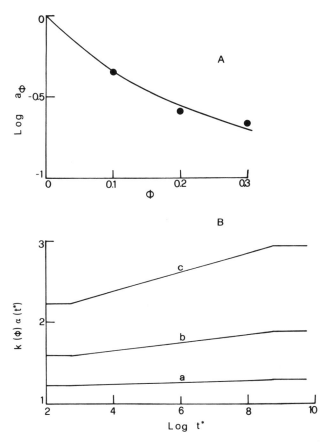

Figure 8.   a) Horizontal shift factor a_ϕ for epoxy resin/glass bead
            composites at different filler content.

         b) Vertical shift factor, $K(\phi)\,\alpha(t^*)$, for epoxy resin/
            glass bead composites at $\phi=0.1$ (a), $\phi=0.2$ (b) and
            $\phi=0.3$ (c).

In conclusion, it has been shown that while the presence of
the filler does not affect the glass transition temperature of the
thermoplastic resin, a significant increase in T_g can be observed
for the epoxy resin based composites. A shifting procedure which
takes into account the reinforcing effect of the filler and the
T_g of the composites permits to superimpose stress relaxation data
at different temperatures in a single generalized stress relaxation master curve.

REFERENCES

1. F.F. Lange, Fracture of brittle matrix, particulate composites,
    in: "Composite Materials", L.J.Broutman and R.H.Krock eds.,
    Academic Press (1974)

2. L. Boltzman, Pogg.Ann.Physik., 7:624 (1876)
3. J.V. Gauchel, DSc Thesis, Washington University, St.Louis (1972)
4. A.V. Tobolsky, "Properties and Structure of Polymers", John
    Wiley and Sons, New York (1960)
5. T.S. Chow, J.Mater.Sci., 15:1873 (1980)
6. E.H. Kerner, Proc.Phys.Soc., 69B:808 (1956)
7. H. Eilers, Kolloid. Z.Z.Polym., 97:313 (1941)
8. K.D. Ziegel and A. Romanov, J.Appl.Poly.Sci., 17:1119 (1973)
9. J.A. Mason and E.H. Chiu, J.Polym.Sci., Symposium No. 41:95 (1973)
10. J.D. Ferry, "Viscoelastic Properties of Polymers", John Wiley
    and Sons, New York (1970)
11. F.R. Schwarzl, H.W. Bren, C. Nederveen, G. Schippert, L. Struik
    and C.W. Van der Wal, Rheol.Acta, 5:270 (1956)
12. L.E. Nielsen, "Mechanical Properties of Polymers and Composites"
    Marcel Dekker, New York (1974)
13. Z. Rigbi, Trans.Soc.Rheol., 9:379 (1965)
14. C. Migliaresi, L. Nicolais, L. Nicodemo and A.T. Di Benedetto
    Polymer Composites, 2(1):29 (1981)
15. A. Apicella, L. Nicodemo and L. Nicolais, Rheol.Acta, 19:291(1980)
16. A.E. Moehlenpah, A.T. Di Benedetto and O. Ishai, Polym.Eng.Sci.
    10(3):170 (1970)
17. Y. Miyano, M. Kanemitsu, T. Kunio "Time and Temperature depen-
    dence of flexural strength in transversal direction of fibers
    in CFRP", presented to the "I International Meeting on Compo-
    site Materials", Milano, Italy (1980)

ON FEASIBILITY OF ACCELERATED CREEP MEASUREMENTS

IN SOME POLYMERIC MATERIALS

A. Franceschini, A. Momo, and P. Campagna

Centro Ricerche Fiat S.p.A.

Str. Torino, 50 - Orbassano (Italy)

INTRODUCTION

It's not easy for the car designer to choose properly a plastic material for a particular application, on account of both the great variety of different materials with similar characteristics, and the paucity of data about their long-term behaviour: this is particularly true in the case of creep resistance.

A satisfactory evaluation of this property generally requires very long tests for each material, under different stress and temperature conditions.

Owing to these reason, we are studying the possibility of foreseeing the creep deformation of a polymer by means of short run tests. First results of this work have already been published[1].

One possibility of cutting down a creep-test time is based on the "time-temperature equivalence" according to the WLF equation[2]. This equation is theoretically applicable only to amorphous polymers above their glass transition; however, it was satisfactorily applied also to crystalline polymers, and in wider temperature ranges [3-7].

Another possibility of reducing the test time comes from the "time-stress equivalence". This equivalence can be defined by stating that a highly stressed sample of polymeric material creeps in an identical but accelerated way as a moderately stressed sample. Studies on this line are few, and not systematic[8,10].

In our previous work[1] we pointed out that it's possible to build up empirically a master curve for a semicrystalline thermoplastic (polypropylene), and for a blend of it with a rubber, by shifting along the time axis the creep curves obtained at different stress levels.

In this work we show that master-curves can be constructed in the same way also for an amorphous thermoplastic (polycarbonate) and a reinforced thermoset (unsaturated polyester).

EXPERIMENTAL

For the tests on polycarbonate we used injection moulded specimens of Bayer Makrolon 3119, annealed for half an hour at 135°C. For the tests on polyester we used compression moulded SMC ("Sheet Moulding Compounds") specimens supplied by IVI (Quattordio, Alessandria). They contained about 25% random oriented short glass fibres and 40% $CaCO_3$; all the specimens were annealed for 2 hours at 120°C.

Tensile creep tests were conducted in thermostatic cells ($\Delta T = \pm 2°C$).

Displacements were measured by an extensometer stuck to the specimen in order to avoid its slippage.

Numerical data were elaborated on a digital PDP 11/70 computer.

RESULTS

Creep curves for the annealed polycarbonate are shown in Fig. 1 as log strain versus log time. In all these curves the temperature is held constant at 80°C, while the stress is varied from 41.5 MPa to 52.4 MPa.

All the curves are quite similar, but they are shifted to lower times at higher stresses. It is possible to superimpose empirically all the curves by moving each along the time axis and letting it move slightly also along the strain axis.

In order to avoid any subjectivity in reducing the data, and above all in order to reduce drastically the very long calculation time, we tried to solve the problem with the aid of a computer. Creep curves are interpolated by analytical functions which allow their numerical superposition.

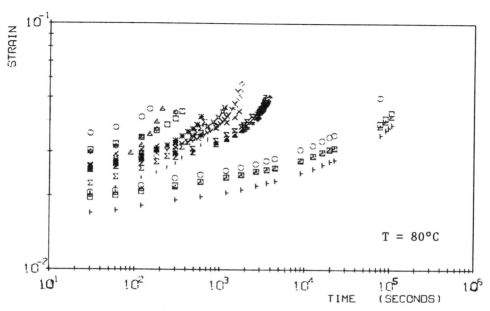

Fig. 1 - Experimental creep curves for annealed polycarbonate

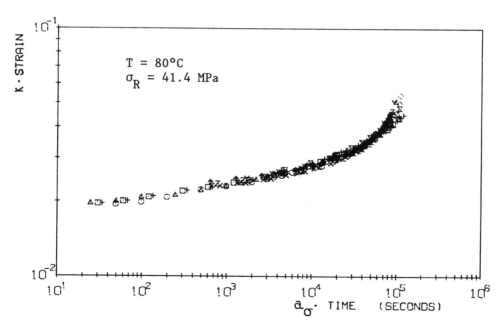

Fig. 2 - Creep master curve for annealed polycarbonate

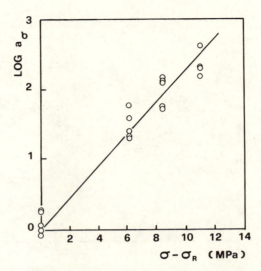

Fig. 3 - Shift factor a_σ as a function of stress

This numerical break-down is quite similar to the one obtained with a long and boring manual elaboration. The master curve drawn directly by computer from the experimental data of Fig. 1 is shown in Fig. 2.

The horizontal shift factor a_σ, i.e. the acceleration factor of the creep time, depends on the stress level σ, as is shown in Fig. 3. The fact that a straight line comes out in a semilog plot means that also in the case of polycarbonate a_σ may formally be expressed by an equation of the type proposed in our previous work for polypropylene[1]:

$$a_\sigma = \exp\left(\frac{\gamma}{RT}(\sigma - \sigma_R)\right)$$

where σ_R is the reference stress, i.e. the stress at which the master curve is built up; T being the temperature, R the gas constant and γ a material constant.

The vertical shift factor k depends of course on the quantity chosen for the y axis. The k values relative to strain

Table 1 - Vertical shift factors for annealed polycarbonate

Stress (MPa)	k in relation to		
	Strain	Compliance	Strain energy
41.4	1.03	1.03	1.03
47.5	0.90	1.03	0.79
49.9	1.10	1.33	0.92
52.4	0.91	1.15	0.72

(used in this work), or compliance (used in [1]), or strain energy, are shown in Tab. 1 as functions of the stress. You can see that all the values are very close to unity, and don't show any dependence on the stress level.

It is important to notice that a_σ isn't at all influenced by the aforementioned choice of the vertical axis.

To show other examples of the good results which may be obtained with the aid of the computer we display in Figs. 4 and 5, both in log-log and semi-log representation, the master curve calculated for a polypropylene, which we already reported in[1] as a results of a standard, manual procedure.

Finally, the master curve of the SMC is shown in Fig. 6, as obtained by manual calculation: actually in this case the shape of the experimental curves doesn't yet lend itself to easy manipulation by the computer program.

Fig. 6 shows a fair overlay of the curves related to different stresses; in spite of this the shift factor a_σ is not a simple function of the stress, as in the cases previously discussed. Till now it isn't clear if this is due either to the cross-linked structure of the material, or to the presence of glass fibres, or to a different deformation and fracture mechanism. We are now working on this point.

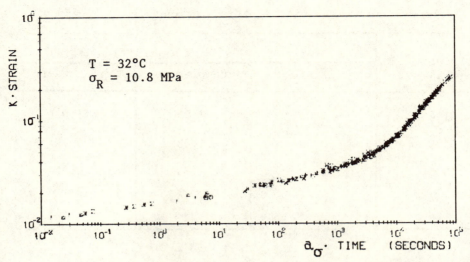

Fig. 4 - Creep master curve for polypropylene MOPLEN EPT 30R
(log-log representation)

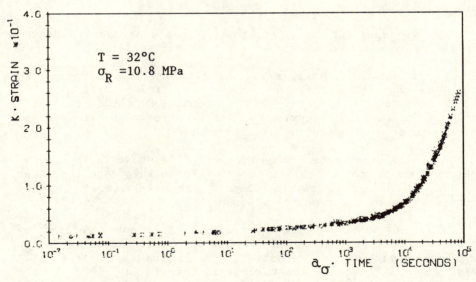

Fig. 5 - Creep master curve for polypropylene MOPLEN EPT 30R
(semi-log representation)

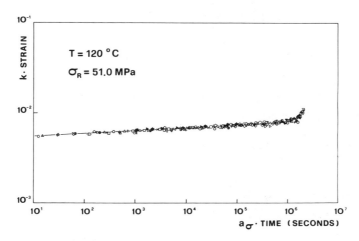

Fig. 6 - Creep master curve for Sheet Moulding Compound

CONCLUSIONS

The principle of time-stress equivalence, previously applied in the case of polypropylene and polypropylene toughened with rubber, is here successfully applied to an amorphous thermoplastic, polycarbonate: the creep curves at different stress levels can here again be superimposed in one normalized master curve.

A computer program is set up, which in many cases allows one to analyse empirical data and to obtain the master curves in a much shorter and easier way.

The master curves are very useful for predicting the material behaviour over long periods by means of short run tests.

Master curves are also obtained in the case of Sheet Moulding Compounds; their interpretation and use as a prediction method is not yet completely developed, because of an odd dependence of the shift factor on the stress. We are now continuing to study this point.

REFERENCES

1. A. Franceschini and A. Momo - Time-stress superposition in some creep experiments - Polymer 21:725 (1980)

2. M.L. Williams, R.F. Landel and J.D. Ferry - Temperature dependence of relaxation mechanisms in amorphous polymers etc. - J. Am. Chem. Soc. 77:3701 (1955)

3. N. G. McCrum - Prediction of creep at long times in polymeric solids - J. Mat. Science 13:1596 (1978)

4. P.K. Agarwal - Effect of moisture on mechanical behaviour of a nylon-epoxy adhesive - Pol. Eng. Science 19:940 (1979)

5. R.J. Crowson and R.G. Arridge - Linear viscoelastic properties of epoxy in dilatation and shear in the glass transition region - 1., Polymer 20:737 (1979)

6. D.B. Bogy et al. - Experimental determination of creep functions for thin orthotropic polymer films - IBM J. Res. Develop. 23:450 (1979)

7. Y. Miyano et al. - Time and temperature dependence of flexural strength in CFRP - in: I st Int. Meeting Compos. Mater, Milano, Italy (1980)

8. E. Catsiff et al. - Generalized creep curves for nylon - Text. Res. J. 23:808 (1953)

9. G. Bauwens-Crowet and J.C. Bauwens - Mechanism of creep behaviour in glassy polymers - J. Mat. Science 10:1779 (1975)

10. L.E. Nielsen pag. 92 in - Mechanical properties of polymers and composites - Dekker; New York (1974)

FATIGUE CRACK PROPAGATION IN SHORT-GLASS-FIBER-REINFORCED NYLON 66:

EFFECT OF FREQUENCY

R. W. Lang, J. A. Manson, and R. W. Hertzberg

Materials Research Center
Lehigh University
Bethlehem, PA 18015

ABSTRACT

The fatigue crack propagation (FCP) response of nylon 66 and composites containing 16 and 31 vol. % short glass fibers was determined at frequencies of 1, 10, 50, and 100 Hz. The sensitivity to frequency was found to depend on glass content and, in the case of the pure matrix, on the value of the stress intensity factor range, ΔK.

Whereas both glass-reinforced compositions were superior to the pure matrix material, an inversion in relative ranking of the former occurred as the cyclic frequency was changed. Also, the strong frequency sensitivity revealed by nylon 66 at low ΔK values disagrees with results reported previously. To explain these phenomena a concept based on the competition between localized and more generalized heating is proposed. For quantitative comparison, values determined for loss modulus as a function of temperature were correlated with the temperature profile measured at the crack tip by means of an infrared microscope. Localized heating at the crack tip is believed to decrease crack growth rates due to a drop in yield strength and the simultaneous increase in plastic zone size which dissipates more energy and effectively blunts the crack. Generalized heating in the bulk material ahead of the crack tip, on the other hand, has an adverse effect on growth rates as a result of an overall decrease in specimen stiffness.

INTRODUCTION

As short-fiber composites are often used in applications that involve cyclic loads, it is important to understand their fatigue response. While failure of typical engineering plastics at a stress range, $\Delta\sigma$, close to the yield stress may involve gross yielding due to hysteretic heating, failure by propagation of a flaw is common at lower stresses.[1] In the case of reinforcement with short fibers of, for example, glass or carbon, fatigue life under constant-stress-range testing is often improved,[2] presumably in part due to a reduction in temperatures attained during cycling. Clearly it is of interest to examine the effects of cycling on fatigue crack propagation rates in such systems. Since the hysteretic heating rate per unit volume, $\dot{\varepsilon}$, under given loading conditions and heat transfer characteristics depends on the frequency both explicitly and implicitly through the loss compliance D" (equation 1)[3], the effect of frequency must be carefully considered. Specifically, $\dot{\varepsilon}$ is given by

$$\dot{\varepsilon} = \pi f \Delta\sigma^2 D''(f,T)/4 = \text{heat dissipated/time} \cdot \text{volume} \tag{1}$$

where f is the frequency, $\Delta\sigma$ the stress range, and D"(f,T) is the loss compliance as a function of frequency and temperature.

Indeed, reported effects of frequency in both unreinforced and reinforced polymers are complex, reflecting the fact that more is involved than just gross damage due to hysteretic heating. Thus extensive studies of FCP in engineering plastics[1] reveal that increasing frequency may increase, decrease, or not affect FCP rates. An explanation for such diverse behavior has been proposed in terms of a competition between local crack-tip heating (which can, in effect, blunt the crack) and generalized heating of the bulk material (which will lower its modulus). Of course, increasing frequency also increases the strain rate, thus increasing the modulus and decreasing creep[4]. Variable effects of frequency have also been noted in adhesive joints[5,6] and in continuous-fiber composites[7-9]. Within the latter systems sometimes more damage (reflected in lower moduli) is seen at lower frequencies[7], while sometimes complex effects of frequency have been observed[8,9].

With short-fiber systems, a few studies of FCP have been reported[10-12], but the performance relative to that of the pure matrix was not explicitly described. It was decided therefore to extend earlier studies of FCP in polymers to examine the effects of fiber parameters (e.g., volume fraction and orientation), interfacial characteristics, matrix, and frequency and environment. Following a preliminary paper on FCP in injection-molded nylon 66 composite systems[13], this paper examines the effect of frequency and glass content on FCP in short-fiber reinforced nylon 66 containing 1.7% water. By combining observations of temperature rise with measurements of damping, it will be shown that otherwise anomalous effects of frequency can be explained.

EXPERIMENTAL

Materials and Specimen Preparation

Three nylon 66 resins reinforced with various amounts of short glass fibers were supplied by the LNP Corporation:

B-N66: unreinforced nylon 66
B-16G: B-N66 containing 16 vol % (30 wt %) of short glass fibers
B-31G: B-N66 containing 31 vol % (50 wt %) of short glass fibers

The number-average molecular weight $\overline{M}_n$ of B-N66 was determined by dilute-solution viscometry using formic acid (88%) as a solvent, and found to be 1.9×10^4; details are described separately[14].

Values of percent crystallinity (by weight) were determined by both differential scanning calorimetry (Perkin-Elmer DSC model 1B) and by a density gradient column (DGC): 31($\pm$1)% and 42% by DSC (average for all matrix materials) and DGC (for B-N66), respectively. These values agree well with those measured in this laboratory for other commercial nylon 66 materials (25 to 35, and 30 to 43% by DSC and DGC, respectively); the difference presumably reflects uncertainty in the values of heat of fusion and density for the 100%-crystalline reference. Since the material had been stored for several years at ambient conditions the moisture content was measured gravimetrically by drying specimens to constant weight at 105-110°C under vacuum. All matrix materials were found to contain 1.7 wt % water. The glass fibers used were of the E-glass type having an average diameter of 10 μm and a length of 6.4 mm before processing.

Specimens of the compact tension (CT) geometry[1] were cut in two different directions from injection-molded plaques so that the direction of the applied load was either transverse (T) or longitudinal (L) with respect to the major flow direction. Dimensions and gating of the plaques are shown in Fig. 1.

Fatigue Crack Propagation Tests

FCP experiments were conducted with an electrohydraulic closed-loop testing machine using standard procedures[1,14]. The applied waveform was sinusoidal with constant load amplitude and a minimum-to-maximum load ratio, R, of 0.1. Environmental conditions were laboratory air at 22-24°C and an average relative humidity of 40%. Tests were conducted at frequencies of 1, 10, 50 and 100 Hz, and the results plotted as log da/dN, the crack growth per cycle, as a function of log ΔK, the stress intensity factor range ($\Delta K = Y \Delta \sigma \sqrt{a}$ where Y is a geometrical variable, $\Delta \sigma$ the applied stress range, and a the crack length).

SERIES B

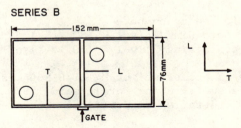

Fig. 1. Geometry of injection molded plaques and orientation and
position of FCP specimens. T: direction of applied load
transverse to major flow direction; L: direction of applied
load longitudinal with respect to the major flow direction
(nominal plaque thickness = 3.2 mm).

Temperature Rise during FCP

Crack-tip temperatures and temperature profiles ahead of the
crack tip were measured using an infrared radiometric microscope,
model RM-2B, with a 15X objective corresponding to a spot size
0.075 mm (Barnes Engineering Company). Temperatures were recorded
after equilibrium was reached, before the FCP test was interrupted to
record the crack tip position. The estimated error increases from
±0.5°C at around 30°C, to ±(1-2)°C at 65°C, to ±3°C above 65°C.

Dynamic Mechanical Spectroscopy (DMS)

Dynamic mechanical spectra were recorded for each of the materials tested using an Autovibron apparatus (Model DDV-III-C, IMASS
Corporation, Hingham, MA). Test procedures are described and discussed by Webler, et al.[15] Specimens for these tests were cut and
loaded in the T-direction according to Fig. 1.

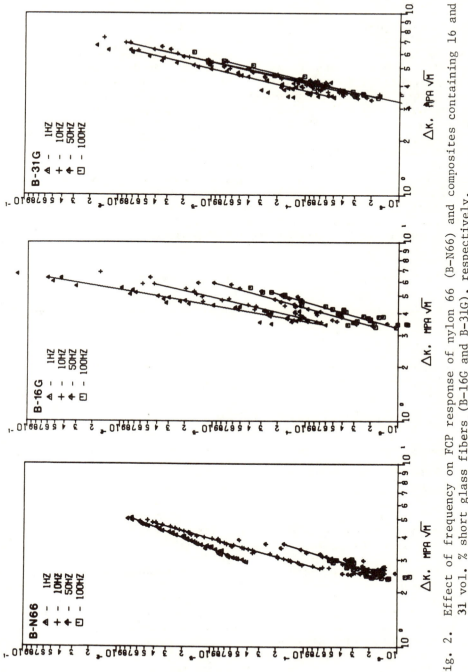

Fig. 2. Effect of frequency on FCP response of nylon 66 (B-N66) and composites containing 16 and 31 vol. % short glass fibers. (B-16G and B-31G), respectively.

EXPERIMENTAL RESULTS

Effect of Frequency and Fiber Content on FCP

   The effect of test frequency on FCP behavior of B-N66, B-16G and B-31G in the frequency range of 1 to 100 Hz is shown in Fig. 2. Although there is some scatter in the data for the glass-reinforced compositions and for the pure B-N66 at 50 and 100 Hz, consistent with a discontinuous crack growth mechanism, the behavior can be described reasonably well by the Paris relationship[16] [$da/dN = A \Delta K^n$, with material variables A and n given in Table 1]; this relationship has been found to hold in several other short-fiber composite systems[5,6,10-13], as well as in some continuous-fiber systems[17,18]. While crack growth rates in all three systems were found to show a positive frequency dependence (i.e., the higher the frequency the lower the growth rate per cycle), the degree of frequency sensitivity varied with glass content. Thus Fig. 3, in which da/dN for different ΔK levels is plotted versus test frequency on a double logarithmic scale, reveals

Table 1. Parameters[a] A and n of the Paris Relationship for B-N66, B-16G and B-31G

Frequency (Hz)	B-N66 A[b]	n	B-16G A[b]	n	B-31G A[b]	n
1	1.6×10^{-6}	5.2	4.8×10^{-11}	11.4	7.1×10^{-9}	7.4
10	5.9×10^{-8}	7.1	1.5×10^{-9}	8.5	5.4×10^{-10}	8.6
50	1.0×10^{-8}	7.6	3.3×10^{-9}	7.1	2.1×10^{-10}	9.1
100	1.3×10^{-7}	5.3	3.0×10^{-9}	6.8	9.9×10^{-10}	8.0

[a] Determined by regression analysis
[b] Units = [(mm/cycle) x (MPa$\sqrt{m}$)$^{-n}$].

that nylon 66 containing 16% (vol.) glass fibers is the most frequency-sensitive composition.

   The high frequency sensitivity of the composition with an intermediate glass content has an important effect on the relative ranking of these materials with respect to FCP. Whereas both glass-reinforced systems are superior to the pure matrix material over the whole frequency range, an inversion in relative ranking of B-16G and B-31G takes place at around 50 Hz (see Fig. 3). At low cyclic frequencies of 1 and 10 Hz, respectively, crack growth rates at any given ΔK value are higher for B=16G than for B-31G, whereas at 50 Hz no significant difference in FCP behavior could be found. Further

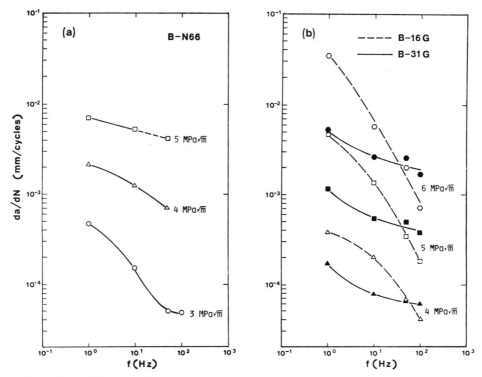

Fig. 3. Effect of frequency on FCP rates for nylon 66 and
         composites. Numbers at right of each curve give ΔK at
         which the rate is measured.

increase in frequency to 100 Hz leads to the above-mentioned inversion, with B-16G being somewhat more resistant to FCP than B-31G.

Effect of Orientation on FCP

The results presented in Figs. 2 and 3 were obtained from
specimens cut out in the T-configuration according to Fig. 1. To
investigate the effect of fiber orientation on FCP, specimens were
also tested in the L-configuration at 10 and 100 Hz; however, no
significant differences with respect to orientation could be found.
A model explaining this somewhat unexpected behavior is discussed
in an earlier paper[13]. Thus the growing crack perceives a quasi-
isotropic fiber array, even though the specimens can be considered
in effect as laminates of plies having different fiber orientations.

Reproducibility of FCP Tests

Reproducibility experiments conducted at 10 and 100 Hz for the
T-configuration showed excellent agreement. Maximum differences in

growth rates were less than 20% for B-16G and B-31G, respectively. In the case of B-N66 the results obtained from two different tests were essentially identical.

Hysteretic Heat Generation

During the course of the fatigue tests significant temperature rises at the crack tip at frequencies higher than 10 Hz were observed and first measured by means of a thermocouple. To record the crack tip temperature more accurately an infrared radiometric microscope was used, a technique applied also by Attermo and Östberg[19] for poly(vinyl chloride), poly(methyl methacrylate) and polycarbonate.

Several attempts have been made to describe the heat generation and the associated temperature rise of polymers subjected to cyclic loading (for a review, see reference 1, ch. 2). While most of the equations proposed cannot predict absolute temperature rises since heat losses to the surrounding environment are not accounted for (for exceptions, see Crawford and Benham[20] and Oberbach[21]) they do show the influence of the major variables associated with hysteretic heating. For example, in reducing equation (1) the temperature rise per unit time (neglecting heat losses), $\Delta\dot{T}$, becomes

$$\Delta\dot{T} = \frac{\pi f D''(f,T,HOH)\Delta\sigma^2}{4c_p\rho} \tag{2}$$

where f is the frequency, D" the loss compliance under constant stress conditions, $\Delta\sigma$ the stress range, ρ the density, and c_p the specific heat. Note that D" itself is a function of frequency and temperature, and, in the case of nylon, of water content; c_p may also vary with temperature and water content.

Effect of Frequency on Hysteretic Heating

In Fig. 4 the crack tip temperature is plotted versus test frequency at a ΔK of 2.6 and 3.1 MPa$\sqrt{m}$ for the unreinforced and glass-reinforced compositions, respectively. For the latter materials, a linear relationship as in eq. 3 fits the data well:

$$\Delta T = C(\Delta K, v_f) \cdot f \tag{3}$$

where ΔT is the temperature rise at the crack tip, f is the frequency and C, the slope, is a function of the applied ΔK and fiber volume fraction, v_f. It should be noted that this first-order dependence of ΔT on frequency is in agreement with equation (2). The deviation from linear behavior observed for B-N66 at higher frequencies (>40 Hz) is possibly related to the higher heat losses to be expected at the higher temperatures attained. Furthermore, as will be seen

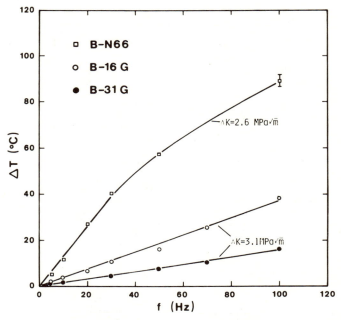

Fig. 4. Effect of frequency f on crack tip temperature rise ΔT for B-N66, B-16G and B-31G [$\Delta T = C(\Delta K, v_f) \cdot f$].

later, D'' reaches a maximum at about 80°C, corresponding to a ΔT of about 55°C at $\approx$50 Hz, and decreases again at higher temperatures. This fact may also have contributed to the observed nonlinearity of ΔT versus f for B-N66 at higher frequencies.

Effect of ΔK on Hysteretic Heating

The effect of the applied stress intensity factor range ΔK on the crack-tip temperature rise is shown in Fig. 5 for B-N66 at 10 Hz and for B-16G and B-31G at 10 and 100 Hz. It was found empirically that the data can be described to a first approximation by an equation of the form

$$\Delta T = B(f, v_f) \ \log \frac{\Delta K}{\Delta K_{T,th}} \tag{4}$$

where the constant B is a function of frequency and glass content and $\Delta K_{T,th}$ is the empirical threshold value for ΔK below which no detectable temperature rise occurs. [In other cases, a simple power dependence of ΔT on ΔK may be more appropriate.[22]] It is interesting that the defined parameter $\Delta K_{T,th}$ has a value in these systems between 2.2 and 2.4 MPa$\sqrt{m}$ (independent of frequency and

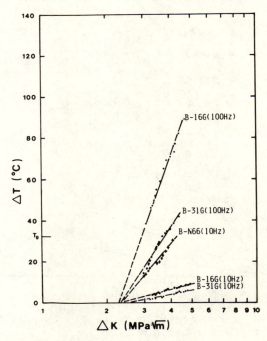

Fig. 5.   Effect of stress intensity factor range on crack tip
          temperature rise in B–N66 at 10 Hz and in B–16G and B–31G
          at 10 and 100 Hz.

glass content) — a range of ΔK corresponding to the passage of most
of the fatigue life. If indeed $\Delta K_{T,th}$ should turn out to be a
true material property, its existence would imply that the concepts
of linear elastic fracture mechanics could be valid for cases in
which most of the fatigue life is passed at values of $\Delta K < \Delta K_{T,th}$.
An analogous transition point separating thermally- from non-
thermally-controlled fatigue failure in unnotched samples has been
characterized by Crawford and Benham (see ref. 1, ch. 2) and defined
as the change-over stress. However, one must distinguish between
thermal failure where actual melting occurs (as in unnotched fatigue
tests) and thermally influenced FCP ($\Delta K > \Delta K_{T,th}$) where mechanical
failure still prevails (as in precracked samples). Clearly more
evidence is needed in order to assess the theoretical significance
of this presently empirical parameter.

Clearly more evidence is needed in order to assess the theoretical significance of this presently empirical parameter.

Temperature Profiles

Temperature profiles ahead of the crack tip were recorded at various ΔK levels. Fig. 6 shows typical curves for the temperature rise as a function of the distance from the back edge of the sample. The location of the crack tip is indicated by arrows. As expected, ΔT decreases exponentially as the distance from the crack tip increases.

Dynamic Mechanical Spectroscopy (DMS)

Since DMS is a useful technique to obtain elastic and visco-elastic parameters of polymeric materials subjected to cyclic loading, the storage modulus E' and the loss modulus E'' were determined over a temperature range from $-120°C$ to $+180°C$ at a frequency of 110 Hz (Fig. 7). With respect to the interpretation of the FCP results it is interesting to note that the glass transition temperature T_g, here taken as the temperature at which E'' reaches a maximum, is in all cases in the range 50 to 60°C.

To evaluate the relative ranking of the investigated materials as to their ability for heat generation in terms of eq. (2), the temperature dependence of D'' is required (see Fig. 8). Equation (4) was applied:

$$D'' = \frac{E''}{E'^2 + E''^2} = \frac{1/E''}{1 + (\tan^2 \delta)^{-1}} \tag{4}$$

The following observations may be made with respect to hysteretic heating during a fatigue test. While dry nylon 66 heats up only slightly[23], extensive heating was observed when 1.7% water was present. However, on adding glass fibers the tendency for heat generation decreases as the fiber content increases. (See Figs. 4 to 6.) The following facts can explain the latter behavior:

(1) the higher the fiber content, the lower the amount of material with high damping.

(2) the higher the fiber content, the greater the thermal conductivity and consequent heat loss.

(3) the higher the fiber content, the lower the average stresses in the matrix at a given applied load (in an isotropic or quasi-isotropic system).

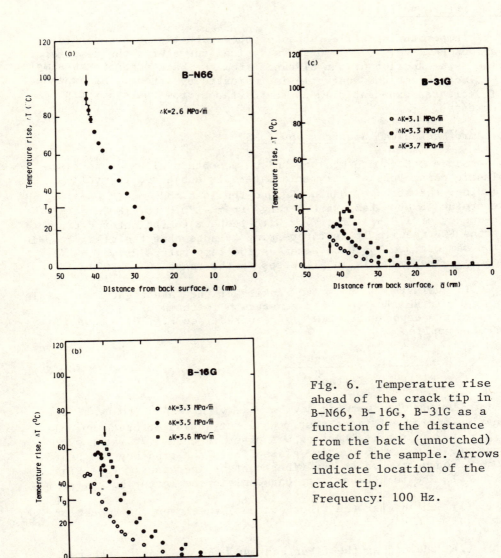

Fig. 6. Temperature rise ahead of the crack tip in B-N66, B-16G, B-31G as a function of the distance from the back (unnotched) edge of the sample. Arrows indicate location of the crack tip.
Frequency: 100 Hz.

(4) the lower the fiber content, the higher D''_{max} and the higher the tendency for autoaccelerated heatup (see Fig. 8).

While the data in Fig. 8 are for a frequency of 110 Hz, the relative behavior at other frequencies is expected to be at least qualitatively similar.

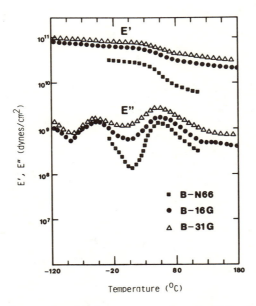

Fig. 7. Dynamic mechanical spectra for B-N66, B-16G, and B-31G.

DISCUSSION

Since cyclic growth rates are frequently related to the plastic zone size at the crack tip[24] or to the crack-opening displacement (COD)[25,26], let us examine the dependence of these parameters on frequency. Irwin[27] derived the size of the plastic zone, r_y, for plane stress conditions from stress field equations by setting the y component of stress, σ_y, equal to the yield strength, σ_{ys}:

$$r_y \simeq \frac{1}{2\pi}\left(\frac{K}{\sigma_{ys}}\right)^2 \quad \text{(plane stress)} \tag{6}$$

For conditions of plane strain r_y is smaller, and given by[28]

$$r_y \simeq \frac{1}{6\pi}\left(\frac{K}{\sigma_{ys}}\right)^2 \quad \text{(plane strain)} \tag{7}$$

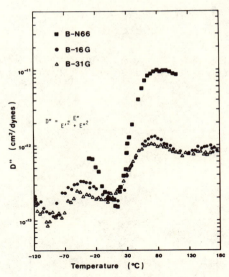

Fig. 8. Dynamic loss compliance D'' as a function of temperature
          for B-N66, B-16G, and B-31G.

The crack-opening displacement, δ, at the crack tip is given by[26,29]

$$\delta \simeq \frac{K^2}{E\sigma_{ys}}$$
(8)

At constant ΔK, clearly factors which decrease σ_{ys} will increase
r_y, while factors which decrease σ_{ys} or E will increase δ. A temperature rise due to a frequency increase will therefore <u>increase</u> both
r_y and δ, while the associated increase in strain rate may be
expected to <u>decrease</u> r_y and δ, at least to some extent. An increase
in frequency can clearly have contradictory effects.

The next question is the role of δ and r_y in determining FCP
rates. In general, it is desirable to keep δ as small as possible
to avoid reaching the critical value required for crack extension.
The case of r_y seems to be not so straightforward. While it has
been suggested that suppressing the plastic zone at the crack tip
should increase the resistance to cyclic crack propagation[4], we
believe that an <u>increase</u> in plastic zone size may be beneficial
because it results in more effective crack tip blunting and dissipates more energy at the crack tip. Consequently, according to
equations 6 and 7 ($r_y \propto 1/\sigma_{ys}^2$), any mechanism leading to a decrease
in σ_{ys} on a localized scale at the crack tip is expected to lead to
an <u>increase in FCP resistance</u>. Furthermore, in general the elonga-

tion at yield, ε_{ys}, and at break, ε_b, also increase with decreasing σ_{ys} (e.g. due to a temperature rise[30]), thus allowing for a larger critical COD, δ_c, for crack extension. The localization of the phenomena concerned should be important, for r_y may be expected to depend on the local value of σ_{ys} at the crack tip whereas δ is controlled by the modulus and yield strength profile of the whole unbroken ligament. However, since the temperature rise of the materials tested does not only occur in a localized region at the crack tip but also can be seen to be significant some distance away from it, Fig. 8, two competitive events occur simultaneously. Hence, with high damping polymers and their composites the overall effect of frequency on fatigue crack growth rates will be determined by a balance between the

- temperature rise at the crack tip resulting in an <u>increase in r_y and δ_c (beneficial)</u>, and the

- temperature rise of the bulk material ahead of the crack tip resulting in a modulus decrease and a simultaneous <u>increase in crack opening displacement (detrimental)</u>.

This concept also is in agreement with the conclusion from experimental findings (see ref. 1, ch. 3) that localized heating at the crack tip leads to attenuated crack growth rates, but generalized heating of the bulk material should be deleterious to FCP resistance.

In the case of the materials studied, it seems likely that the effect of frequency is dominated by consequences of hysteretic heating, for σ_{ys} is only slightly increased by an order-of-magnitude increase in strain-rate but greatly decreased by relatively small increases in temperature[30], and the modulus is nearly independent of frequency in the range studied[31].

The Matrix: B-N66

At first glance, the qualitative trends in FCP rates with frequency (Fig. 3) appear to be consistent with the arguments presented above. Certainly the increase in temperature at the crack tip should favor blunting of the crack tip. However, as shown in Figs. 6 and 8, the matrix material shows a much stronger tendency towards hysteretic heating at and ahead of the crack tip than is the case with the fiber-reinforced specimens. Indeed, with the temperatures noted, one might expect to find a deleterious effect of increasing frequency due to a severe reduction in E and σ_{ys} in the bulk material with a consequent large increase in δ. It should also be noted that the strong sensitivity to frequency seen appears to contradict other findings reported previously[23,31] for nylon 66. [Such conflict is common in the literature[1,4,32,33].] In addition, crack growth rates were found to be lower than those reported by Bretz et al.[14,23], who used the same test procedure and comparable materials with respect to water content, molecular weight and crystallinity. Clearly some factor must exist that can overcome the otherwise deleterious effects of high heat generation ahead of the crack tip.

In fact, there is one difference between this study and those of Bretz et al.[14,23]: the specimen thickness. Bretz et al. used specimens having a nominal thickness of 6.2 mm, whereas our specimens were 3.2 mm thick. As pointed out earlier[13], it seems likely that differences in specimen thickness cause the observed variations in FCP response due to differences in the stress state. Thus the thinner the specimen, the greater is the tendency toward plane-stress rather than plane-strain conditions, the larger is r_y and the higher the expected resistance to FCP. Similar effects of thickness on FCP have been reported for polycarbonate[34,35]. Indeed, a preliminary examination of the fracture surface revealed the presence of shear lips (indicating plane stress conditions) whose width increased with ΔK (at constant frequency) and with frequency (at constant ΔK). This phenomenon (not present in the reinforced samples) reflects an additional contribution to the dissipation of energy, and is consistent with the occurrence of a frequency-induced drop in σ_{ys} ahead of the crack tip, and with the concomitant decrease in FCP rates observed (for the relationship between shear lip width and σ_{ys} see ref. 1, ch. 4).

As a result of the change in stress state the question of its effect on the frequency sensitivity arises. The following reasons can be put forth to explain why the frequency dependence might change with specimen thickness. In general, the temperature rise under cyclic loading conditions will decrease as specimen thickness decreases since the heat transfer to the surrounding environment is enhanced by the increase in the surface-to-volume ratio of the specimen. Thus, in the case of a precracked FCP specimen, the bulk material ahead of the crack tip is expected to heat up less as the specimen thickness decreases. On the other hand, it may well be that the temperature rise on a local scale at the crack tip increases as the specimen thickness decreases due to the larger plastic zone volume associated with plane stress conditions (compare equations 6 and 7). Hence progressively more heat will be generated at the crack tip at higher frequencies with thinner specimens leading to a larger drop in σ_{ys} and in turn to an increase in plastic zone size. As discussed earlier, both effects (larger decrease in σ_{ys} at crack tip and less decrease in stiffness of the bulk material with increasing frequency in thinner specimens) should be beneficial in terms of FCP thus explaining the increased frequency sensitivity of thinner specimens at low values of ΔK.

The effect of ΔK on the _relative_ frequency sensitivity is also of interest. From Fig. 3a it is evident that the frequency sensitivity of B–N66 varies with the applied ΔK range, being highest at the lowest ΔK. An explanation for this finding is the favorable tendency towards the localized temperature increase at smaller crack

lengths (i.e. at low values of ΔK). As pointed out previously
(1, ch. 3), the distinction between localized and generalized heating depends on the size of the hysteretically heated damage zone
relative to the dimensions of the entire specimen. An arbitrary way
to quantify this statement for the case of FCP specimens is to define
a quantity R_L as the ratio of the length of the unbroken ligament
experiencing a temperature rise above a reference temperature T_{ref},
e.g. $T_g \approx 55°C$, to the length of the ligament that remains below
T_{ref}. To gain both the low yield strength at the crack tip and the
high modulus or stiffness of the unbroken ligament, R_L should be
small.*

It is apparent from the trends shown by the temperature distributions in Fig. 6 that the ratio R_L, which reflects the competition
between localized and generalized heating, shifts towards the latter
(i.e. increasing R_L) as the crack length and consequently ΔK
increase. Hence, the benefits gained with increasing frequency at
low ΔK (small R_L) diminish due to the increasing tendency for generalized heating at high ΔK (larger R_L) leading to the decrease in
frequency sensitivity with increasing ΔK. Again we see a complex
balancing of effects.

The Reinforced Specimens

Generally similar arguments may also be used for the glass-
reinforced compositions. However, instead of using the term "plastic
zone" it seems more appropriate to use the term "damage zone" with
regard to the plastically deformed volume at the crack tip. As
with the neat matrix, the positive frequency sensitivity, (see Figs.
2 and 3b) is believed to be caused by crack-tip heating that leads
to a decrease in σ_{ys} and an increase in ε_b, which in turn leads to
an increase in damage-zone size. Visual observations revealed that
the size of the damage zone at the crack tip increased relatively
more with increasing frequency in B-16G than in B-31G in which a
change is barely detectable. Clearly a larger damage zone is
desirable because of both the higher energy necessary to creat it
and the increased blunting effect at the crack tip. In fact, the
diffuse damage zone in continuous-fiber-reinforced plastics is known
to blunt the effect of a flaw; discontinuous crack growth is often
observed (see ref. 1, ch. 5).

The larger increase in damage zone size with frequency for
B-16G is rationalized as follows. From Fig. 4 it can be seen that
the crack-tip temperature rise ΔT varies more with frequency in
B-16G than in B-31G. Moreover, although the dependence of ΔT on ΔK
is only moderate for the composites at 10 Hz, it becomes more
prominent as the frequency is increased (Fig. 5). Most important,

---

*This statement is true only if the region experiencing at tempera-
ture rise is close to the plastic zone in size. It cannot be applied
to the limiting case of $R_L=0$ in which no heatup occurs at all.

however, whereas the crack tip in B-16G at 100 Hz is above the glass transition temperature T_g ($T_g \simeq 55°C \equiv \Delta T \simeq 32°C$) over the whole ΔK range this is not the case for B-31G which heats up much less at the same frequency. Hence, the larger decrease in σ_{ys} with frequency for B-16G causes the observed change in damage zone size (recall equations 6 and 7).

Another way to view this is to consider Fig. 6b, which shows that the temperature at $\Delta K=3.3$ to 3.6 MPa$\sqrt{m}$ drops below T_g at a distance less than 8 mm ahead of the crack tip so that most of the unbroken ligament still retains a high modulus. Hence it is concluded that the higher frequency sensitivity for B-16G and the inversion in the relative ranking of the glass-reinforced materials in terms of their FCP resistance at different frequencies is a result of the higher frequency-induced temperature rise at the crack tip in B-16G and its effect on yield strength and damage zone size.

One final aspect worth consideration in the case of composites experiencing a temperature change is that of thermally induced internal stresses. In cooling a glass reinforced polymer from processing temperature to room temperature, tensile stresses develop in the matrix in a tangential direction around the fibers. These can be caused simply by the mismatch in the thermal coefficients of expansion or by shrinkage of the matrix due to morphological changes, e.g. crystallization. An increase in temperature as observed in our FCP experiments should either decrease these tensile stresses, or if stress relaxation occurs over a long period of storage time (or due to annealing) should result in compressive stresses in the matrix. Accordingly, a temperature increase is expected to be beneficial from the standpoint of internal stresses.

CONCLUSIONS

Several observations and conclusions may be summarized:

1. By testing the FCP behavior of nylon 66, and nylon 66 containing 16 and 31 vol % of short glass fiber, respectively, it was found that the Paris law holds reasonably well.

2. While crack growth rates for all materials decreased with increasing frequency in the range of 1 to 100 Hz, the degree of frequency sensitivity was discovered to depend on the glass content and in the case of unreinforced nylon 66 also on ΔK. Whereas both glass-reinforced compositions were superior to the pure matrix material an inversion in relative ranking of the former occurred as the cyclic frequency was changed. Also, the strong frequency sensitivity revealed by nylon 66 at low ΔK values disagrees with results reported previously.

3. To explain these phenomena a concept based on the competition between localized and more generalized heating has been used successfully. Localized heating at the crack tip is believed to decrease crack growth rates due to a drop in yield strength and the simultaneous increase in strain capability. This in turn increases the plastic zone size which dissipates more energy and effectively blunts the crack. Generalized heating in the bulk material ahead of the crack tip, on the other hand, has an adverse effect on growth rates as a result of the increase in crack opening displacement which is a consequence of the decrease in modulus.

4. The combination of measurements of loss compliance as a function of temperature with direct measurements of temperature profiles at the crack tip provides a powerful approach to explaining effects of frequency on FCP in nylon 66 and short-glass-fiber reinforced nylon 66, and presumably in other polymeric systems as well.

ACKNOWLEDGMENTS

The authors wish to acknowledge partial support by the Office of Naval Research, and are grateful to Mr. John Theberge, LNP Corporation, for providing the specimens.

REFERENCES

1. R. W. Hertzberg and J. A. Manson. "Fatigue in Engineering Plastics," Academic Press, New York, 1980.
2. J. Theberge, B. Arkles, and R. Robinson, Ind. Eng. Chem., Prod. Res. Dev. 15:100 (1976).
3. J. D. Ferry, "Viscoelastic Properties of Polymers," 2nd ed., Wiley, New York, 1970.
4. H. A. El-Hakeem and L. E. Culver, Int. J. Fatigue, 1:133 (1979).
5. S. Mostovoy and E. J. Ripling, Polym. Sci. Technol. 9B, 513 (1975).
6. J. A. Marceau, Y. Moji, and J. C. McMillan, paper presented at the SAMPE Symp., 21st April 6-8 (1976).
7. K. L. Reifsnider, W. W. Stinchcomb, and T. K. O'Brien, ASTM STP 636, p. 171 (1977).
8. W. W. Stinchcomb, K. L. Reifsnider, L. A. Marcus, and R. S. Williams, ASTM STP 596, p. 115 (1975).
9. C. T. Sun and W. S. Chan, ASTM STP 674, 418 (1979).
10. A. T. DiBenedetto and G. Salee, Polym. Eng. Sci. 19:512 (1979).
11. P. A. Thornton, J. Composite Mater. 6:147 (1972).
12. M. J. Owen and P. T. Bishop, J. Phys. D: Appl. Phys. 7:2036 (1974).

13. R. W. Lang, J. A. Manson and R. W. Hertzberg, Org. Coatings and
     Plast. Chem. 45:778 (1981).
14. P. E. Bretz, R. W. Hertzberg, and J. A. Manson, in press, J.
     Mater. Sci., 1981.
15. S. M. Webler, J. A. Manson, and R. Lang, Polymer Preprints,
     22:1, p. 257, March, 1981.
16. P. C. Paris and F. Erdogan, J. Bas. Eng. Trans. ASME Ser. D85:
     4, 528 (1963).
17. M. D. Campbell and B. W. Cherry, in "Fracture Mechanics and
     Technology," (G. C. Sih and C. L. Chow, eds.), Vol. 1, p. 297.
     Sijthoff and Noordhof, Alphen aan den Rijn, The Netherlands,
     1977.
18. D. J. Wilkins, this publication.
19. R. Attermo and G. Östberg, Int. J. Fract. Mech. 7:122 (1971).
20. R. J. Crawford and P. P. Benham, J. Mater. Sci., 9:18 (1976).
21. K. Oberbach, Kunstoffe 63:35 (1973).
22. C. M. Rimnac, Ph.D. research, Lehigh University, 1981.
23. P. E. Bretz, R. W. Hertzberg, J. A. Manson, M. T. Hahn and
     R. W. Lang, "Effect of Test Frequency and H_2O Content on
     Localized Crack-tip Heating in Nylon 66," to be published.
24. A. J. McEvily, R. C. Boetner and T. L. Johnson, in "Proceedings
     of 10th Sagamore Army Materials Research Conference," Syracuse
     University, New York (1963), p. 95.
25. L. R. Hall and R. C. Shah, Eng. Fract. Mech. 3:169 (1971).
26. J. G. Williams, J. Mater. Sci., 12:2525 (1977).
27. G. R. Irwin, in Handbuch der Physik, Vol. VI, S. Flugge, ed.,
     Springer, Berlin (1958), p. 551.
28. F. A. McClintock and G. R. Irwin, ASTM STP 381:84 (1965).
29. G. R. Irwin, Eng. Fract. Mech. 1:231 (1968).
30. M. I. Kohan, "Nylon Plastics," Wiley, New York, 1973.
31. R. W. Hertzberg, M. D. Skibo, J. A. Manson, and J. K. Donald,
     J. Mater. Sci. 14:1754 (1979).
32. S. Arad, J. C. Radon, and L. E. Culver, J. Appl. Polym. Sci.
     17:1967 (1973).
33. S. Arad, J. C. Radon, and L. E. Culver, Eng. Fract. Mech.
     4:511 (1972).
34. J. A. Manson and R. W. Hertzberg, Crit. Rev. Macromol. Sci.
     1:433 (1973).
35. G. Pitman and I. M. Ward, J. Mater. Sci., 15:635 (1981).

CREEP AND FRACTURE INITIATION IN FIBRE REINFORCED PLASTICS

Jan-Fredrik Jansson and Henrik Sundström

Department of Polymer Technology
The Royal Institute of Technology
100 44 Stockholm - Sweden

INTRODUCTION

While most fibres used as a reinforcing constituent in polymer-based composites have reasonably simple deformation and fracture properties, the polymer matrices are anelastic and even non-linear viscoelastic with fracture properties which are strongly dependent on temperature, deformation rate, loading time etc. Therefore the mechanical behaviour is usually characterized as being either fibre or matrix dominated, i.e. more or less determined by only one of the constituents.

In most cases, however, the influence from the matrix has to be taken into consideration and the properties can be regarded as completely fibre dominated only for a few simple cases.

Neither is there any method available at present, capable of predicting the long-time behaviour, based only on the properties and composition of the consituents.

The problems have been observed and discussed by a large number of authors. Yet in spite of its immense practical importance, the connection between creep and fracture initiation has been very insufficiently studied.

We therefore now present a brief review and analysis of the present state of knowledge and some initial studies concerning the relations between the creep behaviour and creep fracture phenomena in thermoset-based fibre composites with special regard to the fracture initiation.

CREEP IN POLYMER- BASED FIBRE COMPOSITES - THE APPEARANCE OF NON-
LINEAR VISCO-ELASTIC BEHAVIOUR

The creep of polymer-based fibre composites has been studied
intensively[1-11]. The behaviour is usually regarded as a result of
the creep in the single constituents (the fibres and the matrix),
rupture in the fibres, matrix or interface and slippage of the
fibres in the matrix.

Kevlar is one exception to the rule that fibres can be regarded as being elastic. Very remarkable is that its creep rate is
even faster than for epoxy at low stresses, being the reverse at
high stresses[2]. Therefore, at low initial strains the load carried
by the fibres is transferred into the matrix resulting in an increasing logarithmic creep rate, whereas at high initial strains
the load carried by the matrix will be transferred into the fibres
giving a decreasing logarithmic creep rate.

Shear loading is the coupling mechanism transferring loads
between individual fibres and thus the general way of spreading
the load through the material. Like the transverse behaviour of
unidirectional laminates the visco-elastic shear behaviour is
dictated by the resign. For laminates built up from woven roving,
chopped strand mat etc. the creep is always substantial and must
be taken into consideration in the design work.

Fig. 1 shows creep curves for a chopped strand mat reinforced
polyester laminat with 33% (by weight) glass fibre at room temperature. A considerable amount of creep is obtained. A very common
way is to redraw the curves for constant loading times in an
isochronous stress-strain diagram. This is shown in Fig. 2 for
two different systems of glass fibre reinforced polyester at 1.000
h. The curves are linear for small stress-strains, where the deformation is reversible as well, showing a linear anelastic behaviour. At a certain stress-strain level the curves turn off,
however, towards the strain axis. The behaviour becomes non-rever-
sible and thus the material non-linear viscoelastic. For all neat,
glassy polyester or epoxy matrices a transition to marked non-
linear anelastic behaviour appears at a strain level of approximately 0,8%, becoming non-linear viscoelastic at still higher
strain-levels. In highly stressed positions of the composite the
matrix will therefore be non-linear resulting in a locally increasing creep and relaxation rate and decreasing stiffness, ending
in cracking and matrix failure. It is therefore interesting to
notice that for one of the systems in Fig. 2, the linear strain
limit is lower for the reinforced material (at 25%, weight, glass
fibre) than for the neat resin without any appreciable increase
in the stress limit. This indicates bad bonding between the fibres
and the matrix, incomplete curing or any other mistake in the
build-up of the composite.

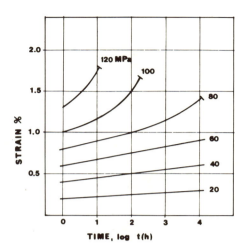

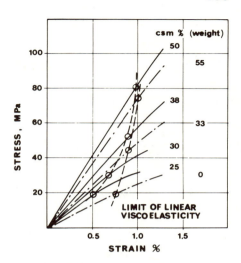

Fig. 1. Uniaxial creep in glass
fibre reinforced unsatu-
rated polyester (UP).
33% (weight) glass,
chopped strand mat (csm).

Fig. 2. Isochronous stress strain
diagram (1000 h, 23°C) for
two different systems of
glass fibre reinforced
(csm) polyester (UP).

The nature of the transition to the non-linear behaviour is
indicated in Fig 3. showing in principal the isochronous stress-
volume strain (stress dilation) behaviour for a polymer-based
fibre composite. At the same level as where the transition to non-
linear stress-strain behaviour appears, the dilation accelerates
and thus an additional increase in volume is introduced. This can
be caused either by the accelerated dilation in the matrix observed
above the limit of linear anelasticity for all glassy polymers, or
by formation of cracks, or by both phenomena. The later explanation
is further strengthed by the observed irreversible deformation in
the non-linear region.

The transition from linear anelastic into marked non-linear
viscoelastic behaviour also coincides with the transition from the
secondary to the tertiary creep region.

CREEP FRACTURE

As can be seen in Fig. 4, the strain at total failure usually
decreases with decreasing stress and increasing time to fracture.
In most cases a linear relation is observed between fracturestress
and logarithmic time to failure. The initiation and growth of
fracture include fiber and matrix fracture and failure of the interface. The situation is extremely complex and in spite of the large
number of very extensive studies reported in literature mainly

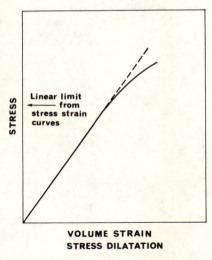

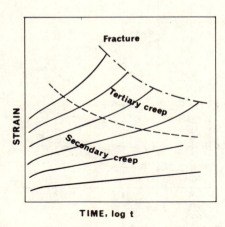

Fig. 3. Isochronous stress dila-   Fig. 4. Creep and creep fracture-
        tion curve for a polymer   strain curves for polymer
        based fibre composite.   based fibre composite.

based on fracture and micromechanics, the detailed knowledge of
the fracture processes and the conditions for the initiation and
growth of cracks is still far from complete. This is especially
true with respect to fracture initiation during long time loading
and creep.

    To give a few examples of the different phenomena that arise, a
very brief review is given in the next few paragraphs describing
the fracture processes in a $\pm 45^\circ$ cross-ply laminate and in a laminate built up from woven roving.

    Fig. 5 shows the crack initiation and growth in the $\pm 45^\circ$
crossply glass/polyester laminate at uniaxial tensile creep in
the 0° direction. At the beginning of the tertiary creep zone intralaminar cracks appear in the single plies perpendicular to the
stress direction.

    The cracks grow gradually through the single plies at distances
which are approximately constant. This causes large shear stresses
in the resin-rich interlaminar layers that crack in the tensile
stress direction.

    A still more complex situation appears in bending of woven
roving laminates as has been analysed by Diggwa and Norman[12].

    A series of fracture modes have been defined at different lo-
cations in the bent specimen and after different loading times. Fig. 6.

- Warp filament fracture occurs either in isolated filaments or in small groups of filaments in the warp strand. A few of this type of cracks appear already in the primary creep region. They are observed in larger numbers and grow rapidly in the tertiary creep zone.

- Transverse weft fracture includes resin cracks between and along the contours of the filaments. The cracks run perpendicular to the warp strands in those zones of the bent sample which are in longitudinal tension. They are present in considerable numbers already at the early part of the primary creep.

- Longitudinal weft fractures appear in the matrix in highly sheared areas of the sample

- Continued weft fractures occur as combinations of the longitudinal and transverse fractures and increase especially in the tertiary creep zone.

- Interstrand fracture occurs between the warp and weft strands and may follow their contour.

- Minor resin fracture appears in the resin-rich areas between the warp and weft strands. They usually occur in the stretched part of the sample and may be strongly affected by the curing conditions and shrinkage during the curing.

- Warp strand shear or buckling occurs in regions of longitudinal compressions. They are likely to be the main failure mechanism on the compression surface of the sample.

The catastrophic crack growth in the tertiary creep zone involves combinations of all modes of fractures defined above.

Thus although some crack initiation occurs in both the primary and secondary creep regions the main crack activities take place in the tertiary creep region.

The transition to non-linear visco-elasticity, the irreversible deformation and the increasing stress dilatation above the linear limit seems therefore mainly to be due to crack phenomena. The strain limit of linear anelasticity might therefore be defined as the "critical strain" for long time fracture, which has been studied intensively by for instance Menges et al. at IKV in Aachen[13-15].

The strain limit of linear anelasticity, the strain limit of the transition from the secondary to the tertiary creep zones (the "critical strain") is dependent on the type of reinforcement and matrix, the stress direction etc, and on the loading time.

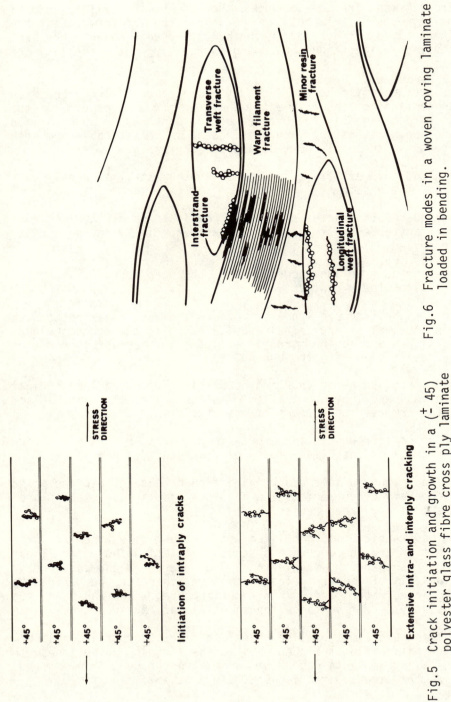

Fig.6   Fracture modes in a woven roving laminate loaded in bending.

Fig.5   Crack initiation and growth in a (± 45) polyester glass fibre cross ply laminate in uniaxial tension.

The fracture activities (initiation and crack growth) can also be detected by acoustic emission. This technique is developing very rapidly at present. The elastically stored mechanical energy is released during fracture, and one part is transformed into sound. Thus the type of fracture etc. will influence the energy of the pulse, its frequency, amplitude etc[16-21].

Very few measurements have been reported relating the acoustic emission intensity to the fracture processes during creep, in spite of the advantages this method might offer for the continuous "non-destructive" studies of the long-time fracture kinetics. Some recent preliminary data are shown in principle in Fig. 7, however. As can be expected the very low acoustic emission intensity in the secondary creep region is followed by a dramatic increase in the tertiary creep zone, corresponding as well to the transition to non-linear visco-elastic behaviour, constituting the initiation and growth of catastrophic cracks.

INTRODUCTARY STUDIES

As an introduction to future more fundamental studies on the physical conditions of the initiation and growth of failures during creep in fibre composites, a series of very slow tensile tests have been worked out on different glass fibre/polyester laminates.

The fracture activity was registered by acoustic emission at different strain rates between 1 and 60% per hour.

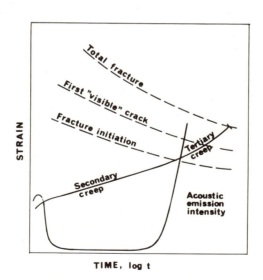

Fig. 7. Fracture phenomena during creep.

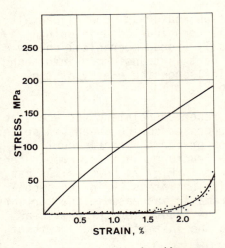

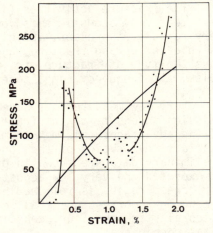

Fig. 8. Stress-strain diagram and acoustic emission intensity in uniaxial tension for a glass fibre/polyester (+45,-45,0,-45,+45) cross ply laminate. Glass content 60% by weight. Strain rate: 1,5% per hour

Fig. 9. Stress-strain diagram and acoustic emission intensity in uniaxial tension for glass fibre/polyester woven rowing laminate, stressed in the warp direction. Glasscontent 50% by weight. Strain rate: 2,4% per hour.

The number of crack events with an energy above a certain level and frequencies in the region 50-150 kHz, was measured in consecutive time intervals of equal length, evenly distributed over the test period. The length of the intervals was chosen inversely proportional to the strain rate. The result therefore shows the fracture intensity, normalized with respect to the strain rate.

In Fig. 8 the stress-strain behaviour for a (+45,-45,0,-45,+45) five ply laminate stressed in the O-direction, is compared with the corresponding fracture intensity. Although the stress-strain curve is non-lienar also for small strains very few fracture events are noticed below approximately 2% strain. The non-linearity may be explained by the non-linear viscoelastic behaviour of the matrix and the redistribution of stresses between the matrix and the fibres.

In Fig. 9 the stress-strain-acoustic emission behaviour is plotted for a laminate built up from woven rowing, stressed in the warp direction. Already at small strains a pronounced fracture intencity peak is noticed, due to cracks formed in the matrix along fibres perpendicular to the stress direction. The intensity of these fracture mechanisms seems to increase with increasing strain rate.

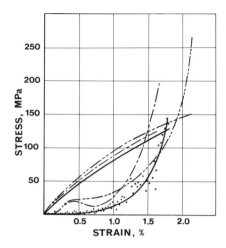

Fig. 10. Stress-strain diagram and acoustic
emission intensity in uniaxial tension for a glass fibre/polyester
csm-laminate. Glass content 40% by
weight. Strain rate:

⎯⎯⎯ 1.7% per hour
⎯ - ⎯12%   "   "
⎯ -- ⎯ 58% "   "

Fig. 10 shows the fracture activity in a chopped strand mat
(csm) laminate. A week crack intensity peak appears already at
strains below 0,5%, plausibly due to matrix failure in highly
stressed regions containing fibres perpendicular to the stress
direction. The peak disappears at low strain rates and seems to
merge into the main fracture processes at high rates.

ACKNOWLEDGEMENTS

This review has been made within the research programme:
Polymerbased Fibre Composites, sponsored by The Swedish Board of
Technical Research and Development.

The author wish to thank Prof Bengt Rånby, the head of the
Department of Polymer Technology.

REFERENCES

1. Sturgeon, J.B., Creep of Fibre Reinforced Thermosetting Resins,
      in Creep of Engineering Materials, ed. POMEROY
      D.D. Mech. Engin. Publ. Ltd. London, (1977)
2. Ericksen, R.H., Room Temperature Creep of Keolar 49/epoxy
      Composites. Composites, 189, July (1976).
3. Sturgeon, J.B., Creep, Repeated Loading, Fatigue and Crack
      Growth in $\pm$ 45° Oriented Carbon Fibre Reinforced
      Plastics. J. Materials Sci., 13, 1490 (1978).
4. Jain, R.K., Goswamy, S.K., Asthana, K.K., A study of the Effect
      of Natural Weathering on the Creep Behaviour of
      Glass Fibre-Reinforced Polyester Laminates.
      Composites, 39, Jan. (1979).
5. Peleg, M.   A Model for Creep and Early Failure. Materials
      Sci. Engin. 40, 197 (1979).
6. Hancox, N.L., Minity, D.C.C., The Torsional Creep of Carbon
      Fibre Reinforced Epoxide Resins. J. Materials
      Sci. 13, 797 (1978)
7. Beckwith, S.W. Creep Evaluation of a Glass/Epoxy Composite.
      Sampe, 8, Jan. (1980).
8. Holmes, M., Rahman, T.A., Creep Behaviour of Glass Reinforced
      Plastics Box Breams. Composites, 79, April (1980)
9. Ehrenstein, G.W. Kriechen, Rest- und Zeigstandfestigkeit bei
      GF-UP. Kunststoffe 66:5, 289 (1976).
10. Holloway, L.   Glass Reinforced Plastics in Construction:
      Engineering Aspects. Surrey University Press,
      Glasgow (1978).
11. Johnson, A.A.   Engineering Design Properties of GRP. The
      British Plastics Federation, London (1979).
12. Diggwa, A.D.S., Norman, R.H., Mechanism of Creep of GRP,
      Plastics and Polymers, 263, Oct. (1972).
13. IKV Veröffentlichungen 1969-75, 3. Kritische Dehnung, Aachen
      (1976).
14. Brintrup, H.   IKV Forschungsdokumentation 1977.
15. Meffert, B.   IKV Forschungsdokumentation 1977.
16. Roeder, E., Crostack, H-A., Der Maschinenschaden 50:4, 121 (1977).
17. Garret, K.W., Bailey, J.E., J. Material Sci., 12, 2189 (1977).
18. Rotem, A.   The Discrimination of Micro-Fracture Mode of
      Fibrous Composite Material by Acoustic Emission
      Technique. Fibre Sci. and Techn. 10, 101, (1977).
19. Rotem, A.   Effect of Strain Rate on Acoustic Emission from
      Fibre Composites. Composites, Jan, 33, (1978).
20. Williams, J.H., Samson, S.L., Acoustic Emission Montoring Fiber
      Composite Materials and Structures. J. Composites
      Mtrls., 12, 348 (1978).
21. Laroche, D., Bunsell, A.R., Stress and Time Dependent Damage in
      Carbon Fibre Reinforced Plastics. Adv. in
      Composite Materials, Ed. Bunsell, A.R. et.al.,
      page 985, Pergamon Press (1980).

EFFECT OF FIBER ASPECT RATIO ON ULTIMATE PROPERTIES OF

SHORT-FIBER COMPOSITES

J. L. Kardos, E. Masoumy, and L. Kacir*

Materials Research Laboratory and
Department of Chemical Engineering
Washington University
St. Louis, MO   63130

INTRODUCTION

The maximum fiber utilization efficiency is achieved for
short fiber composites by aligning the fibers with the directions
of major applied stresses, by controlling the fiber aspect ratio
and volume fraction, and by choosing the appropriate type of fiber
and matrix. In developing design equations to predict mechanical
behavior, it is helpful to begin with a well-aligned short fiber
system. The use of well-aligned systems allows one to isolate
the strong effects of fiber aspect ratio and orientation distribution on the mechanical behavior and fracture mechanisms.

Numerous articles are available in the scientific literature
related to the properties of oriented short fiber composites
(1-7). One major common disadvantage of the published work is
that reported experimental data were obtained on material systems
which were not well characterized and defined. For example, fiber
orientation distribution was not generally specified and the actual
aspect ratio of the short fibers in the composite was not generally
characterized.

Kacir, et al. (8-12) have investigated the dependence of
stress-strain properties on fiber alignment distribution for
brittle epoxy composites containing aligned short glass fiber
bundles. The uniaxial tensile stress-strain behavior of specimens

* Current Address: Israel Aircraft Industries, Ltd. Ben-Gurion
  International Airport Israel

with different orientation patterns, which were determined experimentally, was compared with analytical predictions based on
"Laminate Analogy" methods (11). Good agreement was obtained
between the empirically fit experimental results and the theoretical
predictions in the case of stiffness, stress-strain relationship
and strength.

The "Laminate Analogy" approach was originally introduced by
Halpin and Pagano (13) for the prediction of the elastic stiffness
and thermal expansion properties of a randomly oriented short
fiber composite. Jerina and Halpin (14) extended this approach to
the prediction of the tensile strength of two dimensional, random
or aligned fiber composites by using the laminate analogy in
combination with maximum strain criteria for ply failure. Nicolais
(15), and more recently Halpin and Kardos (16), reviewed the theory
associated with the laminate analogy approach to the prediction of
the stress-strain properties of short fiber composites.

The role of the matrix in protecting the fibers from environmental effects and transferring the external applied stresses to
the short fibers is widely mentioned in the literature. Replacing
the brittle matrix with a ductile one may have a significant
effect on the mechanical properties of aligned short fiber
composites primarily by affecting the crack propagation characteristics. The ability of the ductile matrix to absorb energy is
much higher than that of the brittle matrix, leading to a non-linear
stress-strain curve which has a yield point.

In this paper are summarized the major results from a study
of the effects of fiber aspect ratio and fiber orientation distribution on the stress-strain behavior of aligned composites
containing a ductile epoxy matrix. Comparisons are also made with
the behavior of a similar system having a brittle epoxy matrix.
These results are part of a more comprehensive study in the process
of being published (17,18) to which the reader is referred for
more background, experimental details, and additional related
results.

EXPERIMENTAL

Materials and Specimen Preparation

The ductile epoxy matrix consisted of a 60/40 by weight
mixture of Epon 815 resin and Versamide 140 curing agent. The
glass fibers were obtained from PPG Industries as chopped strands
of E-glass (Type 3450) which had been sized for epoxy with amino
silane coupling agent. The chopped strands yielded fiber bundles
with nominal fiber lengths of 1/4 inch, 1/2 inch and one inch,

which corresponds to fiber bundle aspect ratios of 185, 398, and
557, respectively.

   The major steps in fabricating a well (but not perfectly)
aligned composite specimen are specifically described below.
   1) The glass-fiber bundles were aligned in the mold using
the vibrating V-shaped trough. The stainless steel, window-frame
mold (6" x 6") was previously sprayed with release agent and lined
with aluminum screens on both top and bottom.
   2) The mold was placed in a vacuum oven at 70°C for at least
five hours to dry the glass-fibers.
   3) A prepreg was prepared by pouring onto the fibers an
acetone solution of a resin/curing agent mixture (60 parts by
weight Epon 815, 40 parts of Versamide 140 hardener, 100 parts
acetone). The mold was left at one atm. for about 30 minutes at
room temperature, and then kept in a vacuum oven (30 Torr) at room
temperature for about 60 minutes in order to vaporize the excess
acetone. Most of the air bubbles also came out in this stage.
The vacuum oven was then heated gradually (2°C/min) to 50°C and
held there for 55 minutes in order to B-stage the epoxy. During
the heating under vacuum most of the remaining air bubbles and
acetone were removed.
   4) The screens were removed and the prepreg was then covered
with Teflon sheets and rolled with a roller, to obtain a smooth
surface.
   5) The prepregs were removed from the window frame and
compression-molded at 110°C using a stainless steel mold with a
vent in each side to accommodate the last air bubbles during
compression. After heating the mold to 110°C, the press was
pumped (closed and opened) about five times to vent the last
bubbles. After eight minutes at atmospheric pressure, the pressure
was gradually increased to 46,000 psi and held for about 30 minutes.
The mold was cooled under pressure to room temperature at about
25°C/min. The sample was removed from the mold and post-cured at
110°C for about four hours. This procedure yields 50 v % sheets.

Fiber Orientation Distribution

   In order to analyze the fiber orientation distribution in the
aligned reinforced sheets, red-colored glass fiber bundles were
added to the fiberglass at a two weight percent concentration
prior to the orientation and prepregging process. The colored
fibers were prepared by mixing glass fibers and red Alizarin dye
in water, followed by heating the coated bundles to remove the
water and fix the dye. The number of colored bundles in a typical
sheet was 150-200. The angle of deviation of each colored bundle
from the major alignment direction was measured, and the bundle
orientation distribution characterized.

Mechanical and Physical Properties

Tensile properties (elastic modulus, tensile strength and tensile strain at failure) were obtained for longitudinal (major alignment direction), transverse and off-axis tensile specimens (5°, 10°, 15°, 20°, 30°, 45°, 60° off-axis angles). The specimens were tested in a uniaxial tensile test with an Instron Universal Tester at a crosshead speed of 0.05 in/min. The tensile specimens were straight-sided strips to which epoxy/glass fabric tabs were bonded to ensure uniform distribution of clamping stresses. All tensile experiments were carried out at 23°C and strains were measured with an Instron strain gage extensometer. Ten specimens were tested in the longitudinal direction while five were tested at each of the other angles. The principal Poisson ratio, ν_{12}, was measured with micro-strain gages arranged perpendicular and parallel to the major fiber alignment direction (0°).

Optical microscopy was used to study fiber orientation.

The procedures described above were optimized to produce samples containing 50 ± 4 percent of glass fibers by volume with void contents of less than 2.5%. Volume fractions were measured gravimetrically by burning the resin off at 427°C. Density measurements for both void content and volume fraction calculations were made with a pycnometer. Fiber bundle aspect ratios were measured before molding and after burn-off, with no significant difference due to fabrication.

Control specimens of the matrix were fabricated by casting the 60/40 resin/curing agent mixture between glass plates and curing with the following cycle: ten hours at 25°C, one hour at 40°C, two hours at 70°C, one hour at 110°C.

RESULTS AND DISCUSSION

The Ductile Matrix

A 60/40 Epon 815/Versamide 140 (by weight) matrix system yielded an optimal combination of properties (tensile modulus (3×10^5 psi), strength (6.4 Ksi), ultimate strain (7%), degree of ductility (σ_{yield} = 7 Ksi), and relatively low viscosity of the uncured mixture). The existence of a yield point in the stress-strain curve is a clear indication of the ductile behavior of the matrix compared to the typical nearly linear stress-strain curve which characterizes a brittle epoxy.

Fiber Orientation Distribution

Fiber orientation is one of several important parameters

controlling the mechanical behavior of reinforced plastics. Prediction of mechanical properties of short fiber systems therefore requires an accurate determination of the distribution of fiber orientation. Various methods for the characterization of fiber alignment in opaque composites have been described in the literature and recently reviewed by Kacir, et al. (10). In the present work, red-colored glass fiber bundles were used as tracers. The angle of deviation from the longitudinal axis of the sheet for each colored bundle was measured and recorded in the form of a cumulative percent of fibers vs. fiber orientation. In this case the cumulative percent was taken in 5° intervals from $(-\alpha)$ to $(+\alpha)$ up to 50°, after which 10° intervals were used. Fiber alignment data for 1/4-, 1/2- and 1-inch aligned short fiber composites are presented in Table 1.

In order to test the reproducibility of the fabrication process used in this study, the fiber orientation distribution was characterized for each of three different sheets for each fiber length. Figure 1 presents the results for the 1/4-inch fiber length, which were typical of the other two fiber lengths. It is clear that there is a high degree of reproducibility and ability to experimentally control the fiber orientation distribution using the procedures described above.

Stress-Strain Properties

The effect of fiber length on the longitudinal, transverse and off-axis tensile properties of the aligned short fiber composites was studied, with the off-axis properties being obtained from specimens cut at 5, 10, 15, 20, 30, 45, 60 and 90 degrees to the major alignment direction. Typical stress-strain curves are shown in Figures 2, 3 and 4 for 1/4-, 1/2- and 1-inch aligned bundle composites, respectively (bundle aspect ratios of 185, 398 and 557, respectively). In contrast to the nonlinear, ductile behavior of the matrix, the stress-strain curves for all the longitudinal (0°) specimens are nearly linear to fracture. This phenomenon can be attributed to the elastic deformation characteristics of the fibers and to the stress concentrations near the short bundle ends, which cause premature failure of the composites at failure strains up to 1.57%. The stress-strain relation for the pure matrix is nearly linear up to about 2%, which contributes to the linear behavior of the longitudinal (0°) composites as well as the nearly linear behavior of the off-axis samples for angles below about 15°. This nearly linear nature of the stress-strain curves suggests that the elastic and strength characteristics of those composites can be predicted using laminate analogy methods, based on linear elastic laminate theory, even though the matrix itself behaves in a ductile manner. Nevertheless, the micro-deformation mode of the matrix in the reinforced sheets is not necessarily linear nor can the fracture be characterized as brittle.

Table 1

Cumulative Fiber Orientation Distribution as a Function of Fiber Length

Fiber Length, inch	Bundle Aspect Ratio	Cumulative Percent of Fibers Oriented Between ± α														
± α		1	5	10	15	20	25	30	35	40	45	50	60	70	80	90
1/4	185	7.11	25.94	50.62	64.85	76.56	80.75	85.78	89.12	92.89	94.57	95.41	97.09	97.93	98.77	100
1/2	398	5.41	26.29	48.57	63.22	74.74	81.9	88.67	93.4	94.87	96.65	100	100	100	100	100
1	557	4.85	38.83	65.05	80.58	89.32	92.23	96.11	96.11	98.05	99.02	100	100	100	100	100

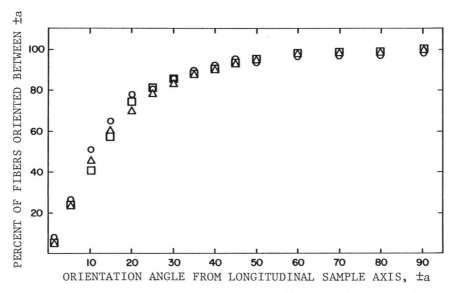

Figure 1.   Cumulative fiber orientation distribution for 1/4-inch
            fiber bundle composites (bundle aspect ratio is 185).

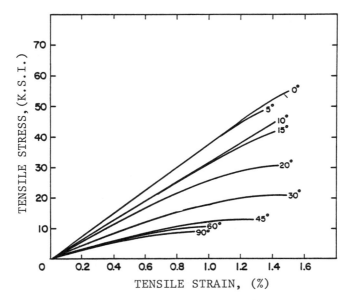

Figure 2.   Typical experimental stress-strain curves as a function
            of off-axis angle for a bundle aspect ratio of 185
            (fiber volume fraction = 0.5).

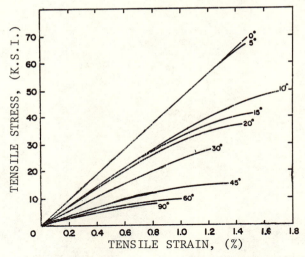

Figure 3.   Typical experimental stress-strain curves as a function
of off-axis angle for a bundle aspect ratio of 398
(fiber volume fraction is 0.5).

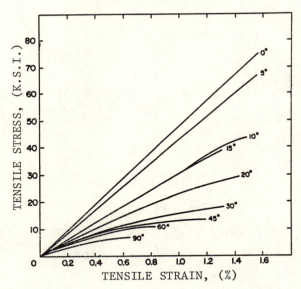

Figure 4.   Typical experimental stress-strain curves as a function
of off-axis angle for a bundle aspect ratio of 557
(fiber volume fraction is 0.5).

The effect of bundle aspect ratio on the elastic modulus in the major alignment direction, $\overline{E}_x$, is shown in Figure 5. Experimental results obtained previously by Kacir et al. (10) for aligned glass fiber bundle-brittle epoxy composites are also shown. As the aspect ratio increases, the experimental elastic moduli approach the modulus for a continuous, longitudinal composite (5.35 Msi), as expected. There is no evidence that the nature of the matrix (brittle vs. ductile) has any effect on the longitudinal (0°) elastic modulus. This is a natural result, since the elastic modulus is a property which is measured at a very low strains where both matrices behave elastically. The elastic modulus of the glass fibers and their volume fraction in the composite control the elastic properties of the aligned composites. The small difference in the elastic modulus between the brittle and ductile matrices (the ductile matrix modulus of elasticity is about 67% of that of the brittle matrix) has negligible effect on $\overline{E}_x$.

The effect of off-axis angle on the tensile modulus, $E_x(\theta)$ is shown in Figure 6 for 1/4-, 1/2-, and 1-inch aligned short fiber bundle composites. The symbols represent the experimental data points while the lines have been constructed using the equation suggested by Leknitskii (19), which was derived from classical elastic theory for anisotropic systems:

$$\frac{1}{E_x(\theta)} = \frac{\cos^4(\theta)}{\overline{E}_x} + \left(\frac{1}{\overline{G}_{xy}} - \frac{2\overline{\nu}_{xy}}{\overline{E}_x}\right) \sin^2\theta\cos^2\theta + \frac{\sin^4\theta}{\overline{E}_y} \qquad (1)$$

where $\overline{E}_x$, $\overline{E}_y$ and $\overline{G}_{xy}$ are the longitudinal, transverse and in-plane shear moduli of the aligned fiber composites, respectively, and $\overline{\nu}_{xy}$ is the major Poisson ratio.

The validity of Equation 1 for the aligned short fiber, ductile epoxy composites was examined in this work using the following approach. First $\overline{E}_x$ and $\overline{E}_y$ were measured experimentally for all three fiber bundle lengths. Next, $\overline{E}_x(\theta)$ was measured for all three fiber bundle lengths at various angles within the 15-45° range where a shear mechanism dominates the failure. Then, using Equation 1, $\overline{G}_{xy}$ values were calculated for all three fiber bundle lengths. Average values thus calculated for 1/4-, 1/2-, and 1-inch composites were 0.685, 0.93, and 0.64 Msi, respectively. The scatter in $\overline{G}xy$ values for each aspect ratio was very small. The rather significant differnce in the calculated $\overline{G}xy$ values arises mainly from differences in the fiber orientation distribution for each

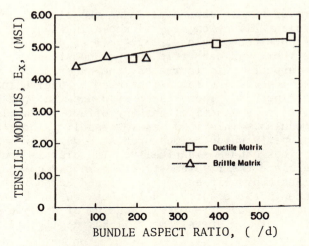

Figure 5.   Dependence of longitudinal tensile modulus (E_X), on the
bundle aspect ratio, ℓ/d, for ductile and brittle epoxy
matrix systems containing 50 v % aligned glass fiber
bundles.

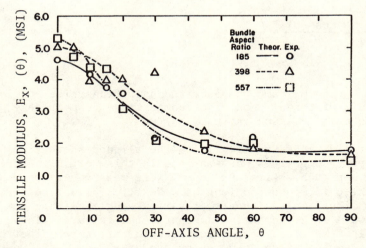

Figure 6.   Dependence of tensile modulus on off-axis angle between
major fiber alignment and tensile loading directions for
various fiber bundle aspect ratios.   Volume fraction is
0.5.   Smooth curves are calculated with Letnitskii
equation.

fiber length. Kacir et al. (11) showed that $\overline{G}_{xy}$ will be highly
dependent on degree of fiber alignment. Equation 1 fits the
experimental data quite well at the angular extremes. It is
important to note that the bundle aspect ratio effect on modulus
diminishes and becomes negligible as the off-axis angle exceeds
about 15°.

The effect of fiber bundle aspect ratio on the longitudinal
(0°) tensile strength of the aligned short fiber composites, $\overline{\sigma}_x$,
is shown in Figure 7. The experimental curve contains the results
obtained in this work for the aligned short fiber ductile epoxy
composites of relatively high aspect ratio, as well as the previously
published results of Kacir et al. (10) for the brittle system. A
smooth best-fit line has been drawn through all the data points.
As in the case of the elastic modulus, there is no evidence that
the nature of the matrix (ductile vs. brittle) has any pronounced
effect on the tensile strength of the longitudinal (0°) specimens.

The tensile strength of aligned short fiber composites is
mainly controlled by the fiber aspect ratio, fiber strength, fiber
volume fraction, fiber alignment distribution, and the degree of
adhesion between the fibers and the matrix, which is related to
the interfacial shear properties. The longitudinal fibers are the
major contributors to the increase in resistance to crack propagation. The mode I fracture toughness parameter, K_{IC}, for longi-
tudinally aligned short glass fiber bundles in a brittle epoxy
matrix was found to be 20.56 Ksi $\sqrt{in}$, while K_{IC} for the transverse
specimens was only 4.23 Ksi $\sqrt{in}$ (unpublished experimental data by
L. Kacir). These results demonstrate the influence of fiber
alignment on the resistance to crack propagation. It seems, from
the experimental results in this study, that the additional
contribution of the ductile nature of the matrix to the arrest of
propagating cracks is small and does not have significant influence
on the tensile strength of the longitudinal (0°) specimens.

On the other hand, in the case of transverse and off-axis
specimens, the role of the ductile matrix in providing resistance
to crack propagation is much more pronounced. For example, the
transverse tensile strengths for the 1/4- and 1/2-inch aligned
fiber bundle lengths were 8200 and 8600 psi respectively, compared
to the pure ductile matrix strength of 7200 psi. Since the
transverse fibers act as stress concentrators, one would expect
the transverse strength of the composite to be below that of the
pure matrix. Due to the ductility of the matrix and to its
resistance to crack propagation, the actual tensile strengths of
the transverse composites were even slightly higher than that of
the matrix. In the case of aligned short glass fiber bundle-
brittle epoxy composites, the transverse strengths of the

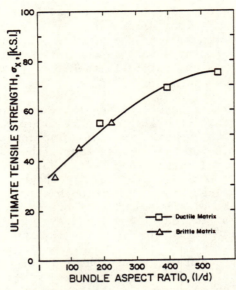

Figure 7.   Dependence of longitudinal tensile strength on the bundle
aspect ratio, ℓ/d, for ductile and brittle epoxy matrix
systems containing 50 v % aligned glass fiber bundles.

composites were about 45 to 55 percent of the tensile strength of
the pure brittle matrix (10).

The strength of a perfectly aligned short fiber composite
approaches an asymptotic value at large aspect ratios.   The
existence of stress concentrations near the fiber ends weakens the
composite and it fails at a reduced ultimate longitudinal strain
when compared with that of a continuous fiber composite.   Thus the
strength limit for the aligned short fiber composite is substantially below that of the continuous fiber reference.   Chen (20)
carried out a stress analysis based on a finite element method for
a unidirectional discontinuous fiber composite which has an
idealized fiber packing arrangement.   He found that at sufficiently
large aspect ratios the strength of short fiber glass/epoxy
composites approaches roughly 60 percent of the strength of the
same continuous fiber composite system, which he defined as a 60%
strength efficiency.   It is interesting to note that in this work
the longitudinal tensile strength efficiencies of 1/4-, 1/2- and
1-inch aligned, glass fiber bundle composites were 43, 55 and 59
percent, respectively.   The 1-inch aligned fiber composite has a
fiber bundle aspect ratio of 557 which seems to be sufficiently
high to achieve a value very close to Chen's plateau value.   The
elastic modulus efficiency for this composite was 99%.   It must be
remembered, however, that the fiber bundles are not perfectly
aligned, even in the case of the 1-inch fiber lengths.

The effect of off-axis angle (from the major alignment direction) on the tensile strengths of the aligned short fiber composites is plotted in Figure 8 for 1/4-, 1/2- and 1-inch fiber bundle composites. At angles beyond about 15°, there is very little effect of bundle aspect ratio which implies that the failure modes at the larger angles are dominated by shear and transverse mechanisms. From an analytical viewpoint Azzi and Tsai (21) adapted Hill's (22) generalized Von Mises-Hencky (23) distortional energy yield criterion to predict the off-axis strength of uni-directional composites for both uni- and multidirectional continuous fiber systems. They reported the experimental data for off-axis continuous unidirectional glass fiber epoxy composites to be in good agreement with theory. The theoretical formulation, suggested by Azzi and Tsai, may be modified for the case of aligned dis-continuous fiber composites in the following way:

$$\frac{1}{\overline{\sigma}_x(\theta)} = \frac{\cos^4\theta}{\overline{\sigma}_x{}^2} + \left(\frac{1}{\overline{\tau}_{xy}{}^2} - \frac{1}{\overline{\sigma}_x{}^2}\right)\ \sin^2\theta\cos^2\theta + \frac{\sin^4\theta}{\overline{\sigma}_y{}^2} \qquad (2)$$

where $\overline{\sigma}_x(\theta)$ is the off-axis tensile strength of the composite, $\overline{\sigma}_x$ is the tensile strength in the major fiber alignment direction,

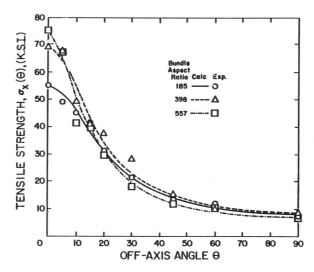

Figure 8.   Dependence of tensile strength on off-axis angle between major fiber alignment and tensile loading directions for various fiber bundle aspect ratios. Volume fraction fibers is 0.5. Smooth curves are calculated with Azzi-Tsai equation.

and $\overline{\sigma}_y$ is the transverse tensile strength. $\overline{\tau}_{xy}$ is the in-plane shear strength along the main orthotropic axis of the material and θ is the angle between the direction of the applied external load and the major alignment direction.

Exactly the same procedure was used to test the validity of Equation 2 as was used in the off-axis modulus analysis; namely $\sigma_x(\theta)$, $\overline{\sigma}_x$, $\overline{\sigma}_y$ and $\overline{\nu}_{xy}$ were measured experimentally and the in-plane shear modulus $\overline{\sigma}_{xy}$ calculated using $\sigma_x(\theta)$ measurements in the angle range 15-45°, where shear failure dominates. Having thus "pinned" the curves with experimental data in the mid-angle range, $\sigma_x(\theta)$ was then calculated for the entire angular range.

The scatter in the calculated $\overline{\tau}_{xy}$ values was very small for each fiber length, indicating the orthotropic nature of the aligned system. Having thus "pinned" the predictive calculation using the mid-angle experimental data to calculate $\overline{\tau}_{xy}$, $\overline{\sigma}_x(\theta)$ was calculated as a function of θ for each fiber bundle aspect ratio. The results are plotted in Figure 8 and fit the experimental data at both small and large angles quite well.

Predictions Using the Laminate Analogy

The longitudinal tensile strength of short fiber composites can be predicted by using the laminate analogy and applying the Halpin-Kardos equation to estimate the failure strains of perfectly aligned short fiber composites. Details of this calculational format may be found in references 16 and 23. Briefly the aligned composite is considered to be a laminate, each ply of which has perfectly uniaxially oriented short fibers. The plies are oriented and given thicknesses (laminate fractions) which correspond to the actual fiber bundle orientation distribution. The stiffness of each ply is calculated with the Halpin-Tsai equation and the laminate stiffness is then calculated by summing up the individual ply contributions. A maximum strain failure criterion is chosen and the failure strains for a continuous fiber, uniaxially oriented ply are obtained experimentally. Next the Halpin-Kardos equation is used to calculate a strength reduction factor which, along with the ply moduli, can be used to reduce the continuous fiber ply allowable strains to those for a perfectly aligned short fiber ply of the same volume fraction. These allowable strains are next transformed from each ply principal direction to the principle laminate directions. If the laminate strain has exceeded any of the transformed ply strains, those plies are considered as having failed. Knowing the laminate strain at which the first ply failed, as well as the laminate stiffness, allows one to calculate the stress in the laminate at the point of first ply failure. The failed ply is deleted, the modulus recalculated, and the strain at which the next ply fails is calculated. This procedure continues until all of the plies

have failed. The laminate strain increments are added up to
produce the failure strain and the stress increments are summed
to provide the failure strength. Details of these calculations
for the system of this study are too space consuming for the
summary nature of the paper and may be found in reference 18.

Table 2 presents a comparison of experimental results for
the failure strength, modulus, and failure strain with the values
predicted utilizing the laminate analogy and the Halpin-Kardos
equation. The failure strain predictions are reasonably good, with
a disagreement of at most 10% with experiment. The modulus
predictions are consistantly low, ranging between 8 and 13% below
experiment. The strength predictions are consistantly above
experiment with the largest discrepancy being about 5%. This
remarkable accuracy is most likely a result on compensating
errors in the stiffness and strain parameters.

Table 2

Comparison of Experimental Stress-Strain Properties with Values
Predicted Using the Laminate Analogy and Halpin-Kardos Equation

Fiber Length, (inches)	Bundle Aspect Ratio	Experimental Results 50% by volume of fiber			Theoretical Predictions		
		$\overline{\sigma}_x$ (Ksi)	$\overline{E}_{xx}$ (Msi)	ε_x (%)	$\overline{\sigma}_x$ (Ksi)	$\overline{E}_{xx}$ (Msi)	ε_x (%)
1/4	185	55.0	4.62	1.45	55.6	4.24	1.32
1/2	398	69.4	5.03	1.49	71.0	4.36	1.64
1	557	75.6	5.30	1.57	79.0	4.77	1.66

CONCLUSIONS

1. Even though a ductile matrix was used, highly oriented
short fiber systems still can produce nearly linear, brittle-like
tensile stress-strain behavior when tested in the major fiber
direction (0°). As the fiber bundle aspect ratio increases, the
tensile modulus approaches that for a continuous fiber system of
the same volume loading, whereas the tensile strength achieves
only 59% of the continuous fiber value for the largest aspect
ratio studied (557).

2. When the aligned fiber system is tested off-axis, the
effect of aspect ratio on modulus becomes very small at angles
exceeding about 15°, and the angular behavior of the modulus is
reasonably well described by the Leknitskii equation. The off-axis

tensile strengths are also relatively unaffected by aspect ratio beyond about 15°. A modified Azzi-Tsai equation adequately fits the off-axis strength data. Transverse tensile strengths were higher than the matrix strength for this ductile matrix system.

3. A calculational format utilizing the laminate analogy with the Halpin-Tsai equation and the Halpin-Kardos equation successfully predicts the ultimate properties of this aligned fiber bundle, ductile matrix system well within design accuracy.

ACKNOWLEDGMENT

The authors gratefully acknowledge the support of this research in part by the Polymers Program in the Materials Division of the National Science Foundation under Grant No. DMR-7812806.

REFERENCES

1. R. M. Anderson and R. E. Lavengood, SPE J., 24, 20 (1968).
2. L. E. Nielsen and P. E. Chen, J. Materials, 3, 352 (1968).
3. W. Rosen in "Fiber Composite Materials", American Society for Metals, Metals Park, Ohio (1965).
4. A. Kelly and W. R. Tyson, J. Mech. Phys. Solids, 13, 329 (1965).
5. J. K. Lees, Polym. Eng. Sci., 8, 195 (1968).
6. B. F. Blumentritt, B. T. Vu and S. L. Cooper, Polym. Eng. Sci., 14, 633 (1974).
7. B. F. Blumentritt, B. T. Vu and S. L. Cooper, Composites, 6, 105 (1975).
8. L. Kacir, M. Narkis, and O. Ishai, Polym. Eng. Sci., 15, 525 (1975).
9. L. Kacir, M. Narkis, and O. Ishai, Polym. Eng. Sci., 15, 532 (1975).
10. L. Kacir, M. Narkis, and O. Ishai, Polym. Eng. Sci., 17, 234 (1977).
11. L. Kacir, M. Narkis, and O. Ishai, Polym. Eng. Sci., 18, 45 (1978).
12. L. Kacir, M. Narkis, and O. Ishai, Composites, 9, 89 (1978).
13. J. C. Halpin and N. J. Pagano, J. Comp. Mater., 3, 720 (1969).
14. K. L. Jerina and J. C. Halpin, "Strength of Molded Discontinuous Fiber Composites", U. S. Gov't Rep., AD 759827 (1972).
15. L. Nicolais, Polym. Eng. Sci., 15, 137 (1975).
16. J. C. Halpin and J. L. Kardos, Polym. Eng. Sci., 18, 496 (1978).
17. E. Masoumy, L. Kacir and J. L. Kardos, Polymer Composites, submitted for publication.
18. E. Masoumy, L. Kacir, J. L. Kardos and S. L. Chang, Polymer Composites, submitted for publication.
19. S. G. Leknitskii, "Theory of Elasticity of an Anisotropic Body", p. 32, Holden-Day, Inc. (1963).

20. P. E. Chen, Polym. Eng. Sci., 11, 51 (1971).
21. V. D. Azzi and S. W. Tsai, Experimental Mechanics, 5, 283
     (1965).
22. R. Hill, Proc. Royal Soc. Series, London, 193, 281 (1948).
23. R. Von Mises, Z. Angew, Mathematik und Mechanik, 8, 161 (1928).
24. J. L. Kardos, J. C. Halpin, and S. L. Chang, Rheology, Vol. 3:
     Applications, Eds. G. Astarita, G. Marrucii, and L. Nicolais,
     Plenum, N. Y., 1980, p. 255.

AN OPTICAL TECHNIQUE FOR MEASURING FIBER

ORIENTATION IN SHORT FIBER COMPOSITES

S. H. McGee and R. L. McCullough

Department of Chemical Engineering and
Center for Composite Materials
University of Delaware
Newark, Delaware

ABSTRACT

A technique for measuring the state of fiber orientation in short fiber composites using a Fraunhoffer diffraction apparatus is described. A theoretical analysis is compared to diffraction patterns produced from masks with known state of orientation and the agreement is found to be excellent. Samples from sheet molding compounds are analyzed to demonstrate processing induced fiber orientation in the material.

INTRODUCTION

The degree of fiber orientation in short fiber composites plays a major role in determining the behavior of the composite. For a composite, such as sheet molding compounds, the degree of fiber orientation can range from planar random to completely aligned. A material with one state of orientation would behave much differently, when subjected to the same mechanical loads, than a similar material which had a different state of orientation. The state of fiber orientation is a complex function of many molding parameters which may or may not be under control of the molder.

Currently there seems to be no simple method of measuring the state of fiber orientation. The process of counting the number of fibers at a given angle in a photomicrograph may be feasible for a few measurements, but is unreasonable for an extended number of measurements and impractical for routine work. It is the purpose of this work to present a new technique for quickly measuring fiber

orientation which utilizes principles of the theory of light diffraction.

APPARATUS

The diffraction apparatus is a typical Fraunhoffer diffractometer and is shown schematically in Figure 1. A discussion of the
design and alignment of similar diffractometers is given by Lipson and
Lipson [1]. Red light (wave length 632.8 nm) from a 2.5 mW helium
neon laser is focused on a 155 micron pinhole by a lens L_0 (focal
length 4.8 cm). Since the light rays passing through the diffraction
mask must be parallel, the pinhole is located at the focal point of
lens L_1. After passing through the diffraction mask, the light is
focused on the observation screen or recording film using lens L_2.
Lenses L_1 and L_2 have identical focal lengths, 61.0 cm.

The diffraction masks were produced from X-radiographs of a
sheet molding compound in which short lengths of copper wire had been
included as tracer fibers. The X-radiographs were photographically
reduced to an appropriate size using a high contrast, black and white
35 mm film, such as Kodak Kodalith graphic arts film. The resulting
negatives, on which the fibers appeared as slits and the resin as an
opaque background, were mounted using standard 35 mm slide mounts
and used as the diffraction masks.

The diffraction patterns were recorded using a Graflex roll film
camera and Kodak Verichrome Pan film. A typical exposure time was
0.10 seconds. A Macbeth optical densitometer was used to measure
the light transmittance of the photographic negative of the recorded
diffraction pattern.

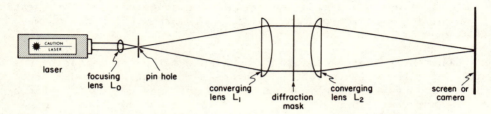

Fig. 1.   Schematic diagram of the Fraunhoffer diffractometer.

ANALYSIS

The principles of Fraunhoffer diffraction are described in detail in most optics texts, see for example Lipson and Lipson [1] or Born and Wolf [2]. The diffraction masks produced from photomicrographs as described above can be modeled as the superposition at various angles of collections of centro-symmetric, randomly space, identical, parallel apertures. The diffraction pattern from this type of diffraction mask can be found from a three step process. First, the diffraction pattern for a single aperture is found using linear Fourier optics theory. Second, using the principles of interference theory, the diffraction pattern from a collection of randomly spaced parallel apertures is found. Finally, the entire diffraction pattern is computed by superimposing the diffraction patterns of the collections of oriented apertures.

It is well known that the amplitude of a Fraunhoffer diffraction pattern, except for a phase constant, is simply the two dimensional Fourier transform of the aperture [1], viz.

$$A(u,v) = \int\int_{-\infty}^{+\infty} O(x,y) \exp[-\frac{2\pi i}{\lambda f}(xu + yv)] \, dx \, dy \qquad (1)$$

where A is the amplitude, $O(x,y)=1$ if (x,y) is located inside the aperture and is zero otherwise, f is the focal length of lens L_2, λ is the wave length of the light, and (x,y) and (u,v) are coordinates in the object (or diffraction) and Fourier (or observation) planes respectively. Equation (1) has been evaluated for several simple geometric apertures by Lipson and Lipson [1] and Smith and Marsh [3].

For a single rectangular aperture oriented parallel to the (x,y) axes, Eq. (1) can be evaluated analytically to yield

$$A(u,v) = ab \, sinc(\pi a \, v/\lambda f) \, sinc(\pi b \, v/\lambda f) \qquad (2)$$

where

$$sinc(x) = sin(x)/x$$

and a and b are the lengths of the sides of the rectangle. This result is given by Lipson and Lipson [1] among others. For a rectangle (representing the individual fibers) oriented at some angle, θ, to the (x,y) axes a simple coordinate rotation gives the amplitude as

$$A(u,v,\theta) = ab \, sinc(\pi a \, u'/\lambda f) \, sinc(\pi b \, v'/\lambda f) \qquad (3)$$

where (u',v') are the rotated coordinates given under a planar transformation by

$$u' = u \cos\theta + v \sin\theta$$

$$v' = -u \sin\theta + v \cos\theta \qquad\qquad (4)$$

Equations (3) and (4) give the amplitude of the diffraction pattern for a single aperture.

For similar, parallel apertures, the function $O(x,y)$ in Eq. (1) can be specified utilizing convolution theory. The following development follows that of Lipson and Lipson [1]. If a single aperture is described by $O_1(x,y)$, a two dimensional collection of similar, parallel apertures is described by the convolution of this function with an array of delta functions, viz.

$$O(x,y) = \int\!\!\int_{-\infty}^{+\infty} \sum_{n=1}^{N} \delta(x'-x_n)\delta(y'-y_n)\, O_1(x-x',y-y')\ dx'\ dy'$$

$$= \int\!\!\int_{-\infty}^{+\infty} g(x',y')\, O_1(x-x',y-y')\ dx'\ dy' \qquad\qquad (5)$$

where (x_n,y_n) are the coordinates of the origin of the n-th aperture and $g(x',y')$ is the summation of the delta functions.

Utilizing the convolution theory for Fourier transforms, the amplitude of any diffraction mask of identical, parallel apertures can be written as the product of the amplitude of a single aperture, i.e. the Fourier transform of the aperture, and the Fourier transform of an array of delta functions representing the aperture locations, i.e. $g(x',y')$, viz.

$$A(u,v) = F(O_1)\ F(g) = A_o A_D$$

Thus, for randomly spaced, parallel apertures, the transform of the array of delta functions, A_D, needs to be determined.

The diffraction pattern (i.e. the Fourier transform) of a set of N randomly arranged delta functions is given by

$$A_D(u,v) = \int\!\!\int_{-\infty}^{+\infty} \sum_{n=1}^{N} \delta(x-x_n)\delta(y-y_n) \exp\left[-\frac{2\pi i}{\lambda f}(xu+yv)\right]\ dx\ dy$$

$$= \sum_{n=1}^{N} \exp\left[-\frac{2\pi i}{\lambda f}(x_n u + y_n v)\right] \qquad\qquad (7)$$

In general, the summation in Eq. (7) cannot be evaluated. However, the intensity of the transform can be computed by writing the intensity as a double summation, viz.

$$I_D(u,v) = A_D A_D^*$$

$$= \left| \sum_{n=1}^{N} \exp\left[-\frac{2\pi i}{\lambda f}(x_n u + y_n v)\right]\right|^2$$

$$= \sum_{n=1}^{N}\sum_{m=1}^{N} \exp\left[-\frac{2\pi i}{\lambda f}\{u(x_n-x_m)+v(y_n-y_m)\}\right] \tag{8}$$

The variables x_n , x_m , y_n and y_m are all random variables and will randomly take on both positive and negative values and so, in general, the summation of Eq. (8) will be zero for large N. There are, however, two exceptions to this, 1) when n = m and 2) when u = v = 0. For the first exception, all the terms in the summation are equal to 1 and there are N of these terms. For the second exception, there are N^2 terms each equal to one. Thus, the intensity of the random array of delta functions is

$$I_D(u,v) = N + N^2\bar{\delta}(u,v) \tag{9}$$

where $\bar{\delta}(u,v) = 1$ for (u,v) = (0,0) and is zero otherwise. Combining Eq. (3) and Eq. (9), the complete diffraction pattern of a collection of N randomly spaced, parallel rectangles is

$$I(u,v,\theta) = [N + N^2\bar{\delta}(u',v')]*[(ab)^2\text{sinc}^2(\frac{\pi au'}{\lambda f})\text{sinc}^2(\frac{\pi bv'}{\lambda f})] \tag{10}$$

Given the diffraction pattern for N rectangles oriented at an angle θ, i.e. Eq. (10), the diffraction pattern for an angular distribution of rectangles is given by summing the collections of oriented rectangles. Since the distribution function is continuous, the summation becomes an integral and the intensity is given by

$$I(u,v) = (a,b)^2 \int_{-\pi/2}^{\pi/2} [N(\theta) + N^2(\theta)\bar{\delta}(u',v')]*$$
$$[\text{sinc}^2(\frac{\pi au'}{\lambda f})\text{sinc}^2(\frac{\pi bv'}{\lambda f})]\, d\theta \tag{11}$$

where $N(\theta)$ is now a distribution function and (u',v') are functions of the angle θ and are given by Eq. (4). Equation (11) is the final experssion which describes the diffraction pattern for randomly spaced, parallel rectangles with an orientation distribution $N(\theta)$.

The validity of Eq. (11) was demonstrated using simulation diffraction masks generated using a Monte Carlo computer routine. The distribution function, $N(\theta)$, utilized was of the form discussed by McCullough [4] and is given as

$$N(\theta) = (\cos\theta)^{2f_p/(1-f_p)} \tag{12}$$

where f_p is the Herman's orientation parameter. The Herman's orientation parameter, which has been identified as a useful descriptor of the state of orientation for micromechanical modeling of the mechanical properties of short fiber composites, is given for a planar orientation as

$$f_p = 2 <\cos^2\theta> - 1$$

with

$$<\cos^2\theta> = \int_0^{\pi/2} N(\theta) \cos^2\theta \, d\theta$$

The values of f_p range from $f_p=0$ for planar random fiber orientation to $f_p=1$ for perfectly aligned fibers. In Figure 2 the computer drawn diffraction masks are given for four states of orientation, viz. $f_p = 0.0$, 0.3, 0.6 and 0.9.

In Figure 3, the experimental diffraction patterns generated from the diffraction masks in Figure 2 and the diffraction patterns calculated using Eq. (11) are given. To produce the contour plots in Figure 3 representing the predicted diffraction patterns, Eq. (11) was integrated numerically using a packaged computer routine. The contour plots were generated using a supplied computer program.

From Figure 3 the changes in the diffraction pattern as the state of orientation changes is apparent. For random fiber orientation, $f_p=0$, the intensity of the diffraction pattern is constant at constant radii. However, as the fibers begin to show preferential orientation (for example $f_p=0.3$), two lobes appear which are perpendicular to the direction of orientation. As the orientation increases, the diffraction lobes become narrower, as shown by comparisons of $f_p = 0.3$, 0.6 and 0.9. Thus the width of the diffraction lobes, or equivalently the angle between them (at constant intensity), is a direct indication of the state of orientation.

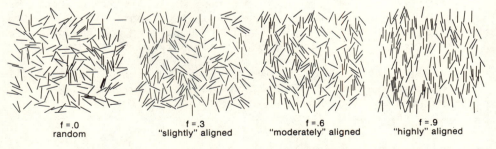

f =.0   f =.3   f =.6   f =.9
random   "slightly" aligned   "moderately" aligned   "highly" aligned

Fig. 2.   Monte Carlo computer simulated fiber distributions for four
          states of orientation, $f_p = 0.0$, 0.3, 0.6 and 0.9.

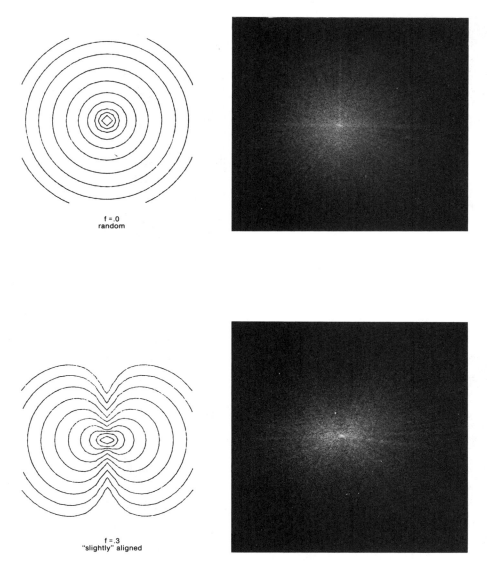

Fig. 3.   Experimental diffraction patterns and predicted diffraction
          patterns for the diffraction masks made from the simulated
          fiber distributions in Figure 2.   (See also page 432)

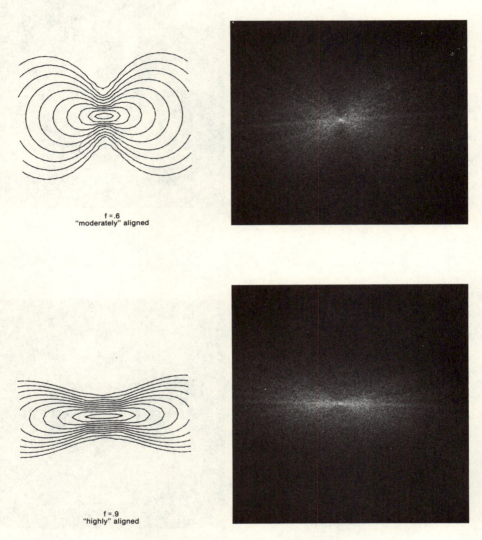

f = .6
"moderately" aligned

f = .9
"highly" aligned

Figure 3 (continued)

These results are confirmed in Figure 4 where the transmittance of the negative of the experimental diffraction patterns in Figure 3 have been plotted. The transmittance tracings were taken at a constant radius of 1.32 cm. The reference angle, i.e. the zero degree angle, is the direction perpendicular to the direction in which the fibers are aligned. It would be informative to compare the measured transmittance with one computed utilizing Eq. (11). Preliminary studies have indicated that the relationship between the intensity of the light incident on the film used to photograph the diffraction pattern and the measured transmittance is not a simple linear relationship. Nonetheless this relationship can be determined through suitable calibrations which will allow a comparison between Eq. (11) and the experimental diffraction patterns.

The transmittance tracings are very useful when used to compare the states of orientation of several masks when the masks are produced in similar manners. It is this feature of the transmittance tracings which will be utilized in the following section.

EXPERIMENTAL

Two sheet molding compounds were produced with copper tracer fibers under two different production conditions. Diffraction masks of the two composites were produced utilizing the method described above. Positive copies of the X-radiographs which were used to produce the diffraction masks and the associated diffraction patterns are shown in Figure 5. Also shown are the transmittance tracings.

The advantage to using a diffractometer in detecting processing induced fiber orientation is apparent from the evidence of Figure 5. The difference in orientation between the two samples is easily seen by comparing the respective transmission tracings. The first sample, though not totally linear, does not show any cyclic variation which would be indicative of fiber alignment. However, the second sample does show a cyclic variation indicating that there is a slight amount of orientation in the sample. It is also interesting to note that the maximum (minimum) in the transmission tracings occurs near 10 (100) and 190 (280) degrees indicating that: 1) the preferential direction along which the fibers, or at least the copper tracers, are aligned is slightly off the machine direction or 2) the sample was machined slightly off the machine direction.

These conclusions, which can easily be drawn from the transmission tracings in Figure 5, can also be deduced, with some difficulty, from the diffraction patterns themselves. However, without the aid of the diffractometer it would be very difficult to make any statements regarding the state of fiber orientation from the X-radiograph directly.

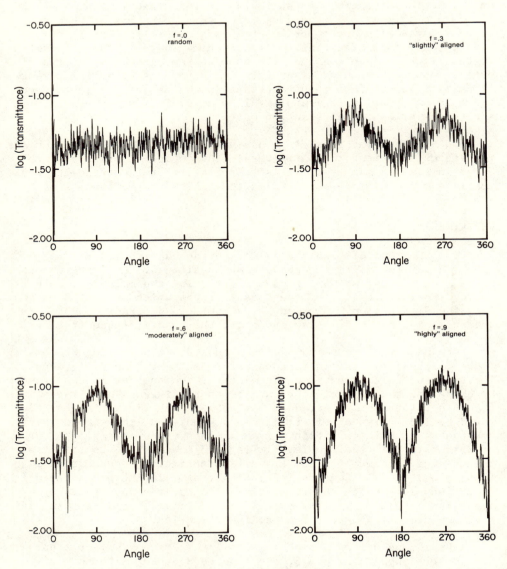

Fig. 4.   Transmission tracings for the diffraction patterns of Fig. 3.

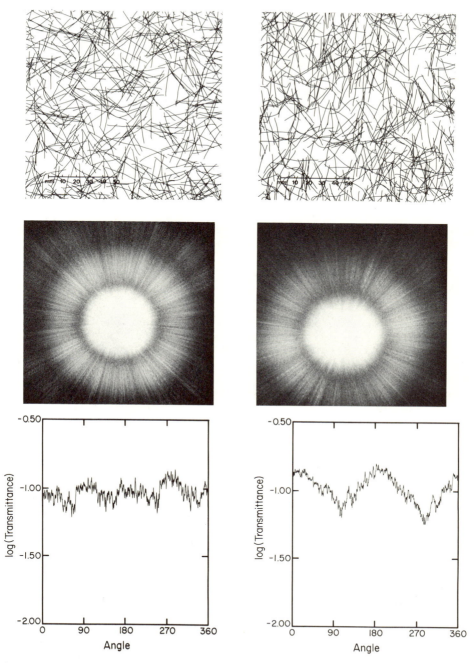

Fig. 5.   Processing induced fiber orientation in sheet molding com-
          pounds.   Positive copies of the X-radiographs used to pro-
          duce the diffraction masks, the diffraction patterns and
          transmission tracings.

From the results shown in Figure 5, the importance of processing parameters in influencing fiber orientation is apparent. Even though the degree of orientation of fibers in the sheet molding compound is slight, partial orientation can introduce anisotropy in the final product.

SUMMARY

A method for measuring fiber orientation in short fiber reinforced composites, using Fraunhoffer diffraction techniques, has been discussed. An analysis based on collections of centro-symmetric, randomly spaced, parallel slits was verified utilizing computer generated diffraction masks with known states of orientation. The technique was applied to actual composites using diffraction masks produced from X-radiographs to demonstrate processing induced fiber orientation.

ACKNOWLEDGEMENTS

The authors would like to thank Dr. Denton of Owens-Corning Fiberglas, Granville, Ohio, for fabricating the copper tracer filled composites and supplying the X-radiographs. The financial support of the Rogers Corporation, Rogers, Connecticut, for this work is greatly appreciated. Steven Fields of the University of Delaware helped with the construction of the experimental apparatus and development of the photographic procedures.

REFERENCES

[1]   S. G. Lipson and H. Lipson, Optical Physics, 2nd Ed., Cambridge University Press, Cambridge, 1981.

[2]   M. Born and E. Wolf, Principles of Optics, 2nd Ed., The MacMillan Co., New York, 1964.

[3]   R. C. Smith and J. S. Marsh, J. Opt. Soc. Am., 64, 798, 1974.

[4]   R. L. McCullough, "Anisotropic Elastic Behavior of Crystalline Polymers", Treatise on Material Science and Technology, Vol. 10, Part B, Academic Press, Inc., New York, 1977.

LOW COST ENERGY STORAGE FLYWHEELS FROM

STRUCTURAL SHEET MOLDING COMPOUND

John F. Kay

Owens-Corning Fiberglas
Research and Development Division
Granville, Ohio, USA

ABSTRACT

Compression molded structural sheet molding compound (SMC) composed of S-2 Glass[R] and polyester resin has been used to fabricate
energy storage flywheel rotors. This technique has the potential of
low cost, high throughput production of rotors for the automobile
industry.

An isophthalic polyester resin and chopped S-2 Glass were used
to mold flat, constant cross section discs 53.3 cm. (21 inches) in
diameter, 2.54 cm. (1.0 inches) thick, and 49.5 kg. (22.5 pounds) in
weight. Materials characterizations have shown a tensile strength of
337 MPa (49 ksi) for the S-2 Glass reinforced rotors, which would
allow the rotor to store 28.6 watt-hours per kilogram (13 watt-hours
per pound) at 330 hertz when a filament wound carbon fiber/epoxy ring
is fitted around the SMC core.

A dynamic test of an SMC flywheel has shown an energy storage
density of 27.7 watt-hours per kilogram (12.6 watt-hours per pound)
at 330 hertz.

S-2 Glass[R]- a registered trademark of Owens-Corning Fiberglas Corp.

INTRODUCTION

Flywheels have been utilized to smooth the cyclic supply and demand of energy for centuries. Although the equations that govern kinetic energy storage in a rotating structure have been known and the notion that a light, strong flywheel spinning very rapidly stores energy more efficiently than a heavy wheel turning at lower speeds has been long understood, it was not until the advent of advanced composite materials that serious thought was applied to actually making an optimally-designed energy storage flywheel.

Fiber reinforced plastic composite materials are ideally suited for energy storage flywheels since their high strengths and low densities provide high specific strengths when compared to metals. Glass fiber is particularly attractive for flywheels since it possesses the highest strength-to-weight ratio (specific strength) of all the common reinforcing fibers.

Filament wound ring-type flywheels have been designed to take advantage of the anisotropy inherent in a composite material and align the continuous hoop-wrapped high strength fibers in the direction of the maximum stresses imposed by rotation. The filament wound rings are, however, utilized to their greatest energy storage potential when placed at the farthest radial distance from the axis of rotation as possible. Among the methods to connect this filament wound ring to the central axis of rotation are spokes, machined hubs, other concentric rings, or solid disc-type cores. Accurate placement of the fibers in a filament wound ring can be expensive and attachment of the ring to a hub using complicated spoke or core systems may preclude the use of such a design for low cost, high volume applications such as the automobile.

Using unidirectional S-2 Glass prepreg material, several quasi-isotropic composite laminate rotor core designs were tested and were found to fail prematurely due to large directional strength variations in the laminated disc.[1,2] While each ply of unidirectional prepreg has very high strength when tested in the 0°, or fiber, direction, this strength is compromised with the addition of other plies at angles up to 90° to its fiber direction.

Interply shear between adjacent plies at different angles also becomes a limiting design factor with the quasi-isotropic disc rotor. Manufacture of the quasi-isotropic disc rotor also is somewhat unattractive since the exact placement of over 100 plies of unidirectional prepreg material and subsequent long cure cycles is labor intensive and, therefore, not conducive to high-volume applications.

In an attempt to provide a flywheel rotor that is isotropic within the plane orthogonal to the axis of rotation, a compression molded disc made from structural sheet molding compound has been considered. While a randomly oriented short fiber reinforced

polyester-based composite does not possess the individual ply
strength of unidirectional prepreg materials, a transversely isotropic
disc can be made which will nearly possess properties approaching
those of the quasi-isotropic disc in its weakest direction. In
addition, the problems of interply shearing effects would be reduced.
The cost of a compression molded flywheel core would be substantially
less than that of a machined metal hub or a series of concentrically
wound inner core rings of filamentary composite material.

Compression molded sheet molding compound prov-des a low cost
method of producing structural composites from glass reinforced plastic.[3] A recent study of structural sheet molding compound composed of
chopped S-2 Glass in E987 isophthalic polyester resin[4] has shown
superior tensile properties listed in Table 2 for 2.54 and 5.08 cm.
reinforced SMC.

The economic advantages of using structural sheet molding compound to form flywheel rotor discs are reduced somewhat by the need to
contain the radial expansion of the core with a carbon fiber/epoxy
filament wound ring placed about its circumference. The circumferential modulus of the continuous filament reinforced ring prevents
radial expansion of the SMC core as the wheel stores energy by rotation, resulting in higher operational speeds and, thus, storage of
more kinetic energy. The prevention of the radial expansion of the
SMC core will also improve significantly its long-term performance
behavior. The matrix for the containment ring is epoxy with "commercial" grade carbon fiber (Thornel 300) as the reinforcement. The
design of the ring is strain limited and serves to contain the greater radial growth of the SMC core under stress. Glass/epoxy (E or S-2)
does not have sufficient tensile elastic modulus ; it would expand
and slip off during spin-up. Kevlar 49/epoxy suffers from poor
transverse properties and would thus limit the radial thickness of
the ring that could be installed; premature failure, due to ring
radial stresses, would result from a thick Kevlar/epoxy ring. Carbon/
epoxy provides more than adequate hoop stiffness with adequate transverse strength so as to provide a system capable of moderate energy
storage density.
Although both constant and tapered thickness profile flywheel
discs were molded, a 10.90 kilogram constant thickness SMC disc with
an 0.85 kilogram carbon/epoxy containment ring was chosen as the
test piece. Flat, constant thickness flywheel discs are more dynam-
ically stable, although slightly less efficient in storing energy,
than tapered thickness rotors. It is thought that the more stable
rotor configuration is better suited for automotive applications
since vehicle motion causing external pertubations are to be greater
than for fixed-base energy storage flywheel units. It is anticipated that the moderate energy storage density of the SMC core-fila-
ment wound ring system, coupled with the low cost, high throughput
SMC process, will make the concept suitable for the automobile industry.

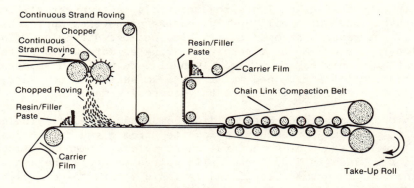

Fig.1. Schematic of SMC machine capable of producing SMC-r or SMC-C/R.
In making R65 SMC for a flywheel disc, a thin layer of polyester paste is deposited on the carrier film, onto which the
chopped glass roving is deposited. No continuous strands are
added, but another layer of polyester paste is placed over the
chopped glass. The material is stored on a take-up roll until
ready to use.

MATERIALS AND PROCESSES

Structural Sheet Molding Compound

Compression molded sheet molding compound provides a low cost
method of producing structural composites from glass reinforced plastic.[3,4] Structural sheet molding compounds are composed of 30 percent
to 70 percent (by weight) fiber reinforcement and are designed to
compete with and/or replace metal parts in load-bearing or structural applications. Figure 1 is a schematic diagram of the equipment
used to make sheet molding compound, which is then compression molded
to form the flywheel discs. The SMC-R process is attractive because
of the material property adjustability that can be obtained by
changing the resin, glass reinforcement type, length and relative
volume fraction within the formulation.

The SMC formulation used in this study consisted of randomly
oriented chopped S-2 Glass in a polyester matrix. E987 resin is a
modified isophthalic polyester resin designed for use in structural
sheet molding compound, chosen for its low linear shrinkage and
excellent dimensional stability. No filler was used because of the
high glass loading. R65 SMC material is composed of 35 weight percent resin paste and 65 weight percent S-2 Glass roving. The high

TABLE 1

S-2 GLASS FIBER REINFORCEMENT
PROPERTIES

Tensile Strength, MPa	4753
Elastic Modulus, GPa	86.8
Tensile Elongation to Failure, %	5.40
Density, g/cm^3	2.49

TABLE 2

S-2 GLASS R65 STRUCTURAL SMC PROPERTIES

	S-2 Glass 2.54 cm.	S-2 Glass 5.08 cm.
Tensile Strength, MPa	253	299
Tensile Modulus, GPa	16.5	16.8
Tensile Elongation at Failure, %	2.27	2.49
Poisson's Ratio	0.26	0.26
Density, g/cm^3	1.728	1.729

virgin fiber tensile strength of S-2 Glass (Table 1) results in high
tensile properties for the SMC composite and, thus, makes it attractive for the flywheel core material. The mechanical properties of R65
derived form flat laminate tests are listed in Table 2 for 2.54 cm.
and 5.08 cm. chopped reinforcements; the longer chop length results
in slight increases in tensile properties.

A typical room temperature tensile stress–strain curve for R65
SMC is shown in Figure 2. S-2 Glass reinforced R65 exhibits two
linear regions in the stress–strain curve. The initial tangent modulus reported in Table 2 is the slope of the line to the proportional
limit. The transition region between the initial slope and the lower
secondary slope (higher strain rate) occurs at 0.5 to 0.7 percent
strain. The transition region location depends both on the full-
scale load used on the test instrument and on the interpretation of
the resultant curve. The transition for structural SMC lasts for
about 0.5 percent strain beyond the proportional limit where a secondary modulus 79 percent of the initial tangent modulus dictates
the behavior to failure.

Mold Design

The constant thickness flywheel rotor was designed by Lawrence
Livermore National Laboratory to the material properties of R65 SMC;

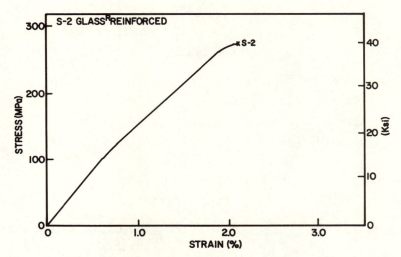

Fig.2. Room temperature tensile behavior of R65 structural SMC. The
       transition between the initial tangent modulus to the propor-
       tional limit and the lower secondary modulus occurs at 0.5 to
       0.7% strain. This transition is considered to be indicative of
       matrix damage.

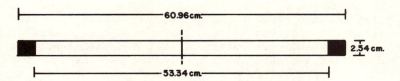

Fig.3. Constant thickness profile rotor. Theoretical weight of the
       disc molded from R65 SMC is 10.98 kilograms. The radial thick-
       ness of the carbon fiber/epoxy ring is 3.8 cm. Total assem-
       bled rotor weight is 12.93 kilograms.

it is a constant thickness profile (Figure 3) 2.54 centimeters thick,
53.3 centimeters in diameter.

     The mold was machined at Gougler Industries*, and features a
hydraulic ejector plate system to facilitate removal of the part
from the mold. The mold is installed in a large hydraulic press. By
varying the steam pressure or oil temperature through the press pla-
tens and the mold base, the temperature of the molding surfaces can
be controlled.

* Kent, Ohio

Fig.4. Individual plies of R65 SMC are stacked together to form the
charge for the flywheel. As each ply is placed on the stack
the protective polyethylene film is removed.

Fig.5. An 18 ply R65 SMC charge placed in the heated constant thickness mold immediately before mold closure. At 88% mold coverage, flow of the material at the edges, with a deleterious
effect on mechanical properties, is minimized.

TABLE 3

R65 SMC MATERIAL SYSTEM FORMULATION

SMC Resin Formulation

Function	Component	Source/Trade Name	Proportions
Resin	isophthalic polyester	OCF E-987	100
Catalyst	t-butyl perbenzoate	Lucidol	0.6
Initiator	t-butyl peroctoate	Lucidol	0.1
Thickener	magnesium oxide	Hatco Modifier M	5.0

Reinforcement

S-2 Glass	OCF 433 x10-114	2.54,5.08 cm. chop length

SMC System – R65

65 weight percent reinforcement (50 volume percent)

35 weight percent resin paste (50 volume percent)

Fig. 6. The assembled flywheel rotor disc, ring, and hub weighs 12.93
kilograms. The five 9.5 millimeter holes have been drilled to
balance the rotor system. (Photograph courtesy ORFEL).

Structural SMC Formulation

The structural SMC resin paste formulation, listed in Table 3,
consisted of a base isophthalic polyester accounting for 35 weight
percent of the composite. This specific SMC system was chosen since
OCF E987 isophthalic polyester resin provides near zero linear
shrinkage and superior surface finish. An external mold release,
carnuba wax, was used so as not to reduce the composite properties
in the interior of the part.

Flywheel Molding

Constant thickness flywheel discs were prepared using proprietary Owens-Corning compression molding technology. From the molded
densities obtained for laminates used for material property characterizations, a charge of 10.98 kilograms was determined to be adequate to fill the mold cavity and maintain a nominal 2.54 centimeter
thickness.

For the constant thickness discs, 18 plies of R65 SMC were cut
to 49.5 centimeter preform size, 88 percent coverage of the mold
(Figure 4). During preform stacking, as each successive ply is

added, the protective polyethylene film is removed and gross air
pockets eliminated by rolling. The result is a stack of consolidated
R65 SMC plies which is placed in the matched metal 120°C oil heated
mold (Figure 5) and held for 50 minutes at 10.3 MPa. The resultant
parts are easily removed due to the ejector plate; molded rotors
need only to be deflashed.

Rotor Assembly

The complete rotor assembly consists of three parts: the SMC
core, the carbon fiber/epoxy containment ring, and an aluminum hub
which allows attachment of the flywheel to the spin test drive tur-
bine. The hub is bonded to the rotor core and allows for the unequal
radial expansion of the rotor and hub during spin-up. This suspension system and hub bonding regime would not, however, be used in
actual practice. It does provide adequate attachment characteristics
for dynamic testing. During testing to failure, the flywheel rotors
are suspended from a single hub fixture on one side only and rotate
about a vertical axis in the spin test chamber. Balance, so as to
align the mass center with the axis of rotation, is critical, and is
symmetric about a plane normal to the axis of rotation and through
the disc center. Stability of the rotor through resonant critical
frequencies and minimization of whirl modes result from careful
balancing.

The carbon fiber/epoxy containment ring was 0.41 millimeters
undersize in inside diameter, providing an "interference fit" between the ring and the core. After placing shim stock in appropriate
positions so as to affect accurate axial alignment of the SMC core
with the filament wound ring. The core was indexed so it could be
exactly positioned for installation. To shrink fit the SMC disc into
the slightly smaller ring, the core was fitted in a Tedlar[R] bag and
placed in a dry ice/acetone bath (67°C) for 20 minutes. The indexed
flywheel disc was then accurately positioned within the ring and the
disc clamped to prevent shifting during warming to room temperature.

Balance and Testing

The finished rotor, ring, and hub assembly (Figure 6), weighing
12.93 kilograms, was sent to the Union Carbide Oak Ridge Y-12 Flywheel Evaluation Laboratory (ORFEL) for spin testing. Balance of
the rotor assembly was attained by drilling five 9.5 millimeter holes
in the SMC core about six millimeters from the outer edge. The hole
depth was approximately 12 millimeters. These holes indicate a substantial system imbalance, prior to drilling, due primarily to a
small offset of the hub from the disc center. The at-rest theoretical energy storage density was calculated at 30 watt-hours per kilogram for 346 hertz (20760 RPM).

The testing procedure at ORFEL consists of a low speed spin-up
to ascertain balance while monitoring run-out of the rim, bottom,

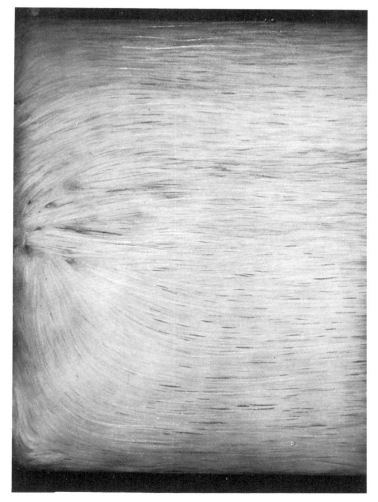

Fig.7. Optical micrograph of a polished section orthogonal to the
       plane of the disc. Note the favorable flow condition at the
       edge of the molded part due primarily to the shear flow of the
       upper and lower layers at 88% mold coverage (X6).

and top surface. Rebalancing, if necessary, occurs between spin-ups;
inspection of the wheel for damage between runs is visual only.
X-ray and ultrasonic inspection occurs only before initial testing.
Sequential spin-ups gradually increase final speed until failure
occurs. After failure, debris is examined in an attempt to gain
evidence for failure analysis.

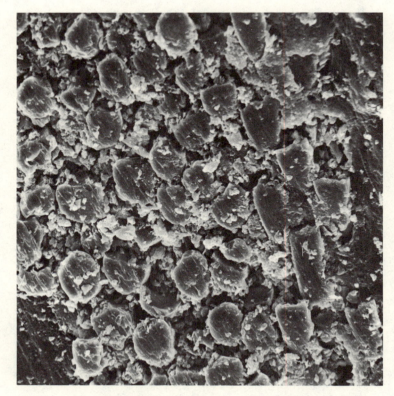

Fig.8. Scanning electron micrograph from the center of an as-cut section of a constant thickness flywheel disc. The fiber bundles
are well coated with resin and no voids exist either among or
adjacent to fiber bundles (x5000).

DISCUSSION OF RESULTS

The nominal diameter of the molded SMC discs was 53.442 centimeters after deflashing with 600 grit emery paper. During the initial moldings, the discs were cross-sectioned to observe flow of the
SMC plies within the molded part, consolidation of the SMC plies, and
to detect any gross interlaminar separations caused by entrapped
exotherm heat.

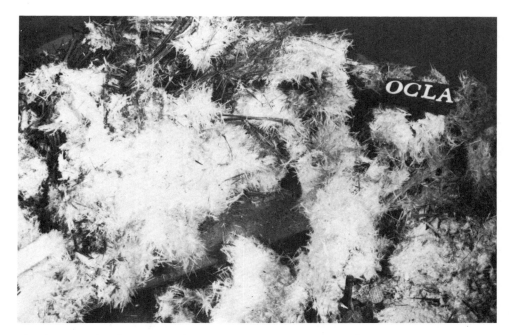

Fig.9. Debris in the spin test chamber after failure was characterized as "furry" clumps of SMC with little pulverized
glass "dust." The finer particles were primarily from the
carbon/epoxy contaminant ring (Photograph courtesy ORFEL).

Ply Consolidation

The use of the polyester resin paste system listed in Table 3
resulted in flywheel discs with a smooth surface and good dimensional
stability. Cross-sections of the constant thickness discs examined
by optical microscopy showed good ply consolidation, and the use of
preforms 49.5 centimeters in diameter resulted in favorable fiber
orientation at the disc edge (Figure 7).

The degree of ply consolidation is also shown in Figure 8 where
fiber bundles can be seen at various angles to the plane of the
photomicrograph. No interlaminar cracks are observed between fiber
bundles across the thickness of the disc. Microscopic analysis provided evidence for adequate wetting of all reinforcing fibers during
the SMC sheet fabrication and good ply consolidation due to flow
during the molding operation.

Energy Storage Density

A summary of the spin test performance is listed in Table 4.
The wheel failed on the twelfth spin-up when an instability was detected at 365 hertz (21,900 RPM), when braking was initiated. At

TABLE 4

DYNAMIC TEST DATA SHEET

Tested: Oak Ridge Y-12 Flywheel Evaluation Laboratory
        (ORFEL) by Dr. Robert Steele, February 13,1981
Maximum RPM: 21,900 (365 hertz)
Failure RPM: 19,980 (failure upon deceleration)
Energy Storage Density: 27.7 wh/kg (12.6 wh/lb.)
Stored Kinetic Energy: 0.358 kilowatt hours
Kinetic Energy Per Unit Volume: 1.367 kwh/ft^3
Radial Growth: 0.33 cm. at 333 hertz (19,980 RPM)
Tip Speed at Failure: 637.7 m/sec (2092 ft/sec)
Maximum Tip Speed: 699.0 m/sec (2293 ft/sec)
Ring Failure Strain: 1.083%

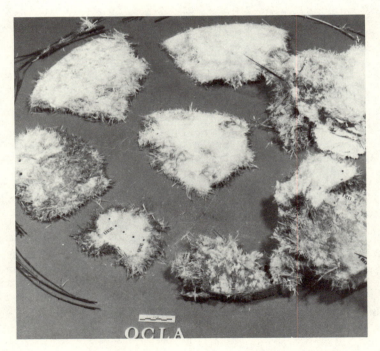

Fig.10. The largest pieces of glass reinforced SMC were approxi-
        mately 15 cm. in maximum dimension. The largest splinters of
        filament wound carbon/epoxy were 15 to 18 cm. long, an indi-
        cation that first failure occurred within the ring (Photo-
        graph courtesy ORFEL).

333 hertz (19,980 RPM) a pressure rise in the test chamber was noted, an indication of composite damage, and failure ensued. Debris was characterized by "furry" clumps of FRP with little "dust". There were several large chunks of SMC 15 to 20 centimeters in maximum dimension (Figures 9,10), but there was no indication that core failure occurred by an internal shear failure. This is an important observation and reflects the establishment of uniform properties through the thickness. There were several carbon fiber/epoxy splinters 15 to 20 centimeters long and 6 centimeters in cross-sectional dimension, implying a hoop stress failure. The large SMC parts would indicate first failure occurred in the carbon/epoxy containment ring. If this is the case, the ring radial thickness could be increased to initiate failure radially at a predesignated speed. Increasing the radial thickness would also increase the energy storage density.

A positive result of the test was the indication that holes drilled in the SMC core did not initiate failure. All five holes

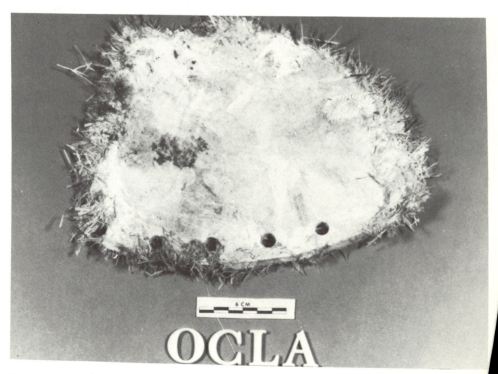

Fig. 11. One section of debris contained all five drilled balance holes, indicating their presence did not decrease the performance of the rotor. This method of balancing may not be applicable to functioning flywheels which must withstand cyclic stresses and prolonged constant stresses (Photograph courtesy ORFEL).

were intact within the same chunk of debris (Figure 11). This implies
that balance may be attained in this manner for future test pieces.
However, due to cyclic performance constraints, holes may prove to be
large enough stress risers to preclude their use in functioning flywheels whose material properties may be lessened somewhat by fatigue
phenomena.

CONCLUSIONS

Structural sheet molding compound of chopped glass fibers in OCF
E987 polyester resin is a viable materials system for low cost energy
storage flywheels. The use of S-2 Glass reinforcement in the SMC process has made possible the low cost flywheel design by providing a
high strength molding compound from chopped fibers. Drilling of holes
to attain a balanced rotor does not cause stress concentrations that
seriously degrade the SMC properties in this application; the effect
of these holes on long term performance is unknown.

The ability of S-2 Glass reinforced structural sheet molding
compound to form a high strength quasi-isotropic composite makes
attractive the potential for the compression molding process to produce low cost, high throughput flywheel discs suitable for automotive
applications.

REFERENCES

1. Kulkarni,S.V.,R.G.Stone,and R.H.Toland,"Prototype Development of
        an Optimized, Tapered Thickness, Graphite/Epoxy Composite Fly-
        wheel", UCRL-52623, Lawrence Livermore Laboratory, November
        1978.
    Duke,J.C.,"A Comparison of Quasi-Isotropic Fiber Reinforced Com-
        posite Laminates", UCRL-15225, Lawrence Livermore Laboratory,
        November 1979.
    utte,R.B.,"Structural SMC- Material, Process, and Performance
        Review", Presented at Cobo Hall, Detroit Michigan-SAE Congress
        and Exposition, February 27-March 3, 1978.
    , J.F.,"Advances in Structural SMC", presented at the ASTM
        Symposium on Short Fiber Reinforced Plastics, Minneapolis,
        innesota, April 13-15, 1980.

REINFORCED POLYESTER STRUCTURAL FOAM

Giancarlo Carignani, and Massimo Mazzola

Technical Center
SNIAL Resine Poliestere S.p.A.
00034 Colleferro (Roma) Italy

INTRODUCTION

In the last few years there has been a great development of many polymer foams with a density varying from 0.1 to 1. named structural foams for their mechanical properties. The interest for this type of expanded polymers is that, in the case very frequent of flexural stress state, to reach an high rigidity and strength is possible to compensate the low absolute value of elastic modulus and strength of these expanded polymers increasing the thickness. Operating in this way it can still be obtained a weight reduction in comparison with the equivalent plate or shell made with compact materials. In fact if the same flexural stiffness between two plates made with different materials is imposed, the following equations can be written:

$$E_e I_e = E_c I_c \qquad 1)$$

$$E_e t_e^3 = E_e t_e'^3$$

$$\frac{t_e}{t_c} = \sqrt[3]{\frac{E_c}{E_e}} \qquad 2)$$

in wich t_e and t_c are respectively the thickness of expanded and E_e and E_c the elastic moduli of the two materials.

The weight reduction in this case is:

$$\frac{We}{Wc} = \frac{\varrho e \ te}{\varrho c \ tc} = \frac{\varrho e}{\varrho c} \sqrt[3]{\frac{Ec}{Ee}} \quad 3)$$

In fig. 1 are reported the relative weights of plates made with different materials at same flexural stiffness. From the fig. 1 it can be noted that if a steel plate weighs 100, with polycarbonate plate we have a weight reduction of 30%. With the aluminium the weight reduction is 45%, with the polyurethane (RIM) the reduction is of 55%, with the reinforced polyester foam the weight reduction is 60%.

If the comparison is made at same flexural strength instead of same flexural stiffness similar results are reached.

The equivalent flexural strength condition can be written:

$$Me = Mc \qquad 4)$$
$$\sigma e \ te^2 = \sigma c \ tc^2$$

$$\frac{te}{tc} = \sqrt[2]{\frac{\sigma c}{\sigma e}} \qquad 5)$$

with the condition 5) the weight ratio can be written:

$$\frac{We}{Wc} = \frac{\varrho e \ te}{\varrho c \ tc} = \frac{\varrho e}{\varrho c} \sqrt[2]{\frac{\sigma c}{\sigma e}} \qquad 6)$$

In fig. 2 are reported the relative weight of plates or shell made with different materials at same flexural strength. The results are qualitatively the same than in the case of equivalent flexural stiffness.

In conclusion in the case of plate and shell the maximum flexural strength, stiffness and elastic stability with the minimum weight can be obtained with the case of an expanded polymer instead of a compact one.

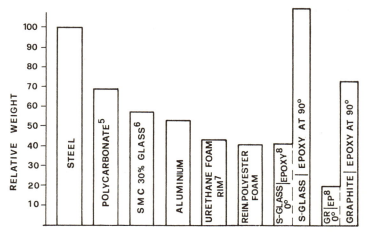

Fig. 1 Relative weight of panels made with different materials with equal flexural stiffness.

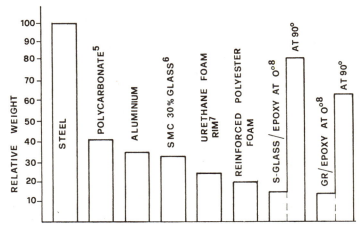

Fig. 2 Relative weight panels made with different materials with equal flexural strength.

ELASTIC PROPERTIES PREDICTION FOR FOAMS AND REINFORCED FOAMS

The problem is approached in two steps. The first is to compute the elastic properties of unreinforced polymer in function of voids content (density), the second is to compute the elastic properties of a composite in wich the continuous phase is the foam

and the fibers (with a length higher than the critical one) are the
reinforcements.

To perform the first step, from many mathematical model
available in the literature, two models have been chosen: the
first is the Kerner[1] modified model that describes the elastic
properties of a matrix with spherical void inclusions, the second
is a model of Cohen-Ishai[2] that describes the elastic properties
of foam with cubic void inclusions.

From experimental microscopic observation one can see
that if the voids content is low (high density) the gas bubbles are
far enough each other and have a spherical shape. When the
density goes down the distance among the bubbles diminishes and
there is same interference among them that cause a shape
deformation. With a decreasing density the shape of the bubble
can be approximated with ideal shape like ellipsoid, cube and
for very low density with dodecahedric shape. In fig.3 are
reported the reduced elastic modulus in function of voids content
computed with the models of Kerner and Cohen-Ishai. In the same
figure are reported also the elastic reduced moduli, experimentally
evaluated in tensile way, for a polyester foam and an epoxy foam.

As can expected for a low and medium voids content less than
40 % the Kerner model is good enough with experimental data,
with a voids content of 40 % we have an error between theoretical
and experimental results of 12 % for very high voids content the
Cohen-Ishai model is more approximeted.

Another observation from experimental data of fig.3 is that
for the voids content range shown in the figure, the modulus as
predicted from the mathematical models is indipendent from the
chemical nature of polymer. For this reason if a very stiff
structural foam is required (with low voids content), it is
necessary to employ a very rigid polymer such as polyester or
epoxy.

Another way to have a very stiff structural foam is to realize
a composite where the high density foam have the function of
matrix and the high propertie fibers like glass fibers or others
are the reinforcment.

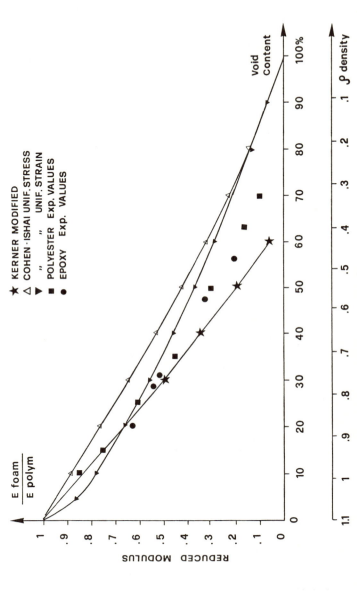

Fig.3 Elastic reduced modulus in function of void content[.] Comparison between Kerner and Cohen–Ishai theoretical models and experimental data for polyester and epoxy foam[.]

Using this composite concept to compute the elastic properties of a reinforced foam it is necessary at first to now the elastic properties of non-reinforced foam (the matrix). This can be performed either in experimental or in theoretical way using for example the Kerner model as seen previously.

Knowing the matrix and the fiber properties, if continuous and parallel fibers are employed the role of mixture and the Halpin-Tsai equation can be used to calculate respectively the longitudinal and transverse properties of ideal lamina.

If the foam is reinforced with chopped fibers randomly distribuited (i.e. with glass fiber mat), the foam can be considered as composed by many laminae with parallel chopped fibers (fig.4). In this case it has to calculate at first the longitudinal and transverse properties of ideal lamina with parallel chopped fibers, using for example the Halpin-Tsai [3] equation, and afterwards with the lamina elastic properties, using the Tsai [4] quasi-isotropic laminate theory it can be calculated the composite elastic properties. In fig.5 is schematically represented this type of calculation.

In fig. 6 are reported the elastic modulus in function of volume glass content for a reinforced polyester foam computed in the way previously exposed. The continuos lines are for constant density of non-reinforced foam, the dotted lines are for constant density of reinforced foam. From the fig.6 it is possible to note that to obtain a certain value of elastic modulus it can be chosen either to use a low density foam with an high glass content, or to use an high density foam with low glass content. In general the choice is made on the basis of other required properties such as thermal coefficient expansion, cost or technological reasons.

REINFORCED POLYESTER STRUCTURAL FOAM (RPSF). MECHANICAL AND PHYSICAL PROPERTIES

The structural polyester foam appears like a microcellular rigid foam with an average diameter of .2 mm, the cells are closed and the foam surface are continuous and repeat exactly the mold surface. The reinforcment used are glass fibers such as continuous filament or chopped strand. The density and the fiber content are constant trough the thickness.

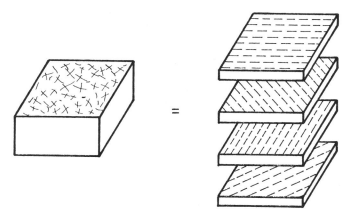

Fig.4   Basic concept of quasi-isotropic laminate theory. A
        laminate made with short fibers randomly distributed
        can be ideally considered as constituted from many
        elemental lamine with short parallel fibers at different
        lamination angles.

In this paper to identify a type polyester foam it has been
used the following classification:
SR – AB – CD
SR it means that is used a rigid polyester foam
AB is the density x 10^2
CD is the glass content (% by weight).

Short term physical and mechanical properties of RPSF

To have a general idea of absolute values of physical and
mechanical properties of RPSF, in table 1 are reported the typical
properties of a RPSF with a density of .64 (kg/lt) and a glass
content of 24% by weight. From the data of table 1 is interesting
to note the high values of moduli and strength and the very low
thermal expansion coefficient (the steel has 12–14×10^{-6} (1/°C),
the aluminium has 24×10^{-6}, the polyurethane has 50×10^{-6}).

Effect of the temperature on tensile properties

In fig.7, fig.8 and 9 are reported respectively the strength,
the elastic modulus and the elongation at break in function of the
temperature for polyester foam with different amount of glass
fibers.

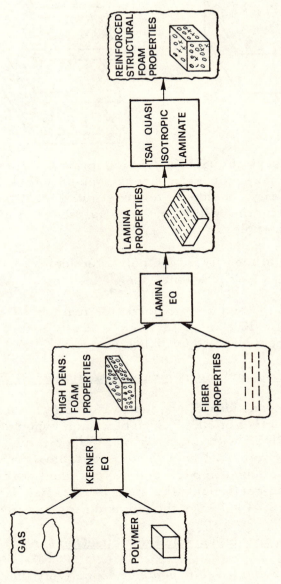

Fig.5 Schematic representation of the procedure to compute elastic properties of a reinforced structural foam.

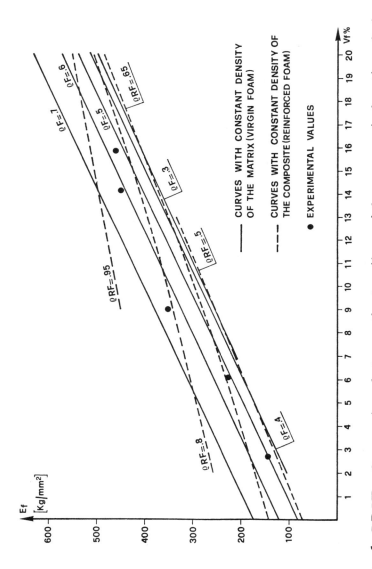

Fig.6 RPSF – Comparison between the tensile modulus curves in function of glass content theoretically evalued with the procedure represented in fig.5 , and experimental results.

Table 1. Typical Properties of a RPSF

Foam Type	SR 64-24	
Density (kg/lt)	.64	
Glass content (% by weight)	24	
Tensile strength (ISO1926-72)	29,4	(MPA)
Tensile modulus (ISO1926-72)	2,22	(GPA)
Elongation at break (ISO 1926-72)	1,9	(%)
Flexural strength (ASTM 1621-64)	51,0	(MPA)
Flexural modulus (ASTM 1621-64)	1,86	(GPA)
Compression strength (ISO 1209-76)	25,5	(MPA)
Compression modulus (ISO 1209-76)	1,17	(GPA)
In plane shear strength (rail shear test)	14	(MPA)
Thermal conducivity coeff.	.03 (Kcal/m/h/°C)	
Specific heat	1300(J/kg/°C)	
Coefficient of thermal expansion	7 x 10-6 (1/°C)	
Mold shrinkage	.08 (%)	

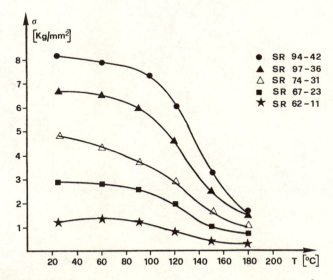

Fig.7 RPSF – Tensile strength v.s. temperature for polyester foams of different density and glass content.

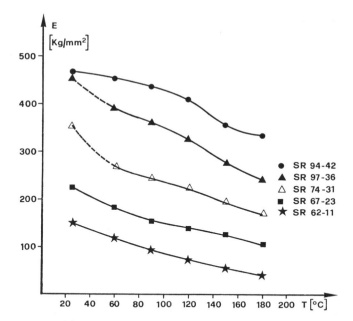

Fig. 8 RPSF – Tensile modulus v.s. temperature for polyester
foam of different density and glass content.

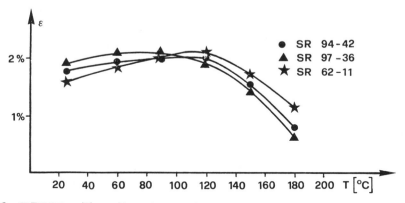

Fig. 9 RPSF – Tensile elongation at break v.s. temperature
for polyester foams of different density and glass content.

The resin matrix is an unsatured polyester resin with a Tg
of 130 °C. We have used the same resin for all type of represented
foams.

It may be observed in fig.7 that the curves are fairly flat up
to temperatures of about 100 °C while the strength becomes
rapidly lower around the glass transition temperature of polyester
matrix wich is in this case about 130 °C. Towards 110 °C the
yellowing of the resin takes place. To separate the influence of
temperature on the matrix apart from the density and glass content
of the foam, the diagram with graph of fig.10 shows the normalized
value of strength, evaluated in respect to its own strength at 60°C,
according to temperature variations for varying densities and
glass contents.

It may be observed in fig.10 that the experimental data are
all fairly near the interpolation curve. From this curve is
possible to deduce immediately the strength of a rigid polyester
foam of any density and glass content, at any temperature once
the strength of the foam at 60°C is know. A second instantaneous
information which the curve in fig.10 gives, is the percentage
variation of strength with the temperature. For instance if we
want to know the highest temperature at which the decrease in
strength is inferior to 30% of that at 60°C apart from the density
and glass content, we deduce from fig.10 that the temperature
is about 120°C.

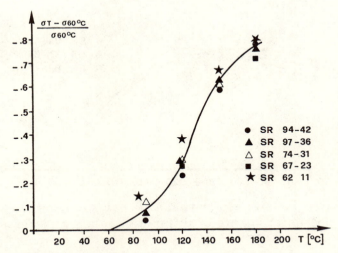

Fig.10 RPSF – Normalized strength curve v.s. temperature
         for polyester foams at different density and glass
         content.

Fig.8 shows elastic modulus curves according to temperature variation for different densities and glass content. Differing from strength behaviour, for modulus the curves show continual decrease, except for the two upper curves in fig.8 and is not possible to identify flexes.

Fig.9 shows tensile elongation at break curves according with temperature variation for all densities and glass content. It may be observed:
1) – All traces are approximatively the same indipendently of density and glass content. Only three curves are shown given that all the others fall in the same area.
2) – The curves show a flat maximum between 60°C and 120°C. (Range of temperature in which the resin becomes more ductile, without any degratation or brittleness).

Impact behaviour of RPSF

Another very interesting characteristic of RPSF is the impact behaviour. In the case of a plate or shell subjected to an impact, if the plate is made with a not expanded composite, the impact energy is absorbed by the creation of few failures that propagate in the plane of plate or shell and through the thickness.

In the case of plate or shell made in RPSF, with the same flexural stiffness than a non expanded composite, the impact energy is absorbed by creation of many and many small craks (cell walls) that are localized only under the impact area on the side in tension. On the side in compression no damage has been observed. This peculiar impact behaviour can be seen easily the falling ball test. As an example of this type of behaviour in the fig.11 are shown the damages caused by a ball of 5 kg falling from 4 m on three specimens with the same flexural stiffness: the first is made with an SMC containing 30% of glass fiber; the second one is a compact polyester reinforced with 30% of glass fibers molded with wet compression molding technology, the third one is a RPSF. One can see that in the case of RPSF the damage area is very small and localized in the impact zone.

Application examples

With RPSF it has been pointed out an industrial technology

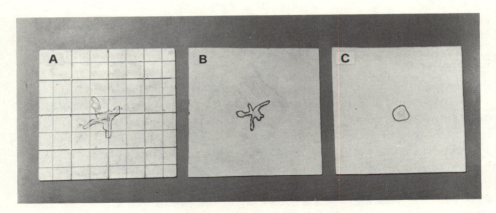

Fig.11 Different failure behaviour, after a ball impact of 20 Kgm,
        on three panels made with: a) SMC, b) GRP,
        c) Reinforced polyester structural foam.

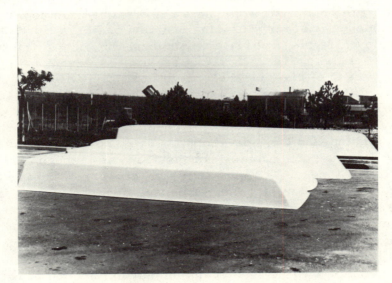

Fig.12   Van Roof prototypes made with reinforced polyester
         structural foam. The first is long 4.5 m and weights
         40 Kg instead of 70 Kg of equivalent standard roof
         made with GRP now in production.

to produce molded articles of big dimension too. It's also possible to realize at low cost prototypes employing the same technology.

The molding techniques are a modification of two well known systems:
1) injection
2) compression

With the first one liquid foam is injected in a closed mould made by metal (steel or aluminium) or fiberglass, using the following operative conditions: pressure from 1.3 to 3 bar, temperature from room temperature to 80°C. The molding time is strictly depend from the temperature: that means higher is the temperature shorter is the cycle.

With the compression technique the liquid foam is poured in the open mould made with a plug and a female part and built with the same material used for the injection technology. Also the molding conditions are the same. As glass reinforcment can be used chopped strand roving and/or continuous filament mat.

An interesting example application of RPSF are the commercial van roof. For this item fifty prototypes have been produced using compression technology (fig.12). These roofs, having the same stiffness of standard GRP and steel ones, weigh 40 kg (88.3 lb) instead of 70 kg (154.5 lb) of GRP version and 100 kg (220.7 lb) of steel version. Furthemore the RPSF van roof present other advantages such as higher thermal insulation, sound deadening and better impact behaviour.

Another realization with RPSF are the car body panels. There have been realized some car body panels for a new experimental medium size car for an italian firm. The following pieces have been molded: the roof, the hood, the external back side parts, the walls and floor of the trunk. The main advantages of this panels are:
1) low weight: there is s weight reduction of 30% compared with SMC parts and 80% compared with steel parts.
2) thermal insulation
3) sound deadening
4) better impact behaviour
5) the surface appereance is comparable with steel and SMC.

CONCLUSION

A composite material consisting of three phase components (resin, gas and continous or long chopped fibers) has the maximum flexural strength and stiffness with minimum weight. It is possible to calculate at the first approximation the elastic modulus of this kind of composite with the quasi-isotropic laminate theory.

With polyester foam reinforced with continous glass fibers it is possible to obtain at the same time high specific flexural stiffness and strength, as well as v.s. the temperature, high thermal insulation and sound deadening, low thermal expansion, good surface appearance, easy and non toxic foaming and molding technology and last, but not least, low cost.

Prototypes and small series made RPSF have confirmed properties and advantages shown at lab level.

REFERENCES

1. L.E. Nielsen,"Mechanical properties of polymers and composites", M. Dekker inc. N.Y. (1974).
2. L.J. Cohen and O. Ishai, The elastic properties of three-phase composites, J. Composite Mat. vol.1 (1967).
3. J.E. Ashton, J.C. Halpin P.H. Petit, "Primer on composite materials: analysis", Technomic Pub.Co.Inc. Stanford Conn. (1969).
4. S.W. Tsai and N.J. Pagano, Invariant properties of composite materials, in: "Composite material workshop", S.W. Tsai, J.C. Halpin, N.J. Pagano ed.,Technomic Pub. Stanford Conn. (1968).
5. General Electric, Technical bull.
6. Snia Viscosa, Technical bull. (1980)
7. B.C. Wendle, "Engineering guide to structural foam", Technomic Pub.Co. Inc., Stanford Conn. (1976)
8. S.W. Tsai and H.T. Hahn, "Composite Material Workbook", AFML TR 78-33 (1978).

DIMENSIONAL STABILITY OF REINFORCED MATRICES

A. Apicella, P. Masi, L. Nicodemo, L. Nicolais, and
S. Piccarolo*

Istituto di Principi di Ingegneria Chimica
University of Naples, 80125, Naples, Italy
*Istituto di Ingegneria Chimica, University of Palermo, Italy

INTRODUCTION

Particulate fillers are often used to improve the dimensional stability of polymeric items obtained through processing techniques such as injection molding, deep drawing, hot stamping, etc. In these operations the material undergoes large multiaxial deformations which result in molecular orientation and remain as frozen-in stresses during cooling of the material. Once the formed objects are exposed to sufficiently high temperatures, various degrees of spring-back take place and subsequent changes in shape and dimension occur (1).

The effect of draw ratio and glass filler on the recoil kinetics of composite sheets is analyzed here. Both amorphous and semicrystalline matrices have been used.

EXPERIMENTAL

For amorphous matrices drawn composites have been obtained with the same processing procedures described in previous papers (2-4). The constituent materials used were Montedison's Edistir FA and NA, general purpose polystyrenes, referred in the text as PS_2 and PS_1, and Silenka 8041 E-glass fibers with 95% of the fibers having a nominal length of 3 mm and 5% having a nominal length of 6 mm. The fiber diameter was 10 microns. Composites obtained with polystyrene and glass beads have also been studied. The glass beads (Ballotini Europe 3000 CP/01) have a diameter between 40 and 50 microns.

The recoil experiments were performed by immersing the drawn

469

samples into a thermostatic bath filled with glycerol, which is not absorbed into polystyrene. Specimens were removed after various residence times at temperatures ranging from $108^{\circ}C$ to $165^{\circ}C$ and their length measured. The initial length of the samples (l_o) cut from the extruded strips was 9 mm. Longer strips gave spiral-like shrinkage due to inhomogeneous deformations (3). As previously discussed (2), although the deformation during extrusion, drawing and recoil is inhomogeneous, the specimens approach homogeneous uniaxial extension in the region around their central line. For these reasons the recoil kinetics of drawn samples have been studied by measuring the length at the central line using a micrometer with a sensitivity of 0.01 mm.

Tensile tests have been performed on rectangular specimens by means of an Instron 1112 tensile machine.

The semicrystalline matrix was a commercial polypropylene (Montedison Moplen MOF41) and the fillers were the fibers and glass beads used for polystyrene. The range of filler volume fraction was 0.05% - 5%.

As reported in ref. 6, long tapes of filled and unfilled polypropylene were prepared with a Negri-Bossi single screw extruder. A temperature of $200^{\circ}C$ was maintained along all the extruder. The molten tapes were hot drawn at constant draw ratio and then rapidly quenched at room temperature in a water bath. The samples so obtained were subsequently drawn using the Instron testing machine at an initial deformation rate of 0.16 $min.^{-1}$ and at temperatures T_S rangin from $25^{\circ}C$ to $150^{\circ}C$, until, for each temperature, the natural draw ratio NDR (7) was attained.

Nearly the same NDR values were reached both for filled and unfilled polypropylene. The results obtained at different T_S are reported in Table 1.

The recoil experiments were carried out as previously described for polystyrene at temperatures, T_A, ranging from $120^{\circ}C$ to $165^{\circ}C$.

T A B L E   1

T_S	25	60	90	120	150
NDR	5.5	6.5	7	7.2	7.5

Table 1. Natural Draw Ratios (NDR) at the drawing temperature (T_S) used in the experiment.

RESULTS AND DISCUSSION

The recoil kinetics for the polystyrene samples PS with a nominal DR=16 at different temperatures are reported in Figure 1.
The data are plotted as l_o/l_t vs t. At high temperatures an asymptotic equilibrium value, l_f, is reached. The data have been superimposed by orizontal shifts giving a single master curve relative
to a reference temperature of 135^oC (3) as l_o/l_t vs t' at fixed DR.

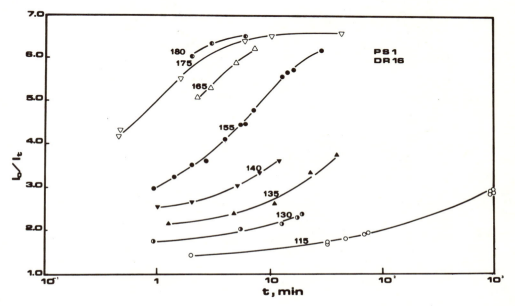

Fig. 1 - Recoil kinetics of drawn PS at $T > T_g$

The variable t' is related to the real time by $t'=t/a_T$, where
a_T is the shift factor given by a modified Williams-Landel-Ferry
(WLF) equation (8):

$$\log a_T = \frac{C_1 (T-T_g)}{C_2+(T-T_g)} - \log a_{T_o} \qquad (1)$$

with $C_1 = 13.3$, $C_2 = 47^oC$, $T_g = 83^oC$ and $a_{T_o} = 1.12 \times 10^{-7}$ for PS_2,
$T_g = 93^oC$ and $a_{T_o} = 5.74 \times 10^{-7}$ for PS_1. The values of C_1 and C_2
are those reported for Polystyrene (8) and used in reference 3.
Log a_{T_o} takes into account that the reference temperature is not
T_g but 135^oC.

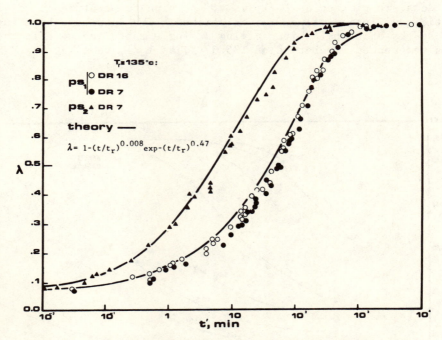

Fig. 2 - Normalized master curves for PS recoil

For different drawn ratios this line of thought has been followed. The quantity $(1_o/1_f)$ represents the deformation due to
orientation and consequently the maximum value of the shrinkage
that can be obtained for a sample with a given processing history.
The rate of recoil at a time t can be defined as a function of the
parameter λ defined as $((1_o-1_t) / (1_o-1_f)) 1_f/1_t$, which is the
ratio $((1_o-1_t)/1_t) / ((1_o-1_f)/1_f)$, i.e. the ratio of the deformation
at time t and the final one, both measured with reference to the
configuration at the time of observation. In Fig. 2, λ vs t' at
different DR are reported.

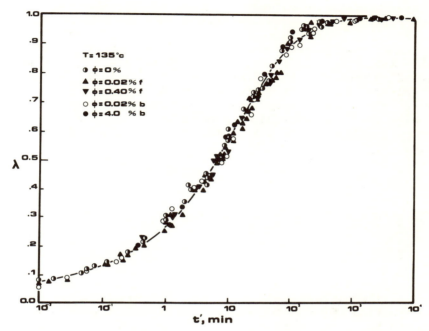

Fig. 3 - Normalized master curve, T, = 135°C, for differently
filled PS$_2$.

The continuous lines through the data are plots of the equation:

$$\lambda = 1-(t/t_r)^{0.008} \exp - (t-t_r)^{0.47} \qquad (2)$$

where t_r is a characteristic relaxation time (84 min for PS$_1$ and
13 min for PS$_2$). The validity of this equation is limited to the
range of variables of Fig. 2.

Data for differently filled and oriented PS$_2$ samples have been
plotted in the same form (4) giving a single master curve (Fig. 3).

The presence of the filler does not influence the recoil kinetics of the polymeric matrix, but has a strong influence on the equilibrium value of the length reversion ratio l_o/l_f (LRR).

Lower equilibrium values of the LRR have been indeed observed for glass fiber and glass bead composites of the same DR at increasing filler volume fractions (4). A possible reason for this diffference could be accounted by a different degree of induced or retained molecular orientation for the filled and unfilled samples subjected to the same thermal and mechanical history. However the data reported in Fig. 4 indicate that this is not the case, at least for glass bead composites. The experimental values of l_o/l_f for bead composites reported as a function of the volumetric filler content ϕ in Fig. 4 (full dots) are, in fact, well fitted by the relation:

$$\varepsilon_{rf} = \varepsilon_{rp} \left[1-(\phi/\phi_{max})^{1/3} \right] \qquad (3)$$

In this equation ε_r is defined as:

$$\varepsilon_r = (1_o-1_f)/1_o = 1-(1_o/1_f)^{-1} \qquad (4)$$

and the subscripts f and p refer to the bead filled composites and to the unfilled polymer respectively: ϕ_{max} represents the maximum volume fraction of beads for random packing of large spherical particles and is expected to be equal to 0.63 (5,9,10). Equation 3 is based on the hypothesis that ε_{rp} represents, under fixed thermal and mechanical history, a property of the drawn polymer both for filled and unfilled samples. Therefore, using the Smith's analysis (11) relative to the strain at break of glass bead composites, equation 3 follows. The good agreement between the experimental data and the predicted values (calculated with $\phi_{max}=0.63$, $(1_o/1_f)=2.3$ in eq. 3) indicates that, for a fixed DR, the molecular orientation in the glass bead filled system and in the unfilled polymer is the same.

The equilibrium values of the LRR for fiber filled composites are also reported in Figure 4. In this system the filler effect on the dimensional stability is much stronger than for bead composites. At very small fiber concentration ($\phi=0.35\%$) l_o/l_f is reduced from a value of 2.3 (i.e. 130% of the deformation of the unfilled PS) to a value of 1.15 (i.e. 15% of the deformation). Also in this case the decrease of the equilibrium LRR as a consequence of filler content could be attributed to the effect of fibers on the recovery process more than on the amount of molecular orientation of the polymeric matrix at a fixed DR. This fact can also be inferred from the data of Young moduli reported below.

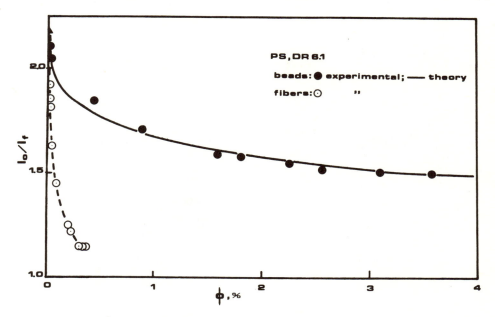

Fig. 4 - Equilibrium LRR for PS composites

The stronger influence on the equilibrium value of the LRR of the glass fibers with respect to the glass beads is probably due to a different recovery mechanism. The fibers oriented in the drawing direction may act as a constrain for a certain amount of polymer surrounding them and consequently the fiber filled specimens partly stress relax and partly creep. In fact, if one puts an unfilled drawn specimen with fixed ends at $T > T_g$, after some time the internal stresses relax without modifying the shape of the sample, while, if the ends are free to move, the specimen creeps until an equilibrium value of LRR is reached. The stress relaxation contribution on the overall process increases at increasing fiber content.

The elastic moduli of unfilled PS and composites containing 0.87% of glass fibers and glass beads respectively are reported in Figure 5 as a function of DR. The modulus increases rapidly at increasing DR, up to DR = 3.5 and then increases slowly. This effect of DR on the elastic modulus is similar to that of the tensile strength (12) or on birifrangence (13) reported in literature for similar systems. However the effect of filler on the modulus is very small as expected for this very low filler content. The data show that the molecular orientation of the matrix in both fiber and bead composites is the same of that of the unfilled polymeric matrix at the same DR.

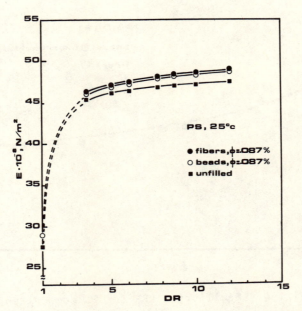

Fig. 5 - Elastic moduli for PS composites of different DR's.

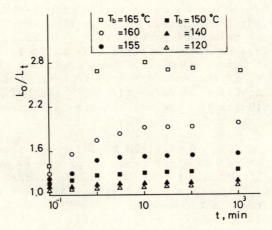

Fig. 6 - Recovery behavior versus time for samples
         drawn at 25°C for various annealing temp-
         eratures T_A.

The results obtained on drawn filled polypropylene are very different from those on polystyrene discussed previously.

The recovery behaviour of unfilled polypropylene drawn at T_S = 25°C and for different annealing temperatures, T_A, is reported in Figure 6 as l_o/l_t vs time, t. It can be noticed that a large amount of total retraction takes place in very short time not experimentally observable and then l_o/l_t increases toward an apparent constant value, l_o/l_{inf} for t larger than 10^3 min. These values are an increasing function of T_A, at a fixed T_S.

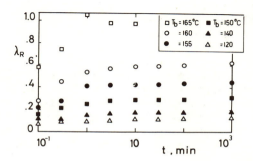

Fig. 7 - Reduced deformation, λ_R, versus time for samples drawn at 25°C for various annealing temperatures.

Similar conclusions can be drawn by plotting the data as λ_R vs t (Figure 7) where

$$\lambda_R = \frac{1 - 1/(l_o/l_t)}{1 - 1/(l_o/l_{inf\ R})} = \frac{l_o - l_t}{l_o - l_{inf}}$$

and $l_o/l_{inf\ R}$ represents the values of l_o/l_t at T_A = 165°C evaluated assuming that the data can be correlated by an exponential function and extrapolating the limiting value with a numerical technique (14). For all practical purposes $l_o/l_{inf\ R}$ coincides with the final value of l_o/l_t. Moreover all the data on unfilled polypropylene can be superimposed to form a master curve of λ_R vs t', where t'=t/a_T and a_T is an experimental shift factor which does not follow the WLF equation (6).

The data reported are similar to others presented in literature (e.g. references 15,16) and can be interpreted on the basis of the key role which is played by the intra and interfibillar taut tie molecules which bridge crystalline blocks to each other (17). However the experimental results obtained do not support the model discussed in reference 15 mainly because λ_R does not follow an equation of the type: $\lambda_R = A + B \log t$, while the large initial retraction can be attributed (15) to the melting out the crystallites.

For filled polypropylene the large amount of data obtained (18), shows a not well defined influence of the filler on the retraction of the unfilled matrix in contrast with the results here reported for filled polystyrene.

The values of $\delta = \dfrac{(l_o/l_f)_c - (l_o/l_f)_m}{(l_o/l_f)_m}$

where c and m are referred to the composites and unfilled polypropylene respectively are reported in Fig. 8, showing that the influence of the filler is very scarce. The scattering of data can be attributed to experimental errors.

These results indicate that the influence of small amounts of filler on the morphology of polypropylene, for what concerns the number of taut tie molecules and number and form of crystallites present, is scarce. Similar conclusions can be drawn for what concerns the necking mechanism as demonstrated by the fact that nearly the same NDR is obtained for each T_S, both for filled and unfilled polypropylene.

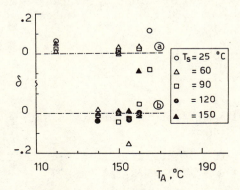

Fig. 8 – Values of δ vs annealing temperature T_A, for various drawing temperatures T_S.

        Part a) : Polypropylene filled with glass fibers at a volume fraction of 5%.

        Part b) : Polypropylene filled with glass beads at a volume fraction of 5%.

In conclusion the recoil of filled polystyrene is largely affected by the presence of filler. In particular the fibers strongly enhance the dimensional stability of the polystyrene at very low concentration. This fact is very interesting for possible applications.

In contrast the recoil of polypropylene is not influenced by the presence of fillers at least for the polypropylene studied. It is probably important to investigate the influence of the degree of cristallinity on the phenomenon of recoil.

ACKNOWLEDGEMENTS

We wish to thank Mr. P. Chiavacci for his assistance in the measurements with the Polypropilene samples and C.N.R., "Programma Finalizzato Chimica Fine e Secondaria" for supporting part of the work under Grant No. 80 02116.

REFERENCES

1. K.M.Kulkarni, Polymer Eng.Sci., 19, 474 (1979)
2. L.Nicolais,L.Nicodemo,P.Masi,A.T.Di Benedetto, Polymer Eng.Sci. 19, 1046 (1979)
3. A.Apicella,L.Nicodemo,L.Nicolais, Rheol.Acta, 19,291 (1981)
4. L.Nicolais,A.Apicella,L.Nicodemo, Polymer Eng.Sci., 21.151 (1981)
5. L.Nicodemo,A.Apicella,L.Nicolais, Appl. Polymer Sci. 26,129(1981)
6. S.Piccarolo, in "Rheology",G.Astarita,G.Marrucci,L.Nicolais Eds. Plenum Press (New York 1980), Vol.3,pg.443
7. L.E.Nielsen,"Mechanical Properties of Polymers and Composites", Vol.2,pg.300 Marcel Dekker (New York 1974)
8. J.D.Ferry, "Viscoelastic Properties of Polymers", J.Wiley, New York (1970)
9. R.F.Fedors, J.Colloid Interf.Sci.,46,546 (1974)
10. G.D.Scott, Nature (London),188,908 (1960)
11. T.L.Smith,Trans.Soc.Rheol.,3,113 (1959)
12. S.S.Sternstein,J.Rosenthal,Adv. in Chemistry Series No.154 "Toughness and Brittleness of Plastics",R.D.Deanin and A.M. Crugnola Eds., (1976)
13. T.T.Jones, "The Effect of Molecular Orientation on the Mecha- nical Properties of Polystyrene",Macromol.Div.,IUPAC,Aberdeen (Sept.10-14,1973)
14. J.Kittrel, Adv.in Chem.Eng.,8,97 (1970)
15. J.Tunnicliffe,D.J.Blundell,A.H.Windle, Polymer 21,1259 (1980)
16. F.J. Balta Calleja,A. Peterlin,Die Makromol.Chem.141,91(1971)
17. A. Peterlin,Polym.Eng.Sci.,18.488 (1978)
18. A.Giovinazzo,Chemical Engineering Thesis, University of Naples, Italy (1981)

DEFINITION OF INTERPHASE IN COMPOSITES

P.S. Theocaris

Department of Mechanics
The National Technical University
Athens (624), Greece

INTRODUCTION

The mechanical behavior of a two-phase composite system depends partly on the filler characteristics such as, the geometry of inclusions, their size, the size distribution, the orientation of inclusions, the filler volume fraction, the relative positions between the inclusions, the physical state of the filler, etc. and partly on the matrix characteristics. These characteristics are the physicochemical state of the matrix, the degree of polymerization of the polymer matrix, the crystallinity, the degree of cross-linking, etc.

These parameters refer to the filler and the matrix separately. However, apart from these parameters, there is another parameter which is of cardinal importance for the characterization of a composite system. This parameter is the effectiveness of the bond between matrix and filler in transferring stress across the interface. Many models have been developed for explaining the mechanical behavior of composite materials. All of them have a common characteristic that they consider the filler-matrix interface as a perfect mathematical surface. However, the real situation that exists is much different from the situation considered by the models.

These differences may be classified as follows:
i) All models assume the surfaces of inclusions as perfect mathematical surfaces, which can ensure continuity of stresses and displacements. However, in real composites these surfaces are rough and this roughness results in stress concentrations near the corners, whilst, on the other hand, the real contact surface

between the two phases is greater than the respective contact
surface predicted by the models
ii) The rate of polymerization, as well as the mobility of
macromolecules at the contact surface is affected by the presence
of filler in the matrix material. This results in a reduction of
the number of possible conformations of molecules in the boundary
layer and the formation of a denser material in the boundary layers
than in the bulk[1-4]. The increase in polymer fraction in surface
layers leads to a general increase of the modulus of the filled
system[1].
iii) Thermal stress concentrations occur in a composite material,
since the thermal properties of each of the components are
different. Thus, the load-carrying capability of the system is
affected by these stresses according to the nature of the stress
field developed by external loads.
iv) Around each inclusion a complex situation develops, which
consists of areas of imperfect bonding, where mechanical stresses
are developed due to shrinkage[5-6] and high stress gradients or
even stress singularities due to the geometry of the inclusion[7,8]
are apparent and finally voids, microcracks, and other
discontinuities may be created.

     Thus, the composite may be considered as consisting of three
phases, that is the matrix, the inclusions and a third one, which
is the area of imperfections surrounding each one of the inclusions
and which is called *interphase*. In this paper the effect of the
boundary interphase on the overall thermomechanical behavior of
polymeric composites will be presented[9-14]. Especially the
mechanism of crack propagation in a composite will be studied and
particularly the role of the straight interphase to the
deceleration and arrest of cracks propagating in the composite will
be developed.

A THEORETICAL MODEL FOR PARTICULATE COMPOSITES

     The representative volume element consists of three phases,
i.e. three concentric spheres, each phase maintaining a constant
volume fraction. The intermediate phase, or interphase, corresponds
to the zone of imperfections surrounding the inclusions and it is
considered as consisting of a homogeneous and isotropic material
of finite thickness. If the thermomechanical properties of the
constituent materials, as well as of the composite, are known, the
respective properties and the volume fraction of the interphase
can be evaluated, under the assumption that, if perfect adhesion
between the two phases of the composite exists, then the thickness
of the interphase vanishes and the interphase layer tends to an
interface.

     Using the equations of thermoelasticity for thermal stresses
and displacements along with the boundary conditions, which account

for continuity of stresses and displacements at the two interfaces, a non-linear system is formed having as unknowns the modulus of elasticity, the thermal expansion coefficient, the radius, the volume fraction and Poisson's ratio of the interphase material.

In order to check the theoretical predictions two series of aluninium-epoxy composite specimens were manufactured. One of them was characterized by good adhesion, termed as *improved*, while the other one was characterized by imperfected adhesion termed as *ordinary*. When these theoretical considerations are taken into account for this simple model, it may be derived that the interphase material in the case of improved specimens has superior mechanical properties when compared to the respective ones of the ordinary specimens.

Moreover, with the help of the same model a prediction for the glass-transition temperature of the composite, T_{g_c}, may be derived yielding the relation:

$$T_{g_c} = \frac{\upsilon_i(\alpha_{i2}-\alpha_{i1})T_{g_i}+\upsilon_m(\alpha_{m2}-\alpha_{m1})T_{g_m}}{\upsilon_i(\alpha_{i2}-\alpha_{i1})+\upsilon_m(\alpha_{m2}-\alpha_{m1})} \qquad (1)$$

where υ and α are the volume fraction and the thermal expansion coefficients, while indices m and i stand for the matrix material and the interphase respectively and indices 1 and 2 hold for temperatures below and above T_g respectively.

Application of Eq.(1) yields glass transition temperatures very close to the values experimentally determined for various composites[11].

MODELS FOR FIBER-REINFORCED COMPOSITES

For composites reinforced with fibers the deformation of the matrix is used to transfer stress by means of shear tractions at the fiber-matrix interface to the embedded high-strength fibers. On the other hand, fibers retard the propagation of cracks and thus produce a material of high strength.

The introduction of short fibers into a polymeric matrix leads to stress concentration at the end of filaments, whose flat ends are held in the matrix by adhesion, whilst, on the other hand, the glass-transition temperature of the matrix material is changing. This kind of behavior appears to depend strongly on the quality of adhesion between matrix and inclusions.

In all models describing the thermomechanical properties of composites, a perfect adhesion between the main phases is conside In reality, the existence of the fiber into the matrix tends to restrict the matrix from freely elongating in the region of the fiber. This results in a strain perturbation at the vicinity of fiber. The strain perturbation in the matrix diminishes with increasing distance from the fiber, until the strain of the bulk matrix is achieved.

In order to explain the thermomechanical behavior of composites reinforced with short fibers we have considered two theoretical models. In the first model the interphase material is considered to be homogeneous and isotropic, whereas in the second model it is considered to be inhomogeneous with properties varying continuously from the fiber to the bulk matrix.

HOMOGENEOUS-INTERPHASE MODEL

In this model the cylindrical fiber is surrounded by a hollow cylindrical interphase and this, in turn, is surrounded by the matrix. The following assumptions hold:
i) Perfect bonding between all interfaces.
ii) Fiber and matrix materials only carry tensile stresses.
iii) Interphase only carries shear stresses.
iv) Matrix, fibers and interphase are homogeneous isotropic materials.
By considering now that the model is subjected to a uniform temperature rise ΔT and simultaneously to a tensile stress σ_c parallel to the z-direction of the fibers, then, from the equilibrium equations for an element of the model and the appropriate boundary conditions, the following expressions may be derived:
The tensile stress in the fiber σ_f is expressed by:

$$\sigma_f = E_f\left[\frac{du_m}{dz}\left(1 - \frac{\cosh nz}{\cosh\frac{n\ell}{2}}\right)+\varepsilon_f^T\frac{\cosh nz}{\cosh\frac{n\ell}{2}}\right] \tag{2}$$

$$= \frac{2G_i}{E_f r_f(r_i-r_f)} \tag{3}$$

u is the displacement, r the radius, G and E the shear and moduli, ε_f^T the thermal strain developed in the fiber, indices m, f and i denote matrix, fiber and interphase

r stress, τ, along the interphase was found to be:

$$\frac{G_i}{-r_f)}\frac{\sinh nz}{n\cosh\frac{n\ell}{2}}\left[\varepsilon_f^T - \frac{du_m}{dz}\right] \tag{4}$$

red.   decreasing variation of E_m with temperature[9] we
                        the displacement in the matrix is varying with
the   rding to the relation:

$$\frac{du_m}{dz} = \frac{\sigma_m}{E_m(T)}\left[1 - \frac{1}{E_m(T)}\frac{dE_m(T)}{dT}\right] + \alpha_m\Delta T \tag{5}$$

where α_m denotes the linear thermal expansion coefficient of the matrix. Finally, the tensile stress in the matrix is found to be:

$$\sigma_m = \frac{\sigma_c - E_f\upsilon_f\left[\alpha_m\Delta T\left(1 - \frac{\cosh nz}{\cosh n\ell/2}\right) + \epsilon_f^T\left(\frac{\cosh nz}{\cosh n\ell/2}\right)\right]}{\upsilon_m + \upsilon_f\frac{E_f}{E_m(T)}\left(1 - \frac{1}{E_m(T)}\frac{dE_m(T)}{dT}\Delta T\right)\left(1 - \frac{\cosh nz}{\cosh n\ell/2}\right)} \tag{6}$$

where υ_m is the real-volume fraction of the matrix given by:

$$\upsilon_m = 1 - \upsilon_f - \upsilon_i \tag{7}$$

INHOMOGENEOUS-INTERPHASE MODEL

An improvement of the previous model is the inhomogeneous-interphase model according to which the interphase material is assumed presenting continuously varying properties with the distance from the fiber surface. This model is capable to evaluate the extent of the boundary interphase and give a quantitative criterion of the quality of adhesion between the two main phases. It results also in a prediction of the longitudinal modulus of elasticity of the fiber-composite.

If we consider the homogeneous interphase model consisting of three concentric cylinders the fiber, interphase and matrix, and suppose that each of the materials of the cylinders is characterized by the stress function Φ given by:

$$\Phi = C_1\log r \pm C_2 r^2\log r \pm C_3 r^2 \pm C_4 \tag{8}$$

where C_1, C_2, C_3, C_4 are constants and r is the radial distance from the fiber axis. From the continuity assumption of stresses and displacements at the interfaces and from energy balance considerations we find that the global elastic modulus of the composite E_c is expressed by:

$$E_c \simeq E_f\frac{r_f^2}{r_m^2} + E_i\left(\frac{r_i^2 - r_f^2}{r_m^2}\right) + E_m\left(\frac{r_m^2 - r_i^2}{r_m^2}\right) \tag{9}$$

which equation is another expression of the law of mixtures.

We assume now that the interphase material consists of an infinite number of concentric cylinders, so that its modulus of elasticity is a continuous function of the distance r from the fiber expressed by:

$$E_i(r) = E_m + E_f\left(\frac{r_f}{r}\right)^{n_1} - E_m\left(\frac{r_f}{r}\right)^{n_2} \tag{10}$$

where the exponents n_1 and n_2 are intimately related to the quality of adhesion. Indeed, if $a = n_2/n_1 = 1$ there is a perfect adhesion between fiber and matrix, whereas for diminishing values of a the quality of adhesion is deteriorating.

According to the above assumption, the rule of mixtures takes the form:

$$E_c = E_m + (E_f - E_m)\upsilon_f + \upsilon_f\left[\frac{E_f}{1-k_1}\left(\upsilon_f^{k_1-1} - 1\right) - \frac{E_m}{1-k_2}\left(\upsilon_f^{k_2-1} - 1\right)\right] \tag{11}$$

where $n_1 = 2k_1$ and $n_2 = 2k_2$.

For the evaluation of the parameter n_1 we may use the following relation given in ref. 13:

$$\left(\frac{\Delta r + r_f}{r_f}\right)^2 - 1 = \lambda\left(\frac{\upsilon_f}{1-\upsilon_f}\right) \tag{12}$$

where:

$$\lambda = 1 - \frac{\Delta C_p, \text{filled}}{\Delta C_p, \text{unfilled}} \tag{13}$$

in which $\Delta C_{p_{unf}}$, $\Delta C_{p_{filled}}$ represent the sudden changes in heat capacity in the temperature position of the transition region of the unfilled and filled polymer respectively. Moreover, Δr is the thickness of the interphase. On the other hand, the thickness of the interphase may be found from relation (10). If we compare these two values we have for n_1, and for the case with $n_1 = n_2$, the expression:

$$n_1 = \frac{2\ln 255.5}{\ln\left(1 + \lambda\frac{\upsilon_f}{1-\upsilon_f}\right)} \tag{14}$$

Then, the modulus of the interphase takes a minimum value given by:

$$E_{r_{min}} = E_m + E_f\left[\frac{E_f}{E_m} \cdot \frac{1}{a}\right]^{\frac{1}{a-1}} - E_m\left[\frac{E_f}{E_m} \cdot \frac{1}{a}\right]^{\frac{a}{a-1}} \tag{15}$$

The value of r_0, corresponding to $E_{r_{min}}$, is given by:

$$r_0 = r_f\left(\frac{E_f n_1}{E_m n_2}\right)^{\frac{1}{n_1-n_2}} \tag{16}$$

We have applied both models to the case of a fiber composite with

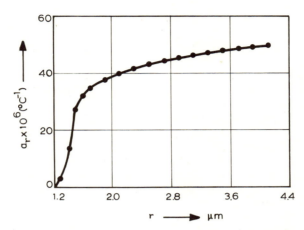

Fig.4: Variation of the thermal expansion coefficient α(r) of the
interphase material against the normal distance r from the
fiber surface.

EXPERIMENTAL EVIDENCE FOR THE INTERPHASE

Application of Eq.(1) yields a glass-transition temperature for
an iron-epoxy composite, filled with iron particles of average
diameter 150μm at a volume fraction u_f=0.05, a value equal to
T_{g_c}=120°C while the experimentally determined value was T_{g_c}=117°C.
On the other hand the homogeneous-interphase model yields for the
thermal expansion coefficient of the abovementioned composite, at
$T<T_{g_c}$, α_{c_1}=48.2×10^{-6} °C^{-1}, against 48×10^{-6} °C^{-1}, which was the
experimental value. However, at $T>T_{g_c}$ it gives 122×10^{-6}C^{-1} against
158×10^{-6} °C^{-1} experimental.

The inhomogeneous-interphase model yields a longitudinal
thermal expansion coefficient, for an E-glass fiber-reinforced
polymer, a value equal to 6.5×10^{-6} °C^{-1} against 6.7×10^{-6} °C^{-1}
experimental. However, the same model, yields a transverse thermal
expansion coefficient for the same material, equal to 12×10^{-6} °C^{-1},
against 25.8×10^{-6} °C^{-1} experimental.

From the above comparison it becomes clear that there is a good agreement between theoretical and experimental results in most cases. The deviations, which in some cases occur, are due to the large number of parameters which affect the overall behavior of the composite material and cannot be taken simultaneously into consideration in a theoretical model.

MECHANISM OF CRACK PROPAGATION IN COMPOSITE MATERIALS

The modes of crack propagation in composite materials are influenced by a great number of factors. Such factors are, the adhesion efficiency between phases, the rate of loading, the mechanical characteristics of each phase and the order of succession of the phases during the initiation and propagation of cracks. In the following, we shall analyse each one of the above parameters separately. The behavior of fiberlike composites in propagating cracks along their phases may be simulated by a bimaterial plate, whose one phase has the characteristics of the matrix and the other phase the mechanical properties of the fiber. If a transverse crack is propagating along the one phase of a bimaterial plate it is characterized by a constant maximum velocity, which is maintained till a very short distance from the interface. Within this short distance from the interface the crack starts decelerating and finally completely stops at the interface. After that, the crack reinitiates in the second phase. In general, at interfaces between a more ductile phase I and a more brittle phase II, the propagating crack tends to deviate from the transverse direction and, extremely to bifurcate. This is due to the fact that when the crack arrives at the interface the displacements which the brittle phase, having higher modulus of elasticity, has to obey are much larger than those the corresponding stresses could produce, because of the continuity of stresses and displacements on the boundary. The criterion of such deviation or bifurcation is the velocity of the stress pulse applied to the composite, in combination with the ratios of the moduli and Poisson's ratios for the two phases. For high stress-pulses and great differences in the moduli of the two phases the arrested crack at the interface starts to propagate along the interface for some distance and when the strain energy necessary for its initiation in both branches is built up it starts to propagate in phase II creating a double branch of independently propagating cracks. The directions of these branches, oblique at the beginning, tend to become transverse and parallel to the crack of phase I.

The above described crack-arrest phenomena may be studied by the method of caustics. Indeed, these phenomena are characterized by a continuous disappearance of the caustic in phase I and a subsequent appearance of it in phase II at the close vicinity of the interface[15,18].

Fig.5 shows the fracture process of a composite specimen

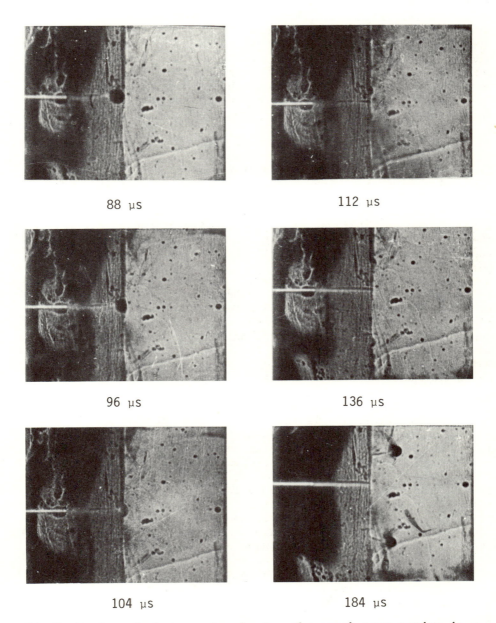

88 μs

112 μs

96 μs

136 μs

104 μs

184 μs

Fig.5: Series of photographs showing the crack propagation in a bimaterial composite specimen. Phase I 30% plasticizer, Phase II 0% plasticizer.

(phase I with 30% of plasticizer, phase II with a pure epoxy polymer). This figure clearly shows the splitting of the unique crack in phase I into two branches. The crack-arrest process at an interface remains unchanged for all possible material combinations having also the common qualitative characteristics of all crack-arrest processes of other definite material discontinuities, such as a longitudinal discontinuity of the material or a void, which resulted in decelerating and arresting the propagating crack and further in deviating the way of propagation of the moving crack.

The crack-arrest process may be viewed as consisting of two distinct periods. During the first period of crack-arrest the singular stress field at the tip of the arrested crack vanishes, although the external load, which caused the singularity, continues to act upon the specimen, while the previously at the crack-tip deposited strain energy is being now absorbed by the material lying in the neighborhood of the crack-tip. At the same time, during the first period of crack-arrest, the caustic (appearing in phase I) submerges in the interface, until it completely disappears, which means that the stress intensity factor diminishes progressively and up to zero. On the other hand, the second period of crack-arrest is characterized by the gradual build-up of a new singularity, this time beyond the "barrier" of the interface, which ends by causing a reinitiation of the crack. At the same time, the caustic emerges subsequently in phase II of the specimen and it is continuously increasing in size, up to the critical size of crack reinitiation. This phenomenon is accompanied by the progressive increase of K_I as the crack propagates in phase II (see Fig.6).

Another factor, which affects the crack-propagation mechanism in composite materials, is the material characteristics and more precisely the succession of phases. The crack propagation velocity is increased when the crack propagates from a more ductile to a more brittle material, while the opposite result is observed when the succession of phases is reversed. Finally, the type of loading is a factor of cardinal importance in determining the crack-propagation behavior of composites. Fig.7 presents the dependence of crack-length on time for three specimens of the same composition (30%-20% of plasticizer) fractured under all three types of loading, that is static, quasi-dynamic and dynamic. From the curves shown in Fig.7 we can observe that, as the strain rate is increased, the crack propagation velocity in phase I is also increased. The same phenomenon is also observed in phase II. However, the important is, that the observed increase in crack-propagation velocity in both phases is of the same order of magnitude. Moreover, in the case of static loading the crack propagation velocity starts with a relatively low value in phase I of all composites to reach, after the crack arrest process, a value of the same order with the fracture velocity in phase II under the higher strain-rate dynamic load.

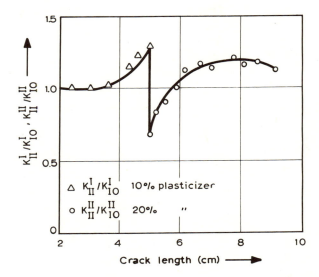

Fig.6: The variation of the normalized value of the stress intensity
factor at each phase of a composite bimaterial plate. Phase
I: 10% plasticizer, and phase II: 20% plasticizer.

The same qualitative behavior is observed when the interface
is inclined to the load axis. The main differences are: i) The
crack-paths are curved, ii) The crack-arrest phenomenon is smoother
in this case, as compared to that of a longitudinal one, iii) The
crack arriving at an interface (with a more brittle phase II)
propagates, for a while, along the interface not symmetrically, but
to the direction of the minimum deviation from its initial path
creating multi-branching (see Fig.8).

These differences are due to the fact that the strong shear-
stress area is not located in the close vicinity of the interface,
but covers almost the whole specimen. The spreading of shear stresses
is produced by an initial bending of the specimen because of its
asymmetric geometry. This bending phenomenon, which depends on the
mechanical characteristics of the two phases, and, mainly, on the

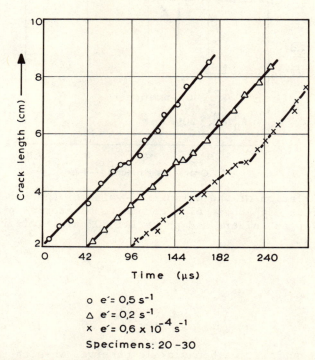

Fig.7: Diagrams of the crack length versus the crack propagation
time for a composite bimaterial plate, and for three
different strain-rates.

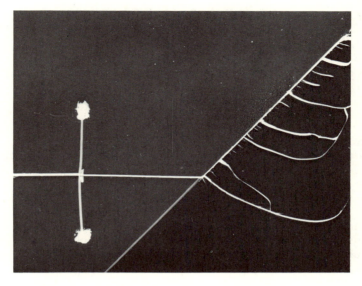

Fig.8: The path of a crack in a bimaterial plate with an angle of
       inclination 45°. Phase I 30% plasticizer, and phase II 10%
       plasticizer.

angle of inclination of the interface, affects also the propagation
velocities in both phases. These velocities do not differ strongly
between the two phases, since the crack-arrest phenomenon is weak
in this case.

CRACK-PROPAGATION BEHAVIOR OF PARTICULATE COMPOSITES

    The crack-propagation behavior of particulate composites may
be viewed as more complicated than the respective behavior of
bimaterial plates. This is due to the fact that, in particulate
composites, the factors which determine their fracture behavior are
numerous. Among these factors may be classified the adhesion
efficiency between phases, the filler volume fraction, the geometry
and the size of inclusions, the mechanical properties of the two
phases and others[19].

    The filler-volume fraction is characterized by the so-called
optimum concentration, which is regarded as the limit of saturation
of the adsorption centers on the surface of the filler by the
macromolecules. For filler concentrations greater than this
optimum value, the continuity of the network structure is upset and
this leads to a deterioration, rather than to an improvement, in the
mechanical properties with filling. This reversal of the
reinforcement is explained by the fact that this difference in

reinforcement is due to the difference of thermal expansivities of the filler and the polymeric matrix and, on the other hand, to the occurrence around the filler-particles of a stress concentration zone, which greatly weakens the adhesion-bonds between filler and matrix. Moreover, the filler does not exhibit any resistance to crack growth and there is a reduction in the content of the continuous phase, which is responsible for the strength of the composite.

The particle-size affects the crack propagation mechamism, since when the particle diameter is small, as compared to the dimension of the fracture zone, the presence of the inclusion at the vicinity of the propagating crack-tip has as a result to change the rheological properties of the viscoelastic matrix, having no effect on the path of the crack.On the contrary, when the particle diameter is large as compared to the dimensions of the fracture zone, then a bifurcation phenomenon is observed, which is more pronounced as the distance of the crack-axis from the inclusion-center is decreasing.

The method of caustics can be succesfully applied in the case of particulate composites. Fig 9 presents a series of photographs showing the crack propagation process in an iron-reinforced epoxy specimen with filler volume fraction 5 percent and particle diameter 0.15mm. The observed variation of dimensions of the caustics created at the crack-tip depends on both the reflection and refraction of the stress waves at the rigid interfaces of filler particles and the variable size of the fracture zone. It is possible to determine from this photograph the variation of crack-propagation velocity, as well as the dynamic stress-intensity factor. It was found that the crack-propagation velocity fluctuates around a mean value, as the crack length is increased, Fig.10, while the dynamic stress intensity factor is increased by steps with increasing crack length,Fig.11. This stepwise mode of increasing of K is due to both the rigid interfaces of the filler particles, which play the role of barriers, and to the different properties of the filler material.

CONCLUSIONS

In the present work, the role of the boundary interphase on the determination of the crack-propagation, as well as the thermomechanical behavior of composite materials, was investigated.

Theoretical models taking into account the presence of a third phase, surrounding each one of the inclusions, were presented and the influence of a large number of parameters on the modes of fracture of composite materials was studied. It was shown that the method of caustics is an experimental method suitable in determining the crack-propagation behavior of composites. On the

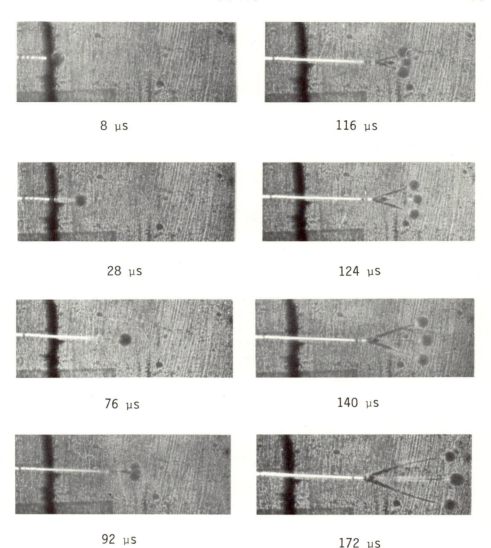

8 μs

116 μs

28 μs

124 μs

76 μs

140 μs

92 μs

172 μs

Fig.9: Series of photographs showing the crack propagation in a
0.15mm iron particulate-epoxy composite with v_f=0.05.

other hand, the theoretical models presented in the present paper
lead to the determination of the thermomechanical properties of
the boundary interphase, which affect the overall behavior of the
composite and the adhesion efficiency between phases.

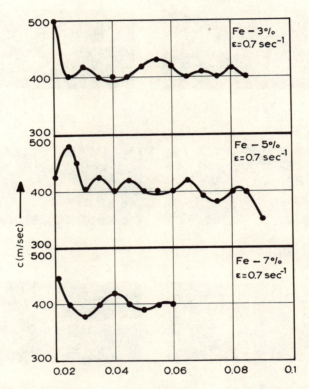

Fig.10: Diagrams of the crack propagation velocity versus the crack
        length for three composite specimens.

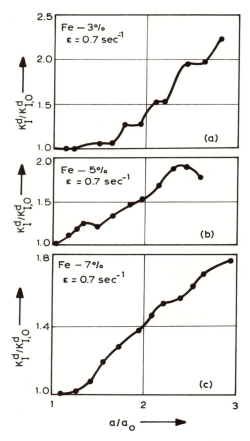

Fig.11: Diagrams of the normalized SIF versus normalized crack
length a/a_0 for three composite specimens.

REFERENCES

1. Y.S. Lipatov, *"Physical Chemistry of Filled Polymers"*, Originally published "Khimiya", (1977). Translated from the Russian by R.J. Moseley, International Polymer Science and Technology, Monograph No.2.
2. Y.S. Lipatov and L.M. Sergeeva, *"Adsorption of Polymer"*, New York (1974).
3. Y.S. Lipatov and T.E. Geller, Vysokomol Soedin, **8**, 592 (1966).
4. Y.S. Lipatov and F.G. Fabulyak, J. Appl. Pol. Sci., **16**, 2131 (1972).
5. P.S. Theocaris and E. Marketos, Fibre Science Techn., **3**, 21 (1970).
6. P.S. Theocaris and S.A. Paipetis, Jnl. Strain Analysis, **8**, 286 (1973).
7. P.S. Theocaris and S.A. Paipetis, Fib. Sci. and Techn., **9**, 19 (1976).
8. P.S. Theocaris and S.A. Paipetis, Int. Jnl. Mech. Sci., **18**, 581 (1976).
9. G.C. Papanicolaou, S.A. Paipetis and P.S. Theocaris, Colloid and Polymer Sci., **256**, 7, 625 (1978).
10. G.C. Papanicolaou and P.S. Theocaris, Colloid and Polymer Science, **257**, 239 (1979).
11. P.S. Theocaris and G.C. Papanicolaou, Fibre Science and Techn., **421**, (1979).
12. P.S. Theocaris and G.C. Papanicolaou, Colloid and Polymer Sci., **258**, 9, 1044 (1980).
13. P.S. Theocaris, G.C. Papanicolaou and G.D. Spathis, Colloid and Polymer Sci., **258**, 11, 1231 (1980).
14. P.S. Theocaris, G.C. Papanicolaou and G.D. Spathis, to be published in Fibre Sci. and Techn. (1981).
15. P.S. Theocaris and J. Milios, Eng. Fracture Mechanics, **13**, No 4, p.599 (1980).
16. P.S. Theocaris and J. Milios, Int. Jnl. of Fracture, **16**, No.2, p.31 (1980).
17. P.S. Theocaris and J. Milios, Jnl. Solids and Structures, **17**, p.217 (1981).
18. P.S. Theocaris and J. Milios, to be published in Jnl. of Eng. Mat. and Techn. (1981).
19. P.S. Theocaris, G.C. Papanicolaou and G.A. Papadopoulos, Jnl. Composite Materials, **15**, p.41 (1981).

IN SITU ANALYSIS OF THE INTERFACE

J.L. Koenig and Chwan-hwa Chiang

Department of Macromolecular Science
Case Western Reserve University
Cleveland, Ohio 44106

INTRODUCTION

Interfacial phenomena are important in all phases of composite technology. The properties of all components are affected by the nature of their surfaces which in turn significantly affect composite properties. Similarly, interfacial phenomena are important in composite fabrication, testing, and evaluation. Finally, interfacial behavior can strongly affect the response of the composite to the thermal, mechanical, and environmental conditions arising from service application [1-4].

From a materials point of view, the details of the nature of composite interfaces can be described to nearly any extent desired. The problem, however, is to relate these details to the behavior of the composites. The macroscopic average properties of a composite are not sufficient for evaluating the materials aspect of composite interfaces. Thus, the micromechanics of composites behavior may provide the only suitable method of analysis for evaluating interfacial effects. Although different interfacial phenomena will occur in different composite systems, and even though the internal mechanics may differ, there are a number of features of both the nature of interface and the micromechanics of composites which are common to all systems.

Applications of coupling agents for surface modification of filler and reinforcements in plastics have generally been directed toward improved mechanical strength and chemical resistance of composites [5-7]. The bonding created by an adhesion promoter occurs by virtue of its having functional groups which can react with the substrate and the organic polymer phase. The most common case, silanes on glass,

entire mechanical-thermal history of the composite as well as the
chemical and structural characteristics of the bulk constituents [18].
The mechanical and water resistance properties of polymer resins are
dependent upon the morphological structure of the resin, although
three-dimensional crosslinked polymer systems have been generally
treated as completely amorphous disordered structures on the molecular
level. The presence of small cavities in the resin, for instance, is
an important determinant influencing the rate of water penetration [19].
Dynamic mechanical and swelling properties of epoxy resins [20] have led
to the suggestion that in fact two separate phases may exist in the
disordered material. Microscopic examination of several glass-reinforced crosslinked epoxy resins have revealed that globular structures
in the resin are affected to various degrees by water. Boiling water
also lowered the T_g of the materials, and the observed decline in
flexural strength was attributed to cracking and changes in the supra-
molecular structure [21].

Although there are some complications, most researchers agree,
water primarily degrades the fiberglass-matrix interface via hydrolysis of the silane coupling agent in addition to attack on the matrix
and filler [22]. However, a basic structural model of interface in composite at the molecular level has not been well established. In order
to improve the moisture resistance of composite, the basic aspects of
these related subjects will be discussed. Especially we will concentrate on the molecular structure, the hydrothermal stability, and the
mechanical properties of the interface between the anhydride-cured
epoxy resin and the amino-silane treated fiberglass.

THE NATURE AND TYPES OF INTERFACES IN COMPOSITES

An interface can be described by specifying the type of discontinuity observed, along with a description of the extent at the interface. This description will usually involve the changes occurring at
a resin-fiber interface, such as state of aggregation, composition,
crystal structure, orientation, and chemical structure. A given
interface may involve several of these features. A rigorous classification would be quite difficult. If chemical interactions occur between the silane coupling agent and polymer matrix, then additional
features of the interface must be considered. What initially was an
interface on a macroscopic scale may become an interfacial zone characterized by multiple interfaces and additional phases (Figure 1). A
complete description of the interfacial zone might require a number
of parameters in composites, i.e., geometry and dimensions, microstructure and morphology, mechanical, physical, chemical, and thermal
properties of different phases or localized areas in the interfacial
zone. The selection of the components of a composite is usually based
initially on the mechanical and physical properties of the individual
constituents. The types and nature of the interfacial interactions
that can occur will vary considerably with the nature of the composite

has been presented as forming siloxane bonds from the reactant to the glass surface and covalent bonds from the coupling agent to the matrix phase. This phenomenon then provides not only strong interfacial primary bonding but also a condition not easily disturbed by environmental conditions [3-10]. The interface in a fiber-matrix composite becomes a surface which is common to both fiber and matrix with the coupling agent acting as an interfacial region. It has physical and mechanical properties which are neither those of the fiber nor the polymer matrix.

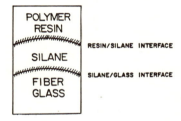

Fig. 1.   Schematic of interfaces in composites.

For the past years several theories have been proposed to interpret the mechanism of reinforcement in composites. The chemical bonding theory is the most widely accepted of these theories and suggests that the coupling agent forms covalent bonds to both the glass surface and the resin [11]. Other theories have been advanced in which the interactions at the glass-coupling agent interface are believed to be hydrogen bonding phenomena occurring between silanols in the coupling agent and glass surface [12]. Still others contend that Van der Waals forces at the interface are sufficiently strong to account for the mechanical strengths of laminates [13].

Kumins and Roteman suggested that the boundary region, of which the coupling agent is a part, between a high modulus reinforcement and a lower modulus resin can transfer stresses most uniformly if it has a modulus intermediate between that of the resin and the reinforcement [14]. It is difficult to reconcile this concept with the need for stress relaxation at an interface because of differential thermal shrinkage between polymer and filler. Hooper proposed that the silane treatment contributed a mode of mechanical relaxation through a deformable layer of silicone resin [15]. However, the layer of silane in a typical glass finish is too thin to provide stress relaxation through mechanical flexibility. A preferential adsorption theory proposed by Erickson et al. [16] is a modification of the deformable layer theory. This theory was based on the assumption that different finishes on glass fibers have, to different degrees, the power to deactivate, destroy, or adsorb out of the uncured liquid resin mixture, certain constituents necessary to complete resin curing. Such a layer would need ductility and strength to provide relaxation and effective transfer of stress between the fibers in load-bearing situations. Plueddemann suggested a reversible hydrolyzable bond theory which is a combination of the chemical bonding theory, the restricted layer theory, and the deformable layer theory [17]. This theory proposes a reversible breaking and remaking of stressed bonds between coupling agent and glass in the presence of water thus allowing relaxation of stresses without loss of adhesion.

Many other functions can possibly be served by coupling agents at a matrix-glass interface. The coupling agent may protect the interface against stress corrosion by water. The nature of the interfaces existing in a composite at any instant is dependent upon the system. In fact, organic polymer matrices are expected to show a marked chemical reactivity with the silane-treated fiber [23].

It is generally held that the matrix and fibers must be bonded together if the desired properties of composites are to be realized. The term "bonding" can be convenient to distinguish three kinds of bonding interactions: mechanical, physical, and chemical. Mechanical bonding refers to the type of interlocking that occurs as a consequence of the geometrical shape of the bodies, or as a result of purely rheological interactions. Physical bonding refers to interactions resulting from physical forces such as gravity, and magnetic fields. Chemical bonding covers all those interactions at the molecular level which result from electronic interactions between atoms or chemical reactions between functional groups. From a practical point of view, the relative importance of mechanical and chemical bonding is uncertain at present, but they are stronger than physical bonding in the composites [24].

Chemical bonds are usually designated according to the type of chemical mechanisms involved, and chemical mechanisms of bonding are customarily grouped into primary and secondary bonds. The primary bonds include the ionic, covalent, and metallic bonds, whereas the secondary bonds include the hydrogen bonds and Van der Waals bonds. The terms primary and secondary emphasize the relative magnitude of the forces involved, but not necessarily their relative importance [2].

As with any composite, the properties depend on the proportions of two components, on their properties, on the degree and nature of the interfacial adhesion, and often on the phase size. If adhesion is poor, strength and modulus will be reduced, and each constituent will exhibit its own T_g . If adhesion is good, the modulus will fall between upper and lower bounds. An exception is the use of block or graft copolymers, which exhibit good interfacial adhesion due to the covalent linkages between the phases [25].

It is clear that chemical bonding can potentially develop across the interface between any two solids if they are in intimate contact. The theoretical strength of such an interface provides a convenient frame of reference for considering the strength of solid-solid interfaces. In general, there are two approaches to the calculation of bond strength: 1: the atomic or molecular approach based upon classical or quantum mechanics, and 2: the thermodynamic approach [26,27]. The theoretical strength of interfacial bonding can be obtained by these two methods.

It is important to understand the type of interfacial regions that exist in the two different components and their surface conditions. The silane modified glass surface generates several different types of interfaces. Chemically, it can be a chemically bonded or non-bonded interface. Physically, it can be a compatible or non-compatible interface, and mechanically, it can be a smooth or rough interface. These interfacial bondings are fundamentally for predicating the load transfer through the interfaces in composites.

THE GLASS SURFACES

There are two fundamental properties which determine the untreated silica or silicate glass appropriate to an end use. These are the surface areas and the extent of hydration. Amorphous silica consists of silicon and oxygen tetrahydrally bonded into an imperfect three-dimensional structure. The most commonly used glass fiber is E-glass which contains 55% SiO_2 as the main component with the remainder being oxides of other metals such as Al, Ca, Mg, Fe, etc. After contact with water, the surface oxides are hydrated and form hydroxide groups which are considered to be adsorption sites for coupling agent molecules [28]. It is evident that surface silanols are a major factor

in determining various structures with different environments. Koenig
and Shih have demonstrated that the structural difference of the surface silanol can be detected utilizing Raman spectroscopy [29]. There
are several peaks appearing in the 1050-950 cm[-1] region due to the
SiO_2 stretching mode of the surface silanols. Glass microspheres display two lines at 980 cm[-1] for a wet sample and at 1005 cm[-1] for a dry
sample. These two frequencies are related to the silanols which are
hydrogen bonded to the adsorbed water molecules. There is an additional peak at 992 cm[-1] for the glass microspheres. This peak is responsible for the silanols which are hydrogen bonded to adjacent silanols. The typical silanol groups on the silica surface are shown in
Figure 2. The roughness of the glass surfaces are also very important
in determining the extent of surface modification.

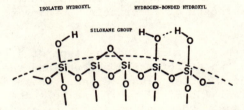

Fig. 2.   Typical groups which can occur on the pure silica surfaces.

THE STRUCTURE OF SILANES IN AQUEOUS SOLUTION

     All commercial silane coupling agents are of the structure X_3Si-R
where X are hydrolyzable groups on silicaon, and R is an organofunctional group with organic chain which is available for reaction with a
given resin. The composition of silane coupling agents in dilute
aqueous solution depends on the nature of the organofunctional group
on the silicaon and the PH value of the solution. Neutral organofunctional silane coupling agents, usually prepared in dilute acetic
solution, hydrolyze rapidly to silane triols, and then condense slowly
to oligomeric siloxanes. The monomer and small oligomers are soluble

in water and large oligomers precipitate from solution. Aqueous solutions of the silanes have only limited stability and must be used
within a few hours [30].

Amino-organofunctional silane coupling agents are the only silanes that hydrolyze almost immediately in water. But solution of
aminosilane coupling agents in toluene or other nonpolar organic
solvents give precipitates upon contact with moist atmosphere. Since
it is well known that cyclic 6-membered chelate rings have extraordi-
nary stability, it is found that an internal cyclic chelate ring structure is formed in solution of aminosilanes [31,32]. The chelate ring
structure is destroyed when the proper ionizing chemical is added to
the solution, such as acid or alkaline chemicals. In the acid solution, there is a chemical transfer of a proton to the amino group and
the amino group becomes an amonium ion in aqueous solution. The silanol group will not react with the ammonium ion and the silane molecule
stays as the linear form in the solutions. In the alkaline (KOH) solution, all the silanol groups lose hydrogen and become SiO_2 groups, and
the silane molecules do not condense with each other. The molecular
structures of the monomeric γ-aminopropyltriethoxysilane (APS) in the
acid, neutral, and alkaline aqueous solutions are shown in Figure 3.

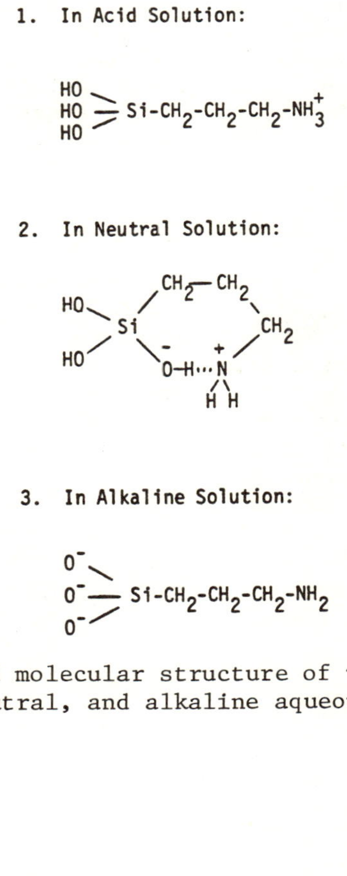

1. In Acid Solution:

$$\begin{matrix} HO \\ HO \\ HO \end{matrix} \!\!\Big\rangle Si\text{-}CH_2\text{-}CH_2\text{-}CH_2\text{-}NH_3^+$$

2. In Neutral Solution:

3. In Alkaline Solution:

$$\begin{matrix} O^- \\ O^- \\ O^- \end{matrix} \!\!\Big\rangle Si\text{-}CH_2\text{-}CH_2\text{-}CH_2\text{-}NH_2$$

Fig. 3.   The proposed molecular structure of γ-aminopropylsilantriols
           in acid, neutral, and alkaline aqueous solutions.

More work must be done in order to understand the detailed structural
composition of silane in solutions as well as the solids. The properties of silane solution and the nature of the silane molecule could
affect the structure and the reactivity of the silane deposited on the
surfaces.

THE NATURE OF SILANES ON GLASS SURFACES

The silane molecules are usually deposited on glass surfaces as a
monolayer film and the amount deposited is dependent on the concentration of the solution (Figure 4). Schrader reported that the coupling
agent when deposited on the glass surfaces usually from heterogeneous
layers consisting of physisorbed and chemisorbed fractions [33]. The
outer fraction, about 95% of the total adsorbed silane, can be extracted by water at room temperature. The second or inner fraction can be
extracted by boiling water, but the third or surface fraction is firm-
ly bound to the substrate and survives extraction in boiling water for
up to 100 min. These results have also been confirmed by Fourier
transform infrared spectroscopy [34]. It was found that γ-MAPS forms
essentially three different structures in the interphase, analogous to
APS interphase. The outermost layers consist of small oligomers which
are simply physisorbed so that they can be washed away by organic solvents. Nearer the glass surface, there is a second region which consists of oligomers similar to the outer layers except for a few bonds
connecting the oligomers. In the region near the glass surface, the
interconnecting crosslinks become extensive and a regular three-dimensional network is found. Thus, a gradient in the structure of the
coupling agent interphase exists and the type of gradient determines
the hydrolytic stability of the interphase.

THE ROLE OF THE EPOXY RESIN IN COMPOSITES

The versatility of epoxy resins has resulted in their use in many
industrial, commercial, military, and consumer applications. Reinforced plastics, produced via laminating and molding methods, have found
applications in the chemical, electrical, and electronics industries.
The various forms and types of epoxy resin, in their thermoplastic or
uncured state, are converted or hardened into useful thermosets by
reaction with a variety of hardeners [35]. Depending upon the resin and
hardener comprising the system, the amount of hardener used can vary
from as low as 1 part hardner per 100 parts of resin to greater than
100 phr. Epoxy resins are hardened into thermoset compounds by any of
three general reactions: 1. self-polymerization forming direct linkages between the epoxy groups, 2. linkage of epoxy groups with aromatic
or aliphatic hydroxyls, and 3. crosslinking with the hardener through
various radicals [36].

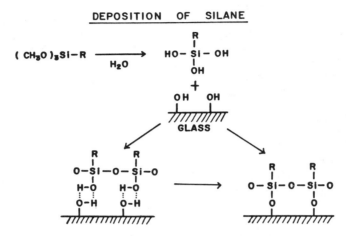

Fig. 4.   The silane coupling agents hydrolyzed in aqueous solution
           and adsorbed on the silica surfaces.

     Resistance to chemical attack is determined by exposure of the
cured systems to the materials of interest, e.g., acid, alkaline, solvents, etc. A weight change as well as the alternation of mechanical
and electrical properties serve to characterize the performance of
epoxy systems in corrosive environments [36]. It is found that long term
exposure of the epoxy matrix to water causes the hydrolysis and leaching of unreacted anhydride molecules. Hydrolysis of the ester linkages in the unstressed resin is significant only in a highly alkaline
medium. However, the application of high tensile stress to the matrix
dramatically accelerates hydrolytic attack in films exposed to neutral
PH or alkaline media. The mechanochemical degradation is modelled by
an exponential dependence of the hydrolysis rate on the applied stress
[37].

THE NATURE OF THE SILANE-RESIN INTERFACE

Although silane coupling agents were first introduced to improve
the water resistance of reinforced plastics, it was soon observed that
they also imparted significant improvement to the initial properties
of laminates. The degree of improvement obtained even under optimum
conditions varied with the resin, the glass content, and the severity
of the test. There is no correlation between the polarity of the silane, or wettability of silane-treated glass, and the effectiveness of
the silane coupling agent [38]. Better wetting of silane-treated glass
will allow more complete displacement of air from the glass surface
and reduce the number of voids in the composite.

The coupling agent-matrix interface is more stable to attack of
water. Thus, it may be important to maximize the extent of the interfacial bonding at the interface. Such bonding is necessary to obtain
high performance of FRP [39]. Bjorksten and Yaeger proposed a chemical
reaction between the vinylsilane and the polyester resin through the
vinyl group [11]. The vinyl group of 3-methacryloxpropyltrimethoxysil-
ane copolymerizes with styrene or acrylate [40]. However, the evidence
was not strong that the vinyl group of silane on the surfaces copolym-
erizated to the vinyl group of the polyester resin. The coupling
agent interphase can homopolymerize through the organic groups or co-
polymerize with the matrix resin [41].

Direct evidence for chemical bonding at the interface between an
anhydride-cured epoxy resin and the aminosilane treated fiber has been
obtained by using Fourier transform infrared spectroscopy [42]. It was
found that the nadic methyl anhydride can react with γ-aminopropyl-
triethoxysilane (APS) and N-methylaminopropyltrimethoxysilane (MAPS).
In comparing the relative reactivities of these two coupling agents
with the epoxy resin, the secondary aminosilane has a higher reactivity than the primary aminosilane. The primary aminosilane forms a
cyclic imide with the nadic methyl anhydride at the interface of the
composite [43]. The formation of this imide group inhibits the further
copolymerization of the matrix and the silane layers, so interfacial
bonding does not occur. The molecular structure of the interface in
MAPS treated fiberglass reinforced composites is different from that
of the APS-treated fiber composites. The silane-resin interface con-
stitutes copolymers of the epoxy resin and the aminosilane coupling
agents. The penetrating ability of the uncured resin and the nature
of the siloxane interphase on the fiber surfaces determines the structure of the resin-silane interphase [44]. In addition, the silane-
induced esterification incresaes the curing density of the epoxy matrix about 5-10% relative to the bulk resin. Thus, there is a gradient
of matrix composition and structure from the fiber surface to the bulk
resin. This interface region could protect against the moisture
attack at the interface in composites.

THE ROLE OF INTERFACIAL BONDING IN HYDROTHERMAL STABILITY

Exposure of a composite to cyclic temperature and humidity variations induces cyclic stress within a composite [22]. The degradation is
accompanied by debonding of the resin from the fiber reinforcement due
to the difference in the thermal expansion coefficients between the
matrix and the fiber. Silane coupling agents are applied to the fiber
reinforced composites to stabilize the interface by chemical bonding
of the matrix to the glass. In order to account for the enhanced
effect of the silane coupling agent at the interface of the composites
several methods have been developed to measure fiber-matrix interfacial strength [45-50]. All methods typically suffer from a lack of
reproducibility and produce data with coefficients of variation greater than 10 percent. One of these methods is to measure the force
required to pullout a fiber embedded in a matrix. This method, in
addition to its relative simplicity of sample preparation and measurement, is expected to give realistic information when one considers the
pullout of the fibers from the fracture surfaces of composites. A
general relationship between the fiber strength and the interfacial
strength has been established theoretically as a function of the embedded length of fiber by Greszczuk [51]. Lawrence has developed a
theory on the effect of partial debonding on the maximum debonding
stress, including the effect of friction [52]. However, most of these
experiments have dealt with steel filaments (not a real fiberglass) and
epoxy, polyethylene, and polypropylene resin [53].

In order to measure the interfacial strength of the fiber reinforced composites, an improved pull-out test has been developed in our
laboratory. The joint shear strengths between 25 mil diameter fiberglass filaments and epoxy resins are tested in a transparent media to
display the fracture. It was found that the adhesion is dependent on
the curing conditions of epoxy resins. Three major types of failures
of the specimens have been found when the fibers pulled-out of the
cured epoxy resins. These failures include fiber failure, matrix
failure, and the interface failure. For the pre-cured sample, the
matrix deformation can be observed and the result shows only the elastic behavior of the incompletely cured eopxy matrix. For post-cured
samples, the fiber broke inside the matrix because of the large residual pressure in the matrix. Thus, only the pulled-out fiber sample
can provide the value of the maximum shear stress at the interface in
composites. These tests make it clear that there are many variables
influencing the rate of bond degradation and these include the nature
of silane, amount of silane, temperature of water, PH of the water,
resin formulation, surface constitution, and processing condition of
composites.

It is certain that the water migration is sensitive to the resin structure [54]. The water molecules can reach and weaken the resin-fiber interface by diffusion through the resin or along the interface itself. Because the resin-silane interfacial bonds are more stable to the attack of water, the major degradation should appear at the coupling agent interphase and the glass-silane interface. Therefore, the reaction of the siloxanes with the water controls the interfacial strength of the composites. As long as the interface is rigid or has flow, bond forming and scission in the presence of water is reversible [1]. An increase in the amount of water in the interface region causes a shift of the equilibrium to be less favorable for bond formation and decreases the interfacial strength of the interface. The acid and alkaline chemicals generated from the surface would influence the rate of hydrolysis and the state of equilibrium and so would result in poor mechanical performance of composites.

The long term mechanical performance of glass reinforced composites depends on their stability in various environmental atmospheres. Glass reinforced plastics adsorb moisture after extended periods of exposure to high humidity and this moisture often degrades their mechanical and structural integrity [55]. There is a strong indication that degradation mainly occurs by a weakening of the fiber-resin interface, with a subsequent loss in the shear strength of the composites.

CONCLUSIONS

The great improvement in properties imparted to the fiberglass reinforced composite by traces of appropriate reactive silanes at the interface suggests that an understanding of the molecular structure of the interface might be the key to understanding the mechanical behavior of composites. The effect of water at the interface of composites is quite complex, depending upon the nature of the polymer system and that of the fibers. Thermosetting resins like epoxy absorb water with swelling and reduction in modulus. Molecular water diffuses readily through the resin phase and attacks the interface of composites. Liquid water can leach soluble materials from the interface and the matrix phase. Silane coupling agents may partially prevent water from attacking the fiber-resin interface. So the strength retention of the silane-treated fiber composite is better than that of non-silane treated fiber composites.

REFERENCES

1. E.P. Pluddemann, Interface in Polymer Matrix Composites, Acedemic
    Press, Vol. 6, 1974, Chapters 1 and 6.
2. D.L. Harroad and R.T. Bagley, Sci. of Adv. Materl. and Proc. Eng.,
    Vol. 10, E1 (1966).
3. R.T. Schwartzand, H.S. Schwartz, Fundamental Aspects of Fiber
    Reinforced Plastics Composites, Interscience Publishers, 1968,
    p. 163.
4. D.T. Clark and W.J. Feust, Polymer Surfaces, John Wiley & Sons,
    1977, p. 47.
5. R.L. McCullough, Concepts of Fiber-Reinforced Composites, Marcel
    Dekker, 1971, p. 62.
6. P.E. Cassidy, J.M. Johnson, and G.C. Rolls, Ind. Eng. Chem. Proc.
    Res. Develop., 11, 170 (1972).
7. S. Oswitch and R.F. Golownia, Reinforced Plastics, 252 (1970).
8. T.B. Husbands, C.F. Derrington, and L. Pepper, Gorv. Rept., AD-
    730744 (1971).
9. W.A. Jemian, R.C. Wilcox, and A.C.T. Hau, Grov. Rept., AD-A011289/
    6st (1974).
10. C.E. Browning, Proc. 28th Ann. Tech. Conf., Reinforced Plastics
    Div., SPI, 15-A (1973).
11. J. Bjorksten and L.L. Yaeger, Mod. Plast., 29, 124 (1952).
12. J.G. Vail, Soluble Silicates, Vol. 1, Reinhold, 171 (1952).
13. W.A. Zisman, Proc. 19th Ann. Tech. Conf., Reinforced Plastics
    Div., SPI, 21-B (1964).
14. C.A. Kumins and J. Roteman, J. Polym. Sci., 1, 527 (1963).
15. R.C. Hooper, Proc. 11th Ann. Tech. Conf., Reinforced Plastics
    Div., SPI, 8-A (1956).
16. P.W. Erickson, Proc. 27th Ann. Tech. Conf., Reinforced Plastics
    Div., SPI, 13-A (1970).
17. E.P. Plueddemann, J. Adhesion 2, 184 (1970).
18. K.J. Brookfield, Reinforced Plastics, 136 (1972).
19. J.L. Parham, Permeability of Epoxy Systems, Grov. Rept., AD-
    726930 (1971).
20. A.S. Kenyon and L.E. Nielson, J. Macromol. Sci., A3, 275 (1969).
21. K.E. Hoffer, M. Stander, and L.C. Bennett, Proc. 32nd Ann. Tech.
    Con., Reinforced Plastics Div., SPI, 11-I (1977).
22. C.E. Browning, "The Mechanisms of Elevated Temperature Property
    Losses in High Performance Structural Epoxy Resin", Ph.D. Thesis,
    University of Dayton (1976).
23. E.P. Plueddemann, Filler and Reinforcements of Plastics, 1, 86
    (1978).
24. H.W. Yip and J.B. Shortall, J. Adhesion, 8, 155 (1976).
25. J.A. Manson and H. Sperling, "Polymer Blends and Composites",
    N.Y., 1978, p. 62.
26. P.J.W. Debye, Adhesion and Cohesion, Elsevier Publ., 1962.
27. D.D. Eley, Adhesion, Oxford Univ., 1961,
28. M.P. Wagner, Rubber Chem. and Tech., 49, 703 (1977).

29. J.L. Koenig and P.T.K. Shih, Materials Sci. and Eng., 20, 127 (1975).
30. E.P. Plueddemann, Additives for Plastics, 1, 123 (1978).
31. C-h. Chiang, H. Ishida, and J.L. Koenig, J. Colloid & Interface Sci., 74, 396 (1980).
32. C-h. Chiang, Nan-I Liu and J.L. Koenig, J. Colloid & Interface Sci., (accepted).
33. M.E. Schrader, I. Lerner, J.L. D'Oria, and L. Deutsch, Proc. 22nd Ann. Tech. Conf., Reinforced Plastics Div., SPI, 13-A (1961).
34. H. Ishida and J.L. Koenig, J. Polym. Sci., Physics, 18, 193 (1980).
35. H. Lee and K. Neville Handbook of Epoxy Resins, McGraw Hill, New York, 1967.
36. M.K. Antoon and J.L. Koenig, J. Polym. Sci., Chem. Ed., 18, 480 (1980).
37. M.K. Antoon, "FT-IR Investigation of the Structure and Moisture Stability of the Epoxy Matrix in Glass-Reinforced Composites", Ph.D. Thesis, Case Western Reserve University, 1980.
38. W.D. Bascom and J.B. Romans, Ind. Eng. Chem. Prod. R & D, 7, 172 (1968).
39. E.P. Plueddemann and G.L. Stark, Modern Plastics, August, 1976 (1977).
40. S. Sterman and J.G. Marsden, Proc. 18th Ann. Tech. Conf., Reinforced Plastics Div., SPI, 1-D (1963).
41. H. Ishida and J.L. Koenig, J. Polym. Sci., Physics, 17 615 (1979).
42. C-h. Chiang and J.L. Koenig, Polymer Composites, 2, 88 (1980).
43. C-h. ChiangProc. 36th Ann Tech. Conf., SPI, Reinforced Plastics Div., 2-D (1981).
44. C-h. Chiang and J.L. Koenig, "Spectroscopic Characterization of the Matrix-Silane Coupling Agents Interface in Fiber Reinforced Composites", (Unpublished Paper).
45. K. Kendall, J. Materials Sci., 10, 1011 (1975).
46. H.A. Clark and E.P. Plueddemann, Modern Plastics, 133 (1963).
47. J.F. Mandall, J.H. Chen, and F.J. McGarry, International J. of Adhesion and Adhesives, 40 (1980).
48. A. Takaku and R.G.C. Arridge, Appl Phys., 6, 2038 (1973).
49. L.J. Broutman, Interfaces in Composites, ASTM, 27 (1968).
50. W.A. Fraser, F.H. Achker, and A.T. DiBenedetto, Proc. 30th SPI Annual Tech. Conf., Reinforced Plastics Div., SPI, 22-A (1975).
51. L.B. Greszczuk, Interfaces in Composites, ASTM, 49 (1969).
52. L. Lawrence, J. Materials Sci., 7, 1 (1972).
53. R.R. Mayers and J.S. Long, Treatise on Coatings, Marcel Dekker, Vol. 1, Part 3 (1961).
54. B.S. Mekta, A.T. DiBenedetto, and J.L. Kardos, Proc. 31st Ann. Tech. Conf., Reinforced Plastics. Div., SPI., 21-A (1976).
55. D.A. Scola, Proc. 31st Ann. Tech. Conf., Reinforced Plastics Div., SPI, 14-A (1976).

INTERNAL STRESSES IN FIBRE REINFORCED PLASTICS

K.H.G. Ashbee, J.P. Sargent and E. Walter

University of Bristol
H H Wills Physics Laboratory
Tyndall Avenue
Bristol BS8 1TL
England

SUMMARY

Three experiments are reported. The first two investigate two stages in the development of self-stress during fabrication of a composite, and the third investigates the generation of interfacial pockets of osmotic pressure associated with the uptake of water. The techniques developed for these experiments include a combination of optical interferometry and thin plate elasticity theory to study cure stresses, the application of oblique incidence polarising optical microscopy to uniquely determine the magnitudes of radial and tangential stresses present in the resin after cooling from the cure temperature, and an analysis of the equilibrium of pressure-filled cavities in order to investigate the phenomenon of loss of load transfer attributable to osmosis during weathering.

CURE SHRINKAGE

Both cross-linking and the evolution of volatiles (arising from condensation reactions in polyamides and from unreacted epichlorhydrin in epoxies) produce resin dimensional changes and, if each phenomenon occurs homogeneously, it might be expected that neither will give rise to internal stress. However, the formation of chemical bonds between resin and fibres inhibits relative displacement parallel to resin/fibre interfaces. This includes displacement due to resin cure shrinkage, the consequent inhomogeneous distribution of which will give rise to internal

517

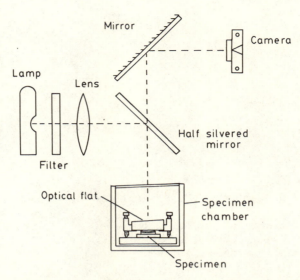

Fig. 1.   Schematic   representation of the apparatus showing   the
optical path.

stresses. The problem is compounded if the degree of cross-linking
is affected by differences in the rate of exothermic heat   transfer
since fibres will usually have thermal diffusivities different from
those of matrix resins. It is no doubt further compounded by any
retention of volatiles in resin remote from the free surface.

     To investigate the nature of the resultant dimensional changes
during curing of an epoxy/glass system, the following <u>in situ</u>
experiment was devised. 19mm diameter soda-lime glass cover slips
of three different thicknesses, namely 120μm, 150μm and 230μm, were
used to sandwich a 17mm x 200μm thick layer of Redux 312/5 to 1mm
thick soda-lime glass microscope slides. Both the cover slips and
the microscope slides had been thoroughly cleaned by   ion
bombardment. Each sandwich was mounted so that the free surface of
the cover slip is in close proximity to an optical flat. Figure 1
is a schematic of the optical set-up. The air gap between cover
slip and optical flat is small enough to permit optical
interference and changes in its size and shape, due to deformation
of the cover slip arising from restricted radial contraction of
resin adjacent to either glass surface, will introduce changes in
the pattern of interference fringes. Figure 2 shows a sequence of
photographs recorded during a 120C cure of a sandwich manufactured

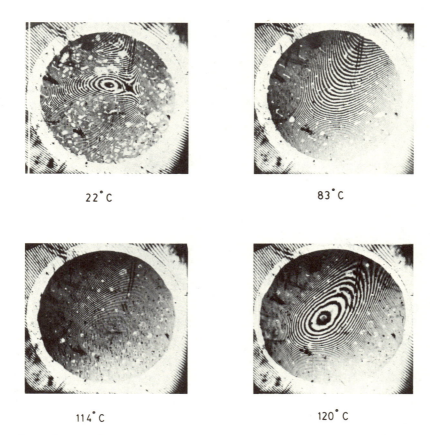

22°C

83°C

114°C

120°C

Fig. 2.   Part of a sequence of interference patterns photographed
           during   the 30 minute warm-up period for a glass
           slide/adhesive film/150 μm thick cover slip sandwich.

with   a 150μm thick cover slip. The time taken for the specimen to
reach   the cure temperature was 30 minutes and most of the shape
change occurred during this heating-up period.   The subsequent much
smaller changes have been further examined in Figure 3 by creating
Moiré patterns between photographs of the pattern of interference
fringes,   each recorded after progressively longer times at   120C,
and the pattern as it was when the specimen had completed its   1/2
hour cure. The Moiré fringes are the circumferential fringes, each
of which is the locus of identical displacement normal to   the
sandwich and adjacent members of which differ in this displacement
by half a wavelength.

      Figure 4 shows the displacement field across a diameter of
this same specimen that has been determined from the Moiré pattern

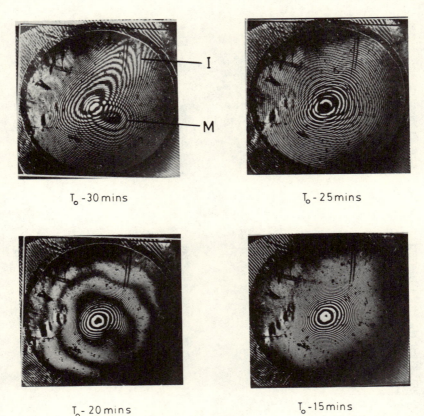

T_o -30 mins T_o - 25 mins

T_o - 20 mins T_o - 15 mins

Fig. 3. Moiré fringes (M) generated by superimposition of
interference patterns (I) in order to reveal   the
deformation   of the cover slip after reaching the   cure
temperature (120C).

after 5 minutes at the cure temperature.   The sign of the
displacements was established by applying an external   axial
compressive stress to the specimen. As expected,   radial shrinkage
in resin adjacent to the cover slip has been significantly less
than radial shrinkage elsewhere in the resin,   as a consequence of
which the cover slip has been deformed into a parabaloid.

To a first approximation, the deformation of the cover   slip
can be regarded as identical to that of a circular membrane,
rigidly supported at its edge and subjected to a pressure drop (p)
across its surfaces. For such a membrane, Love (1959) has shown
that the fourth differential of the axial displacement (w) is a
measure of p.

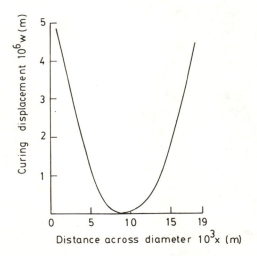

Fig. 4.   Normal displacement across one diameter of the cover slip
           after 5 minutes at the cure temperature i.e.(T_o-25mins).

$$D \nabla^4 w = -p$$

where the flexural rigidity $D = 2Eh^3/3(1-\nu^2)$. $2h$ is the membrane
thickness   and E and ν respectively   are   Young's   modulus   and
Poisson's ratio for the membrane material.   The displacement field
shown in Figure 4 is described by the equation $w = 0.073(x -$
$0.0095)^2$, the fourth differential of which is zero. It is concluded
therefore that there are no stresses acting normal to the sandwich.
The deformation of the cover slip is due entirely to in-plane shear
stresses caused by restricted radial shrinkage of the interfacial
resin.

CURE SHRINKAGE PLUS DIFFERENTIAL THERMAL CONTRACTION

     Measurements of the transverse components of the residual
stress   system in a model uniaxial glass fibre/epoxy resin
composite,   that has undergone curing followed by cooling from the
cure temperature to room temperature, have been reported by
Cunningham, Sargent and Ashbee (1981).   Their specimen was cast
inside a thick-walled glass tube so that, when sectioned into
transverse slices for examination in a polarising microscope,
relief of stresses by way of radial displacements would be
minimised.

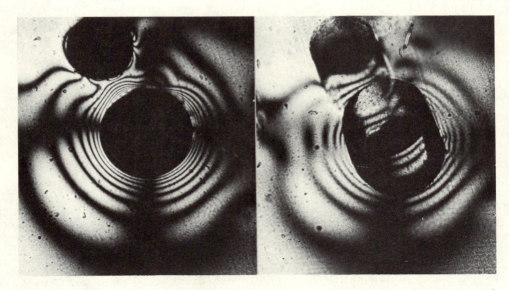

Fig. 5.   A   transverse   section   from   a   model   unidirectional
composite   photographed   in   the   polarising   microscope
using   both   normal   and   oblique   incidence.   The   diameter
of the central fibre is ~0.5mm.

In   order   to   separate the tangential   and   radial   principal
stresses, optical   retardation measurements were made   at   various
locations   in the resin using normal and oblique incidence so   that
the   photoelasticity   analysis   due   to   Drucker   (1943)   could   be
employed. A   pair   of   normal   and   oblique   incidence   images   is
reproduced in Figure 5.   It was found that, for a square array   of
parallel   fibres in a 30% fibre volume fraction uniaxial composite,
the   largest principal stress in the resin is $\sigma_{\theta\theta}$ the magnitude of
which is ~100MPa tensile near the resin/fibre interface falling   to
~20MPa   tensile   at a radial distance equal to one fibre   diameter.
The   radial principal stress in the resin is everywhere   very   much
smaller at <10MPa compressive. This latter   observation   is   at
variance   with   finite element calculations reported by   Crane   and
Adams   (1981)   for   a square array of parallel glass fibres   in   an
epoxy composite of similar fibre volume fraction.   The   calculations
predict   a   radial   stress   at the resin/fibre   interface   which   is
~30MPa compressive at the line joining adjacent fibre centres (θ =
0) changing sign at $\theta = \pi/13$ and reaching a maximum tensile stress
of ~10MPa at $\theta = \pi/8$.

OSMOSIS

"In   order   clearly   to realize the quantity   referred   to   as

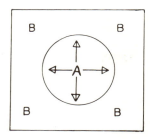

Fig. 6.   After Van't Hoff (1888), see text.

osmotic pressure, imagine a vessel, A (Figure 6), completely full
of an aqueous solution of sugar, placed in water, B. If it be
conceived that the solid walls of this vessel are permeable to
water, but impermeable to the dissolved sugar, then, owing to the
attraction of the solution for water, water will enter the vessel A
up to a certain limit, thereby increasing the pressure on the walls
of the vessel. Equilibrium then ensues, owing to the pressure
resisting further entry of water. This pressure we have termed
osmotic pressure." - J. van't Hoff, (1888).

     In FRP, vessel A corresponds to interfacial pockets of water
(or other solvent) that has diffused in from the outside
environment and has dissolved water solubles located at the
interface, vessel B corresponds to the in-service environment
(humid) in which the FRP component resides, and the solid walls of
vessel A that are permeable to solvent but not to solute correspond
to the matrix resin. Photoelastic evidence for the occurrence of
interfacial pressure pockets due to osmosis was first reported by
Ashbee and Wyatt (1969). Examples in a carbon fibre/epoxy resin

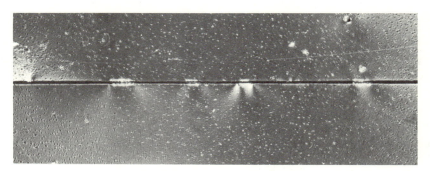

Fig. 7.   Photoelastic contrast adjacent to pressure pockets on the
            surface of an AS graphite fibre in a MY 750 epoxy matrix
            composite after 600 hours exposure of the composite to
            distilled water at 80C.

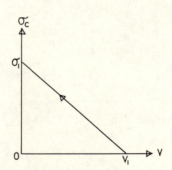

Fig. 8.   External stress as a function of volume
          for an interfacial cavity.

composite are shown in Figure 7.

When formed, each interfacial pocket closely resembles a penny-shaped crack that has grown in a solid tensioned between fixed grips. The overall change in energy is from U_O, corresponding to the state of strain energy prior to fracture, to $U_O - U_G$, where the Griffith energy U_G given by the Zener approximation is

$$U_G = \frac{1}{2} \frac{\sigma_1^2 \cdot 4\pi}{E \quad 3} ab^2 = \frac{2\pi}{3} \sigma_1^2 \frac{ab^2}{E}$$

$\sigma = \sigma_1$ is the tensile stress applied externally and perpendicular to the crack, E is Young's modulus, a and b respectively are the half thickness and radius of the crack. Referring to Figure 8, if the interfacial cavity could be closed by application of traction on its wall, the stress σ_c acting across the cavity would rise from 0 to σ_1 and, correspondingly, its volume would decrease from V_1 to 0.

The Griffith energy gained by this process would be

$$U_G = \frac{1}{2} \sigma_1 V_1 = \frac{2\pi}{3} \sigma_1^2 \frac{ab^2}{E}$$

Hence

$$V_1 = 4\pi \cdot \sigma_1 \frac{ab^2}{3 \quad E}$$

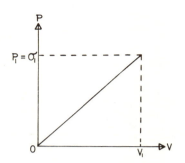

Fig. 9.   Volume   as   a   function   of   internal   pressure   for   an
           interfacial cavity.

In reality, creation of the interfacial crack occurs with zero
external stress. Its presence is due, instead, to creation of an
internal pressure and an internal pressure $p = \sigma_1$ gives the same
V_1, see Figure 9.

$$V_1 = \frac{4\pi \cdot p \cdot ab^2}{3 \quad E}$$

and the elastic stored energy is

$$E_{el} = \frac{1}{2} p_1 V_1 = \frac{2\pi}{3} \frac{p_1^2}{E} ab^2$$

Therefore $$p_1 = \frac{3EV_1}{4\pi \, ab^2}$$

and $$E_{el} = \frac{3E \, V_1^2}{8\pi \, ab^2}$$

A solute exerts an osmotic pressure equal to the pressure that
would be exerted by a gas having the same number of molecules in a
volume equal to that occupied by the solution. Consider a change in
volume of the interfacial cavity from V_o to V_1 resulting from
dissolution of n moles of solute.

$$pV = nRT$$

The strain energy release $= \int p dV$

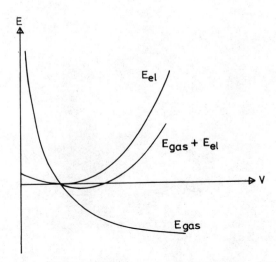

Fig. 10. Energy of a pressure-filled cavity as a function of its volume.

$$= \int \frac{nRT}{V} \, dV$$

$$= nRT \, \log_e \frac{(V_o + V_1)}{V_o}$$

$$= nRT \, \log_e \frac{p_1}{p_o}$$

p_o and p, are the values of the osmotic pressure corresponding to volumes V_o and V_1 respectively.

The overall energy $E = E_{gas} + E_{el}$

$$= -nRT \, \log_e \frac{(V_o + V_1)}{V_o} + \frac{3 \, E \, V_1^2}{8\pi \, ab^2}$$

This function is sketched in Figure 10.

The value of V_1 corresponding to minimum energy release may be found by differentiation

$$\frac{dE}{dV_1} = -nRT \, \frac{1}{V_o + V_1} + \frac{3E}{4\pi} \frac{V_1}{ab^2}$$

$$= 0 \text{ when } V_1^2 + V_1 V_o - \frac{4\pi}{3E} ab^2 nRT = 0$$

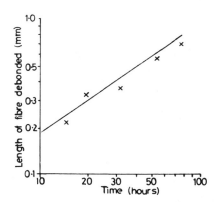

Fig. 11. Axial propagation of an individual region of interfacial debonding in an AS graphite fibre/MY 750 epoxy composite during immersion in distilled water at 80C. This particular region was completely debonded around the fibre circumference.

i.e. when $V_1 = -\dfrac{V_o}{2} \pm \sqrt{\dfrac{V_o^2}{4} + \dfrac{4ab^2nRT}{3E}}$

We must take the +ve root, hence

$$V_1 = -\frac{V_o}{2} + \sqrt{\frac{4ab^2nRT}{3\pi}}\left(1 + \frac{3EV_o^2}{32ab^2nRT}\right)$$

$$\sim -\frac{V_o}{2} + \sqrt{\frac{4ab^2nRT}{3\pi}} \qquad \text{for large values of b}$$

Interfacial pressure pockets of the kind seen in Figure 7 grow slowly and presumably maintain their volumes at minimum energy values predicted by the above equation, i.e. at volumes determined by the square root of the number of moles of dissolved solute. In cases where the solute is leached from the fibre, its availability is presumably governed by the laws of diffusion which, for the simplest of models, predicts a dependence on the square root of time. This in turn suggests that the volume of interfacial pressure-filled cavities should increase as the fourth root of time. If it is assumed that the change in cavity thickness during cavity growth is negligible, then changes in cavity volume are approximated by the more easily measured changes in interfacial debonded area. Figure 11 shows a few measurements for a single debonded area in a carbon fibre/epoxy composite immersed in distilled water at 80C. As anticipated, the slope of the $\log_e(V)$ versus $\log_e(t)$ plot is close to 1/4.

ACKNOWLEDGEMENTS

This work was supported by the US Army, grant no. DA-ERO-78-G-117, monitored by Dr. N. Schneider. The authors are also grateful to Professor Sir Charles Frank for stimulating discussions.

REFERENCES

Ashbee K H G and Wyatt R C W, 1969, "Water damage in glass fibre/resin composites", Proc. Roy. Soc. **A**312:553.

Crane D A and Adams D F, 1981, "Finite element micromechanical analysis of a unidirectional composite including longitudinal shear loading", U S Army Materials and Mechanics Research Centre TR 81-7, Final Report, Contract No. DAAG46-79-C-0076.

Cunningham B, Sargent J P and Ashbee K H G, 1981, "Measurement of the stress field created within the resin between fibres in a composite material during cooling from the cure temperature", J. Mat. Sci., 16:620.

Drucker D C, 1943, "Photoelastic separation of principal stresses by oblique incidence", J Appl Mech 10:A156.

Love A E H, 1959, "Treatise on the Mathematical Theory of Elasticity", (Cambridge University Press), Chap. XXII, p. 455.

Van't Hoff J, 1888, "The Function of Osmotic Pressure in the Analogy between Solutions and Gases", Phil. Mag. 26:81.

STRESS AND STRENGTH ANALYSIS IN AND AROUND COMPOSITE

INCLUSIONS IN POLYMER MATRICES

A. Pavan

Politecnico di Milano

Piazza L. da Vinci, 32 – 20133 Milano, Italy

INTRODUCTION AND SCOPE

Even confining our consideration to particulate inclusions and to rigid polymer matrixes, the cases of composite materials of practical interest that contain particles that are themselves composites are numerous. They range from the all-polymer cases of toughened plastics, including (i) conventional types, such as high-impact polystyrenes and ABS resins, which contain soft inclusions of rubber which in turn occlude a portion of the hard matrix material (i.e. many hard sub-inclusions) (1), as well as (ii) the recently-developed block-copolymer systems, which form particles having a capsule or core-shell structure (i.e. one single hard sub-inclusion) (2,3), to the cases (iii) of elastomer-coated mineral or glass fillers (4) and (iv) hollow glass beads. Somewhat related are also the various instances of interlayers or "inter phases" that form between matrix and filler as a result of strong interaction. Here, more typically, the filler is inorganic (v), but instances of particulate polymer-polymer composites can also be included, such as may result from (vi) the blending of semi compatible resins (5), or (vii) the casting "in situ" of a resin filled with preformed rigid polymer particles, as used for orthopaedic or dental restoration (6), or (viii) the grafting of rubber particles to promote interfacial adhesion and dispersion in toughened plastics (1).

All these materials are ternary composites comprised of a continuous matrix phase and dispersed two-phase inclusions; the inclusions can be described schematically as having spherical shape and – with the exception of case (i) – spherical structure: a central globule encapsulated within a spherical shell.

Theoretical analysis of three-dimensional composites

becomes increasing complex, of course, with an increasing number
of constituent phases and structural factors requiring specifica
tion. Only a limited number of studies have attempted to clarify
the effect of particle structure and properties on the properties
of the overall composite. Most studies deal with the effective or
macroscopically-equivalent moduli of the composite, a few cover
stress analysis around the inclusions (e.g. 7-13), and very few
treat strength analysis (e.g. 8, 10-12).

        This paper examines an elementary model for composites
with composite inclusions, and describes a micromechanical analysis
of stress and strength performed on it. The model consists of a
single spherical inclusion embedded in an infinite matrix and
itself embedding a single spherical sub-inclusion. Consideration
of an isolated inclusion is explained by interest in highly-loca
lized phenomena, such as the onset of craze-yielding. On the other
hand, the choice of a single sub-inclusion structure providing
spherical symmetry is mainly motivated by the possibility of
performing a rigorous micromechanical analysis of this idealized
geometry/[1].

        The numerical results presented here are for the
specific case in which the particle core and matrix are of the
same (or mechanically-equivalent) glassy polymer, and the particle
shell is of rubber. Strictly, then, the model adheres to the real
cases of toughened plastics having particles of the capsule- or
core-shell-type normally produced from block copolymers (ii).
However, in view of the anyhow large difference in stiffness between
the two component phases of the particle, the results very
probably hold good, approximately, for polymers filled with
elastomer-coated rigid particles of any kind (iii). Moreover, in
the author's view, this model can be usefully applied to toughened
plastics with a particle structure of the cellular type (i), even
if the multiplicity of their sub-inclusions is not reflected in the
model. Reasons for this will be discussed in the next sections.

THE MODEL

        The configuration of the model examined here and the
pertinent reference system are depicted in Fig. 1.
        The constituent phases are assumed to have the ideal
properties of homogeneity, isotropy, and linear elasticity. On
these assumptions, the only relevant material constants are three,
say the ratio between the two shear moduli G_1 and G_2, and
the Poisson's ratios ν_1 and ν_2 of the two component materials,
1 and 2. Since the analysis was carried out numerically, we had
to assign specific values to these constants. The following values

---

Note/[1]. A layered sphere model was also examined by Matonis and
        Small (8), Wang and Schonhorn (9), Broutman and Agarwal
        (13) with different constituent properties and aims.

were taken as representative of a glassy plastic (index 1) and rubber (index 2): $G_1/G_2 = 10^3$, $\nu_1 = 0.35$, $\nu_2 = 0.49982/^2$

Furthermore, we postulate perfect adhesion at the interfaces. When adhesion is provided, in practice, via chemical grafting, this may result in the formation of transitional zones between different phases. The presence of these interphases could be readily accomodated in a more refined version of this model by introducing additional spherical layers with appropriate properties. No account of any such transitional zones between different materials is taken in the present study, however.

This model can be worked out exactly by methods based on the classical theory of elasticity. In the course of this treatment each constituent phase maintains its individual characteristics: the topological features that may be envisaged as relevant in the real composite are thus preserved.

An alternative approach was followed by Bucknall (7, 16) and by others (12, 17). Being specifically concerned with

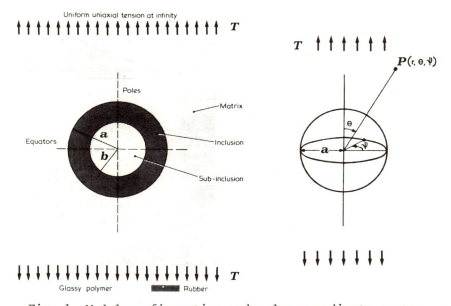

Fig. 1. Model configuration and polar coordinate system.

Note/^{2}. The Poisson's ratio of rubber, ν_2, being close to the limiting value of 0.5, is more critical and needs to be realistically specified, which does not seem to be recognized in some papers (e.g. 8, 10). The value taken here is of best credit (7, 14, 15).

conventional rubber-modified plastics, whose rubbery inclusions
are actually dispersions of numerous glassy sub-inclusions in a
continuous matrix of rubber, Bucknall proposed treating the
particle as a quasi-homogeneous body according to one of the
current theories for macroscopically isotropic, quasi-homogeneous
composites. The composite inclusion is thus replaced with a simple
homogeneous one of a "mechanically equivalent" material. According
to this approach, the only effect of the presence of hard sub-
-inclusions in the rubber particles is to modify the effective
elastic constants of the (composite) rubber, i.e. mainly to raise
its shear modulus /[3].

From these premises, it is predicted that stress
concentrations are lower for composite rubber particles: the
latters' efficiency as stress-concentrating agents is thus
expected to be reduced, and hence also the capacity of the rubber-
-modified plastic to deform and absorb energy through multiple
craze formation (16). There has as yet been no direct experimental
confirmation of these predictions.

In fact, the concept of a mechanically-equivalent
homogeneous particle is not wholly satisfactory: while it has been
proven to be adequate for predicting average macroscopic properties
of the overall composite (as in ref.s 16 and 17), it appears
questionable when highly-localized phenomena are investigated (as
in ref. 12). Suffice it to note that the effective or equivalent
elastic constants of a composite are largely determined by
composition (phase moduli and volume fractions) according to the
prevailing theories: several of the mathematical theories proposed
for isotropic particulate composites do not even specify the
topology of the constituent phases, e.g. do not distinguish
between matrix and dispersed phase. The fact that local effects
governing the ultimate properties are independent of structure,
is not, however, convincing.

Different predictions are obtained with the core-shell
model. A sample is presented in Fig. 2, which compares some local
stresses calculated according to the two different approaches. As
indicated in the insets, full lines are results obtained with the
core-shell model examined in this paper as described in the next
section; broken lines are calculations based on Bucknall's
"homogenized" particle model. Accordingly the former were

Note/[3]. The presence of the sub-inclusions also increases the
        total volume of the (composite) rubber particles. This
        effect has a major influence upon the average macroscopic
        properties of the overall composite (16), but does not
        affect the stress concentrations locally developed by
        an isolated particle: in relative terms, these are
        independent of particle size (see later). Since we are
        concerned with a single particle model and local
        phenomena, we ignore this effect.

performed as follows: for any given volume fraction of sub-
-inclusions, the effective elastic constants of the composite
rubber were first calculated and then introduced into Goodier's
equations (18) for the local stresses around a simple inclusion.
Calculation of the effective composite rubber moduli poses a
problem: either exact bounding expressions are used, or
approximate single-valued equations have to be chosen, to avoid
the complication of discussing bounds. This choice introduces
some arbitrariness, of course. For the calculations given in
Fig. 2, we applied Kerner's equations for the shear and bulk
moduli (19). It is to be noted that Kerner's equation corresponds
to Hashin-Shtrikman's lower bound equation for the shear modulus,
representing the minimum predictable effect of the sub-inclusions
upon the stiffness of the (composite) rubber.

      Without going into the details of Fig. 2, we would
just mention that the two approaches lead to decidedly different
predictions. Of course the mere comparison of these results cannot
"sort out" the two models: only experimental checks could do so.

      On purely conceptual grounds, we note that the layered
sphere model deserves consideration for at least the following
good points:

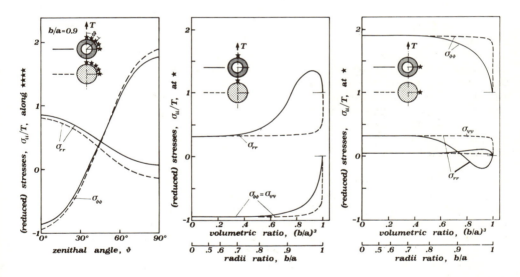

Fig. 2. Stresses at sites on the outer surface of the composite
     particle, calculated according to the layered sphere
     model (solid lines) and the "homogenized" particle model
     (dashed lines). Sites and models are indicated in the
     insets, where white areas represent glassy polymer, dark
     shaded areas represent pure rubber, and light shaded
     areas represent "homogenized" composite rubber.

(I) It preserves the relevant topological features of the composite
    particle, each phase maintaing its individual characteristics;
(II) Its analysis remains significant even with high proportions of
     sub-inclusions in the composite particle, while, on the other
     hand, composite theories err significantly.
(III) With this model, stress and strength analysis can be
      meaningfully extended to the interior of the composite
      particle — a point that will be touched upon later.

STRESS ANALYSIS

        As an elementary case, the problem of the stress field
developed by — or stress-field perturbations brought about by —
the composite inclusion under the action of a uniform uniaxial
tension, T, at infinity (Fig. 1) appears to be of interest. (Other
sources of internal stresses, namely thermal, were considered
separately (19)). Under these conditions the mechanical problem
is axisymmetric. On the basis of symmetry considerations, the
superposition principle, and dimensional analysis, the following
a-dimensional expression can be predicted for any stress tensor
component σ_{ij} at a point $P(r, \vartheta, \psi)$:

$$\sigma_{ij}/T = \text{function } (P(\frac{r}{a}, \vartheta); \; \frac{b}{a}; \; \frac{G_2}{G_1}, \; \nu_1, \; \nu_2)$$

        This indicates that: (i) in terms of factors of stress
concentration or reduced stresses, σ_{ij}/T, the results will not
depend on the level of applied tension T;   (ii) only the relative
dimensions b/a of the sub-inclusion and inclusion and (iii) the
relative position r/a of the point P with respect to the inclusion
radius count, which means that with particles of different sizes
the stresses are the same at similarly-situated points.
        The specific form of the above function depends on the
particular stress tensor component and material phase region
considered. General solutions of the axisymmetric boundary-value
problem of the elasticity theory are available in literature, for
region both inside and outside a spherical boundary (18, 21), or
comprised between two concentric spherical boundaries (22). We
used Goodier's solutions (18), extending them also to the shell
region by a simple linear combination (23). The values of the
twelve arbitrary multipliers involved have to be determined from
continuity conditions at the interfaces. This requires the
solution of a set of twelve simultaneous equations, which it is
practical to perform numerically via a digital computer. This was
done for a number of values of b/a in the interval 0 < b/a < 1;
for each of them, each stress component was the computed for any
point.
        As an example of the results obtained, Fig. 3 shows
the spatial distribution of the four non-zero stress tensor
components in the region of the glassy matrix around the inclusion
for the case b/a = 0.9. Each map represents stereographic projec-

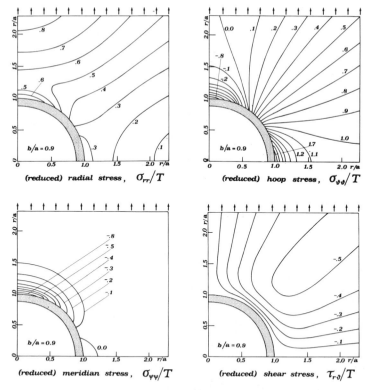

Fig. 3. Distribution of stresses in the matrix around the
composite particle (layered sphere model). White
areas represent glassy polymer, shaded areas
represent rubber. Arrows indicate the direction
of the tension T applied at infinity.

tions of some contour levels on a quadrant of a generic meridian
plane: owing to the model's symmetry, the same distribution is
repeated in every other quadrant and on any meridian plane. Values
of the local stresses, reduced with respect to the applied tension
T, are indicated by labels on each line. With the usual sign
conventions, positive values indicate tractive (compressive)
normal stresses, while negative values indicate compressive
(tractive) normal stresses if the applied stress is tractive
(compressive).

Inspection of these maps revealed the following
noteworthy features:

(a) Different stress components have quite different spatial distributions. Thus consideration of stresses alone can give no indication of strength. Correct failure predictions, based on criteria properly formulated in invariant tensorial terms, must consider all the stress components involved.

(b) Stresses do not vary monotonically with the radial distance r, in general (This appears also in other authors' results (e.g. 9, 10, 24)). Thus, even if it is true that the stress field perturbation as a whole decays rapidly away from the inclusion, this does not justify confining attention, a priori, to matrix stresses at the inclusion surface.

(c) Shear stresses $\tau_{r\vartheta}$ at the inclusion-matrix interface are almost nil. They are kept away from the inclusion by its rubber shell owing to the low shear modulus of rubber compared with the shear modulus of the matrix and continuity of shear stress through their interface. This result remains valid for very thin rubber shell thicknesses.

(d) Distribution and magnitude of all stresses do not vary appreciably when b/a varies from zero up to about 0.8 (cf. Fig. 2), i.e. they are insensitive to the presence of the hard sub-inclusion of that size: the composite particle behaves as if it were entirely made of rubber. This result, in agreement with the "homogenized" particle model, provides theoretical support to the experimental finding that the behaviour of rubber-modified plastics is determined by the volume fraction of the total dispersed rubbery phase rather than by the actual rubber content (25, 26, 16).

(e) When the sub-inclusion is a large proportion of the total inclusion (b/a > 0.8), on the other hand, the stress distribution is vastly different compared with the case of a simple, homogeneous rubber particle (b/a = 0) (cf. again also Fig. 2). Thus, from a theoretical point of view, the behaviour of the composite particle may be expected to be different.

STRENGTH ANALYSIS

The analysis of stresses *per se* cannot, of course, help us very much to understand the performance of a composite material. It is of fundamental value, however, in providing the necessary link between structure and performance if the state of stress can be related to the onset of failure, i.e. if a failure criterion is available.

A generally-valid criterion for failure does not exist, of course: different materials under given conditions or a given material under different conditions (temperature, strain rate, etc.) may obey different stress criteria, depending on the prevailing deformation mechanism or mode of failure. Conversely, it depends on the state of stress which mode will prevail in a given material under given conditions. A complete strength analysis ought thus to consider all possible modes of failure and apply the corresponding criteria. This is especially important in particulate composites, where the local, triaxial state of stress may be highly variable.

We will limit ourselves here to consideration of failure by crazing, i.e. the dilatational mechanism of heterogeneous deformation, which is believed to be both the major source of toughness in particulate polymer composites and a possible precursor of fracture. Indeed, while it provides a means of dissipating energy, at the same time the strength of a specimen is lowered.

Of the several craze initiation criteria proposed in literature, none is generally accepted. Perhaps the largest consensus has been won by the critical strain criterion proposed by Oxborough and Bowden (27) as a more general (i.e. triaxial) formulation of the stress-bias criterion originally set forth by Sternstein and Ongchin (28) (the validity of this very generalization was, however, recently questioned (29)). The Oxborough-Bowden criterion states that crazing will occur when the largest tensile strain in any direction reaches a critical value, ϵ_c, which depends on the isotropic component of the stress tensor:

$$\epsilon_{max} = \epsilon_c \equiv \frac{1}{E}\left[C + \frac{D}{3\,\sigma_m} \right] \qquad \text{(eq. 1)}$$

or alternatively

$$\epsilon_{max} = \epsilon_c \equiv \frac{1}{E}\left[C + \frac{D'}{\Delta} \right] \qquad \text{(eq. 1')}$$

where ϵ_{max} is the maximum (principal) tensile strain, σ_m is the mean normal stress and Δ the related volume strain or dilation, E is the tensile or Young's modulus of the as yet uncrazed material, and C and D (or $D' = D/3K$, K being the bulk modulus) are two other material constants.

To apply this criterion to our model, let us note that both ϵ_{max} and Δ can be expressed in terms of stress tensor components, and so can be calculated at any point P from the results of our previous stress analysis (30). It is convenient to normalize both these quantities in relation to their respective values at infinity, $\epsilon_{max\,\infty}$ and Δ_{∞}; their relative ratios are concentration factors, then, independent of applied tension T:

$$\epsilon^*_{max}(P) \equiv \frac{\epsilon_{max}(P,T)}{\epsilon_{max\,\infty}(T)} \qquad \text{(eq. 2)}$$

$$\Delta^*(P) \equiv \frac{\Delta(P,T)}{\Delta_{\infty}(T)} \qquad \text{(eq. 3)}$$

Samples of their spatial distribution in the matrix region around the model particle are given in Fig. 4. It is just worth observing here that their pattern is rather complex and totally different for the two quantities. Since it is a

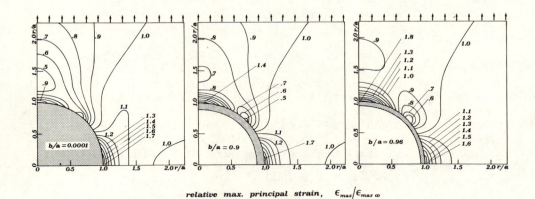

relative max. principal strain, $\epsilon_{max}/\epsilon_{max \, \infty}$

relative dilation, Δ/Δ_∞

Fig. 4. Distribution of maximum principal strain (above) and
dilation (below) in the matrix around the composite particle
(layered sphere model), for three values of the radii ratio
b/a. Signs and shadings as per Fig. 3.

combination of the two that determines crazing according to the above criterion, any attempt at predicting failure of this composite on the basis of intuition appears to be hopeless.

In order to apply the Oxborough-Bowden criterion (eq.1 or 1') numerically (31), we needed first to assign numerical values to the two constants, C and D, involved. We took those measured in "crystal grade" polystyrene at 20°C in a short-term (1 s) test (27): $C = 30.2$ MNm^{-2}, $D = 1400$ (MNm^{-2})^2. This additional selection will of course restrict the validity of the results even further.

At infinity, i.e. under uniaxial tension T, $\epsilon_{max}\infty = T/E$ and $\Delta_\infty = T/3K$. Introducing these expressions into (eq. 1') gives an equation in T that can be solved to obtain the critical value of it required to produce crazing under this standard state of stress, i.e. the *(tensile) craze-yield strength of the glassy polymer:*

$$T^\circ_{cy} = \frac{C}{2}\left[1 + \sqrt{1 + 4\frac{D}{C^2}}\right] \qquad \text{(eq. 4)}$$

At point P, it is convenient to express ϵ_{max} and Δ as a function of applied tension T through their respective concentration factors (eq. 2 and 3) so as to separate the variables:

$$\epsilon_{max}(P,T) = \epsilon^*_{max}(P) \cdot T/E \qquad \text{(eq. 5)}$$

$$\Delta(P,T) = \Delta^*(P) \cdot T/3K \qquad \text{(eq. 6)}$$

Introduction of these expressions into (eq. 1') again gives an equation in T, which can be solved to yield now the critical value of the tension T that would be required for crazing to occur at point P — should it not occur somewhere else first:

$$T_{cy}(P) = \frac{C}{2\ \epsilon^*_{max}(P)}\left[1 + \sqrt{1 + 4\frac{D}{C^2}\frac{\epsilon^*_{max}(P)}{\Delta^*(P)}}\right] \qquad \text{(eq. 7)}$$

Note that T_{cy} is a function of P, though it acts at infinity, not in P: it is the *craze-yield strength of the composite* as predicted by an observer whose attention in focused on point P.

In Fig. 5 the relative craze-yield strength T_{cy}/T°_{cy} is mapped. Values of this ratio higher than 1 mean reinforcement[4], values lower than 1 mean reduction in strength produced by the

---

Note[4]. The Oxborough-Bowden condition for craze formation (eq. 1 or 1') implies positive dilation. Regions where dilation is negative should be assigned an infinite value of T_{cy}/T°_{cy}. These regions are hatched in Fig. 5.

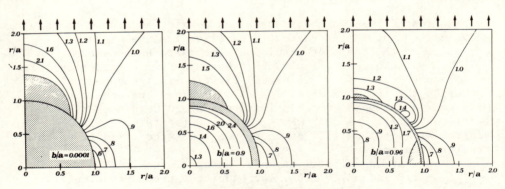

RELATIVE CRAZE-YIELD STRENGTH, T_{cy}/T_{cy}^{o}

Fig. 5. Distribution of craze-yield strength in the matrix around
and in the sub-inclusion within the composite particle
(layered sphere model). Signs and shadings as per Fig. 4.
Hatched areas are regions where conditions for crazing are
not met.

particle. The minimum value of T_{cy} over the whole space is the
(tensile) craze-yield strength of the composite, and the corresponding point is the site where crazing is predicted to start.

This analysis was extended to the interior of the composite particle. In fact, there are experimental indications (32-35)
that crazes have started or occurred within the glassy occlusions
or sub-inclusions in conventional toughened plastics. This possil-
ity would be of particular practical interest, because an
appropriate design of the material could offer a means of absorbing
energy within the composite inclusions even before the matrix
reaches the critical point of its incipient failure. Fig. 5 shows
that reduction in the craze-yield strength of the composite is
indeed produced in the central region of the sub-inclusion as well
as in the matrix region that surrounds the particle equator, when
b/a is high. More specifically, Fig. 6 shows the combined effect
of the two quantities ϵ_{max} (left) and Δ (center) in determining
the craze-yield strength (right) observed at four particular sites
(see inset), over the relevant b/a range. A huge peak in dilation
comes out at the sub-inclusion pole when b/a approaches 1, but is
not backed up by a sufficiently large value of the maximum
tensile strain in promoting craze initiation. In comparison to
that, more favourable appears the combination of a relatively large
maximum strain with a modest level of dilation, which occurs at the
sub-inclusion center. . The most favourable site for craze initiation remains the equatorial region outside the particle, as a
matter of fact, as generally found by experiment. The craze-yield

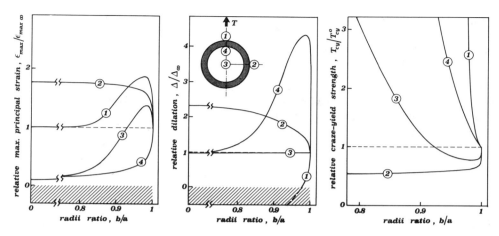

Fig. 6. Maximum principal strain (left), dilation (center), and
craze-yield strength (right) at four sites in and around
the composite particle (layered sphere model) as indicated
in the inset, as a function of b/a.

strength computed at this site is the absolute minimum over the
whole space, at any b/a ratio: therefore that is *the craze-yield
strength of the composite*. Its value appears to be almost constant
over the entire range of b/a.

CONCLUDING REMARKS

We have considered an elementary model for composite
inclusions in polymer matrixes, and have carried out a micromechanical analysis subjected to some symplyfying assumptions and
based on numerical values applicable to rubbery inclusions
comprising glassy polymer sub-inclusion(s). Results of this analysis
provide some insight of the effects that the composite particle
structure has on stress distribution and craze-yielding.

In the course of the presentation we have underlined
the good points of this model. In closing we must point out its
limitations: (a) As a single particle model it neglects inter-
particle effects or interactions; the results would only apply with
reasonable accuracy to materials containing less than about 10% by
volume of disperse particles (18). (b) Particle shape is idealized/[5]

Note/[5]. A review of the effect of particle shape on the mechanical
properties of filled polymers can be found in ref. (36).
Z.Hashin has pointed out that excentricity of the sub-
inclusion might produce a remarkable effect.

and (c) particle size is not given relevance. (d) While the model
assumes perfect interfacial adhesion, debonding can occur in real
materials and alter stress distribution and yielding behaviour.
(e) The assumption of linear elasticity makes the analysis valid
only for small strains. In any case, the analysis will be valid up
to the point of incipient craze formation, at most: as crazing at a
site occurs, the hypothesis of linear elastic behaviour throughout
the composite material fails. (f) Any time dependence of viscoelastic nature is neglected: the behaviour of viscoelastic composites
is complicated by the presence of interactive effects (38). (g) The
foregoing analysis is for a particular applied stress system, i.e.
uniform uniaxial tension. Results for any uniform stress system
could be derived by superposing the effects of three simple tensions
or compressions of magnitudes corresponding to any three principal
stresses at infinity (18). (h) No source of internal stresses but
applied loads is considered. As quoted previously, thermal stresses
were considered separately (20). (i) The present results are controlled by a set of material   constants, representative of a glassy
plastic and rubber. Both the glassy plastic and the rubber were
treated as retaining in the composite the same properties as they
would have individually in the bulk. This assumption can be
questioned on several grounds (7), however. An extension of the
analysis to a number of different sets of values is in progress (39).
Finally, (j) a continuum approach to the problem is valid only while
the size of the inclusion is relatively large, as pointed out by
Bucknall (7).

REFERENCES

(1)   For a comprehensive review see C.B.Bucknall, "Toughened
          Plastics", Applied Science, London (1977).
(2)   G.Riess, S.Marti, J.L.Refregier, and M.Schlienger, in "Polymer
          Alloys", D.Klempner and K.C.Frisch, eds., Plenum, New York
          (1977), pp. 327-335
(3)   A.Echte, Angew. Makromol. Chem. 58-59:175 (1977).
(4)   J.B.Donnet, Z.Kessaissia, and E.Papirer, Proc. 2nd Int.1 Conf.
          SIMPLAST-79, Warsaw, Nov. 26-28, 1979. See also L.E.Nielsen,
          "Mechanical Properties of Polymers and Composites", M.Dekker,
          New York (1974), vol.2, p.497; S.Newman, in "Polymer Blends",
          D.R.Paul and S.Newman, eds., Academic Press, New York (1978),
          p.84.
(5)   See for example in J.A.Manson and L.H.Sperling, "Polymer Blends
          and Composites", Heyden, London (1976), ch.12.
(6)   R.W.Hertzberg and J.A.Manson, "Fatigue of Engineering Plastics",
          Academic Press, New York (1980), refs.48-57, p.212.
(7)   C.B.Bucknall, J. Mater. (ASTM) 4:214 (1969).
(8)   V.A.Matonis and N.C.Small, Polym. Eng. Sci. 9:90 (1969); V.A.
          Matonis, ibidem 9:100 (1969).
(9)   T.T.Wang and H.Schonhorn, J. Appl. Phys. 40:5131 (1969).
(10)   T.T.Wang, M.Matsuo, and T.K.Kwei, J. Appl. Phys. 42:4188 (1971).

(11) O.Buynkorturk, A.H.Nilson, and F.O.Slate, J. Eng. Mech. Div. EM3 (June 1973).
(12) R.J.Oxborough and P.B.Bowden, Phil. Mag. 30:171 (1974).
(13) L.J.Broutman and B.D.Agarwal, Polym. Eng. Sci. 14:581 (1974).
(14) L.Bohn, Angew. Makromol. Chem. 20:129 (1971).
(15) B.P.Holownia, Rubb. Ind. 8:157 (1974).
(16) C.B.Bucknall and M.M.Hall, J. Mater. Sci. 6:95 (1971).
(17) R.A.Dickie, J. Appl. Polym. Sci. 17:45 (1973).
(18) J.N.Goodier, J. Appl. Mech., Trans. A.S.M.E. 55:A-39 (1933).
(19) E.H.Kerner, Proc. Phys. Soc. 69B:808 (1956).
(20) A.Pavan and T.Riccò, J. Mater. Sci. 11:1180 (1976).
(21) K.Sezawa and B.Miyazaki, Proc. Jap. Soc. Mech. Eng. (Tokyo) 31:No.136,625 (1928).
(22) E.Sternberg, R.A.Eubanks, and M.A.Sadowsky, Proc. First U.S. Nat.1 Congress of Appl. Mech., A.S.M.E., New York, N.Y. (1952), p.209.
(23) T.Riccò, A.Pavan, and F.Danusso, Mecc. Ital(Milan) 108:33 (1977)
(24) A.E.Oberth, Rubber Chem. Technol. 40:1337 (1967).
(25) E.R.Wagner and L.M.Robeson, Rubber Chem. Technol. 43:1129 (1970)
(26) G.Cigna, J. Appl. Polym. Sci. 11:1781 (1970).
(27) R.J.Oxborough and P.B.Bowden, Phil. Mag. 28:547 (1973).
(28) S.S.Sternstein and L.Ongchin, ACS Polymer Prepr. 10(2):1117 (1969).
(29) H.Breuer and J.Stabenow, Angew. Makromol. Chem. 78:45 (1979).
(30) T.Riccò, A.Pavan, and F.Danusso, Polym. Eng. Sci. 18:774 (1978).
(31) A.Pavan and T.Riccò, to be published.
(32) R.J.Seward, J. Appl. Polym. Sci. 14:852 (1970).
(33) K.Lawrence, BPCo., quoted by R.P.Kambour, J. Polym. Sci., Macromol. Rev. 7:1 (1973).
(34) D.Lee, J. Mater. Sci. 10:661 (1975).
(35) A.S.Argon, Pure and Appl. Chem. 43:247 (1975); A.S.Argon, R.E. Cohen, B.Z.Jang, and J.Vander Sande, Int.Conf. "Toughening of Plastics", London, 4-6 July 1978.
(36) T.S.Chow, J. Mater. Sci. 15:1873 (1980).
(37) Z.Hashin, oral comment.
(38) J.R.M.Radok, "Effects of Stress Raisers in Linear Viscoelastic Bodies", in "Fracture Proccesses in Polymeric Solids", B.Rosen, ed., Interscience, New York (1964), pp.747-771.
(39) A.Pavan and G.Tieghi, forthcoming.

DESIGN OF CONTINUOUS FIBER

COMPOSITE STRUCTURES

I. Crivelli Visconti

Istituto di Tecnologie – University of Naples

Piazzale Tecchio – Napoli – ITALY

1- INTRODUCTION: HISTORY OF DESIGN AS A GUIDE LINE

Not differently from any human new adventure, the history of composites begun once upon a time (or at least it seems so now) certainly with the idea to provide something with new properties, new aspects, new challenges, able to solve old problems.

But for the same reason of being a new adventure, also without knowing where the path was leading to, and what would have been the best equipment and facilities, and language necessary to rapidly reach the end of the way.

Necessarly, then, the initial attempts in the use of composites were made trying to realize a structure in a manner very similar in shape, geometry, dimensions, to the one up to then made with previous materials: there was, in other words, no or a mininum amount of composite design.

In the majority of cases, though, that shape, that geometry, those dimensions, were a result of a natural development along the years following the needs, the technological possibilities, the properties of the previous material, and were not suitable to the same structure made in composite material.

This work was performed under CNR and CRF research contracts.

After a ripening time, it was clear that a composite can be used exploiting all its possibilities only if we match its intrinsic properties with technological means, and with the mission of the structure.

This is then what must be accepted as a definition for "design" with composites: to solve on the same line with horizontal inputs and feed-backs the aspects related to mission requirements, technological possibilities, material properties.

The fundamental conclusion that is here assessed as the basic design rule when using composite, is that the traditional division between material man, design man, technological man must be overcome. The composite design must be done either by a super-man who knows a lot of everything, or (and this is more feasible) by a team of people working really together with actual exchanges of knowledge and experience.

The final result of such an approach to composite design will certainly show up with a structure that is by no means similar to the previous one: the mission requirements will be satisfied by a structure whose shape and geometry are compatible with composite properties and with available technologies of fabrication, testing, control.

2 - DESIGN CYCLE

The basic rule of a composite design can be expressed in a more workable way with reference to Fig. 1 [1] where a simple diagram of a design cycle is shown, with the indication of the horizontal exchanges: this will result in the possibility of using in more than one step of the cycle inputs and informations coming from steps of the cycle.

What is important to underline here is that this type of composite design can be approached at various levels of consciousness.

In other words, it is possible to use the design cycle in a superficial manner: and this is what actually happens in a large number of cases when either the consequences of a large oversizing or of a failure in service are not catastrophic; or , it is possible to use the design cycle analyzing each step of the cycle in the deepest manner, trying to reach optimal conditions in terms of safety, weight, cost, endurance of the structure.

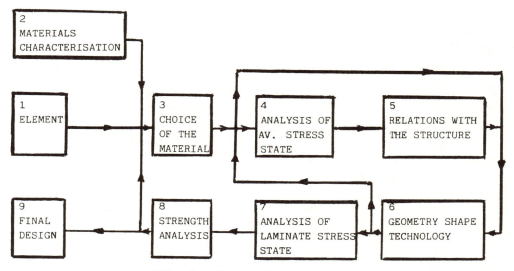

Fig. 1 – Basic design cycle

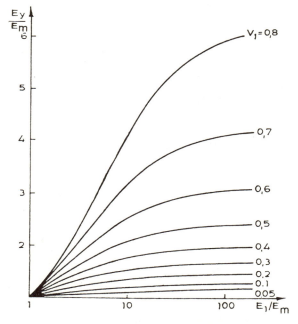

Fig. 2 – Normalized transervse Young's modulus in unidirectional
composites [6]

A qualitative description of the first design mode is easily made, showing what is the rule in many fields of application of GFRP, where:

- fibres are in general in the form of rough fabric and mat
- properties of consequent composite are strongly variable
- technologies are of low level of automation and therefore is not possible to make previsions on type and density of defects in the structure
- quality control is almost ignored during fabrication
- external load are not known exactly and stresses are computed in very rough ways
- the unique and important goal to meet is to have a structure moderately reliable and with the lower direct cost.

When the above conditions are met, the design operation is limited to just calculate the number of fabric and mat layers such to account, on the basis of experience, for the load probably present in service, serving all with a large "factor of safety."

It is possible, however, to use the design cycle trying to know what each step represents, and evaluating the consequence of a particular choice. Of course, how deeply must the evaluations be pushed is a matter of how necessary is to correlate the knowledge of the final result (reliability, endurance, confidence) to the characteristics of the structure (weight, cost, technological feasibility).

Two basic design criteria are now extensively used for advanced composites:

1 - Classic deterministic design
2 - Probabilistic design

The first method makes use of precise values of all parameters entering into the design, such as strength or applied stress, rigidity of the element, service conditions, and safety or reliability of design is controlled by the definition (or determination) of proper design limits and safety factors.

In the second design method the concept of "reliability" and of "probability of failure" is the idea upon which the design is based, replacing the previous concept of "safety factor" [2].

Basically, the necessity of a probabilistic design is bound to the following unavoidable aspects [3] :

- limitations in overall analysis
- inaccuracy in design
- lack of control of fabrication steps
- imperfect knowledge of materials strength
- variability of applied loads
- possible synergistic effects due to cumulative damages.

All these features cannot be precisely defined, but have rather a certain degree of probability to be actually present in the specific case, and therefore also the final result, that is the final "reliability" of the design, can be defined only with a known degree of probability.

3 - DETERMINISTIC DESIGN

Although a probabilistic design, due to what has been said, is the most close to the actual behaviour of nature, a deterministic design is preferred or sufficient in most cases. We can show, however, that many aspects of composites normally neglected should be considered and would be of great help for better design, even neglecting probabilistic or statistical behaviours.

With reference to the design cycle, we can briefly comment on some of the steps there indicated.

ELEMENT

What is to be known here is the mission that the element must fulfill and support. This is normally thougt to be coincident with the loads applied on the element. Indeed this is the most important aspect of the mission to be fulfilled, but in almost all the actual cases other sources of "loads" are neglected or not fully considered.

One source of important efforts for the material is the environment, and a great deal of research has been done in the last years in order to better understand the behaviour of composites after exposures to corrosive atmosphere or humidity, and changes of temperatures [4,5].

It is not clear, though, what the actual behaviour is under simultaneous presence of humidity and stresses, and while this is certainly a topic of necessary research, it is not always considered as an important "load" on the structure.

MATERIALS PROPERTIES AND TECHNOLOGY

This point should be investigated in far deeper details that this note will permit. What is to be stressed, here, is that before attempting to go further into the design we should know what the properties of the material are and how they can vary under given circumstances.

It is well known that we can in general distinguish between:
1)- fibre dependent properties
2)- matrix dependent properties

Among type 1) are the properties closely expressed by rule-of-mixtures type laws, like:

$$E_c = E_f V_f + E_m V_m$$

$$\sigma_c = \sigma_f V_f + \sigma_m V_m \tag{1}$$

where E is the young's modulus, σ the stress (and strength), c, f,m, refer to composite, fibres, matrix respectively. Eq (1) are good expressions for properties along the fibre direction.

An example of type 2) properties is the Young's modulus in direction transerve to the fibre. Fig.2 is a diagram showing that no matter how stiff the fibres are (or high the ratio E_f/E_m is) the upper bound for the tranverse Young's modulus cannot be larger than a few times the matrix Young's modulus [6].

A tremendous amount of work has been in the past years trying to yield expressions easily applicable to composites in order to evaluate all the composite elastic properties from a micro-mechanical point of wiew [7]. While in the fibre direction matrix and fibres are approximately working in a parallel mode, it is not so clear what are the connections between parallel and serial mode of working of the two components, when the fibre are at a variable angle with the measuring direction.

From the designer point of wiew, therefore, experimental deterministic properties are of prime importance. He must be aware, though, that in spite of the large amount of informations found in the literature since the large spectrum of experimental procedures used by different people, not all the results can be actually compared.

This is in fact true because both test procedures, materials and technologies are not standardized in different countries. In the majority of cases this will imply that new experimental tests must be organized to know the properties of the material designed for that particular application, using that particular technology.

The choice of proper technology to be used is also somewhat underestimated with regard to the possible structural defects present within the material: they can affect considerably the actual properties of the material in the final form.

A sistematic study of defects, and their relationship with material properties is a research topic very poorly examined in the literature and a larger attention should be payed to it.

A large scatter of properties of composites is to be ascribed to defects of the material, and could be avoided with a better understanding of the type of defects created by the fabrication method.

The importance of the defect structure in a composite is also related to another aspect frequently ignored: the resistance to cumulative damage; if we know the resistance of the material to the damage due to "load"1, and to "load" 2 separately, it may happen that the damage created by 1 will let the material behave differently, with regard to "load" 2 , from what we know.

Fatigue properties are among the properties that have undergone a deep and broad analysis in the last years, and the literature is full of results [8-18] to be examined, however, with a certain attention, if the results must be compared, due to the lack of testing and materials standardization.

What the designer needs to know is:

− Does a fatigue limit exist?
− Is there a method to design for fatigue?   (2)

The answer is not very simply and clearly statable, but is based on a number of considerations leading to some positive conclusions.

First of all, it is necessary to realize that composite fatigue is influenced by a large number of parameters; among these are the multiple failure modes, that are:

− debonding
− matrix failure

- delamination
- fibre pull-out
- fibre buckling
- fibre failure

it follows that every source of variation of any characteristic of the matrix, of the fibre, and of the interface like fibre and matrix strength, interface bonding, strength to crack advance, variation of defects in the matrix or at the interface, notches, and so on, will affect the fatigue strength, even if it will not affect the static strength.

The parameters of primary importance in the fatigue strength are:

- Constituent materials
    type and volume of fibre
    type and volume of matrix
    characteristics of the interface

- Fibre position
    fibre orientation
    lamination sequence

- Service conditions
    type of applied load
    frequency
    environmental conditions   (3)

- Element geometry
    laminate thickness
    holes
    junctions
    notches and discontinuities

- Fabrication technology

    porosity
    bonding defects
    curing defects
    orientation defects
    surface defects.

While a complete description of the influence of all the parameters are reported elsewhere [12] we shall report only a few examples showing that the answer to the basic questions (2) can be found only after a deep analysis of the behaviour of the composite in a specific situation.

An easy way to show the complexity of the fatigue behaviour of composites is to try to consider the role played by the type of stress, type and dimension of fibres, type and properties of matrix, properties of the interface.

In the case of constant influence of all other parameters, we can in fact assume that if the load is applied (in tension) mainly along the fibre direction, if the fibres are long, parallel and continuous, if the interface bonding is strong enough, if the Young's modulus of the fibres is high enough and if the matrix strength is high enough, then the behaviour of the composite will be satisfaying in comparison with a similar behaviour of a metal component and the S/n curve will decrease in a rather monothonic way. Although it is difficult to distinguish a real fatigue limit, the percentage of decrease in strength compared to the static strength is quite low.

This effect is shown in Fig.3 [20] where the results of tests of axial fatigue of composites using high modulus carbon fibres are reported. Fig. 4 [21] reports similar results, but also shows the decrease observed for compression fatigue. Fig.5 [22] shows the results obtained when the specimen is a cross plied 0/90 laminate: the decrease of properties is evident.

If we compare these results with those of Fig.6 [21], that shows fatigue results for type HT C fibres, we can notice a twofold effect:

I)- The fatigue strengths are higher for HT fibres, due to higher strength of this type of fibres, as long as the applied load does not create damage at the matrix, that is for the initial part of the curve (low number of cycles).

II)-The curves decrease more rapidly than the analogous for HM fibres. This is due to cumulative cracking, induced by the deformation, within the matrix, and is due to the large deformation of the composite with lower Young's modulus fibres. This effect is more evident for composites with untreated fibres, indicating that micro-debonding is present with time (or with cycles) at the interface.

Using the above indications, the designer can be guided in using different types of composites for different loading situations, and gather informations on the fatigue behaviours. Composites with glass fibres, for instance,or in general with lower values of Young's modulus are apt to show fast decreasing fatigue S/n curves,

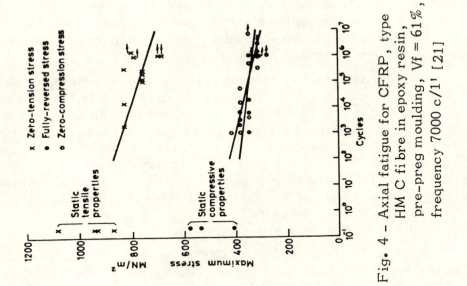

Fig. 4 – Axial fatigue for CFRP, type HM C fibre in epoxy resin, pre-preg moulding, Vf = 61%, frequency 7000 c/1' [21]

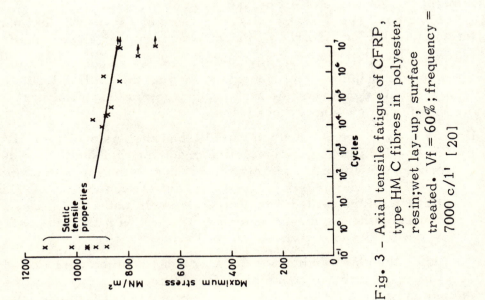

Fig. 3 – Axial tensile fatigue of CFRP, type HM C fibres in polyester resin;wet lay-up, surface treated. Vf = 60%; frequency = 7000 c/1' [20]

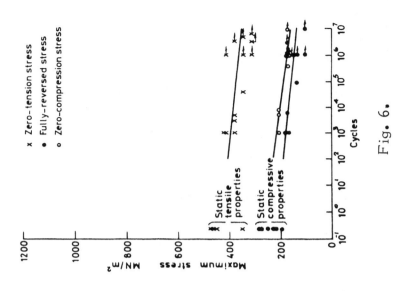

Fig. 6.

Zero tension axial fatigue for cross plied CFRP, type HT, pre-preg moulding epoxy resin, Vf=60%, frequency 100 c/1', curve 1 = surface treated fibres; curve 2 = untreated fibres [21]

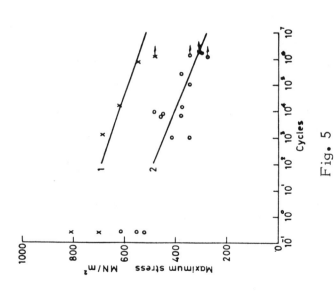

Fig. 5

Axial fatigue for cross-plied CFRP, type HM, surface treated, 11 laminae, pre-preg moulding, epoxy resin, Vf = 63%, frequency 7000 c/1' [22]

unless the resin has higher deformation possibilities. This is a
field of research where further results will be always welcome.

Moreover, and for a reason strictly related to the above observation, better fatigue properties are expected if the orientation
of the laminate is such that the longitudinal fibres support the majority of the load.

RELATIONSHIP BETWEEN ELEMENT AND ENTIRE STRUCTURE: CHOICE OF GEOMETRY, SHAPE AND TECHNOLOGY

This is the step where maximum attention should be payed by the
composite designer because here the true potential of the use of the
material should prove to be effective.

In other words the use of the composite material (if the application is suitable for a composite) should take into account all the
interface with the structure that it belongs to resulting in a new
shape, new geometry of the element considering also the possible
available way of fabrication.

In this step all the features like junctions, notches, special
shapes should be carefully estimated with the aim to improve the
behaviour of the structure still fulfilling the mission and so making
the choice for the most proper technology. It is important to emphasize that the element and therefore the materials for the element, and therefore the fabrication methods to be used must be
tailor-made on the basis of current available technologies; as a
function of the degree of perfection, properties of the materials,
tolerances, type and density of defects allowable, the degree of
sofistication of the fabrication method can be chosen, starting
from manual operation and wet lay-up to automatic laying and moulding of pre-preg goods.

Singularities in the element where connections with other parts
or junctions are present reserve particular attention: notch sensitivity can be designed with particular lamination to artificially
induce micro-delamination absorbing energy or by putting insert
pieces to locally withstand higher stresses   [23-25].

Of course the particular technique must be evaluated as a function of the available technologies.

STRESS ANALYSIS AND STRENGTH ANALYSIS

At this step of the cycle the duty of the design team is to concentrate on the three following aspects:

- determine the stresses acting on each lamina of the composite laminate
- rationalize the strength of the material
- demonstrate the capacity of the element to withstand and fulfill the mission required.

It is obvious, from what has been said previously, how all the aspects mentioned both about the element (or mission) and about properties and technology must now found a quantitative and determined expression in terms of load on the structure and actual foreseen strength of the composite material.

When this is done we can actually say to be almost through.

It is may be clarifying at this point to recall what are the general definitions used to indicate the level of stresses compared with the different levels of loads acting upon a structure. We shall do this with reference to the operational systems commun to aeronautical design.

The procedure is as follows: once the totale load system have been determined, we choose the most critical and call it "limit load" (LL).

Because, as we mentioned, there are uncertainties in the operation of setting the value of the limit load, we also define a more conservative value of "ultimate load" (UL) taken as the limit load multyplied by a "safety factor" (SF), usually of 1.5. Therefore it is:

$$UL = SF \times LL = 1.5 \times LL \qquad (4)$$

$$LL = 0.66 \times UL \qquad (5)$$

The basic design philosophy now is:

- No significant permanent deformation must occur in each component of the structure at the limit load; deformation (pseudoplastic) can be present only at loads higher than LL, as long as the structure maintain its capabilities.

- No structural failure must be present prior to the ultimate load.

- The structure must be dimensioned with reference to either one of the above basic criteria while checked to verify the other one.

The application of such criteria to metals are quite straight forward, for the values of load can be easily translated into values of stress, and being usually:

$$\sigma_y^* \geqslant 0.66 \; \sigma_u^* \qquad\qquad (6)$$

where σ_y^* is the yield and σ_u^* the ultimate stress (strength) of the metal, the criterion to apply is to design for ultimate load, and so doing the criterion for limit load will be automatically satisfied.

If eq. (6) holds, it is then sufficient that the stress calculated by UL, σ_{uL}, be less than the strength σ_u^* ; it must be, therefore:

$$\sigma_{uL} \leqslant \sigma_u^* \qquad\qquad (7)$$

and since from (5):

$$\sigma_{LL} = 0.66 \; \sigma_{uL} \qquad\qquad (8)$$

from (6),(7) and (8) we have:

$$\sigma_{LL} = 0.66 \, \sigma_{uL} \leq 0.66 \; \sigma_u^* \leq \sigma_y^* \qquad (9)$$

or:

$$\sigma_y^* \geqslant \sigma_{LL} \qquad\qquad (10)$$

In case of composites the reasoning cannot be so simple because:

-Eq.(6) is not usually verified by experience. Initial presence of "pseudoplastic" deformation in a laminate begins sometimes at very low levels of stress and strain.

-It is not so well established what the relationships are between linear behaviour and post-linear behaviour of a composite laminate [26-28].

It is more convenient, therefore, to use a limit load criterion, basing all analytical evaluations upon linear elastic lamination theory which holds only up to the "yield" point of the material.

Once the design loads acting on the element have been found the normal procedures is to make use of lamination theory (LT) to finally find all the stresses acting on each lamina of the laminate.

Although LT is the normal and well experienced way to evaluate stress and strains it must be pointed out that it does suffer of intrinsic limitations so that is a very appropriate theory for elasticity calculation, but not for evaluation of actual state of stresses.

The main reasons for such unespected results are due to the following aspects:

A = LT is based on plane state of stress
B = LT does not consider fracture mechanics (FM).

The first assumption A is sufficiently correct for most applications, but is now assumed to be the reason for unexpected failures where thickness is not very low, and in fatigue. Stacking sequence is in fact responsable for the source of normal stresses in the third direction, particularly at the edge of the element [29-33].

While compressive stresses can easily be withstood and can even enhance the interlaminar strength, tension stress can be relatively large such to give rise to very dangerous accumulation of overall stress in the laminate.

The first glance to this effect dates back in 1971, when Pagano and Byron Pipes made an interesting study on the actual stress state present in a symmetric and balanced laminate under uniform tensile load in the x direction.

They were able to show that LT correctly yield elastic solutions for the entire volume of the material, except for a zone close to the edge of the laminate where other stresses arise involving in particular the z direction.

This can be explained with the aid of Fig. 7 [29] referring to a $(\pm 15/\pm 45)_s$ laminate.

The uppermost lamina, taken as a free body, is subjected to the lateral Poisson's stresses σ_y only at the inner side, but not at the free edge (this can be the actual border of the element, or an internal hole, or a similar); the counterclokwise moment must then be equilibrated by a distribution of σ_z.

As shown in Fig. 8 [29] the σ_z distribution presents a strong gradient at the edge.

The maximum value of σ_z at the very edge of the laminate is of difficult exact evaluation, but it can be stated that his dimension is a function of the dimension of the σ_y, that can be calculated by LT.

In cases when σ_z is negative, a compressive state occurs at the interface the two laminae (uppermost and subsequent, but the same effect is present at each interface within the laminate), thus

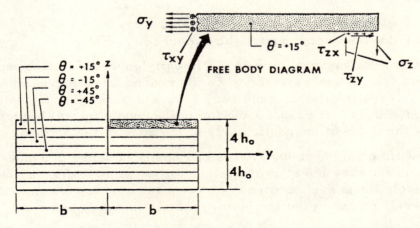

Fig. 7 – Actual stresses present in the upper lamina of a
laminate subjected to a force along x–direction [29]

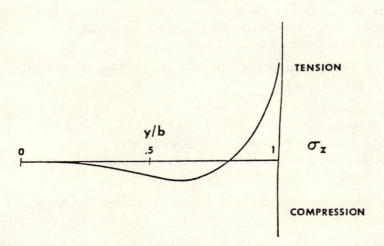

Fig. 8 – Distribution of σ_z along the interface of uppermost
and subsequent lamina [29]

enhancing the overall strength of the laminate preventing delamination and initial failure.

If the stress is positive, the tensile state is indeed an element in favour of early delamination not detectable by simple lamination theory.

The actual distribution of σ_x, τ_{xy}, and τ_{xz} are reported in Fig. 9 [34] of a (± 45) SC-epoxy laminate, high modulus fibres.

It can be seen that the LT result are not more valid for a region close to the edge of the specimen. The approximate distance where edge effects are present is of the order of the totale laminate thickness t.

It is clear, from what has been said, that the use of a proper lamination sequence can produce (for a given signe of load in x direction) tensile or compressive σ_z stresses at the edge; delamination can then be enhanced or prevented.

The second intrinsic limitation of lamination theory (LT) is the exclusion of any-fracture mechanics (FM) consideration.

It is well known that FM is the discipline that concerns with the failure of a material owing to the initial presence and subsequent growth of a crack.

It is clear the relationships that should bound the technology used to make the final structure, the techniques use for NDI, and design through FM.

Since it is not conceivable to realize a structure without any defect or imperfection we should rather attempt to classify (and standardize) the above steps in terms of "structural defects" present in the material, considering though the singularities (called "defects") rather as a "characteristic" of any particular process. Attention should then be payed to the effects, to be evaluated through FM, on the actual properties of the material.

To better explain this let us consider the fundamentals of FM [35 -40]. The basic consideration in FM is that when a crack is present within a stressed material, the crack growth can be effective only if the energy stored at the crack tip, or the driving force due the stress state applied to the body, G, reaches a critical value, G_c that depends mainly on the mechanisms involved in the crack advance.

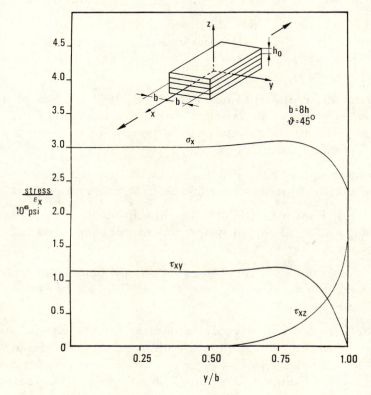

Fig. 9 – Stress distribution in a $(\pm 45)_S$ laminate with b = 8h, at the interface $z = h_o$. Graphite CI (high modulus) in epoxy [34]

For composites, the crack advance mechanisms are quite different from what a designer can be accostumed to with traditional materials, and have been mentioned earlier in this paper.

Fig. 10 [35] shows a model of a crack tip in a composite, with various mechanisms involved in crack propagation (pull out, fibre delamination, breakage, plastic zone etc).

It is true that a noticeable amount of work has been done in assessing the relationship between various mechanisms and the value of G_c, the energy involved [41-44]; however, the synergistic presence of multimechanism effect, as can be due to the different possible "characteristic" or "defects" obtainable by given technologies, appears not be fully estimated.

It has been shown that the value of G, the driving force to crack advance, can be expressed in terms of the applied load. If σ is the uniform stress applied at infinity, around the crack tip the stress state is intensified by a factor K (stress intensity factor), given by:

$$K = \sigma \sqrt{\pi a} \tag{11}$$

where 2a is the length of the crack perpendicular to the direction of the stress.

As for the G's values, it can be said that the propagation of the crack initiates when K reaches a critical value, K_c, called fracture toughness of the material.

If we can assume that material behaviour is brittle, or linear elastic up to fracture, like can be assumed for composite, linear elastic fracture mechanics (LEFM) can be applied, permitting an analytical relationship between G_c and K_c. Assuming further that crack propagation proceeds only by the "opening mode" or mode I, we define a critical value of K_I as [45]:

$$G = K_{Ic}\left(\frac{a_{11}\, a_{22}}{2}\right)^{-1/2}\left[\left(\frac{a_{22}}{a_{11}}\right)^{1/2} - \frac{2a_{12} + a_{66}}{2\,a_{11}}\right]^{-1/2} \tag{12}$$

where a_{ij} is an element of the matrix [a] of:

$$\varepsilon_i = a_{ij}\, \sigma_j \tag{13}$$

In a notched composite material, with a notch transverse to

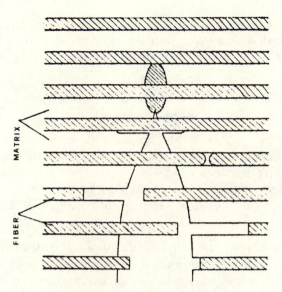

Fig. 10 –Model of the mechanism involved
in crack propagation in a composite
[35]

Fig. 11 – CFRP (90/0/±45)$_S$,HYE-1034-C-H,
Fiberite 3.6 mm thickness = negative value of NR [52]

Fig. 12 – CFRP (45/90/-45/0)$_S$, HYE-1034-C-H,Fiberite,
3.6 mm, thickness: higher value of N.R. [52]

the stress of length 2a, eq.(11) becomes:

$$K_{Ic} = \sigma_c \sqrt{a\,\pi}$$ (14)

and fracture should appear under extremely low loads. The ability to withstand load in presence of notches is explained by the concept of the presence of a "pseudoplastic" zone ahead of the crack tip, where energy is dissipated by microdelamination and microcracking. This zone is supposed to grow to a characteristic dimension, and only at this point failure proceeds, or the crack advances [46].

With this assumption, if a_o is the increment of the length 2a (at the crack tip), necessary to reach the critical situation, it is:

$$K_{Ic} = \sigma_c \sqrt{(a+a_o)\,\pi}$$ (15)

From the above assumption we can now assess that also an un-notched material, due to unavoidable "defects" present, will have a strength which relates to K_{Ic} by:

$$K_{Ic} = \sigma^* \sqrt{a_o\,\pi}$$ (16)

where a_o has been defined previously. If K_{Ic} is constant for that material, the value of a_o can easily be determined using notched and un-notched specimens:

$$a_o = \frac{a}{\left(\dfrac{\sigma_c}{\sigma^*}\right)^2 - 1}$$ (17)

where 2a is the notch length, σ_c the strength of the notched specimen, σ^* the strength of the un-notched specimen.

The possibility to relate technological parameters to FM indications for material strength, via the eq.(16) and (17), implies the verification that a_o is a characteristic constant of the material. However, owing also to what has been shown with respect to stacking sequence, more study is necessary to lighten this aspect, although some work has been made to demonstrate the validity of eq.(17) and to correlate the FM parameters K_{Ic} with technological aspects [47-49].

Variability with stacking sequence and with thickness of the laminate has not yet received full exploration , athough a great effort has been made in this direction [39-50] .

A recent work has attempted to correlate stacking sequence and thickness with the FM parameters, trying to define also a more general parameter for notch sensitivity referred to the value of a_o [31,51] .

The basic approach consists in the observation that while it is true that a given stacking sequence of the laminate could give rise to a positive (tensile) stress under a load in the x direction, and this can be source of delamination it is also true that this delamination effect can be the key to the formation of the pseudoplastic zone a_o.

On this rationale a value of notch resistance (NR) can be presented as a function of the lamination sequence and laminate thickness. A recent work [52] performed to ascertain the validity of such an approach, has shown that the delamination zone varies with thickness and stacking sequence, following tha variation of the NR parameter.

This effect is shown in Figs. 11-12, where T.B.E. radiographs are reported for 2 CFRP laminates: $(90/0/-45)_s$ in Fig. 11, and $(45/90/-45/0)_s$ in Fig. 12.

The laminate of Fig. 11 has a negative value of NR and the delamination zone is limited to an extension close to the notch; in the laminate of Fig. 12, characterized by a positive NR value, the zone extends far away from the notch.

CONCLUDING REMARKS

The discussion of the last paragraph is devoted to rationalize the strength of the material. It should not be forgotten here that the actua use of the material in the form of the final structure must give guarranties on the damage tolerance of the element, as we mentioned before.

This aspect implies the complete knowledge of the growth rate of flaws or "defects" in the material, and the ability of detecting damages induced by external accidental loads.

Once again, then, we find a close relationship between the mechenical and technological aspects in the design, that is between technologies and possible "characteristics" or "defects" of the material.

However, for complete safety of design, it must be stressed that specific techniques of service inspection need necessarly to be included in the operational lay out.

The aim of continuously monitoring the "defects structure" of the material is to predict and program the maintenance procedures, once the effects of partiuolar shapes and dimensions of flaws are known.

In this regards it is of primary importance that more study and research be made on NDT, and on quality control techniques, in general.

The consequence of improvements of this point will be two-fold for not only the knowledge of the structural behaviour is increased and so the design safety is increased but also the durability of the element can be known with greater accuracy, and the cost of mainteance could be lowered, programming the inspections at given and longer periods of time [53].

4- PROBABILISTIC DESIGN

In this section we try to lighten the reasons why probabilistic concepts are convenient in composite design.

It has been shown that because of some considerations like inaccuracy in determining load conditions and strength properties lead to either increase the supposed loads, or to decrease the expected strength a strictly technical reason is offered to approach the safety problem from a different point introducing statistical concepts.

We are now realizing, though, that equally important economical reasons must be considered in solving safety problems in the best way. We do know that the high performances of composites with organic matrices overcome their present higher costs for it is more efficient on a life cycle basis to invest energy in ligthweight advanced structures rather than in burning it to carry heavier structures.

It is clear that this can be true on a total life cycle basis, that is taking into account all the operational costs during the total life of the structure: this point as far as composites are concerned, is totally dependent on the questions of system reliability and durability.

As long as we succeed in lowering the costs involved with reliability and durability we can assume that composites must be preferred.

Due to safety reasons the deterministic approach to design, cannot rely on given properties of materials and conditions of loading unless a proper safety factor is considered.

We have also seen how durability is related to the possibility of inspecting the structure with available technological means, and to correlate the defect structure to analytical tools , via some failure mode.

It follows that the high performance of composites can be obscured by the presence of "uncertainties" or safety factors, so destroying all economical benefits predictable on the total life-cycle basis.

The aim of a probabilistic design is therefore to give a more physical meaning to the safety factor, and to the concept of reliability.

The basic idea of a probabilistic design is the consideration that each event or parameter connected with the element to be designed cannot be determined or assigned exactly, but can be defined by a statistical law.

The main steps in design are then:

I – Find the analytical form of distribution of "loads"
II–Find the analytical form of distribution of "strengths"
III– Compare distribution of "loads" with distribution of "strengths."

Undoubtedly, many difficulties are encountered in performing what very simply is reported in I, II, and III. First of all, the choice of the analytical form of the distribution density, or frequency of appearence, of the values of strength, and the values of applied loads (clearly, in the term "load" we must include any possible external action on the structure to be designed).

The most commun distribution laws are the normal (Gaussian) and Weibull distributions.

The normal distribution of a given property X is characterized by the following expressions.

The frequency or density distribution of the value X is:

$$f(X) = \frac{1}{\sigma \sqrt{2\pi}} \, exp\left[-\frac{1}{2} \left(\frac{X - \bar{X}}{\sigma} \right)^2 \right] \qquad (18)$$

where σ is the standard deviation of the population of n samples and $\bar{X}$ is the mean value of the population; σ and $\bar{X}$ can be estimated by:

$$\sigma = \sqrt{\frac{1}{n-1} \sum_{1}^{n} (X_i - \bar{X})^2} \qquad (19)$$

$$\bar{X} = \frac{1}{n} \sum_{1}^{n} X_i \qquad (20)$$

where n is the sample size.

Integration of eq. 18 gives the distribution function F(X), as:

$$F(X) = \frac{1}{\sigma \sqrt{2\pi}} \int_{-\infty}^{\infty} exp\left[-\frac{1}{2} \left(\frac{X-\bar{X}}{\sigma} \right)^2 \right] dX \qquad (21)$$

It follows that the probability of failure (if X is a possible strength of the material), that is:

$$Pr\left\{ strength < X_o = stress \right\}$$

can be expressed by:

$$F(X_o) = \frac{1}{\sigma \sqrt{2\pi}} \int_{-\infty}^{X_o} exp\left[-\frac{1}{2} \left(\frac{X-\bar{X}}{\sigma} \right)^2 \right] dX \qquad (22)$$

while the reliability R, (probability of survival), that is:

$$Pr\left\{ strength > X_o = stress \right\}$$

can be expressed by:

$$R(X_o) = \frac{1}{\sigma \sqrt{2\pi}} \int_{X_o}^{\infty} exp\left[-\frac{1}{2} \left(\frac{X-\bar{X}}{\sigma} \right)^2 \right] dX \qquad (23)$$

The Weibull distribution is more adequate to represent variations of properties that cannot be taken as normal.

The Weibull distribution function is expressed by a two parameters relationship:

$$F(X) = 1 - \exp\left[-(\frac{X}{\beta})^{\alpha}\right] \qquad (24)$$

where α is called shape parameter and β scale parameter of the distribution; they can be estimated, for a given sample size n, by using the following expressions:

$$\frac{\sum_{1}^{n} X_i^{\alpha} \ln X_i}{\sum_{1}^{n} X_i^{\alpha}} - \frac{\sum_{1}^{n} \ln X_i}{n} = \frac{1}{\alpha} \qquad (25)$$

$$\left[\frac{1}{n} \sum_{1}^{n} X_i^{\alpha}\right]^{1/\alpha} = \beta \qquad (26)$$

It is worth to note at this point that in presence of a normal distribution we define a "design allowable" on A basis or B basis not only as a function of the probability, but also as a function of the degree of "confidence" we have on the given population of samples. In particular:

A basis Allowable : is the value of strength presented by at least 99% of the samples tested out of a population with a confidence of 95% upon the testing and distribution procedures.

B basis Allowable : is the value of strength presented by at least 90% of the samples tested out of a population with a confidence of 95% upon the testing and distribution procedures.

The concept of confidence implied the concept of sample size, i.e. the number of samples on which the statistic is based.

The sample size should be high enough to permit the choice of the analytical distribution laws closest to the actual distribution of density of the properties measured during the tests.

The exact choice of the distribution laws is very important insofar the main differences among the various curves is on the "tails" of the curves and the behaviours of the tails of the distribution curves are rather important in the assessment about re-liability in a probabilistic sense, as we will show.

We report some basic indications available in the literature, with the aim to show what is possible now to do using probabilistic methods in design, and what is only possible to foresee for more complex cases.

It is also worthwhile to know that present terminology existing in the field of probabilistic design is sometimes confusing. For a complet understanding and learning of methods and terms used in Reliability Analyses, we refer to specific publications where the principles of particular systems like weak-link, parallel rendundant bundle theory, wear-out models, and so on, are well documented 54-59.

It is sufficient here to recall the following conclusions, for systems concerned with assessing total statistical properties of a structure as a function of the properties of all its constituents.

Weak-link: the system fails when any of its component fails. The failure probability can be calculated as follows:

$$P_{WL} = 1 - (1-p_1)(1-p_2)\cdots\cdots(1-p_n) = 1 - \prod_1^n i \; P_i \quad (27)$$

where p_i is the failure probability of the it component.

Parallel redundant: The system fails when all its components have failed. The failure probability is:

$$P_R = p_1 p_2 \cdots\cdots p_n = \prod_1^n i \; P_i \quad (28)$$

If the component probabilities p_i are not expressed numerically, but with a distribution function of the density of failure, the respective laws must be inserted in eqs. 27 and 28 that will give expressions for the total strength distribution $F(S)$.

Fiber bundle theory

Assuming that there are n fibres in a bundle, each with identical strength distribution $F(X)$, if X is a level of the strength:

$$F(X) = P_r \left\{ \text{strength of fiber} \leq X \right\}$$

and density function expressed by:

$$f(X) = \frac{dF(x)}{dx} \qquad\qquad (29)$$

and that the failure is characterized by a sequence of failure of each fibre in any order, the total strength distribution of the bundle is expressed by , if s is the total load:

$$F_B(s) = (m!) \int_c^{s/m} f(x_1) \int_{x_1}^{s/m-1} f(x_2) \cdots \int_{x_{m-1}}^{s} f(x_m)\, dx_m\, dx_{m-1} \cdots dx_1 \qquad (30)$$

RANDOM LOADING FOR STRUCTURAL RELIABILITY

If the applied load stress is also assumed to vary its variation should be included as a distribution function $F_A(s)$. It is usually assumed a normal distribution such that:

$$f_a(s) = \frac{1}{\sigma_s \sqrt{2\pi}} \exp\left[-\frac{1}{2}\left(\frac{s}{\sigma_s}\right)^2 \right] \qquad (31)$$

with $-\infty < s < \infty$, $\sigma_s > 0$, $\bar{s} = 0$ where s is the variable load, σ_s is the standard deviation of the normal distribution, with normal mean value $\bar{s} = 0$ (variation with zero mean); σ and $\bar{s}$ can be in general estimated as follows:

$$\sigma_s = \sqrt{\frac{1}{n-1} \sum_1^n (s_i - \bar{s})^2} \quad ; \quad \bar{s} = \frac{1}{m} \sum_1^n s_i \qquad (32)$$

if n is the sample size, and s_i is the variable parameter.

Integration of eq. 31 to obtain the distribution function $F_A(s)$ is not easy, but can be performed with good approximation by using tables on text books.

STRENGTH AND LOADING BOTH VARIABLE

When both strength and loads are supposed to vary statistically an expression can be derived for the total failure probability [60] once the distribution function of strength F(s) is known either by the weak-link (WL) method, or the bundle (B) method, and the applied load (A) is supposed to have a density function $f_A(s)$. It is:

$$P_{WL} = \int_0^{\infty} F_{WL}(s)\, f_n(s)\, ds \qquad (33)$$

or:

$$P_B = \int_c^\infty F_B(s) f_A(s) ds \qquad (34)$$

When both load and strength vary with a normal distribution, the density functions can be espressed as:
for the strength:

$$f(S) = \frac{1}{\sigma_S \sqrt{2\pi}} \exp\left[-\frac{1}{2}(\frac{S-\bar{S}}{\sigma_S})^2\right] \qquad (35)$$

for the load:

$$f_A(s) = \frac{1}{\sigma_s \sqrt{2\pi}} \exp\left[-\frac{1}{2}(\frac{s-\bar{s}}{\sigma_s})^2\right] \qquad (36)$$

where $\bar{S}$ is the mean value of strength, and $\bar{s}$ the mean value of applied stress, estimated like in eqs (32).

We can calculate the reliability R as the probability that no failure occurs for the entire ranges of values of S and s or, that for any possible combination of load and strength, it is:

$$S > s \quad \text{or} \quad S - s > 0$$

Graphically, the probability of failure is therefore proportional to the overlapping area of the tails of the curves of Fig. 13 61 representing the density distributions of load f(s) and of strength f(S).

For normal distribution, letting $\lambda = S-s$ and $\begin{cases} \bar{\lambda} = \bar{S} - \bar{s} \\ \sigma_\lambda = \sqrt{\sigma_S^2 - \sigma_s^2} \end{cases}$

we can define the difference function $f(\lambda)$ as:

$$f(\lambda) = \frac{1}{\sigma_\lambda \sqrt{2\pi}} \exp\left[-\frac{1}{2}(\frac{\lambda-\bar{\lambda}}{\sigma_\lambda})^2\right] \qquad (37)$$

and defining the realiability as the area under the difference distribution curve between O and $+\infty$, we have:

$$R = \frac{1}{\sigma_\lambda \sqrt{2\pi}} \int_o^\infty \exp\left[-\frac{1}{2}(\frac{\lambda-\bar{\lambda}}{\sigma_\lambda})^2\right] d\lambda \qquad (38)$$

Eq (38) can be solved by approximate methods [62].

Referring to Fig. 14 it is worthwhile to note the difference in assuming deterministically the safety factor as the ratio of the

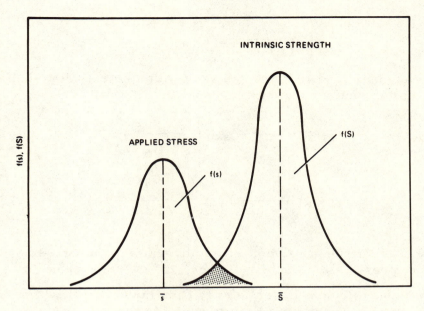

Fig. 13 – Probability of failure as proportional to the overlapping of the tails of the distribution curves [61]

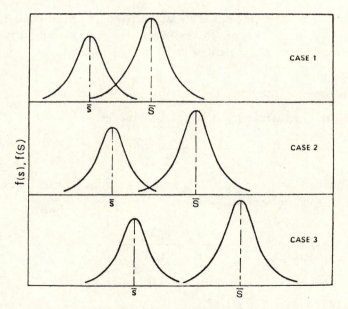

Fig. 14 – Difference between deterministic safety factor and
reliability from distributions curves [61]

mean values of strength to the mean value of load in comparison
to assuming the reliability as proportional to the overlapping areas
of the two density distribution curves. Again, the difference is
clear also for cases of normal distribution with equal mean values,
but with different standard deviations, as shown in Fig. 15. The
latter is clearly indicative that for better reliability, is preferable
to have distribution curves with lower standard deviation rather
than higher mean values.

If strength and applied stress are characterized by distribution functions other than normal, the realiability can then be expressed calculating the failure probability by eqs (33) and (34); graphically the failure probability is then proportional to the overlapping
of the upper tail of the applied stress distribution density $f_A(s)$,
with the lower tail of the strength distribution function $F(s)$.

EXAMPLES OF APPLICATION OF PROBABILISTIC DESIGN

An increasing number of examples can be found when probabilistic concepts are being applied to actual cases of design.

We will now mention some of the simplest ones, trying to
evidentiate the main problems still waiting for answer in more
complex situations.

We have already mentioned how characteristic flaws or defects due to the particular technology employed do play the pre-
minent role in assessing durability and damage tolerance in a
composite structure and how the knowledge of the dynamics of
crack advance rate is affecting maintenance cycles and therefore
the total or life-cycle basic cost of the structure.

Two criteria are now extensively used to analize the damage
tolerance with regard to crack advance under cyclec fatigue with
constant amplitude, namely the stable flow growth model and
the wear-out model.

The first model controls the dimension of the crack at the
applied stress s by:

$$\frac{da}{dn} = C s^{2r} a^r \qquad (39)$$

where a = instant crack length
s = stress range = $s_{max} - s_{min}$

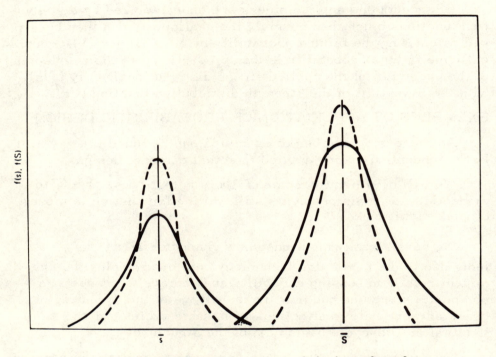

Fig. 15 – Difference between deterministic safety factors
        and reliabilities from different distribution
        curves   [61]

r = growth rate exponent $> 1, \simeq 2$
c = material constant
n = number of fatigue cycles
r is experimentally determined

If a_0 and a_n are the crack length for n = 0 and after n cycles respectively, we have:

$$a_0^{1-r} - a_n^{1-r} = (r-1) C n s^{2r} \qquad (40)$$

If we assume not a deterministic value of a_0 but a statistically variable value, that is a distribution function $F(a_0)$ such that

$$F(a_0) \simeq P_r \left\{ \text{initial flaw size} \leq a_0 \right\}$$

and

$$f(a_0) = \frac{dF(a_0)}{da_0}$$

we can derive then a density distribution function for the flaw size after n fatigue cycles.

The wear-out model [40] tends to by-pass flaw growth treating directly the variation of strength during cyclic stressing, still basing the variation of strength on the flaw growth rate.

If we call F(n) the strength after n cycles, F(o) the initial strength, s_{max} the maximum applied stress, we can derive the following expression:

$$F(n)^{2(r-1)} = F(o)^{2(r-1)} - (r-1) A s_{max}^{2r} n \qquad (41)$$

where r is again the experimental crack growth rate exponent, A a material constant, n the number of fatigue cycles.

The wear-out model can be used for probabilistic design introducing a distribution function for the initial strength and deriving a distribution function for strength after n cycles. This model has been applied, for instance, to design adhesive joints [63] and bolted scarf joints [64].

An example of probabilistic design taking into account variability of both strength and applied stress has been considered with reference to burst strength of filament wound tubes [65], showing that is possible to use eqs 33 or 34 after choosing the most appropriate distribution and density function for applied stress and strength when only a limited number of tests is available 40 .

APPLICATION TO COMPLEX STRUCTURES

When the stress analysis of a complex structures cannot be performed with close solution, a ripetitive sequence of operations must be iteratively performed using numerical methods with automatic computation, in order to obtain a solution for stress state and strains on the structure. Obviously methods using finite elements (FE) analysis will not yeld other than deterministic solution, for deterministic inputs must be given to the computer.

The idea to treat FE models with parametric imputs and distribution functions will lead to very expensive run.

It has been proposed to use an approximate method, and an example is reported supposing a uniform distribution of applied loads [66], where a unique failure mode (buckling, in the example) is analyzed by calculating the possible effects of each parameter affecting, with its distributions functions of properties, the critical buckling load.

The method utilizes a computerized simulation technique, based on the Monte Carlo method, to simulate all possible situations of presence of the single parameters each with its own probability of occurrance.

The random variables considered, in the example shown, are:

elastic constraints = Uniform distribution
EI   = Normal distribution
applied load   = Uniform distribution

In the evaluation of EI, the single influence of sub-parameters, like:

Ply thickness = Uniform distribution
Ply angular error = Uniform distribution
Elastic modulus for composites = Uniform and normal distributions

were evaluated by a Variance Separation method, to indicate what are the single probabilistic effects of each parameters, with the given distribution of values.

The method, although the example was not of particular complexity, is applicable to other structures: the limit is, however, the presence of a single failure mode.

The next step would be the application of the model as a single step in a more complex Fault Tree structure [67] , where each single step would consider the possibility (or probability) of a single event which will eventually lead to failure.

Fault tree analysis is a well known method and the application to composite reability is highly recommanded, in spite of the apparent complexity of its structure [68-70] .

5 - CONCLUSIONS

The aim of the present paper has been to show what are the most peculiar aspects on designing a composite structure, considering the properties of the material from a more general point of view taking into account the influence of technology on the design.

In this regard, we have shown how the design stage is actually encompassing all the different faces of a composite realization, and must reflect all the knowledge and all the unknowledge about composite behaviour.

In particular it has been shown that a proper design must consider element mission, technological aspects, material properties at the same time, and that there is still a large field of uncertainty, which much research work must be focused upon, if composite design has to became a familiar language.

Among the principal points that do need further analysis we include:

- further standardization of materials, testing procedures, technological procedures

- influence if environment upon composites under stress

- influence of matrix strain-to-failure upon laminate stress-strain curve, and research on high-strain matrices

- influence of standardized technological parameters upon defect structure

- composite quality control basic concepts and technological means of NDI

- relationships between FM/defects/technology

- probabilistic design application to FE techniques.

REFERENCES

[1]   I. Crevelli Visconti, "Basic Design for structural Applications
                                 of composite" in Adv. in Composite Materi
                                 ed. Piatti, Applied Science Publ. England
                                 p. 75, 1977

[2]   I. Crivelli Visconti, S. Mignosi, T. Sgobba,"Elementi di
                                 statistica per prove di laboratorio e am-
                                 missibili" Giorn. Aeritalia, CMC ,Naples
                                 nov. 1980

[3]   C.C. Chamis   "Design of composite structural component
                             in Composite Materials, vol. 8 ed. by C.C
                             Chamis, p. 231, Accademic Press.N.Y.
                             1975

[4]   A. Kelly   "European SAMPE meeting , London,Heath
                            Airport, Nov. 1980.

[5]   E.M. Wu,D.C. Rhumann"Stress rupture of glass epoxy Com-
                            posites: Environmental and stress effects"
                            Composite Realiability, ASTM, STP 580
                            p. 263, 1975

[6]   A. Maceri, I. Crivelli Viscont, "Sui moduli elastici di compositi
                                 con fibre unidirezionali" Ing. Mecc.31,1,9

[7]   I.Crivelli Visconti"Materiali Compositi" Tamburini,Milano,197

[8]   Toschio Tenimoto and Sadao Amiime"Progressive nature of
                                 Damage of Glass Fiber Reinforced Plastics
                                 Journal of Composite 1975 pp.380-390

[9]   D. Harris,   "Fatigue and accumulation of Damage in
                            Reinforced Plastics" Composite Vol.8
                            (1977),pp. 214-219

[10]   N.J.Owen   " Fatigue Testing of fibre reinforced pla-
                            stics!' Composite 1970 , pp. 346-355

[11]   Bharan, C.K.H.   " Fatigue failure in fiber reinforced ma-
                            terials" Proceeding of the Internat. Conf.
                            on Composite Materials - E. Scala,
                            E. Anderson, I. Thot, P.R. Noton, Ed.
                            Metallurgical Society, N.Y. 1976,p.
                            830-839.

[12] Hahn, H.J. and Kim, R.Y., "Fatigue behaviour of Composites
                  Laminate"Journal of Composite Materials
                  Vol. 10 (1976) p.156

[13] Beadmore, P. and Kulkarni, H,. "Cyclic Softening Phenomena
                  in Fiber Reinforce Composites" ICCM/2
                  1978, p. 611-21

[14] Yang,J.N.,Miller R.K.,and Sun,C.T."Effects of High Load
                  on Statistical Fatigue of Unotched Graphite
                  Epoxy Laminates", J. Composite Materials
                  Vol.14 , 1980, p.82

[15] Yang J.N., and Sun,C.T.,"Proof Test and Fatigue of Unnotched
                  Composite Laminates",J.Comp. Mat.,
                  Vol. 14, aprile 1980, p. 168

[16] M.J. Owen,   "Fatigue damage in GFRP"Composite
                  Materials, Ed. Broutman-Krock, Vol. 5,ch.7
                  Acc. Press, N.Y. 1975

[17] M.J. Owen,   "Fatigue of CRP"Composite Materials,
                  Ed. Broutman - Vol. 5 ch.8, Acc. Press
                  N.Y. 1975

[18] I.Crivelli Visconti, M. Di Ilio, L. Carrion, G. Macchia
                  "Materiali Compositi: Tecnologie e Pro-
                  gettazione" final report, FIAT CRF re-
                  search contract, Ist. Tecnologie,Napoli,
                  1980

[19] I. Crivelli Visconti, M. Di Ilio, to be published

[20] J. Owen, S.Morris, Ann. Techn.Conf., 25th SPI Conference
                  Washington, DC, Set. 8E, 1970

[21] S. Morris, Ph. D. thesis , Univ. of Nottingham, England,970

[22] M.J. Owen, S. Morris, Int. Conf. Carbon Fibres, Comp.
                  and Appl. Plastics Inst. n° 51,London,971

[23] K. De Booy,   "Battle damage tolerant Wing Structural
                  development"NASA TMX 3377, April 1976
                  Third Conf. on fibrous Comp. in Flight
                  Vehicle design.

[24] S.L. Huang, T.E. Hess,"A hibrid Composite Fuselage de-
                  sign with integral Crack Arresters" Third

Conf. on Fibrous Comp. in fligth Vehicle
Design. NASA TMX-3377, April , 1976

[25] T.E. Hess, S.L. Huang, H. Rubin,"Fracture control in Composite materials using integral crack
arresters",AIAA/ASME/SAE 17th
Structure, StructuraDynamics and Materials Conference, May, 1976

[26] P.H. Petit, M.E. Waddoups,"A method for predicting for
non linear behaviour of laminated com-
posites",J. Comp.Mat.3,2,1969

[27] H.T. Hahn,S.W. Tsai,"On the behaviour of composite laminates after initial failures",J. Comp.
Mat. 8,288, 1974

[28] G.P. Sendeckyi, M.D. Richardson, J.E. Pappas,"Fracture
behaviour of thornel 300/5208 Graphite,
epoxy laminates, Part. 1 : unnotched
laminates" Comp. Reliability, ASTM
STP, 580, p. 528, 1975

[29] N.B. Pagano, R. Byron Pipes,"The influence of stacking sequence on laminate strength",J. Com.
Mat.5,50, 1971

[30] R. Byron Pieps, M. Daniel, "Moirè analysis of the interlaminar shear edge effects in laminated
composites" , J. Comp.Mat.5,255,1971

[31] G. Caprino, L. Carrino,"Influenza degli effetti di bordo sul-
la resistenza all'intaglio di laminati in
carbo-resina" IV Nat. I° Intern. Meeting
on Comp. Mat.,Milano,19-21 nov.1980
CMC - Naples

[32] R.L. Foye, D.J. Becker, " Design of orthotropic laminates"
11th Ann. Conf. AIAA, Denver, Col.,
Aprile, 1970

[33] B.E. Kaminski,   "On the determination of the failure
surface for an orthotropic quasi-homo-
geneous material",Master thesis Georgia
Inst. of Technology, 1972

[34] R. Byron, N.B. Pagano, "Interlaminar stresses in composite laminates under Uniform axial extension", I.Comp.Mat. 4,538, 1970

[35] M.F. Kanninen, E.F. Rybicki, H.F. Brinson, "A critical look at current applications of fracture mechanics 1 the failure of fibre reinforced composites", Composites, 8,1,p.17,1977

[36] C. Zweben, "Fracture mechanics and composite materials a critical analysis" ASTM STP 521,p.65 1973

[37] G. Caprino, I.Crivelli Visconti"Meccanica della frattura e fatica nei materiali compositi" Convegno Scuola su "Prop. mecc. e frattura dei mat. polimerici", Tirrenia, Italy, 2-5 June, 1980 Ass. Ital. Sc. Tecn. Macromolecole.

[38] G.C. Sih, P.D. Hilton, R. Vadaliance, P.S. Shenberger, G. Villarreal, "Fracture mechanics for fibrous composites", ASTM STP 521, p.98,1973

[39] G. Caprino, J.C. Halpin, L. Nicolais,"Fracture mechanics in Composite materials"

[40] J.C. Halpin, K.L. Jerina, T.A. Johnson,"Analysis of the test methods for high modulus fibres and composites" ASTM STP 521,p.5,1973

[41] G.A. Cooper, A. Kelly, "Role of the interface in the fracture of fibre composite materials", ASRM STP 452, p. 90, 1969

[42] A. Kelly, "Interface effects and the work of fracture of a fibrous composite",Proc. Roy.Soc. A 319, p. 95, 1970

[43] G.A. Cooper, "The fracture toughness of composites reinforced with weakaned fibres" J.Mat. Science, 5,p.645,1970

[44] M.R. Piggott, "Theoretical estimation of fracture toughness of fibrous composites" J.Mat. Sc. 5, p.669, 1970

[45] G.C. Sih, P.C. Paris, G.R. Irwin, Int. J. Fracture Mechanics, 1, 1965

[46] M.E. Waddoups, J.R. Eisenmann, B.E. Kaminski, "Macroscopic fracture of advanced composite materials", J. Comp. Mat., 5, 1971

[47] J.A. Werbuch, M.T. Hahn, "Crack tip damage and fracture toughness of B-Al Composites", J. Comp. Mat., 13, p. 82, 1979

[48] P.W.R. Beaumont, A.S. Teleman, "The fracture and toughness of fibrous composites" Failure mode in cor posites, I.I. Toth ed., Met. Soc. of AIME p. 49, N.Y., 1973

[49] R.B. Sanford, R.F. Stonesifer "Fracture toughness measurement in unidrectional GRP", J. Comp. Mat., 5, p. 244, 1971

[50] R. Byron, R.C. Wetherhold, J.W.Jr. Gillespie, "Notched strength of composite materials", J. Comp. Mat. 13, 1, p. 148, 1979

[51] G. Caprino, unpublished work

[52] G. Caprino, V. Renta, S. Mignosi, private comunication

[53] J.E. McCarty, R.W. Johnson, "Durable and damage tolerant composite commercial aircraft structure desic approach" J. Aircraft, 15, 1, p. 33; 1978

[54] L. Breiman,   "Statistics with a view toward applications Houghton Mifflin, Co. Boston, Mass. USA, 1973

[55] R.E. Barlow, F. Proschan, "Mathematical theory of realiability", Wiley, N.Y. 1967

[56] D.D. Bien,   "Optimum allocation of redundancy among subsystems connected in series", NASA TN D-7764, March, 1973

[57] H.E. Daniels,   "Statistical theory of the strength of bundles of fibres", Proc. Roy. Soc., A, Vol. 8 N° A 995, 1945

[58] C.A. Cornell,   "A probability based structural code"

paper n° 66-85, J. American Conc. Inst.
Dec. 1969

[59] J.F. McCarthy, O. Orringer,"Some approach to assessing
failure probability od redundant structures",
Composite Realiability, ASTM, STP 580,
p.5,1975

[60] M. Shinozuka, J.T.P. Yaò, A. Nishimura" A note on the re-
liability of redundant. structures" TR n°27
Inst. for study of fatigue and reliability
Columbia, Univ. N.Y. ,nov. 1965

[61] B.H. Jones,   "Probabilistic design and reliability",
Comp. Mat., Broutman-Krock ed., Vol.8
n°8 p.33 Chamis ed. Acc. Press. Publ.
N.Y. 1975

[62] C. Hastings,   "Approximation for digital computers,
Princ. Univ. Press., 1955

[63] A.P. Berens ,B.S. West,"Evaluation of an accelerated characterization technique for reliability asses-
sment of adhesive joints", Comp. Reliability, ASTM, STP 580,p.90,1975

[64] D.L. Reed,J.R. Eisenmann,"Reliability aspects of a composite bolted scarf joint", Composite re-
liability ASTM STP 580, p.105,1975

[65] E.M. Lenoe ,D.Neal, "Structural Integrity assessment of
filament wound components", Comp.
Reliability, ASTM, STP 580,p.54,1975

[66] R.E. Maxwell,R.H. Toland, W.C. Johnson, "Probabilistic
design of composite structures"Comp.
Reliability, ASTM STP 580,p.35,1975

[67] J.E. Masters, Y.T. Yeow,M.R. Louthan,K.L. Reifsmider,
H.F. Brinson,   "A qualitative fault tree analysis for the
tensile failure of fibrous laminated com-
posites", Composites, 8,n°2,p.111,1977

[68] D.F. Haasl,   "Advanced concepts of fault tree analysis "
System Safety Symposium, Boeing Compa-
ny, Seattle, Wash., June 1965

[69] F.A. Scerbo ,J.J. Pritchard, "Fault tree analysis; a technique for product safety evaluation", ASME
publ. 75 -SAF-3.

[70] J.M. M chels   "Computer evaluation on the safety fault
tree model" System safety Symposium,
Boeing Co. , Seattle, Wash.,June, 1965.

COMPRESSION FATIGUE BEHAVIOR OF NOTCHED COMPOSITE LAMINATES*

Richard M. Walsh, Jr. and R. Byron Pipes

Center for Composite Materials
University of Delaware
Newark, Delaware, 19711

ABSTRACT

The influence of compression fatigue on graphite/epoxy laminates containing circular holes of 0.25 inch diameter was investigated. The laminate stacking sequences were $[0_2/\pm45]_{5S}$, $[0/45/0/-45]_{5S}$, $[0/\pm45/90]_{5S}$, and $[90/0/\pm45]_{5S}$. Specifically this study examined the nature and extent of induced compression fatigue damage and the determination of the effects of this damage on the laminate residual failure mechanisms. Two modes of compressive failure were found to occur: the diagonal shear and the net compression modes. Both failure modes were characterized by local instability of individual lamina or small laminae subgroupings, with diagonal shear predominant in the fiber-dominated laminates, and net compression predominant in the quasi-isotropic laminates. The mode and direction of failure were dependent upon the nature of the specimen delamination. It was also found that the laminate stacking sequence influenced the intraply crack development in the laminates as well as the failure mode. The failure mechanisms were essentially the same for two different material systems which were studied (Narmco 5208 and 5209).

INTRODUCTION

The increasing use of graphite/epoxy laminates in the aerospace industry, along with the potential for more widespread structural applications in the future, places a significant emphasis on understanding the fatigue characteristics of these materials. Such under-

---

*Supported by Materials Sciences Corp., Blue Bell, PA, USA, and U.S. Naval Air Systems Command, Contract No. N00019-79-C-0633.

standing is essential in facilitating the efficient and safe design
of structures utilizing these laminates. Fatigue studies dealing with
tension-tension cycling have resulted in favorable behavior which is
superior to that of metals[1,2]. Other studies of composites in tension-
compression and compression-compression fatigue[2], however, indicate
a substantial reduction in the fatigue life as compared to the tension-
tension case. This behavior underscores the added importance of fully
characterizing the fatigue behavior and failure mechanisms resulting
from compressive loading.

The mechanism of failure in compressive fatigue[1] has been observed as beginning with local matrix failure in the vicinity of a
stress concentration. Failure of the matrix then leads to fiber
splitting followed by progressive delamination and local instability.
In view of this deleterious mechanism of failure, an area of particular interest is that of compression-compression fatigue of notched
laminates. It is this topic which is the subject of this work.

This experimental study focused on the characterization of the
nature and extent of the induced fatigue damage in the graphite/epoxy
laminate, specifically in the vicinity of a circular hole. Of particular interest were the mechanisms of failure corresponding to compressive residual strength tests. Specimens of each laminate were
tested statically to establish the initial static strength, and identical specimens were then subjected to compression-compression fatigue
loading to induce damage in the area of a circular hole. Throughout
the program, nondestructive evaluation (NDE) of the damage was performed by means of ultrasonic C-scan techniques. Test samples were
also dissected for micrographic analysis. The failure mechanisms of
the fatigued specimens were observed by means of subsequent static
compressive loading to failure.

TEST PROGRAM

Fabrication of the test specimens was performed using Narmco
(T300/5208 and T300/5209) graphite/epoxy prepreg material. Four laminate configurations were examined for each resin system and are denoted as laminates A, B, C, and D as follows:

A: $[0/0/45/-45/0/0/45/-45/0/0/45/-45/0/0/45/-45/0/0/45/-45]_s$

B: $[0/45/0/-45/0/45/0/-45/0/45/0/-45/0/45/0/-45/0/45/0/-45]_s$

C: $[0/45/-45/90/0/45/-45/90/0/45/-45/90/0/45/-45/90/0/45/-45/90]_s$

D: $[90/0/45/-45/90/0/45/-45/90/0/45/-45/90/0/45/-45/90/0/45/-45]_s$

Panels with these stacking sequences were laid up and cured in an
autoclave using a two-step cure cycle with maximum process tempera-

tures of 350°F (177°C) for 5208 and 250°F (121°C) for 5209, and a
pressure of 85-100 psi (0.586-0.689 MPa) with no post curing. The
endtab material was a glass-reinforced, epoxy resin material. The
adhesive used to bond the entabs to the graphite/epoxy specimens was
Hysol Aerospace Adhesive EA 9309 which was selected due to its high
peel strength.

     The test specimen configuration along with the dimensions are
shown in Fig. 1. The specimens were 1.0 inch (2.54 cm) in width with
a 0.250 inch (0.635 cm) diameter circular hole centrally located in
the unsupported span. In determining the length of the unsupported
span (gage length), the objective was to maximize this dimension, and
thereby minimize any grip effects, without reaching a state of instability which would result in gross buckling of the laminate. A gage
length of 2.0 inches (5.08 cm) was selected, and found to exhibit no
signs of gross laminate buckling based on strain gage readings made of
selected specimens in static tests. All tests were performed with an
Instron Model 1321 closed-loop, servo-hydraulic test machine. The
test fixture was the modified IITRI compression fixture shown in
Fig. 2.[4]

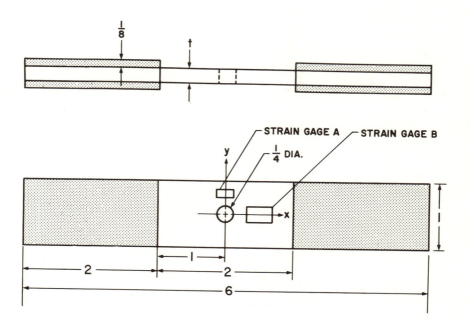

MICRO-MEASUREMENTS
STRAIN GAGE A: EA-06-125BZ-350
STRAIN GAGE B: EA-06-125AC-350

Fig. 1. Notched Fatigue Specimen

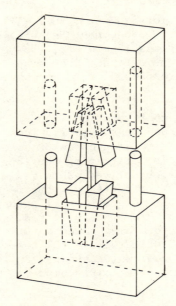

Fig. 2. Modified IITRI Compression Fixture

Specimens of each laminate were tested statically to establish
the initial static strength. Other identical specimens were subjected
to fatigue loading followed by residual strength evaluation. The
fatigue schedule is summarized in Table 1. Certain of the test samples were sectioned after fatigue loading in order to conduct micrographic analyses. The section orientations are shown in Fig. 3.

Table 1. Fatigue Program

(R = 0.1, frequency of 10 Hertz, ambient conditions)

Specimen	S-Level	Cycles
1	(Static)	--
2	(Static)	--
3	(Static)	--
4	0.50	1×10^6
5	0.50	1×10^6
6	0.50	1×10^6
7	0.60	1×10^6
8	0.60	1×10^6
9	0.60	1×10^6
10	0.60	5×10^5
11	0.60	5×10^5
12	0.60	5×10^5

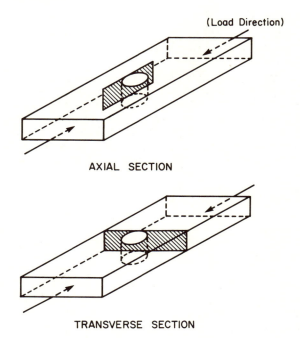

Fig. 3. Micrograph Section Orientations

Progressive damage characterization of specimens subjected to fatigue loading was monitored through the use of ultrasonic C-scan techniques which were utilized as a means of nondestructive evaluation (NDE). The specific technique which was employed is the peak amplitude based C-scan.[5] In order to monitor the initiation and subsequent growth of the fatigue-induced damage, all specimens were removed from the test machine at regular intervals of 2.5×10^5 cycles and inspected via the C-scan. Upon completion of the fatigue tests, a series of C-scans was assembled from which the damage sequence could be characterized.

FAILURE MECHANISMS

Since the two resin systems (5208 and 5209) possessed different maximum process temperatures, it was expected that resin controlled failure mechanisms might be different for the two systems. In actuality, the static failure mechanisms for the two material systems were indistinguishable. However, the failure mechanisms for the four laminates exhibited significant differences.

Two distinct failure modes were observed for laminate "A". They
were the diagonal shear (DS) failure and the net compression (NC)
failure modes which are illustrated in Fig. 4.

Fatigue damage in laminate "A" was manifest as cracking of the
+45° and. -45° laminae in regions adjacent to the central notch. In
regions where the +45° laminae were cracked, the -45° laminae were
not, and in regions where the -45° laminae were cracked, the +45° laminae were not, as illustrated in Figs. 5 and 6. As can be seen, the
-45° laminae exhibited cracking in the first and third quadrants while
the +45° laminae cracked in the second and fourth quadrants.

The origin of the fatigue cracking in the ±45° laminae must be
attributed to the shearing stresses which act tangent to the circular
notch and parallel to the load direction. The distribution of the
shearing stress in this region is an antisymmetric function as shown
in Fig. 7. The direction of the shearing stress may be determined by
a free body diagram as in Fig. 8. Acting as the remote end of the
free body is the constant, applied stress σ, while at the section
adjacent to the hole, the compressive stress varies from $K_T^\infty \sigma$ [†] at the
edge of the hole to σ at some distance from the hole. Since the two
resultant forces are not equal, a third force is required for equilibrium. The third force is the resultant of the shearing stress and

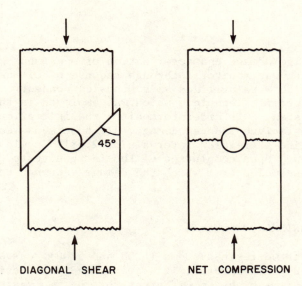

DIAGONAL SHEAR   NET COMPRESSION

Fig. 4. Notched Compression Failure Modes

[†]where K_T^∞ is the elastic stress concentration factor.

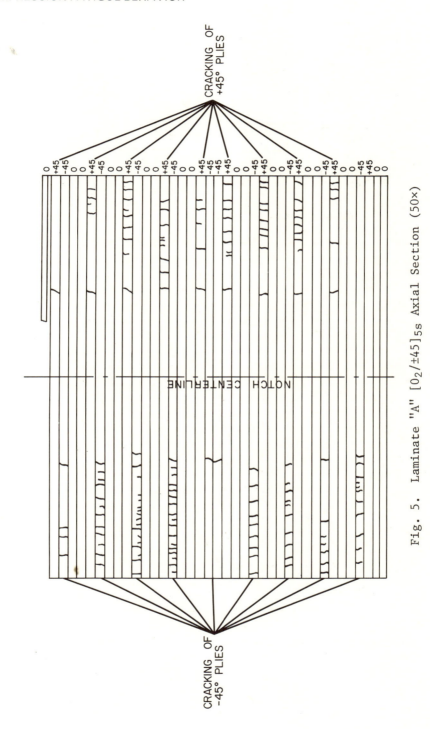

Fig. 5. Laminate "A" $[0_2/\pm45]_{5s}$ Axial Section (50×)

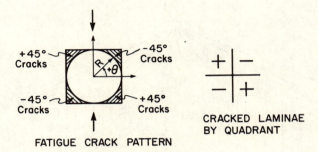

Fig. 6.   Fatigue crack pattern.

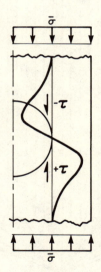

Fig. 7.   Antisymmetric Distribution of the Shearing Stress, τ

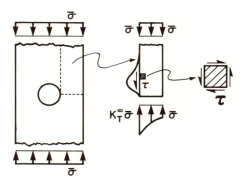

Fig. 8. Direction of the Shearing Stress, τ

its direction is the same as
the remote applied stress since
the resultant force at the section adjacent to the hole will
always be greater than the resultant of the remote stress.
Hence, when the shearing stress
is transformed to a coordinate
system rotated 45 degrees from
the axis of the specimen, transverse tensile stresses act on
the -45° laminae and transverse
compressive stresses act on the
+45° laminae in the first and
third quadrants. This behavior
is reversed in the second and
fourth quadrants as shown in
Fig. 9.

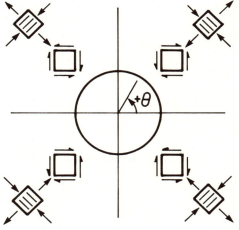

One significant observation
of the residual strength failure
of laminate "A" was that the diagonal shear failure appeared to be
primarily responsible for laminate failure, although net compression failure mode was also
present. Further, the diagonal
shear failure mode always occurred

Fig. 9. Shear Induced Transverse
Tensile Stresses

parallel to the +45° laminae. No -45° diagonal shear failures were
observed for laminate "A". Since the +45° laminae in laminate "A"
are located nearest the surface, it might be expected that cracks in
these laminae, developed during fatigue cycling, may play a role in
the ultimate failure mode.

Another mechanism characteristic of the "A" laminate was fatigue
growth of cracks in the 0° laminae parallel to the load direction and
tangent to the edge of the circular notch. The cracks occurred in all
0° laminae only, with their surfaces contained in a single plane tangent to the notch as shown in Figure 10. The delamination which grew
at the intersections of the axial cracks and the near-surface 0/45
interface were mapped in planform by ultrasonic techniques. As can
be seen in Fig. 11, the delamination progresses monotonically with
increasing cycles or stress amplitude.

The delamination progression was such that it initiated on both
sides of the notch at the axial tangent points and propagated towards
the endtabs as shown in Fig. 11. The regions of delamination initially developed in four quandrants about the notch in a uniform and
symmetric manner. Growth of each region progressed along the axial
tangent line to the notch, as well as along the circumference of the
notch towards the center of the specimen. This circumferential progression eventually resulted in the bridging of the delamination in
the upper two quadrants and, similarly, the lower two quadrants.
Upon completion of the joining of these regions, the delamination
continued towards the endtabs, contained within a region bounded by
the axial tangent lines to the notch.

The laminate "B" failures were similar to those of laminate "A"
in that they also exhibited characteristics of the diagonal shear and
the net compression failure modes. Diagonal shear was the predominant
mode of failure with net compression appearing in a small number of
instances. In contrast to laminate "A", the shearing direction was
not always parallel to the near-surface angle ply (+45°). The failures occurred in the -45° direction as well as the +45° direction.
Another distinct diagonal shear failure which was observed was the
"dual shear" failure in which the shear failure was in the +45° direction on one side of the notch, and in the -45° direction on the other
side of the notch as illustrated in Fig. 12. In characterizing each
of these shear failures, it was found that the mode of delamination
determined the direction of failure. In all instances, the specimen
delamination was such that subgroups of two 0° plies with an angle
ply "sandwiched" in between were formed. This resulted in two types
of delamination which determined the diagonal shear failure direction.
The first case (I) is that in which the subgroups had a configuration
of [0/+45/0]. In this case, the direction of shear failure was that
of the +45° lamina. The subgroups of the second case (II) had a
[0/-45/0] configuration resulting in a -45° shear direction. Dual
shear failure took place when both delamination cases were present,

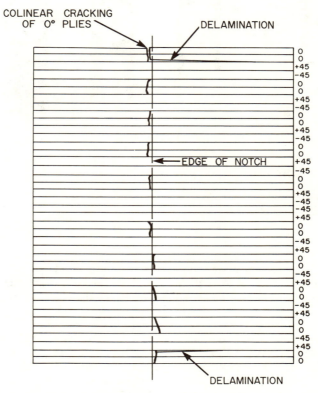

Fig. 10. Laminate "A" $[0_2/\pm45]_{5s}$ Transverse Section (50×)

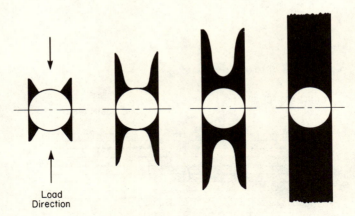

Load
Direction

Fig. 11.   Near-Surface Delamination Growth: Laminate "A"

with case I on one side of the notch and case II on the other side.
A similar effect was present in the "A" laminate in which the 0/+45
interface consistently remained intact, resulting in a +45° shear
direction.

A further observation pertaining to the diagonal shear failure
mode was that the origin of the crack appears to be at the 45° tangent
point to the hole as shown in Fig. 13.   Also, microscopic analysis of
laminate "B" revealed that while intralaminar cracking of the angle
plies was present as in laminate "A", it occurred to a lesser degree.
This result is apparently linked to the difference in stacking sequence between laminates "A" and "B".   Despite this, however, the
cracking of the angle plies and its subsequent relation to the diagonal shear failure mode observed in laminate "B" is consistent with
the previous explanation of this failure mode in laminate "A".

The net compression failure mode was present along with the diagonal shear failure mode in only a small number of specimens tested.
In those instances, it was generally confined to the near-surface
layers with the shear failure extending through most of the specimen
thickness.   The net compression failures were characterized by delam-

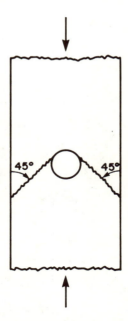

Fig. 12. Dual Shear Compression Failure Mode

ination between the angle plies and the 0° plies, leaving individual
0° plies unsupported and prone to local instability. This is in contrast to the subgroupings of the 0° plies and the angle plies in the
diagonal shear mode. Between the zones of diagonal shear and net
compression failures were large interlaminar cracks which extended
past the damage zone parallel to the loading direction, as well as
from edge to edge in the transverse direction.

The overall failure sequence in laminate "B" was one dominated
by the angle plies as has been shown. In this sequence, out-of-plane
deformation causes delamination of the laminate into subgroups and
induces bending stresses which fail the 0° plies parallel to the
cracks in the "sandwiched" angle ply, with the direction being determined by that of the angle ply. Net compression infrequently followed
the shear failure due to unsymmetric loading which resulted after the
shear failure stopped at the previously mentioned large interlaminar
crack.

The mode of near-surface delamination in laminate "B" proved to
be the same as that described for laminate "A" with the significant
contrasting element being the degree with which the delamination

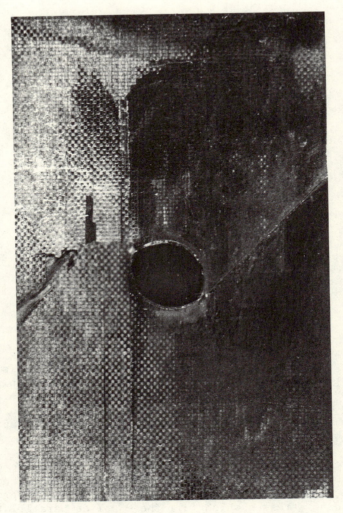

Fig. 13.   Laminate "B" Diagnol Shear Failure

took place. Specifically, the delamination monitored in laminate "B"
did not develop to the extent that it did in laminate "A" despite comparable stress levels and number of fatigue cycles. Delamination in
laminate "B" was generally confined to small regions in the four quadrants, often with no significant delamination in a number of the quadrants.

The quasi-isotropic laminates "C" and "D" exhibited failure mechanisms common to both. The predominant failure mode was that of net
compression with the fracture line running transverse to the loading
direction, and directly through the center of the notch. Extensive
interlaminar cracks were observed extending past the damage zone
parallel to the loading direction and from edge to edge in the transverse direction. These were similar to those observed in laminates
"A" and "B" which were present between zones of diagonal shear and
net compression failure. In most, cases, the cracks developed at the
0/+45 degree interface.

Microscopic inspection of laminates "C" and "D" revealed minimal
intralaminar cracking of the individual laminae, particularly the
angle plies, which contrasts with the results found in laminate "A".
An explanation for this may be made based on the presence of both 0°
and 90° plies in the quasi-isotropic laminates which effectively
resist both components of shear stress by strengthening the laminate
in both the axial and transverse directions. Their presence results
in a reduction of the transverse deformation of the angle plies and,
hence, a reduction in the cracking as well.

Also discovered in the microscopic inspection was the presence
of fatigue-induced delamination at the [0/+45] degree interface ini-
tiating in the near surface layers. An additional factor to be considered in the failure of the quasi-isotropic laminates is that the
presence of a 90° ply in the stacking sequence results in a reduced
axial stiffness in the quasi-isotropic laminate. Such a decrease
leads to an increased tendency for local and global instability.

The various factors observed in the failure of the quasi-isotropic
laminates indicate that failure is the result of local instability of
the individual 0° plies. This is supported by the previous findings
which reveal a failure sequence involving delamination of the 0° plies
and their adjacent, uncracked 45° plies, leaving unsupported individual 0° plies which subsequently fail in a mode I buckling failure.
Such a sequence initiates in the outermost plies and progresses
towards the center of the laminate. An example of such a configuration is shown in Fig. 14.

The fatigue-induced delamination in the near surface layers of
laminate "C" developed in essentially the same manner as that described for laminate "A" occurring also at the [0/+45] degree interface nearest the surface. The one difference which was found was that

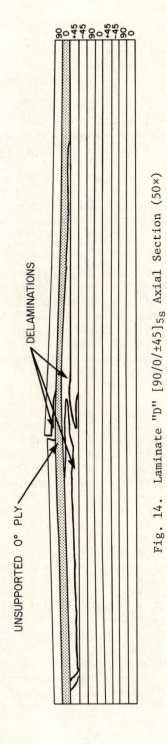

Fig. 14. Laminate "D" [90/0/±45]₅ₛ Axial Section (50×)

in laminate "C" the regions of delamination did not remain confined
within the two axial tangent lines to the notch as was the case in
laminate "A". The delamination was observed to extend in the transverse direction beyond these lines, thereby encompassing the entire
perimeter of the notch as in Figure 15. This transverse development
was relatively small, however, with the primary direction of delamination progression being the axial one, as in laminates "A" and "B".

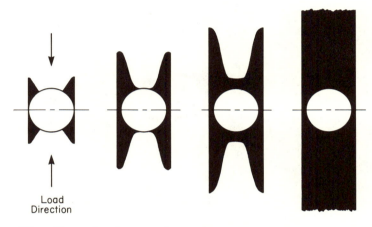

Fig. 15. Near-Surface Delamination Growth: Laminate "C"

    The near-surface delamination found in laminate "D" was interest-
ing in that while it also initially formed in the four quadrants as
described for laminate "A", the primary direction of delamination
progression was the transverse direction. The regions of delamination
were not confined within the transverse tangent lines to the notch,
but instead extended in the axial direction, encompassing the notch
perimeter as in laminate "C". This near-surface delamination took
place at the outmost [90/0] interface and is illustrated in Fig. 16.

    Based on these results, and those of the ultrasonic examinations
of the other three laminates throughout the fatigue process, it can
be concluded that the primary direction of the near-surface delamina-

Fig. 16. Near-Surface Delamination Growth: Laminate "D"

tion progression is determined by the fiber direction of the laminate surface ply.

Comparison of the failure mechanisms of the fatigued specimens with those from preliminary tests on undamaged specimens showed some differences in the fiber-dominated laminates, and no significant difference in the quasi-isotropic laminates. In the case of the fiber-dominated laminates the indication was that the diagonal shear mode was not dominant in the undamaged specimens, with the net compression mode playing a more significant role than was the case in the fatigued specimens. It should be noted that these results are, as yet, incon-

clusive due to the small data base involved. However, the implication
here is that there may be an important link between the fatigue-
induced cracking of the laminate and the predominant mode of failure.
This is an area which warrants further study.

CONCLUSIONS

     The objective of this research was the investigation of the ef-
fects of compression-compression fatigue on graphite/epoxy laminates
containing a 0.25 inch diameter circular hole, with the focus of this
paper on the mechanisms of failure of the laminates subsequent to the
fatigue cycling. The results of the tests on the four different laminate configurations revealed the existence of two modes of compressive failure which were the diagonal shear (DS) mode and the net compression (NC) mode. The diagonal shear failure was the predominant
mode in the fiber-dominated laminates, while the net compression
failure was predominant in the quasi-isotropic laminates. The diagonal shear mode was characterized by local instability of small lamina subgroupings which developed as a result of progressive delaminations of the composite. The direction of the laminate shear failure
was that of the "sandwiched" angle ply in the lamina subgroupings.
The net compression failure was characterized by local instability
of individual 0° laminae which were separated due to similar progressive delaminations of the composite.

     An important factor related to the modes of failure for the lami-
nates was the nature of the intralaminar cracking which resulted from
the fatigue cycling. The laminate stacking sequence played an important role in determining the nature of this cracking. By analyzing
the stress distribution about the notch, it was found that the positive or negative angle plies were subjected to transverse tension
depending on their quadrant location. In laminates "A" and "B", significant cracking of the angle plies resulted, which is consistent
with the predominance of the diagonal shear failure mode in these
laminates. While the nature of the cracking was very similar for
these two laminates, there was also a distinction between the two
stacking sequences in that the cracking occurred to a lesser degree
in laminate "B", and laminate "A" also exhibited colinear cracking of
the 0° plies tangent to the notch. The quasi-isotropic laminates "C"
and "D", however, did not exhibit any significant intralaminar cracking, which is consistent with the predominance of the net compression
failure mode in these laminates. Therefore, the laminate stacking
sequence was found to affect the nature of the cracking as well as the
mechanism of failure. Preliminary results also indicate the possibility of a direct relation between the nature of the fatigue-induced
intralaminar cracking and the predominant failure mode of the laminate.

     Throughout the fatigue cycling, nondestructive evaluation (NDE) was carried out by means of ultrasonic C-scan techniques. Based on those results, it was concluded that the primary direction of near-surface delamination is the same as the fiber direction of the laminate surface ply.

An additional factor considered in this study was the effect of the resin system on the failure mechanisms. The results showed no discernable difference between the T300-5208 and T300-5209 resin systems at room temperature.

REFERENCES

1.   Rosenfeld, M. S. and Huang, S. L., "Fatigue Characteristics of Graphite/Epoxy Laminates Under Compression Loading," Structures, Structural Dynamics and Materials Conference, March 1977, 423-427.
2.   Ratwani, M. M. and Kan, H. P., "Compression Fatigue Analysis of Fiber Composites," NADC-78049-60, September 1979.
3.   Preston, J. L. "USAF Advanced Composites Design Guide," Air Force Flight Dynamics Laboratory, 1978. Wright-Patterson AFB, Ohio.
4.   Pipes, R. B., "Experimental Methods for Composite Materials," Center for Composite Materials, University of Delaware, Newark, Delaware.
5.   Blake, R. A., Jr., "Ultrasonic NDE of Composite Materials," Center for Composite Materials, University of Delaware, Newark, Delaware, 1-11.
6.   Vinson, J. R., and Chou, T.-W., "Composite Materials and Their Use in Structures," Applied Science Publishers, Essex, England, 1975.
7.   Wetherhold, R. C. and Pipes, R. B., "CMAP-1, Composite Materials Analysis of Plates," Center for Composite Materials, University of Delaware, Newark, Delaware, 1979.
8.   Shames, I. H., "Introduction to Solid Mechanics," Prentice-Hall, Inc., Englewood Cliffs, New Jersey, 1975.
9.   Pipes, R. B., Kulkarni, S. W., and McLaughlin, P. V., "Fatigue Damage in Notched Composite Laminates," Materials Science and Engineering, Blue Bell, Pennsylvania, 1977, 113-120.
10.   Hahn, H. T. and Tsai, S. W., "Fatigue Behavior and Life Prediction of Composite Laminates," AFML-TR-78-43, April 1978.
11.   Ramkumar, R. L., Kulkarni, S. V. and Pipes, R. B., "Definition and Modeling of Critical Flaws in Graphite Fiber Reinforced Epoxy Resin Matrix Composite Materials," NADC-76228-30, January 1978.

CONSIDERATION ON THE FATIGUE DAMAGE OF SPECIMENS USED

FOR COMPOSITE CRITICAL COMPONENTS QUALIFICATION

A. Brivio, G. Parenti, G. Samanni, V. Wagner, and
C. Zanotti

Costr. Aer. G. Augusta
L.S. and L.T.S.
Via Schaparelli, 8, 21013 Gallarate, VA

INTRODUCTION

The fatigue qualification of critical components in Glassfiber Epoxy Composite materials, which are going to be installed on a new design helicopter is carried out according to methods which are also typical for the same components made in metal.

As part of the qualification of a new main rotor blade, a series of fatigue tests on full size prototype components and on simple elementary specimens[1] is performed (fig. 1).

The assessment of the material fatigue performance in terms of the well known S-N (or Whoeller) curves is a critical milestone as the material itself is not a conventional one and its basic characteristics need to be derived.

Because the primary purpose was to carry out the S-N curves evaluation of all the composite systems to be applied, a series of fatigue tests was started in 1979 on specimens in two particular lay-ups which were going to represent the blade main constituents. Unidirectional 0° laminates in R/S and +45°/-45° laminates in E Glassfiber/epoxy systems have been subjected to fatigue tests and the relevant S-N curves were derived.

The results obtained in such tests have been considered enough satisfactory to allow to proceed with the anticipated qualification program. However, a more fundamental interpretation of the fatigue tests incurs always into the difficulty of fatigue damage interpretation. Particularly on 0° laminates such damage seems to be completely masked by different degradation processes taking place below the end tabs[2]. A part of this study is dedicated to the investigation of such phenomena with the aim of better understanding them. This contributes now to the justification of the validity of the experimental data and may help in the future to find methods to reduce such unwanted end-tabs effects.

EXPERIMENTAL PROCEDURE AND RESULTS

   The Glassfiber/Epoxy laminates examined are listed in table 1.
The Resin Content determination was carried out by digestion or
calcination according to standard procedures[3].
   The specimens' geometry is shown in figure 2. The specimens
were cut from a plate obtained by bonding or co-curing the end tabs
to the laminate.
   All fatigue tests were performed under cyclic synusoidal load
in load control mode, measuring the number of load cycles to complete

Fig.1 Test bench for main rotor blade segment fatigue test

Table 1: Laminates Characteristics

LAMINATE	GLASSFIBER TYPE	LAY UP	RESIN CONT
CIBA 920 GR7	R	$[0°]_7$	32%
CIBA 920 GR7	R	$[0°]_7$	24%
3M SP 250 E	E	$[\pm45°]_7$	33%
NARMCO 5216 S2	S	$[0°]_7$	33%
NARMCO 5216 GE5	E	$[\pm45°]_7$	33%

failure . In some cases during cycling the displacement of the ram
at minimum and maximum load was recorded by direct reading of the
operator. The cycling frequency was 6 Hz.
    Some stiffness measurements were performed on fatigue cycled
specimens statically by applying the maximum load of the fatigue
test at a crosshead speed of 0.5 mm/sec and measuring the strain
with a clip-gauge applied on a 50 mm gauge length in the central
part of the specimens, instrument full scale was 5 mm extension.

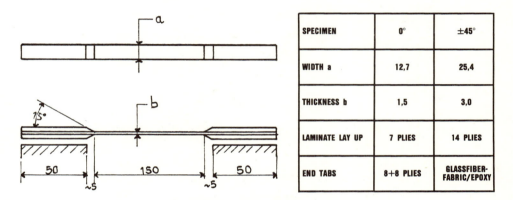

SPECIMEN	0°	±45°
WIDTH a	12,7	25,4
THICKNESS b	1,5	3,0
LAMINATE LAY UP	7 PLIES	14 PLIES
END TABS	8+8 PLIES	GLASSFIBER-FABRIC/EPOXY

Fig.2 Specimens geometry and size

Fatigue tests performed on various Glassfiber/Epoxy laminates

    The experimental S-N curves relevant to the rotor blade main
constituents are reported on fig.3 and fig.4 showing respectively
monodirectional 0° laminates of R/S Glassfiber Epoxy and ±45°
laminates of E Glassfiber/Epoxy. Laminates of different sources have
shown S-N curves close enough to justify their interchangeability in
application as constituents of the rotor blade. A wide scatter of
data points was observed compared to the same tests on metallic materials and is to be ascribed to the intrinsic fatigue characteristic of composite material as well as to the fact that several fami-
lies of specimens cured in different times were tested.

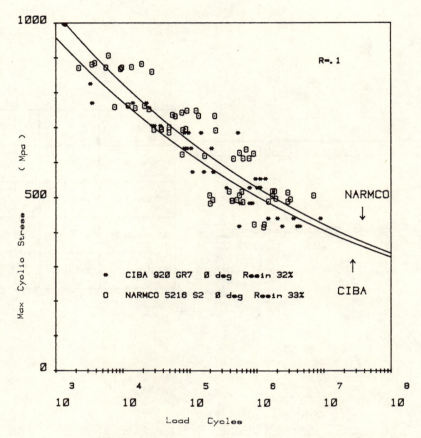

Fig.3 S-N curves (three parameters best fit) for R/S Glassfiber/
Epoxy laminates.

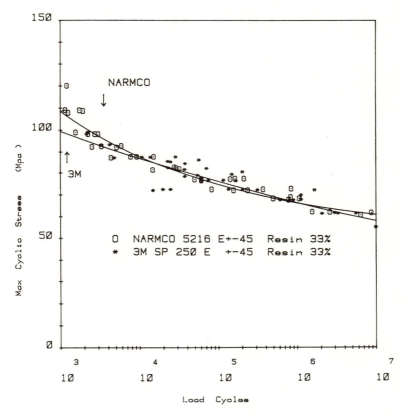

Fig.4 S-N curves for E-Glassfiber /Epoxy laminates

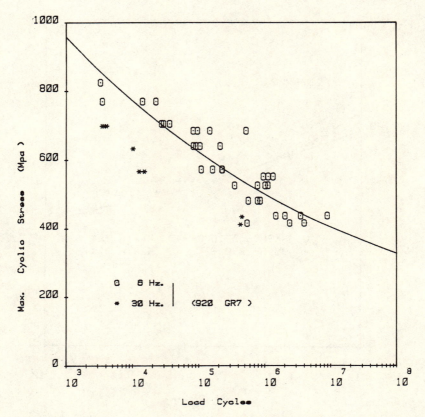

Fig.5 30 Hz tests on R/S Glassfiber/Epoxy laminates

In order to assess the sensitivity of the obtained fatigue data
to the frequency of synusoidal load , some tests were run at 30 Hz.
The relevant data points are shown in fig.5, superimposed on the
6 Hz. S-N diagram for the same material.It may be extimated that the
30 Hz S-N curve is shifted of one decade showing that the test frequency does have an effect on fatigue tests [4].

With the purpose of a better insight into the mechanisms of
the fatigue failure of Glassfiber/Epoxy laminates, a series of fatigue tests on 0° monodirectional laminate was carried out with the
use of some additional equipment and monitoring. This particular
laminate had a higher Glassfiber content as shown on table 1.

Fig.6 X Ray picture of 0° specimens

Before carrying out the tests an X-Ray picture was taken to
show the degree of fiber alignment and is reported on fig.6. The
fiber mis-alignment could be responsible of the local processes
leading to matrix degradation and subsequent local failure in fatigue.
The actual S-N curve obtained in the fatigue tests is shown on fig.7
together with experimental points. One feature which is worth to be
stressed is that in this case all points correspond to specimens
failures : for the lowest load levels, because of unavoidable data
scattering, tests up to 19 million cycles had to be performed .

At the same time, the ram positions corresponding to minimum
and maximum loads were recorded in function of the fatigue cycles,
as indication of damage accumulation. The relevant diagrams are
shown on fig.8. They show an almost constant rate of growth in the
middle part and terminate with a sudden jump immediately before
failure.

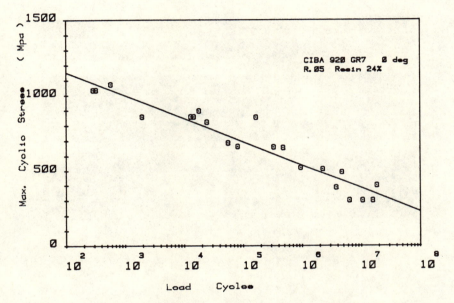

Fig.7 S-N Diagram of 920 GR 7 0° laminate, 24% resin content

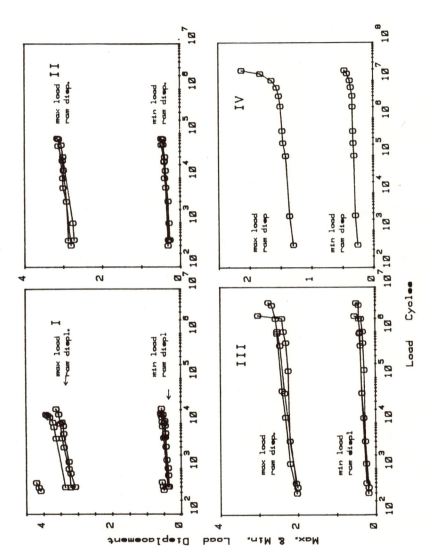

Fig.8 Specimens elongation at min. and max. loads

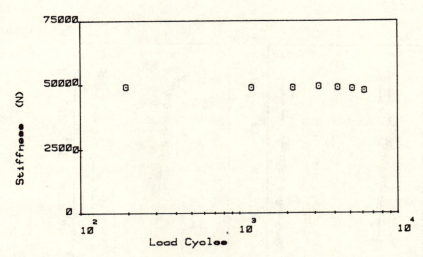

Fig.9 Stiffenss Digrams relevant to the specimens central portion

Specific Tests to localize Fatigue Damage

    The above reported results suggest that besides the damage
taking place in the final part of the test, some continuous degradation seems to be always present. In order to assess whether such
degradation is localized in the end tabs region or spread throughout
the specimen length, some specific tests were conducted measuring
statically the stiffness inside the zone of constant cross section
of the specimens[5],[6].The stiffness was measured several times after
given numbers of load cycles. The data plotted on the diagrams of
fig.9 and 10 show that the stiffness losses are very low (near to
the experimental errors). It was then concluded that the damage
responsible of the overall stiffness loss observed on the specimens
full length was localized mainly, if not completely, under the end
tabs, where different loading conditions are present and failure
can be attributed to processes different from pure axial tension fatigue.

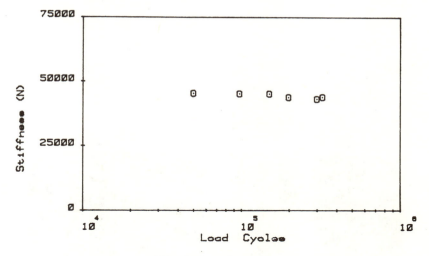

Fig.9 Continued

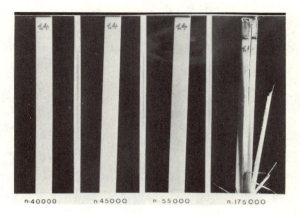

Fig.10 Failure of specimen tested at 642 Mpa

Macro and Thermographic Documentation

To achieve better information on the damage growth, some specimens were photographed at different stages of their life. The
photos are shown on fig.10 and 11. From the pictures it can be
observed that the specimen behavior was different depending on the
load level. Namely highly stressed specimens started to show some
damage just immediately before total failure, whereas specimens
tested at lower stress levels showed important crack growth and
damage well before their failure.

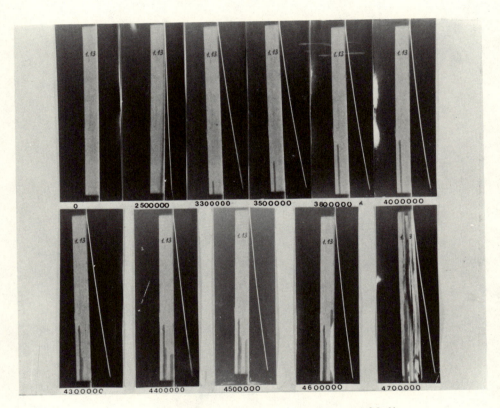

Fig.11 Failure of specimen tested at 489 Mpa

It is possible to see that the splitting starts either from oblique
cut fibers (spcm 1.13) or more frequently in the highly stressed
region under the tabs. This process is particularly evident on specimen 1.13 where at 3.3 M cycles a limited delamination is observed.
The delamination was originated below the end tab, at the interface
between the composite laminate and the tab, where friction and
wearing can be present. Other delaminations which propagated in the
same way followed later.

The importance of the friction and wearing processes at the
interface between the laminate and the tab material is further evidenced by thermovision. Some pictures taken during a test are shown
in fig.12 and 13. In those pictures the dark shades indicate higher
temperatures, except for the white points which outline the isothermal curve at the temperature indicated. As the test proceeds the
temperature field grows as consequence of the heat production under
the grip.

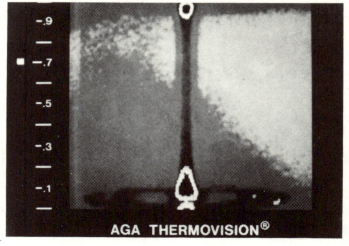

Fig.12 Specimen Thermo-diagram (general view)

A clear indication of the wearing processes taking place under
the tabs is given by the pictures reported in fig.14 which is a
typical fractographic aspect for this kind of tests. The unfailed
and of the specimens presents debonding between the laminate and the
tab material. It was observed the progression of the debonding,
after an initial jump,was continuous until failure.

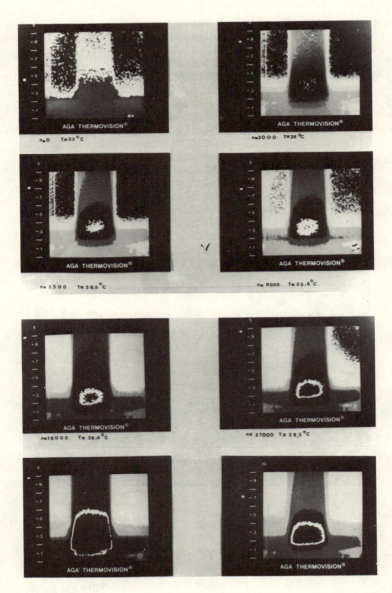

Fig.13 Thermo-diagrams of the tabs region at various cycling stages

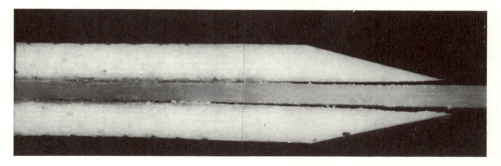

Fig.14 Debonding of the tab from laminate at the end of a test

DISCUSSION OF EXPERIMENTAL RESULTS

From the above mentioned experimental results it is possible
to outline a very simple model explaning the fatigue damage and
failure of the 0° unidirectional specimens.
Debonding of the end tabs propagates suddenly for some millimeters. Such propagation was always observed during the earliest
load cycles and is due to the shear forces concentrated at the
tapered part of the end tabs.
The debonding cracks which are started in this way grow slowly
in the grip area, under the external pressure. Every crack propagation step reduces the stiffness of the specimen, when this is
measured over the whole specimen length (see diagram of fig.11).
Local wearing taking place under the grips, between the laminate and the tabs, leads to the specimen failure; when the load is
low enough , the wearing affects primarily the outer laminate
layers which brake first (fig.11), whilst the inner ones have
still enough " load carrying capacity " to avoid the complete
failure.

CONCLUSIONS

The actual way of performing fatigue tests is adeguate for
material qualification and design purposes, because the derived
S-N curves are conservative with respect to the true fatigue

characteristics of the laminate, also in relation with the simplicity of the specimen design and necessary testing equipment[7]. On the other hand some misunderstanding of the real fatigue characteristics of the laminates may happen.

The on line monitoring of the specimen deformation in the central region, which is a promising extension of the stiffness tests above described, should be useful to show the degradation processes happening outside the end tabs region during a fatigue tests.

Only in this way a real understanding of fiber, matrix, and fiber/matrix interface deteriorations can be achieved[8] with the final goal of technology improvement in respect of matrix modification and both material and processing upgrading, independently from the specimen's shape.

ACKNOWLEDGMENTS

The analysis of fatigue data are carried out by methods developed in cooperation with Pisa University by Mr. Raggi and Mr. Aldinio.

Part of the experimental work herein described was carried out by the Swiss Federal Material Testing and Research Institute EMPA (Duebendorf) under Agusta contract.

REFERENCES

1. M.Raggi, G. Aldinio, P. Alli, Progetto " FITPISA " per tracciamento delle curve di Woehler con best fit lineare, Rel.Agusta 100-45-15
2. C.K.H. Dahran, Fatigue failure mechanisms in a unidirectionally reinforced composite material,"ASTM STP 569", Philadelphia,1975
3. ASTM D 3171-76 Standard test method for fiber content of reinforced resin composites,"ASTM Annual Book of ASTM standards", part 36 Philadelphia, 1977
4. W.W.Stinchcomb, K.L.Reifsnider, L.A.Marcus and R.S.Williams, Effects of frequency on the mechanical response of two composite materials to fatigue loads,"ASTM STP 569",Philadelphia 1975
5. H.T.Hahn and R.Y.Kim, Fatigue behaviour of composite laminate, J.Comp.Materials,vol.10,(Apr.1976)p.156
6. W.C.Kim,L.J.Ebert,Fatigue life-limiting parameters in fiberglass composites,J.of Mat.Science, 14 (1979)
7. C.T.Herakovich,M.P.Renieri,J.G.Davis Jr.,Influence of specimens features on the strength of Boron-Epoxy reinforced metal,J.of Eng.Mat.Technology,paper N°75-Met-C
8. K.L.Reifsnider and A.Talug,Analysis of fatigue damage in composite laminates,Int.J.of Fatigue, (Jan.1980)

APPENDIX

Fig. A1.Full size prototype components tests
Main Rotor Blade Root Fatigue Test

Fig. A2. Test Benches for fatigue tests on Composite Material Specimens

UNRESOLVED STRESS ANALYSIS PROBLEMS IN

KEVLAR COMPOSITE PRESSURE VESSELS

J. Edmund Fitzgerald

School of Civil Engineering
Georgia Institute of Technology
Atlanta, Georgia, U.S.A.

INTRODUCTION

Three major unresolved problems associated with the use of Kevlar/epoxy composite solid rocket motor cases were presented. This paper is a brief summary of the presentation.

DEFLECTIONS

The usual practice in the United States for analyzing deflections, strains, and stresses in helically wound solid rocket motor cases is to use a two step analysis.

The first step utilizes one of several available anisotropic shell computer codes. The results of this first analysis are used for determining the deflections, strains, and stresses in the cylinderical portion of the pressure vessel. Measurements on both sub-scale and full-scale cases have shown that the cylinderical portion of the composite case responds in close agreement to the analytical predictions.

The second step in the analysis consists of using the above calculated deflections and rotations (generally taken at the termination of the cylinderical portion of the case) as input values for the helically wound dome segments. The dome segment is then subdivided into from one thousand to three thousand finite elements. It is generally assumed that the dome segments are axially symmetrical permitting one of several available 2-D computer codes to be used. Thus only one-half of the dome segment cut by the plane containing the axis of symmetry and a radius is required.

The result of the above assumptions and the many finite elements available allows essentially each Kevlar tape in the cut
through the helically wound dome to be represented by one finite
element.

In order to account for the fact that the tapes are generally
not normal to the cut section but traverse the section at various
angles, the elastic modulus of each tape and hence each finite
element is modified and reduced according to the usual cosine power
law.

Full scale hydrotest measurements have not confirmed the
applicability of the above analysis method. On the contrary, in
several instances the direction of the predicted deflection of the
dome closure boss as well as its magnitude was in error. It is
obvious that the predictions for strains are highly suspect since
they essentially are based on the derivative of the calculated
displacements.

It should not be too surprising that the predictions are not
borne out by the full scale measurements. The modeling procedure
described above in essence converts the helically wound structure
into a circumferentially wound structure whose wrap elements have
differing elastic moduli from element to element.

Because each finite element is generally considered isotropic
in its modulus, this assumption implies that the elastic modulus
along the Kevlar tape equals the elastic modulus normal to the wrap
direction (even though the cosine power reduction has been used
on tapes entering the cut plane at an angle).

Even if one utilized anisotropic elements to compensate for
the well known fact that the cross-ply elastic modulus of Kevlar
tape is much less than that in the wrap direction, the 2-D finite
element modelling still loses the fact that the tensile forces are
helical in nature.

Two potential solutions may be considered. First, the use of
a 3-D computer code. Let us suppose that a rough modelling could
be achieved by using, say, 100 cuts either along radis or normal
to the axis and that the same one-to three-thousand elements were
used in each cut plane. The result would be up to several hundred-
thousand elements of up to twenty-one degrees of freedom each.
It is doubtful if even the use of a CRAY computer could handle such
a system. Even if it would, the cost would be exhorbitant.

A second and more promising approach would be to develop

special finite elements wherein each element knew that it was the cross-section of a helix. To the writer's knowledge, such an element has not been developed but should be capable of development by incorporating the differential geometry describing the helical structure.

In summary, the use of 2-D modelling for a helically wound dome is onologous to attempting to describe the behavior of the Chinese Finger toy remembered from ones childhood.

INTERLAMINAR SHEAR AND TENSION

A second problem associated with Kevlar composite structures is that associated with determining the allowable values for inter-laminar shear and tension.

Generally, these values are determined indirectly from hydro-test and hydroburst measurements rather than by direct laboratory test. However, the believibility of these derived values suffers from the same reason as the previous discussion on deflections, namely, the lack, of a credible analytical modelling technique.

Further, the problem is compounded because the calculated interlaminar stress values in the helically wound dome segment can vary several-fold due to relatively minor fabrication variations in the local geometry.

CUMULATIVE DAMAGE

Cumulative damage, particulary in the epoxy matrix material is evident from the pronounced acoustic emission from a Kevlar composite pressure vessel under initial hydrotesting. Cumulative damage, in the Kevlar fibers is inferred from the fact that the cases are frequently observed to fail on subsequent testing at hydrostatic test pressures less then originally withstood. In addition, failures have resulted after some period of time while simply holding a case at a fixed hydrostatic pressure.

Attempts to reduce sources and causes of failure are often fruitless. As opposed to metal case post-failure examination, a failed Kevlar case is indeed the situation of looking for a needle in a random haystack of broken delaminated fiber ends.

Much more detailed, highly instrumented experiments are needed to determine the accumulation of damage and the subsequent degra-dation of strength in Kevlar Composites.

DELAMINATION IN GRAPHITE-EPOXY

Dick J. Wilkins

General Dynamics
Fort Worth, Texas   76101

This summary is intended to provide the perspective and philosophy of the presentation made in Capri. The substance of the presentation is published elsewhere[1].

Delamination growth is the fundamental issue in the evaluation of laminated composite structures for durability and damage tolerance. When primary composite structures are evaluated originally with respect to durability (fatigue) and damage tolerance (safety), they are judged against the design service usage, and tests are performed to demonstrate their structural integrity. However, after the parts enter service, they are exposed to a variety of operational usages so that techniques are needed to estimate the behavior of defects as a function of time under changing service conditions.

The experience of General Dynamics and other contractors has been that only rarely does a composite part fail in a realistic fatigue test. But when test conditions are extended to explore failure mechanisms, delamination is observed to be the most prevalent life-limiting growth mode. An examination of the fracture toughness of glass-, boron-, and graphite-epoxy shows that interlaminar forces (through the thickness) are operating against an inherent weakness

1. Wilkins, D. J., Eisenmann, J. R., Camin, R. A., Margolis, W. A., and Benson, R. A., "Characterizing Delamination Growth in Graphite-Epoxy," Damage in Composite Materials, ASTM STP 775, American Society for Testing and Materials, 1982.

of the resin. An additional observation is that common design
features in laminated composites (free edges, notches, ply drops,
bonded and bolted joints) give rise to interlaminar shear and normal
forces that depend on local details.

Composites are sensitive to applied compression loads because
local and overall buckling are very important mechanisms. Coupon
compression test results can't be scaled up to hardware because
buckling is a geometric situation that can't be readily scaled.
Unfortunately, materials scientists are not generally trained in
buckling theory, so they don't view the prepreg as a structure. The
other realization is that internal compression at the molecular level
doesn't really hurt you except as it causes instability or yielding.
But internal tension and shear forces can cause a fracture or rupture
that tears things apart. So, under an applied compression, one must
be sensitive to internal defects that act as stress concentrations,
and the tension or shear stress concentrations that are associated
with the defects.

These facts suggested a fracture mechanics approach to delamination growth based on the concept of strain-energy release rate.
The technology already being developed for debonding of adhesives
was readily suited to the study of delamination in composites.

A quantitative measure of the effects of defects in composite
structures is a key issue in material selection, manufacturing
procedures, quality assurance plans, structural design, structural
integrity predictions, maintenance planning, and repair decisions.
Experience suggests that the inherent lack of interlaminar toughness
is the root cause of such problems as poor in-plane compression
properties, edge (and internal) delaminations, and low impact
resistance.

Characterization of the behavior of delaminations has been
approached by adapting and developing techniques for coupon design,
static and fatigue testing, data analysis, fracture analysis for
separation of modes, spectrum life prediction, and spectrum truncation. Critical strain-energy release-rate values have been obtained
for Mode I and Mode II delamination. Corresponding delamination
growth rates have been measured.

These test methods are applicable to other fiber-reinforced
composite materials, and a data base needs to be built to compare
various materials on an interlaminar toughness basis. Such comparisons have already begun, but much more work is required.

These interlaminar toughnesses are more likely to be influenced by processing variables than the usual mechanical tests used for quality control and process development. Such variables as prepreg viscosity, cure advancement, layup technique, cure-cycle parameters, cured resin content, number of fiber ends per tow, fiber-sizing effects, and many others could significantly affect interlaminar toughness. A careful test program could help sort out some of the manufacturing unknowns to arrive at quantitative relationships among manufacturing cost, inspectability, and structural integrity.

In a similar way, structural design detail concepts should be analyzed with the available techniques to develop relationships among the in-plane strain field, the geometry of the concepts, and the interlaminar strain-energy release rates. Free edges, open holes, ply terminations, bonded joints, and bolted joints all produce interlaminar normal and shear stress states that vary with geometry, stacking sequence, and ply properties. These design details can be analyzed for producibility, inspectability, and delamination resistance to aid in the selection of structural concepts and meaningful accept/reject criteria.

FINALIZED RESEARCHES ON POLYMERIC MATERIALS:
SUB-PROJECT ON POLYMERIC MATERIALS OF THE FINE AND SECONDARY
CHEMISTRY RESEARCH PROGRAM OF THE ITALIAN NATIONAL RESEARCH COUNCIL

Ezio Martuscelli

Scientific Manager of the Sub-project "Polymeric Materials" of the Fine and Secondary Chemistry Research Program
of Italian National Research Council*

INTRODUCTION

The year 1972/73 was a decisive turning point for the public
research policy in Italy. The Italian Research Council (C.N.R.) at
this time, responding to strong political pressures which criticized
its role as the principal co-ordinator and subsidizer of the public
research, decided to organize "finalized research programs" whose
goals were essential for the socio-economic progress of the country.
The basic premises underlying the finalized project-plans are
as follows:
- to identify through a rigorous analysis of our country's necessi-
  ties, the sectors and areas where it is necessary to press research
  interventions in accordance with the developing plans
- to concentrate the potential research resources of our country
  in these areas.
By pointing out the structural deficiencies of the research-
-apparatus (public or private) and making the necessary corrections,
the finalized project should contribute substantially to introduce
innovations both in industry and in agricolture. They would also
encourage the development and modernization of social transport,
health and education and the safety of the territory and the environment.
The first finalized research programs proposed by the C.N.R.
were approved by the Ministry of Scientific and Technological Research and by the Interministerial Committee of Economical Program-
mation (CIPE) on 9/10/75. The CIPE resolution containing the approval of the first finalized projects, indicated some sectors for which

*c/o Istituto di Ricerche su Tecnologia dei Polimeri e Reologia -
- C.N.R. - Arco Felice (Napoli) - Italy

incentivation was considered necessary. Among these sectors the fine
and secondary chemical sector was expressely mentioned.

This indication derived from the heavy crisis in the chemical
sector in Italy (see later).

Consequently the President of the C.N.R. on October 12, 1976
appointed a commission composed of experts from industry and the
academic community who were to study a finalized research project
in the field of fine and secondary chemistry.

In June 1978 the feasibility study was completed and on July
5, 1979 the CIPE in a resolution authorized the commencement of the
finalized project and invited the C.N.R. to prepare the executive
project.

The elaboration of the executive project was completed in a
few months by a special commission composed of many experts. The
final documentation approved by qualified organs of the C.N.R. was
transmitted to the Ministry for Scientific Research and to the Ministry of Industry which approved it. Following this approval, the
finalized project "Fine and Secondary Chemistry" (PFCFS) became operative in July 1980.

The PFCFS project whose lasting is of 5 years is divided in
four major sectors or subprojects:
a) New Synthesis
b) Polymeric Material
c) Biologically active products
d) Methodologies
each of which has been subdivided into central themes and research
lines.

The goal of PFCFS is to contribute to reducing the deficiencies
in research which causes the poor rate of development and the low
competitivity of Fine and Secondary sectors of the Italian Chemical
Industry.

This goal can be reached through:
a) Increase of coordinated research activities
b) Overcoming the situation of marked separation between University
   & C.N.R. research on the one hand and industrial research on the
   other
c) Involving university researchers and the C.N.R. in programs oriented and finalized to the solution of industrial problems.

ECONOMIC ENVIRONMENTS

a) Industrial Chemistry Sector

The reasons for a finalized project on Secondary Chemistry,
derive from the heavy crisis of the industrial sector. The competi-

tion loss of italian chemical industry, that began in the 1970, has been identified by the negative balance of payments, by the state of economic and financial structures of Industries and by the progressive reduction of the utilization degree of production plants.

The commercial balance of payments of the entire chemical sector (years:1977, 78, 79) has been reported in tab.1. From this tab. it emerges that the balance deficit passed by 571 billions of 1977, to 1.266 billions of 1978 and 2.285 of 1979. The 1980 data seem to confirm this negative trend.

For what concerns 1979 on fig.1 the balance of italian chemical industry has been compared with that of more advanced western countries, and Japan.

The data show that in those countries the chemical sector positivitely contributes to their economy.

As evidenced in tab.2, the research expenses in Italy concerning the chemical sector, are in absolute lower than in the western countries and Japan (the ratio of research-expenses invoiced is, for Italy, lower than the 50% of the other countries; 1,7% against an average value of 4%).

This deficiency of research certainly contribute to the deficit of commercial chemical balance of the overall sector including the secondary and fine one.

So, the incentivation of research activities represents one of the necessary conditions for the equilibrium of the italian chemical inter-exchange with foreign countries.

b) Polymeric material sub-sector

The polymeric material sub-sector (plastics, elastomers and fibers) represents about the 23% of the total italian chemical industry businesses (see fig.2).

As it can be noticed in tab.3, in the 1979, the three sub-sectors showed a deficit of: Lit. 66 billions for elastomers; 142 billions for fibers and 279 billions for plastics.

This negative trend shows that the competitivity of the italian products is lowered.

SUBPROJECT "POLYMERIC MATERIAL"

a) Goals and organization

The consumption development and production in industrial field of macromolecular materials, will be guaranted in the near future, principally through a diversification of products and by the passage

from a production of a quantitative type, to one where the quality
of the product and its specific characteristics, are the most important factors.

The production will be addressed towards materials whose fundamental characteristics are long life and high performance and that
can be used for specific sectors.

This new concept of production, based on quality and performance
could only be affirmed if: an incentivation of scientific and technological research, suitably coordinated, in the field is achieved.

The results of these actions should bring about the development
of new polymeric materials, characterized by a high value rate and
a low cost/performance ratio, by introducing substantial technological innovations in the production process and thus contributing
to an increase in the sectors, where they could also be used.

Consequently studies and researches must be increased, developed and coordinated in the following sectors:
1) Synthesis and development of new monomers, new catalysts and polymerization techniques
2) Combination of monomers and polymers for the synthesis of copolymers (random, alternated, block, grafts, stars, ...) with specific
   characteristics:
a) functionalization of macromolecules
b) formulation of polymeric systems with more components (blends-
   -composites) at high adhesion between the phases and low interfacial energy
c) development and use of new additives
d) development of new transformation methods.

The realization of new macromolecular materials requires the
development and setting up of new and more sophisticated chemical
processes and formulations that will utilize rare products opportu-
nely synthetized.

It is expected that these methods and the sophisticated processes of Fine Chemistry will be more used in the near future for
the formulation of new polymeric materials.

The general goals of the sub-project "Polymeric Materials" are:
- To supply Italian Industry for new instruments (typically of Fine
  and Secondary Chemistry), for increasing and ameliorating the range
  of polymeric materials which have specific uses
- To encourage studies and researches whose goals are the realization
  of polymeric materials "ad hoc" characterized by a high innovation
  degree and a high value rate.

The realization of new polymeric materials will require also
the setting up of new transformation process and planning of more
and more sophisticated machineries.

The sub-project "Polymeric Materials" has been subdivided in
the following research-themes:
a) Polymeric composites
b) Polymeric systems: characterization, compatibility and applied
   studies
c) Polymer degradation mechanism

d) Membranes and separation processes
e) Catalysis of polymerization
f) New polymers and functionalization
g) Valorization of cellulose-other polysaccarides and their derivates.
    Each research theme has been subdivided in research-lines with homogeneous objectives.

b) Goals and main scientific contents of research themes

1) Polymeric composites

    The goals we intend to reach are:
- the realization of composites with short fiber at high-mechanical properties
- individualization of the most important rheological parameters connected with processing of composites with polymeric matrix
- to establish the necessary backgrounds for more advanced projects in order to realize special items.
    The choice of mechanical property studies of composites with short fibers derives from the observation that these materials have been, up to now, studied very little, contrarily to composites with contineous fibers.
    Furthermore it is to be noticed that composites with short fibers, are very interesting being obtained, at competitive prices, with more traditional processing similar to the ones used for the thermoplastic resins.
    Particularly it will be studied composites constituted by poly-carbonate, polypropylene, polyester, epoxy resins, matrixes and glass, carbon and kevlar fibers.
    The theme is divided in two research-lines entitled as follows:
- Study on fiber-matrix adhesion and on composite properties
- Suspension rheology.
    For the second semester of the 1980, 26 researchers were in-volved corresponding to 13.5 full-time researchers (equivalent re-searchers) subdivided in 9 operative teams (4: Industry, 5: Univer-sity). The financial coverage of the second semester of the 1980 is Lit. 110 millions (see tab.4).
    For what concerns 1981, the equivalent researchers involved are 18 and the operating teams 10 (5: University, 5: Industry). The financial coverage is for the 1981 Lit. 250 millions (see tab.5).

2) Polymeric systems: characterization, compatibility and application studies

    The goals of this theme can be resumed as follows:
- to obtain basic informations on the relationship between: proces-sing, structure, morphology and properties of polymeric systems
- to obtain specific morphologies for large production polymers and polymeric blends relating to new characteristics
- to obtain new materials having prefixed properties, through the

mixing of two or more polymers, with a better value of costs performances ratio.

Referring to these goals, new polymeric materials having characteristics "ad hoc" will be realized, through the study of the influence of the formulation and the process conditions on the morphology and their properties.

These materials will be realized through the formulation of
blends and/or the obtaining of particular morphologies.

The theme is subdivided in the following lines, strickly inter-
dependent among them:
- Preparation of interphacial agents and their functionalization
- Preparation, and characterization of particular polymers with "ad
  hoc" distribution of constitutional units, tacticity and molecular
  mass
- Interactions polymer/additive
- Preparation and processing of blends and polymeric materials
- Mechanical and end use properties of polymeric materials
- Characterization of rheological properties relating to the proces-
  sability
- Structural and morphological characterization related to mechanical
  and end use properties.

For the 1980 the teams involved are 20 (9: universitary, 6:
C.N.R., 4: industrials, 1: other Ente). The researchers involved are
96 corresponding to 33.5 equivalent researchers. Concerning the
second semester of 1980 the theme has received a financial coverage
of Lit. 234 millions (see tab.4).

For what concerns 1981 the researchers involved are 40.5 and the
operating teams 23 (9: University, 7: Industry, 1: others). The financial coverage for the 1981 is: Lit. 505 millions (see tab.5).

3) Mechanism of polymer degradation

The principal goals of this theme can be summarized as follows:
- the definition of "Photooxidation mechanisms" for principal polymers in the presence of inhibitors
- the determination of factors influencing the flame behaviour of
  polymers and determination of suppression mechanism of smokes produced during the combustion process
- the determination of variations induced by various ignifuge additives on degradation mechanism of polymers. Flame retardant polymers via chemical modification.

The themes has been subdivided in the following research-lines:
- studies on polymers photooxidation mechanisms
- studies on action mechanisms concerning flame retardant and smoke
  suppressors agents.

For the 1980 about 35 researchers, corresponding to 19.5 equivalent researchers belonging to 12 operating teams (4: universitary,
3: C.N.R., 4: industrial, 1: others) are involved. The financial coverage for the semester of the 1980 is: Lit. 150 millions (see tab.
4).

For what concerns the 1981, the equivalent researchers involved are 23, and the operating teams 13 (4: University, 3: C.N.R., 5: industrial, 1: others). The financial coverage for the 1981 is Lit. 315 millions (see tab.5).

4) Membranes and separation processes

The basic goals can be resumed as follows:
- to deepen basic knowledges necessary for the preparation of selective membranes in different configurations (planar, capillary, tubular) to be used during separation processes
- preparation and study of structures and transport properties of the membranes and their characterization during separation, concentration and purification of liquids and gas.

The program should contribute to increase and develop the aspects relating to the separation problems of the industrial processes.

The researches will be concentrated on the following polymeric systems: fluorinated polymers, aromatic polyamides, polysulphones and polyurethanes.

The themes will be divided into the following three research lines:
- asymmetrical membranes, formation mechanisms
- preparation of new exchanging membranes
- study on diffusion, and transport phenomena.

In 1980 10 operating teams will be involved (9: universitary, 1: others). The total of researchers is 91 corresponding to 15.5 equivalent researchers. For the second semester of the 1980 the project has received a financial coverage of 90 millions (see tab.4).

For what concerns 1981, the equivalent researchers involved are 18.5 and the operating teams 11 (10: University, 1: others). The financial coverage for the 1981 is: Lit. 293 millions (see tab.5).

5) Catalysis of polymerization

The goal of this theme will assure research engagement in the field of Ziegler-Natta catalysis developing the ionic and polycondensation catalysis studies too.

The goals can be subdivided as follows:
- to reach larger knowings on titanium-vanadium catalysts
- to clarify factors determining microstructures; to individuate new catalysts
- to utilize carboanion-carbocation chemistry for new applications on ionic catalysis
- to determine action-mechanism of polycondensation catalysts.

The research has been subdivided in the following lines:
- Ziegler-Natta catalysis
- ionic and polycondensation catalysis.

In 1980 about 20 researchers for a corresponding number of 15 equivalent researchers subdivided in 6 operating teams (5: University, 1: C.N.R.) are involved. Financial coverage - second semester

1980 - has been Lit. 95.5 millions (see tab.4).

For what concerns 1981, the equivalent researchers involved are 18.5 and the operating units 9 (6: University, 1: C.N.R., 2: Industry). The financial coverage for the 1981 is: Lit. 253 millions (see tab.5).

6) New polymers and functionalization

The goals of these researches are:
- functionalization of polymers to introduce tensioactive groups reactive towards different substrates, adhesive, cross-linkings and compatibilizing groups
- formulation of new polymers and additives to be used for electrophoresis
- production and characterization of new thermoplastic elastomers and new compatibilizing materials.
    The research has been subdivided in two main lines:
- polymers with functional groups and new polymers
- cross-linkable polymers for paints and resins.

The operating units are 11 (3: University, 4: C.N.R., 4: Industry) for a total of 21.5 equivalent researchers (about 53 researchers involved). Financial coverage for the second semester of 1980 has been Lit. 110 millions (see tab.4).

For what concerns 1981, the equivalent researchers involved are: 38.4 and the operating teams 17 (6: University, 3: C.N.R., 8: Industry). 1981 budget is Lit. 357 millions (see tab.5).

7) Valorization of cellulose other polysaccarides and their derivates

The goals of these researches are the following:
- to optimize the extraction process of the cellulose and characterize from the point of structural and morphologic view, the resulting samples
- to find new solvents of the cellulose and to study the properties of diluted and concentrate solutions
- to obtain high modulus fibers and to study the morphology and the structure of regenerated forms
- to set up techniques and derivative reactions without modifying the backbone of polysaccaride chains.

The main goal of this theme is boosting researches, that will encourage high technological level innovations, in the field of cellulose and other polysaccarides and their derivates.

For the 1981 the research will be subdivided in the following lines:
- to optimize cellulose extraction processes and morphological structure characterization of resulting samples
- cellulose solutions new solvents; properties of diluted and concentrated, solutions; high modulus fibers; morphology and structure of regenerated forms
- derivation without backbone modifications of polysaccaride chains

(synthesis and study of ionic derivates).

For the second semester of the 1980, the operating teams, the equivalent researchers and the total of people involved were respectively 9 (2: University, 3: C.N.R., 2: Industry, 2: others), 9 and 28 with a semestral budget of Lit. 55 millions (see tab.4).

For what concerns 1981, the equivalent researchers involved are: 18 and the operating teams 9 (3: University, 1: C.N.R., 3: Industry, 2: others). The financial coverage for the 1981 is 288 millions (see tab.5).

c) Participation of Public and Industrial Research Institutes, to the Sub-project "Polymeric Materials"

For the 1980 more than 300 researchers and technicians are involved in the project, and their partecipation corresponds to 127.5 full time researchers (174.9 for the 1981) subdivided into 77 operating teams (92 for the 1981). Most of the equivalent researchers and the operating teams come from the University (respectively 66 (51.8%) and 37 (48.1%) for 1980 - for 1981 90.8 (51%) and 43 (46%). The industrial research laboratories are involved with the (23.3%) of operating teams and 18% of equivalent researchers for the 1980. For what concerns 1981: 32% - 26.5%. As regards C.N.R. operating teams and equivalent researchers are respectively 17 (22%) and 34 (26.7%) for the 1980; for the 1981: 14 (15%) and 32 (18%) (see tabs. 4 and 5).

So, it is to be noticed that year 1981 shows a larger industrial partecipation to the project.

This larger partecipation will contribute to boost the finalization of the entire program.

CONCLUSIONS

The institution of efficient systems for data retrieval and of data banks was already acknowledged at the elaboration stage of the finalized project.

These data banks located in research Institutes and managed by qualified persons, could be the nucleus of sectorial transfer system of innovations.

The finalized project should also give indications necessary for the realization of new initiatives, such as:
- the institution of new research institutes
- the constitution of research agencies
- the constitution of qualified centres at high technological content for scientific equipments.

The success of the project will depend on the co-operation degree between the Researchers of Public Institutions and the Private ones.

The big effort in co-ordination never attended before, if well managed, could have a decisive role in the future of the entire italian chemical sector.

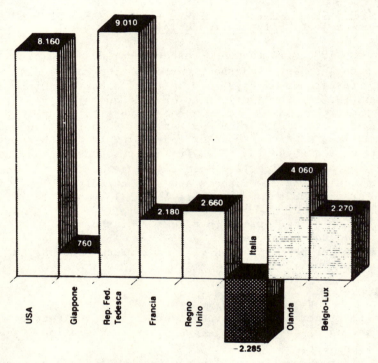

Fig.1. Commercial balance of payments of chemistry sector
for different western countries and Japan.
(billions of Lit.)
Source: Aschimici – Compendio Statistico 1979

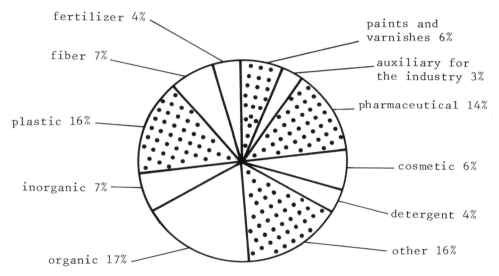

Fig.2. Composition of the total Italian chemical industry
business (1979). Source: Aschimici - (March 1981)
"L'Industria Chimica in Italia"

Table 1. Balance of payments for chemistry sector
(billions of Lit.)

	1977	1978	1979
Import	3,668	4,580	6,592
Export	3,097	3,314	4,307
Saldo	- 571	-1,266	-2,285

Source: Elaborazione Servizio Studi Aschimici su dati
ISTAT

Balance for secondary chemistry: (1979) -1,054

Table 2. Research expenses in chemistry sector in
different western countries and Japan (public
and private).(Billions of Lit.).(Year 1976).

Country	Total	Invoiced	Research invoiced %
D	1,400	26,750	5.2
F	570(x)	14,800	3.8
I	206	12,000	1.7
NL	290	7,500	3.8
UK	478	14,750	3.2
USA	2,577	91,000	2.8
J	1,213	26,900	4.5

(x) only private research

Table 3. Balance of payments for the "Polymeric
Materials" sector (Lit. billions)

Sub-sector	1977	1978	1979
Elastomers	- 46	- 48	- 66
Fibers	+110	+ 24	-142
Plastics	+ 66	- 47	-279
Total	+131	- 71	-487

Table 4. Operating units, equivalent researchers and budget
for each of the themes of research (1980: semester)

Research theme	Operating units					Equivalent researchers					Budget (Mil. Lit.)
	Univ.	CNR	Ind.	Oth.	P.Tot.	Univ.	CNR	Ind.	Oth.	P.T.	
Polymeric Composit.	5	–	4	–	9	8	–	5,5	–	13,5	110
Polymeric systems: characterization and compatibilization	9	6	4	1	20	14	13	5,5	1	33,5	234
Polymer degradation mechanism	4	3	4	1	12	11	3	5	0,5	19,5	150
Membranes and separation process	9	–	–	1	10	14,5	–	–	1	15,5	90
Catalysis of polymerization	5	1	–	–	6	13	2	–	–	15	95,5
New polymer and functionalizat.	3	4	4	–	11	3,5	13	5	–	21,5	110
Valorization of cellulose-other polysaccaride their der.	2	3	2	2	9	2	3	2	2	9	55
TOTAL	37	17	18	5	77	66	34	23	4,5	127,5	844,5
%	48,1	22,1	23,3	6,4	100	51,8	26,7	18,0	3,5	100	

Table 5. Operating units, equivalent researchers and budget
for each of the themes research (year 1981)

Research theme	Operating units					Equivalent researchers					Budget (Mil. Lit.)
	Univ.	CNR	Ind.	Oth.	P.T.	Univ.	CNR	Ind.	Oth.	P.T.	
Polymeric composit.	5	–	5	–	10	9,5	–	8,5	–	18	250
Polymeric systems: characterization and compatibilization	9	6	7	1	23	14	16	9,5	1	40,5	505
Polymer degradation mechanism	4	3	5	1	13	12	3	7	1	23	315
Membranes and separation process	10	–	–	1	11	17	–	–	1,5	18,5	293
Catalysis of polymerization	6	1	2	–	9	16	1	1,5	–	18,5	253
New polymer and functionalization	6	3	8	–	17	16,9	9	12,5	–	38,4	357
Valorization of cellulose-other polysaccaride their der.	3	1	3	2	9	5,4	3	4,5	5,1	18	288
TOTAL	43	14	30	5	92	90,8	32	46,5	8,6	174,9	2261
%	46	15	32	5	100	51	18	26,5	4	100	

INDUSTRY-UNIVERSITY COOPERATION IN RESEARCH

G. Astarita
University of Naples

C. D. Denson
University of Delaware

FOREWORD

This report is based on the panel discussion which was held at the Symposium on June 19, 1981. However, many of the opinions expressed are the personal ones of the authors, and participants in the panel discussion should not be held responsible for them unless quoted explicitly.

Although we both work in the University at the time of writing, one of us (C.D.D.) has worked for many years in industry. We therefore believe to have, between the two of us, a balanced view of the realities of both academic and industrial life.

Both of us have contributed to the writing of the whole report, and we both take responsibility for the opinions expressed. However, for obvious reasons, one of us (C.D.D.) has done most of the work concerned with the situation in the U.S.A., and the other (G.A.) that concerned with the Italian situation.

THE SITUATION IN THE U.S.A.

It is useful to think of university-industrial interactions in the United States along three rather broad lines: gifts or unrestricted grants from industrial sources; consulting arrangements between faculty and industrial firms; and, broad-based university-industrial cooperative research programs of an institutionalized nature.

Gifts and unrestricted grants represent the most widespread
form of academic-industrial interaction, yet they reflect the least
intense level of such interaction. Almost always, the policy is
one of laissez-faire and consequently personal interactions are
minimal. Usually, there are no restrictions on the part of the
industrial corporations with regard to the utilization of the grant
by the academic institutions, and therefore these grants are not
intended to focus on the solution of any particular problem which
may be of industrial importance. There are, however, expectations,
and these are related to assuring that a given department maintain
a high level of excellence and achievement. Thus, such grants or
gifts are generally interpreted as encouragement of the type of
education and research that is being conducted within a given
(engineering) department. Of course, from the University perspective, these gifts are welcomed, and they collectively provide a
department with increased flexibility regarding additions in terms
of faculty, equipment, services, and in aiding new faculty to
establish a research program. From the industrial perspective,
these grants serve to establish a line of communication with a
given department; and through this, some advantage is obtained in
the recruitment of students for future employment. Of some
importance, perhaps, is the fact that these grants can have a
favorable impact insofar as the public image of an industrial
corporation is concerned.

Consulting arrangements between faculty and industrial
corporations bring university-industrial interactions into much
sharper focus. Most often, this involves a one-to-one relationship between faculty, and scientists and engineers from industrial
corporations. Although the consulting effort may be concerned
with either basic or applied research, it generally revolves around
obtaining solutions to specific problems; consequently, it is
proprietary in nature and direct student involvement is rare.

Consulting arrangements provide a number of advantages for
both the university and the industrial corporations. From the
academic perspective, the proprietary problems which the faculty
are asked to address often serve as a basis for graduate research
problems of a non-proprietary nature. The relevance of the research
is thus guaranteed and the published results serve to contribute
to our understanding in a meaningful way. A second benefit which
accrues from this type of interaction is that faculty who consult
can continually revise and update the content and material of both
undergraduate and advanced level courses and thus avoid obsolescence.

Insofar as advantages for the industrial corporations are
concerned, a consultant will bring a different point of view to a
problem by virtue of having knowledge in areas which may be outside

the scope of the industrial corporation for which the work is being done. A second advantage that accrues is related to the recruitment of students by the industrial corporation. The knowledge gained about the "personality" of a given corporation is often passed on to students who are closely associated with the faculty who consult for that corporation, and this can be a positive force in the recruitment of students.

It is interesting to observe that cooperative research programs of an institutionalized nature often come into being as a result of a previous consulting agreement between faculty and industrial scientists or engineers. Indeed, all the case histories which were reported at the panel discussion support this statement. The one-to-one interaction, with mutual trusting and appreciation, which develops in consulting arrangements is an important basis for successful cooperative research programs.

The cooperative research program is by far the most intense and highly focused interaction of the three types we have identified. To understand the nature of these programs, how they are initiated, and the particular advantage that these programs offer to both the academic and industrial sectors requires first an understanding of how important technological innovation is to the industrial sector in the United States, and an understanding of how these innovations are both implemented and integrated into the industrial sector.

In a broad sense the industrial sector may be thought of as being roughly divided into two categories. In the first one, profitability is heavily dependent on being able to employ the economies of scale in the production process. The automotive, household appliance, steel, and mining industries in the United States are examples where this has been the situation. In these industries the half life of any new technology that is integrated into the operation is long. That is to say, major changes and rapid turnover in technology are not feasible since the capital investment in these industries is usually quite large. Therefore, there is little incentive for these industries to have in place or conduct extensive research programs, and certainly no incentive to participate in any cooperative venture with the universities on basic research programs. This is true <u>unless</u> a major dislocation occurs in the marketplace wherein profitability or market position is seriously eroded and radically new technology is required in a relatively short period of time in order to reestablish profitability. In these circumstances it is clearly necessary for these corporations to develop new technology as rapidly as possible; and, one mechanism which is now emerging for doing this is to establish cooperative research programs of a basic nature with various universities.

A key example of this situation is the automotive industry where the necessity for having lighter weight automobiles has created a demand for developing a basic understanding of the processing and performance behavior of polymeric composite materials in a relatively short period of time. This, in turn, has led to the establishment of cooperative research programs with universities which have basic research programs in place in the area of composite materials, as for example, the University of Delaware, Georgia Institute of Technology, and Renssalaer Polytechnic Institute. Insofar as the cooperative research program at the University of Delaware is concerned, a number of industrial corporations participate. And, even though some of these corporations are competitors, no problems have arisen. The success of the program is due largely to the fact that the research results provide fundamental knowledge and a basic understanding which can be used by any one of the individual corporations to develop proprietary technology in any manner which each deems appropriate.

The fact that industrial corporations in this first category start cooperative programs with universities on basic research when a major dislocation of the market occurs is proof of the fact that these corporations believe in the plausibility that such programs may lead to the development of new technology. When such programs were not conducted by these corporations, the reason was that the incentive to develop new technology was small, and not a lack of belief in the ability of the programs to lead to new technology

The second category consists of industries where the economies of scale do not play a key role in profitability--at least in the same sense as they do for, say, the automotive industry. Examples of industries in this category include computers, pharmaceuticals, the aeronautical industry, and industries which manufacture artificial limbs and organs. The business in these industries require highly specialized technology and is highly competitive; and, here, profitability must come from an alternate strategy which usually involves the implementation of a proprietary, technological innovation often based on some new scientific discovery. There is thus a rapid turnover in technology. Consequently, the half life of any technological innovation is often quite short.

The incentive for industrial corporations in this category to participate in university-industrial cooperative research programs is, perhaps, not as clear or straightforward as it is for the first category. Because technological innovation plays such a vital role for corporations in the second category, these corporations tend to conduct most research in-house, and many have extensive, highly specialized corporate R & D efforts. For those that do not, universities could become a major extension of existing R & D efforts. For those that do, cooperative programs with universities could provide these corporations with additional avenues for uncovering new technology. In any event, cooperative research programs with corporations in this category would most likely have to be inventive as well as explicatory in nature, and, of course, proprietary.

One of the distinguishing features of the university-industrial cooperative research program, as opposed to the other types of interactions described earlier, is that students are intimately involved in these programs. This means that students can be formally trained in what may perhaps be viewed as non-traditional areas of science and engineering where no formal educational structure has previously existed. Consequently, these students can begin to make significant contributions immediately upon employment without having to develop new technology by the costly trial-and-error procedure borne of ignorance.

The advantage to the universities of the cooperative research programs, simply stated, is this. These programs provide the faculty with problems, and the financial support to solve these problems, which are relevant and focused; and, they serve to insure that the faculty are productive in doing what they are charged to do: train and develop students to the point where they can independently identify and define problems, and develop and implement programs which will resolve these problems in a significant and meaningful way.

One of the pitfalls which must be avoided in any university-industrial cooperative program is that the program not degenerate to one in which the university is used simply as a testing laboratory and the student as a technician. It is imperative that the university sector be intimately involved in all stages of the program from the initial planning through implementation and be viewed as an equal partner.

In an effort to encourage the interaction between the academic and industrial sectors, the United States Government through the National Science Foundation instituted a program about five years ago to provide financial support for selected research programs which were joint ventures involving a university and an industrial corporation. Dr. Frederick Betz, manager of NSF's program on industry-university collaborative research, pointed out that NSF's role is viewed as one which is catalytic in the triangular relationship between government, universities and industrial corporations. Such an arrangement was also viewed as one which would serve to maximize technological productivity in a resource limited environment.

In the time that the industry-university cooperative program has been in operation, Dr. Betz has judged the program to be a success, although one unexpected aspect of the program has emerged.

All of the industrial firms with a common product or technological
interest (all firms in one segment of an industrial sector) did
not interact with the university which was conducting research in
that particular area of technology. Usually, one or two firms would
have a strong interaction, and these tended to be the firms which
placed the greatest emphasis on developing and integrating new
technology into their business. The reason for this is not entirely
clear, for certainly in the case of the Center for Catalytic Science
and Technology and the Center for Composite Materials at the
University of Delaware, groups of corporations with common product
and technological interests participate.

Probably the most serious question which comes to mind
regarding the operation of NSF's industry-university cooperative
research program is the question of exactly how priorities are
established with regard to which research programs are to be funded.
Insofar as technical feasibility is concerned, proposed research
programs are judged on the basis of the peer review process, and
this seems to be quite appropriate.

The question of establishing priorities and the criteria used
in establishing priorities was not addressed in any detail during
the symposium, and clearly, this leaves the question open to a
wide range of speculation. In the absence of any well-defined
criteria, the program could degenerate to one which serves the
parochial interests of a given industrial firm when, in fact, the
purpose of the program is perceived to be one of encouraging long-
term contributions toward product and/or process innovation, or
encouraging the development of knowledge as a foundation for new or
improved technologies. The assumption that the corporation which
is the leader of a given segment of the industrial sector will
choose to conduct cooperative basic research in an area that will
serve the national interest on the whole may be in error.

THE SITUATION IN ITALY

The situation of industry-university cooperation in research
appears to be strongly different in Italy from what it is in the
United States, and the reasons for this being so are far from clear,
though some of them can be identified.

Of the three types of industry-university interactions discussed previously, the gift or unrestricted grant from industry to
the university is extremely rare in Italy. This may be related to
the fact that in Italy, contrary to the U.S.A., there is no shortage
at all of technically skilled manpower (at the graduate level),
and, therefore, industries see no advantage in programs that make
recruitment of graduates easier. Consequently, only the minor

incentive of improvement of the public image of the corporation is left.

The second type of interaction, consulting by faculty for industrial firms, though not uncommon in Italy, is certainly less frequent than in the U.S. In this regard, however, the attitude of Italian universities and of the Italian Government, as was pointed out by one of us (G.A.) at the panel discussion, is very peculiar. By law, Italian university professors are forbidden to do any consulting, unless they opt for a sort of second-rank professorship, where their salary is 40% lower and they cannot hold any adminis- trative position (such as dean, chairman, etc.) within the academic structure. This seems to clearly indicate that the Italian Government and universities not only do not regard consulting as beneficial to the quality of the academic activity of the faculty, but in fact regard it as detrimental. The law has been enforced only recently, and--curiously enough--it has not been lobbied against by either industry or the university community. This atti- tude is typical of the traditional conception of the instructional function of the university as a static one, one which is not regarded as liable to obsolescence on a short-time scale, and, therefore, one where research plays only a minor role. This traditional conception is reflected in the fact that there is no graduate education in Italian universities, no Ph.D. program; also, curricula are established by act of Congress and, therefore, changing them is a difficult and time-consuming operation. This is perhaps the price that needs to be paid when universities are seven hundred years old.

The viewpoint of Italian industry can be appreciated by considering the contribution given at the panel discussion by E. Colombo, President of the Istituto Donegani, which is the Research and Development firm of the Montedison Corporation, the largest Italian chemical firm. Colombo argued that, with the cost of money being 25% or more in Italy, long-term investments are not attractive to industry, and, therefore, research--which Colombo regards as a long-term investment--is not pursuable by industry unless it is financed by the Government. It is debatable whether this argument is based on sound economical principles; research is characterized by a high risk and a potentially very high rate of return, so that decreasing only marginally the former and/or increasing only marginally the latter may easily offset the high cost of money. Be that as it may, the very fact that the argument is brought for- ward by the President of the research institution of the largest Italian chemical firm (not by a manager responsible for financial planning) seems to indicate that the Italian chemical industry has little belief in the potentiality of research as a basis for the development of new technology.

Neither can it be argued that the lack of interest in basic research is related to the fact that in the chemical industry the economies of scale are operative, since even if that was true ten years ago a major dislocation of the market has occurred since then. After World War II, the chemical industry grew in Italy very rapidly, at an average rate of more than 12% per year, up to the late '60's, when Italy was the seventh largest producer of chemicals in the world, with about 4.5% of the total world production. At that time, the balance of payments for chemicals was even in Italy, but exports were mainly low-price, mass-produced chemicals, while imports were high price chemicals. With low-priced chemicals becoming more and more difficult to sell abroad, since the production technology becomes available also to underdeveloped countries, the balance of payments for chemicals in Italy became worse and worse after 1970, up to negative balances of 571, 1266 and 2285 billion liras in 1977, 1978 and 1979. In spite of this, the Italian chemical industry invests only 1.8% of total sales into research and development, as contrasted to about 4% in the rest of the industrialized world.

It is interesting to remark that Colombo also pointed out that, with the cost of labor in Italy having reached, if not surpassed, that in other industrialized societies of the Western World, the only way for industry to stay competitive is to switch to the production of high price, small volume market chemicals and materials. Such production is typical of industries where the economies of scale are not operative, and, therefore, cooperation in research with universities would appear to be crucially important; yet, the two arguments brought forward by Colombo lead to the conclusion that the Italian industry plans to move to new high technology ventures, provided, however, that the government finances the research effort required to do so.

The Italian Government appears to be responsive to this position, through the programs of the C.N.R., which is roughly the Italian equivalent of N.S.F.; these programs were presented at the panel discussion by E. Martuscelli, Director of the Research Institute for Polymer Technology and Rheology of C.N.R. The Italian Government has identified, in the last ten years, a few broad areas of research in which most of the financial resources are concentrated. One of these is the area of fine chemistry, i.e., the production of high-price chemicals and materials characterized by a large contributed value (the difference between the value of the finished product and the raw materials needed to produce it). The C.N.R. program on fine chemistry finances research work done in industries, in the university, and in the C.N.R. itself, in a supposedly coordinated effort; in this sense, it would be an example of an institutionalized cooperative program. However, while Betz described the N.S.F. program of 10 million dollars per year as one which is supposed to have at most a "catalytic" effect,

the C.N.R. program on fine chemistry has a budget of less than 2
million dollars per year: and this in a situation where industry
is unwilling to do any basic research unless it is financed by
the government.

It was also remarked at the panel discussion that the C.N.R.
program involves the equivalent of about 180 full-time professionals,
which boils down to the almost insignificant budget of about 10,000
dollars per year and per full-time professional. This makes it
unlikely that research of relevant impact to industry can be carried
out, except perhaps in rather exceptional cases. The holding of
this joint symposium, which was partly financed by the C.N.R.
program, may turn out to be such an exceptional case.

The landscape which has been briefly sketched above seems to
lead to a somewhat general conclusion: the three corners of the
triangle industry-university-government in Italy all seem to lack
a strong belief in the possibility that profitable new technologies
may be developed as an outcome of well-organized programs of basic
research. Therefore, even when there is a strong incentive for the
development of such new technologies, there does not correspond a
strong effort in basic research.

It was also remarked at the panel discussion that the triangular structure industry-university-government is somewhat anomalous
in Italy, since the universities are all run by the State (with all
members of the staff being civil servants), and the State also owns
either directly or through a controlling interest a large fraction
of industry. The triangular structures could still be viable if
the three corners, though possibly all branches of government,
kept well-distinguished roles: the promotion of research for C.N.R.,
education for the university, production for industry. This, however, is not entirely accomplished in Italy: for instance,
research proposals to C.N.R. are not subjected to peer review, but
are decided upon by committees, the members of which are elected
largely by university professors. This may be the reason why
C.N.R. resources are often spread so thin: committee members may
not entirely disregard the need to maintain a constituency.

In the rather bleak situation described above, there are,
however, a few healthy signs of evolution. At least one successful
case history of a cooperative industry-university research program
was described by L. Nicolais at the panel discussion; the need to
move into high technology ventures has come to be recognized by
both industry and government, and sooner or later the requirement
of programs of basic research as springboards will become clear;
programs of graduate study will shortly be introduced in Italian
universities, and although at the moment they are still regarded
essentially as training programs for new faculty, they may increase
the role of research in the educational function of the university.

In this regard, it is interesting to observe the existence of cooperative programs between U.S. and Italian industries, and between U.S. and Italian academic groups, and, the fact that they are in general successful. Lovell illustrated the joint venture of Boeing and Aeritalia, who are partners in the design and construction of the new 767 plane; Boeing has a cooperative program in operation with Professsor J. C. Seferis at the University of Washington, and Aeritalia is planning to have a similar arrangement with Professor L. Nicolais of the University of Naples; the holding of this symposium is an outcome of the cooperative program of research of the two academic groups at the universities of Washington and of Naples. It is also interesting that cooperative programs in Italy tend to involve industrial firms which either have international ties such as Aeritalia has with Boeing, or are the Italian branch of a multinational corporation: this seems to indicate that such firms acquire a belief in the potentialities of basic research through their international contacts. This more modern viewpoint will hopefully spread also to the Italian industrial community as a whole.

ACKNOWLEDGMENTS

We are grateful to all the participants to the panel discussion for having brought forward the seeds of many of the ideas discussed in this paper; to the Italian Ministry for Public Instruction for supporting one of us (C.D.D) during the period when this work was done; and to the organizers of the Symposium, Professors L. Nicolais and J. Seferis, for asking us to report on the panel and to develop our own line of thought in doing so.

AUTHOR INDEX

Apicella, A., 215, 469
Ashbee, K. H. G., 517
Astarita, G., 649
Brivio, A., 607
Browning, C. E., 231
Campagna, P., 369
Carfagna, C., 215
Carignani, G., 453
Castro, J. M., 189
Chiang, C., 503
Christie, M. A., 319
Chu, H. S., 53
Cohn, D., 245
Crivelli-Visconti, I., 545
Darlington, M. W., 319
Denson, C. D., 649
Fitzgerald, J. E., 625
Franceschini, A., 369
Gillham, J. K., 127
Goettler, L. A., 289
Gonzalez, V. M., 189
Halpin, J. C., 3
Hertzberg, R. W., 377
Hinrichs, R. J., 147
Jansson, J. F., 397
Jarzebski, G. J., 261
Kacir, L., 407
Kardos, J. L., 407
Kay, J. F., 437
Koenig, J. L., 503
Lang, R. W., 377
Lovell, D. T., 19
Macosko, C. W., 189

Manson, J. A., 377
Marom, G., 245
Martuscelli, E., 633
Masi, P., 469
Masomy, E., 407
Mazzola, M., 453
McCullough, R. L., 261, 425
McGee, S. H., 261, 425
Migliaresi, C., 357
Momo, A., 369
Morgan, R. J., 207
Nicodemo, L., 469
Nicolais, L., 215, 469
Parenti, G., 607
Pavan, A., 529
Piccarolo, S., 469
Pipes, R. B., 587
Samanni, G., 607
Sargent, J. P., 517
Scola, D. A., 159
Seferis, J. C., 53
Sternstein, S. S., 39
Sundstrom, H., 397
Theocaris, P. S., 481
Thuen, J. M., 147
Ting, R. Y., 171
Voto, C., 33
Wagner, V., 607
Walsh, R. M., Jr., 587
Walter, E., 517
Wilkins, D. J., 629
Yang, P., 39
Zanotti, C., 607

PARTICIPANTS

Dr. A. Addeo
Istituto Donegani
Via Nuova delle Brecce, 150
NAPOLI

Prof. F. Alfani
Istituto di Principi di
 Ingegneria Chimica
Universita di Napoli
P.le Tecchio
80125 NAPOLI

Dr. R. H. Andrews
Plenum Publishing Co.
88/90 Middlesex Street
London, E1 7 EZ U.K.

Prof. G. Astarita
Istituto di Principi di
 Ingegneria Chimica
Universita di Napoli
P.le Tecchio
80125 NAPOLI

Dr. M. Avella
ITPR-CNR
Via Toiano
Arco Felice
80072 NAPOLI

Prof. F. Bellucci
Istituto di Principi di
 Ingegneria Chimica
Universita di Napoli
P.le Tecchio
80125 NAPOLI

Mr. J. P. Bentejac
Societe Nationale Industrielle
Aerospatiale - Centre Technique
 de Suresnes
12, rue Pasteur - B.P. 76
95152 Suresnes Cedex
FRANCE

Dr. F. Betz
National Science Foundation
WASHINGTON, D.C. 20550

Mr. J. P. Botman
Hexcel S.A. 4840
Welkeuraedt, BELGIUM

Prof. L. Caglioti
CNR Prog. Fin.
Chimica Fine e Secondaria
Viale Nizza
00100 ROMA

Dr. B. Calcagno
Direzione Ricerche
Sviluppo e Processi
Industriale Pirelli
Viale Sarca 202
20125 MILANO

Dr. G. Campanile
Istituto "G. Donegani"
Via Nuova delle Brecce, 150
NAPOLI

Prof. F. Caputo
Istituto di Costruzioni
  di Macchine
Universita di Napoli
80125 NAPOLI

Prof. G. C. Caprino
Istituto di Tecnologie
Universita di Napoli
80125 NAPOLI

Prof. A. Cardon
Vzije Universiteit Brussel
Pleinlaan 2 ÷ 1051 Brussels
BELGIUM

Dr. C. Carfagna
Istituto Donegani
Via Nuova delle Brecce, 150
NAPOLI

Mr. F. Cecchini
P.F. Chimica Fine e Secondaria
C.N.R.
Via Nizza
00100 ROMA

Dr. A. Chudnovsky
Dept. of Civil Engineering
Case Western Reserve University
CLEVELAND, OH 44106

Prof. P. Corradini
Istituto Chimico
Universita di Napoli
Via Mezzocannone, 8
80134 NAPOLI

Dr. L. Cohen
IMI
Viale dell'Arte, 25
00144 NAPOLI

Ing. E. Colombo
Istituto G. Donegani
Viale Rosellini, 17
20124 MILANO

Dr. R. E. Coulehan
Ciba-Geigy Corporation
ARDSLEY, NY 10502

Prof. I. Crivelli-Visconti
Istituto di Tecnologie
Universita di Napoli
80125 NAPOLI

Prof. M. W. Darlington
Cranfield Institute of
  Technology
Bedford MK 430 AL U.K.

Dr. G. De Luca
Vetrotex
Via Romagnoli, 6
20146 MILANO

Prof. C. D. Denson
Dept. of Chemical Engineering
University of Delaware
NEWARK, DE 19711

Dr. A. De Ruvo
Swedish Forest Products, Research
  Lab.
Box 5604
S 0 11486 Stockholm, SWEDEN

Prof. A. T. DiBenedetto
Vice President
University of Connecticut
STORRS, CT 06268

Prof. J. E. Fitzgerald
Georgia Institute of Technology
School of Civil Engineering
ATLANTA, GA 30332

Dr. A. Franceschini
Centro Ricerche FIAT
Via Torino, 50
10043 ORBASSANO

Prof. C. Garbuglio
Istituto Donegani
Via Nuova delle Brecce, 150
NAPOLI

Dr. S. Gargiulo
Aeritalia
Pomigliano d'Arco
80038 NAPOLI

Dr. F. Gerstle
Sandia National Laboratories
Division 5814
ALBUQUERQUE, NM 87185

Ing. L. Giamundo
Regla Fiberglass
Angri SALERNO

Dr. H. H. Gibbs
E. I. du Pont de Nemours & Co.
Polymer Product Dept.
Building 323
WILMINGTON, DE 19898

Dr. J. K. Gillham
Dept. of Chemical Engineering
Princeton University
PRINCETON, NJ 08544

Prof. F. Gioia
Istituto di Principi di
  Ingegneria Chimica
Universita di Napoli
P.le Tecchio
80125 NAPOLI

Prof. P. Giusti
Istituto di Chimica Generale
Facolta di Ingegneria
Universita di Pisa
56100 PISA

Dr. L. A. Goettler
Monsanto Company
260 Springside Drive
AKRON, OH 44313

Ing. R. Greco
ITPR - CNR
Arco Felice
80072 NAPOLI

Dr. G. Guerra
Istituto Donegani
Via Nuova delle Brecce 150
NAPOLI

Dr. J. C. Halpin
U.S. Air Force
Wright-Patterson Air Force Base
DAYTON, OH 45433

Prof. Z. Hashin
School of Engineering
Tel Aviv University
Tel Aviv, ISRAEL

Dr. B. J. Hayes
Ciba-Geigy
Duxford
Cambridge CB 2 4 QD- U.K.

Dr. R. Hinrichs
Appl. Polymer Tech.
P.O. Box 11950
COSTA MESA, CA 92627

Prof. H. B. Hopfenberg
Dept. of Chemical Engineering
North Carolina State University
RALEIGH, NC 27650

Dr. M. Iannone
Aeritalia
Pomigliano d'Arco
80138 NAPOLI

Dr. J. F. Jansson
Royal Institute of Technology
Dept. of Technology
10044 Stockholm, SWEDEN

Prof. J. L. Kardos
Dept. of Chemical Engineering
Washington University
ST. LOUIS, MO 63130

Dr. J. F. Kay
Owens - Corning Fiberglas
Technical Center
P.O. Box 415
GRANVILLE, OH 43023

Dr. J. Kolarik
Institute of Macromolecular
    Chemistry
Academy of Science
Prague 6 - CZECHOSLOVAKIA

Dr. G. Kumar
Johnson and Johnson Dental
    Product Co.
20 Lake Drive
EAST WINDSOR, NJ   08520

Dr. P. Laconte
University of Louvain La Neuve,
Louvain La Neuve, BELGIUM

Dr. R. F. Landel
Jet Propulsion Laboratory
4800 Oak Grove Ave. MS122-123
PASADENA, CA   91103

Dr. J. K. Lepper
Lawrence Livermore National
    Laboratory
P.O. Box 808
LIVERMORE, CA   94550

Dr. D. T. Lovell
Boeing Commercial Airplane Co.
P.O. Box 3707, MS 73 - 43
SEATTLE, WA   98124

Dr. G. Lovo
Ciba-Geigy
(Div. Mat. Plastiche e Additivi)
21040 ORIGGIO

Prof. C. W. Macosko
Dept. of Chemical Engineering
University of Minnesota
421 Washington Ave.
MINNEAPOLIS, MN   55455

Prof. J. A. Manson
Materials Research Center
Lehigh Univerity
BETHLEHEM, PA   18015

Prof. G. Marom
Hebrew University of Jerusalem
Casali Institute of Applied
    Chemistry
Jerusalem   91904, ISRAEL

Prof. G. Marrucci
Istituto di Principi di
    Ingegneria Chimica
Universita di Napoli
80125 NAPOLI

Prof. E. Martuscelli
I.T.P.R.   C.N.R.
Via Toiano
Arco Felice
80072 NAPOLI

Dr.   P. Masi
Istituto di Principi di Chimica
Universita di Napoli
80125 NAPOLI

Ing. M. Mazzola
SNIAL
Resine Poliesteri
00034 COLLEFERRO

Prof. E. Meinecke
Dept. of Polymer Science
University of Akron
AKRON, OH   44325

Dr. C. Migliaresi
Istituto di Principi di
    Ingegneria Chimica
Universita di Napoli
80125 NAPOLI

Ing. S. Mignosi
Aeritalia
Pomigliano d'Arco
80030 NAPOLI

Dr. B. Minieri
FIAT AUTO S.P.A.
D. T. Laboratori Centrali
C.so G. Agnelli 200
10135 TORINO

Dr. A. C. Moloney
Laboratoire de Polymeres
Ecole Polytechnique de Lausanne
32, CH de Belledrive CH 1007
Lausanne, SWITZERLAND

Dr. R. L. Morgan
Lawrence Livermore Lab.
University of California
P.O. Box 808
LIVERMORE, CA 94550

Prof. L. Nicodemo
Istituto di Principi di
  Ingegneria Chimica
Universita di Napoli
80125 NAPOLI

Prof. L. Nicolais (Cochairman)
Istituto di Principi di
  Ingegneria Chimica
Universita di Napoli
80125 NAPOLI

Ing. G. Ombra
Aeritalia Tern
Pomiglaino d'Arco
80030 NAPOLI

Dr. S. Pagliarulo
Aeritalia
Pomigliano d'Arco
80038 NAPOLI

Dr. S. Pagliuso
Aeritalia SVTN
Pomigliano d'Arco
80038 NAPOLI

Dr. P. Passerini
Aeritalia
Pomigliano d'Arco
80038 NAPOLI

Prof. A. Pavan
Istituto di Chimica Industriale
Politecnico di Milano
20100 MILANO

Prof. M. Pegoraro
Istituto di Chimica Industriale
Politecnico di Milano
20100 MILANO

Dr. P. Pelagalli
Aeritalia
Pomigliano d'Arco
80038 NAPOLI

Prof. A. Penati
Istituto di Chimica Industriale
Politecnico di Milano
20133 MILANO

Prof. S. Piccarolo
Istituto di Ingegneria Chimica
Universita di Palermo
90100 PALERMO

Ing. V. Renta
Aeritalia S.p.A.
Pomigliano d'Arco
80038 NAPOLI

Dr. P. Rizzo
Aviointeriors
Largo I. Antonelli, 4
00145 ROMA

Dr. G. Salee
Hooker Chemical & Plastics
2901 Long Road
GRAND ISLAND, NY 14072

Mr. E. Scafora
ITPR – CNR
Arco Felice
80072 NAPOLI

Dr. D. A. Scola
United Technologies Research
  Center
United Technologies Co.
EAST HARTFORD, CT 06108

Ing. M. Scolaris
Aeritalia
10100 TORINO

Prof. J.C. Seferis (Cochairman)
Dept. of Chemical Engineering
University of Washington
SFATTLE, WA 98195

Prof. G. Sironi
Vitrofil
17047 VADO LIGURE

Ing. A. Staffa
Aeritalia MALN
Pomigliano d'Arco
80038 NAPOLI

Prof. S. S. Sternstein
Rensselaer Polytechnic Institute
Materials Engineering Department
TROY, NY 12181

Mr. P. Sullivan
Owens-Corning Fiberglas Corp.
Route 16
P.O. Box 415
GRANVILLE, OH 43023

Prof. P. Theocaris
Chair of Applied Mechanics
Nat'l Tech. Univ. of Athens
University City, Zografou
ATHENS, Greece

Ms. J. Thuen
Narmco Materials Inc.
600 Victoria Street
COSTA MESA, CA 92727

Dr. R. Ting
5975 Naval Research Laboratory
P.O. Box 8337
ORLANDO, FL 32856

Prof. G. Titomanlio
Istituto di Ingegneria Chimica
Universita di Palermo
Viale delle Scienze
90100 PALERMO

Dr. M. Turner
Ciba-Geigy
Duxford
Cambridge CB2 4 QD - U.K.

Dr. A. Ummarino
Aeritalia TEPN
Pomigliano d'Arco
80038 NAPOLI

Prof. A. Valvassori
Istituto G. Donegani
Viale Rosellini, 17
20124 MILANO

Dr. G. Zanotti
Lab. Tecnologie Sperimentali
"G. Augusta"
Viale Schiapparelli, 8
21013 GALLARATE

Dr. P. L. Wagner
E. I. du Pont de Nemours & Co.
Bldg. 701, Chestnut Run Loc.
WILMINGTON, DE 19898

Ing. V. Wagner
Lab. Tecnologie Sperimentali
Costruz. Aeronautiche "Augusta"
Via Schiapparelli, 8
21013 GALLARATE

Dr. G. Weinand
Hexcel S.A. 4840
Welkeruaedt, BELGIUM

Prof. R. A. Weiss
Institute of Materials Science
The University of Connecticut
STORRS, CT 06268

Dr. D. J. Wilkins
General Dynamics
Fort Worth Division
P.O. Box 748
FORT WORTH, TX 76101